D1103917

HANDBOOK OF ENVIRONMENTAL ECONOMICS
VOLUME 1

HANDBOOKS
IN
ECONOMICS

20

Series Editors

KENNETH J. ARROW
MICHAEL D. INTRILIGATOR

ELSEVIER

AMSTERDAM • BOSTON • LONDON • NEW YORK • OXFORD • PARIS
SAN DIEGO • SAN FRANCISCO • SINGAPORE • SYDNEY • TOKYO

HANDBOOK OF ENVIRONMENTAL ECONOMICS

VOLUME 1
ENVIRONMENTAL DEGRADATION
AND INSTITUTIONAL RESPONSES

Edited by

KARL-GÖRAN MÄLER
Swedish Academy of Sciences

and

JEFFREY R. VINCENT
University of California

2003

ELSEVIER

AMSTERDAM · BOSTON · LONDON · NEW YORK · OXFORD · PARIS
SAN DIEGO · SAN FRANCISCO · SINGAPORE · SYDNEY · TOKYO

ELSEVIER SCIENCE B.V.
Sara Burgerhartstraat 25
P.O. Box 211, 1000 AE Amsterdam, The Netherlands

First edition 2003

British Library Cataloguing in Publication Data
Handbook of environmental economics
  Vol. 1: Environmental degradation and institutional
  responses. - (Handbook in economics; 20)
  1. Environmental economics
  I. Mäler, Karl-Göran II. Vincent, Jeffrey R.
  333.7
  ISBN 0444500634   ᛌᚻ

Library of Congress Cataloguing in Publication Data
A catalog record from the Library of Congress has been applied for.

ISBN: 0-444-50063-4
ISSN: 0169-7218 (Handbooks in Economics Series)

⊗ The paper used in this publication meets the requirements of ANSI/NISO Z39.48-1992 (Permanence of Paper).

Printed in The Netherlands.

INTRODUCTION TO THE SERIES

The aim of the *Handbooks in Economics* series is to produce Handbooks for various branches of economics, each of which is a definitive source, reference, and teaching supplement for use by professional researchers and advanced graduate students. Each Handbook provides self-contained surveys of the current state of a branch of economics in the form of chapters prepared by leading specialists on various aspects of this branch of economics. These surveys summarize not only received results but also newer developments, from recent journal articles and discussion papers. Some original material is also included, but the main goal is to provide comprehensive and accessible surveys. The Handbooks are intended to provide not only useful reference volumes for professional collections but also possible supplementary readings for advanced courses for graduate students in economics.

KENNETH J. ARROW and MICHAEL D. INTRILIGATOR

PUBLISHER'S NOTE

For a complete overview of the Handbooks in Economics Series, please refer to the listing at the end of this volume.

v

CONTENTS OF THE HANDBOOK

DEDICATION

Allen Kneese

If anyone should be called the founding father of environmental economics, it must be Allen Kneese. He was a pioneer as a researcher, and he was a pioneer as a research organizer. He inspired a vast number of younger environmental economists. His studies of water issues in the 1960s induced many, including one of the editors of this handbook, to look at environmental problems through the eyes of an economist. His enduring fight for the use of economic instruments in environmental policy had impacts even outside his own country. He was the first to recognize the need for economists to learn from other disciplines – physics, hydrology, ecology, political science – in order to enable us to produce good and relevant policy recommendations. Allen was an editor of the North-Holland *Handbook of Natural Resource and Energy Economics*. He had promised to write an essay describing his personal perspective on the evolution of environmental economics for this handbook. Unfortunately for all of us, he passed away after a long illness. We dedicate these volumes to his memory.

PREFACE TO THE HANDBOOK

Elsevier published a 3-volume *Handbook of Natural Resource and Energy Economics* in 1985 (the first two volumes) and 1993 (the third volume). Why is it now publishing a 3-volume *Handbook of Environmental Economics*? Is it not true that economic development in Europe and North America during the last thirty years has proved that there is no resource scarcity? After all, prices of minerals have not increased (in real terms), despite the enormous economic expansion that has occurred in these regions. Moreover, air quality has improved substantially in Europe and North America. Are not all environmental problems solved? Many "experts" argue that this is the case, and if they were right, there would be no need for a new handbook!

However, here there is a paradox. On the one hand aggregate data seem to indicate that we have overcome most environmental problems. On the other hand, if we look at a micro level, it is easy to find contrary evidence.

Most environmental problems share the following two characteristics: they are intertemporal, and they are local. Soil erosion may cause severe economic losses in the future, but a long time might pass before the soil is so much eroded that its productivity is affected. And when its productivity is affected, the economic damage will fall primarily on the nearby village of farmers and might be barely felt on a national or international level. Thus, there will be no sign of economic damage until later, and because of the lack of appropriate information and the lack of appropriate property rights, there will be no immediate impacts on agricultural products and their prices.

This parable about soil erosion possibly applies to most environmental problems, which are often invisible unless we look for them. Human-induced climate change is a case in point. Without knowledge of thermodynamics, humans would not have launched the research that uncovered empirical evidence of global warming.

Of course, there are examples of continued environmental deterioration at the aggregate level. Global climate change is perhaps the most dramatic one. Another is the depletion of the world's marine fisheries. But some problems, e.g., biodiversity, are mainly analysed and discussed from a global perspective when the real problem is arguably on a local level. Reduction of biodiversity implies a reduction in ecological resilience, which increases the risk that local human communities will lose essential ecosystem services.

These points are relevant for both rich and poor countries, but if we focus our interest on the poor countries, the magnitude of welfare losses due to environmental degradation is even greater. Urban pollution, soil erosion, reduction both in the quality and quantity of potable water, etc. are the rule, not the exception, in these countries.

Economics, which is about the management of scarce resources, offers the tools needed for a rational analysis of environmental problems. The rapid development of economic theory and methods as applied to the environment is the first reason a new handbook is needed. Several chapters in the earlier Elsevier handbook are outdated. The most obvious example pertains to valuation methods, which economists use to measure environmental changes in monetary terms. The *Handbook of Natural Resource and Energy Economics* had two chapters on valuation, one on theory and one on methods applicable to recreation demand. In contrast, and as a consequence of the explosion of valuation research since the 1980s, the new handbook devotes an entire volume, Volume 3, "Valuing Environmental Changes", to valuation theory and methods. Valuation research has extended into areas, such as experimental economics, that were scarcely imagined in 1985.

Another example is market-based instruments for controlling pollution. An influential chapter by Peter Bohm and Clifford Russell in the earlier handbook made the case for using economic principles to guide the design of pollution policies. Although examples of economic approaches for controlling pollution, such as effluent charges and emissions trading programs, existed in the 1980s, they were so few and so new that experience with them could barely be evaluated. Now, many countries have experimented with pollution charges of various types, and at least one (the United States) has created emissions trading programs at the national level. Economists have analysed the experience with these programs, and the new handbook presents the lessons of their research.

The more important reason for the new handbook, however, is the emergence of entirely new lines of research that either did not exist 10–20 years ago or were so sparsely investigated as to preclude chapter-length reviews. Economic research on the environment today includes much more than studies that estimate the value of particular non-market environment goods or the cost-effectiveness of particular pollution control instruments.

Some of the new research is new because it applies microeconomic theory much more carefully to understand institutional aspects of environmental management (or mismanagement). Volume 1 of this handbook, "Environmental Degradation and Institutional Responses", presents much of the new research in this area, especially as it applies to environmental degradation at a local level. It includes chapters on common property management regimes; population, poverty, and the environment; mechanism design and the environment; and experimental evaluations of environmental policy instruments – chapters that have no counterparts in the earlier handbook.

Other research is new because it examines environmental externalities and public goods at larger economic scales: an entire national economy, several countries in a given region, or all the countries of the world. Volume 2, "Economywide and International Environmental Issues", summarizes advances in this area. New areas of research that are covered in it include environmental policy in a second-best economy (the "double dividend" literature), empirical studies on economic growth and the environment (the "environmental Kuznets curve" literature), national income accounts and the environment, international trade and the environment, and international environmental agreements. One chapter in the Volume 3 of the earlier handbook touched on environmental applications of computable general equilibrium models and the economics of climate change, but both topics receive much more extensive coverage in Volume 2 of this handbook.

Due to the expansion of economic research on the environment, in one sense the scope of this handbook is, ironically, narrower than that of its predecessor. This difference is signalled by the change in title: the *Handbook of Environmental Economics*, not volumes 4–6 of the *Handbook of Natural Resource and Energy Economics*. Unlike the earlier handbook, this handbook does *not* include chapters on the supply of and demand for energy resources, minerals, timber, fish, and other commercial natural resources.

Instead, this handbook focuses on environmental goods and services that, due to property rights failures stemming from externalities and public goods, are not allocated efficiently by markets. Indeed, these environmental resources often lack markets altogether. They include air and water quality, hydrological functions of forests and wetlands, soil stability and fertility, the genetic diversity of wild species, natural areas used for recreation, and numerous others. They are in principle renewable, but in practice they are often subject to excessive degradation and depletion, sometimes to an irreversible degree.

Commercial natural resources appear in this new handbook only in an incidental way. For example, the development of comprehensive measures of national income and wealth requires consideration of all forms of capital, including all forms of natural capital. So, the chapter on national accounts and the environment discusses adjustments to conventional measures of national income and wealth for not only the degradation of environmental quality but also the depletion of stocks of commercial natural resources. Commercial extraction and utilization of natural resources are also sources of many of the environmental externalities discussed throughout the handbook. A prime example is damage from emissions of greenhouse gases, which are released primarily by the burning of fossil fuels.

For these reasons, this handbook is best regarded as a complement to the *Handbook of Natural Resource and Energy Economics*, not a replacement for it. This handbook is intended to be an updated reference on environmental economics, not natural resource economics.

This handbook does share two important features with the earlier one, which we have attempted to accentuate. First, both handbooks draw upon research conducted by not only economists but also natural and social scientists in other disciplines. The chapters in this handbook on common property management regimes and population, poverty, and the environment draw extensively on the anthropological literature, while the chapter on political economy of environmental policy draws on studies by political scientists and legal scholars. Some chapters in this handbook are written by noneconomists, from the earth sciences, ecology, and psychology. External reviewers of chapter drafts were drawn from an even broader range of disciplines.

Second, both handbooks emphasize dynamic considerations. Natural resource economics is inherently about efficient allocations over time, but many textbooks present environmental economics in an entirely static context: the valuation of current use of an environmental resource, or the short-run cost-effectiveness of market-based instruments compared to command-and-control instruments. In fact, environmental economics, properly done, must consider several dynamic issues, which the chapters in this handbook highlight.

One is the dynamics of natural systems. The build-up of greenhouse gases in the atmosphere reminds us that pollution involves stocks as well as flows. The same is true for environmental resources other than air quality. Research by natural scientists has revealed that the dynamics of natural systems can be far from continuous and smooth; they can be nonlinear, complex, and chaotic, subject to abrupt and irreversible (or effectively irreversible) "flips" from one state to another. The first two chapters in Volume 1 highlight the dynamics of natural systems, which economists ignore at the risk of constructing economic models with feet of clay. These chapters complement the excellent chapter on dynamics of natural resources by James Wilen in the earlier handbook.

A second dynamic consideration follows immediately from the stock nature of environmental resources: optimal management of environmental resources is no less intertemporal than the optimal management of commercial natural resources. Indeed, the time frame for economic studies of climate change is much longer – centuries instead of decades – than the time frame typically considered in studies on the optimal management of mineral reserves or timber stocks. Hence, although the same questions arise – what welfare function should we use? what discount rate? – answering these questions is harder and more consequential. Several chapters in Volume 2 address these dynamic welfare issues.

Third, a static perspective could cause environmental economists to overlook important impacts of environmental regulations on technological change – and the impact of environmental degradation on empirical estimates of rates of technological change. Chapters in this handbook address these issues. In particular, a chapter in Volume 1 looks exclusively at the impacts of environmental regulations on technological change. This issue was treated only in passing in the earlier handbook.

A final important dynamic area concerns institutional evolution. Like other fields of economics, environmental economics has been heavily influenced by the "New Institutional Economics". Several chapters in both Volumes 1 and 2 of this handbook examine the forces that shape institutional responses to environmental change at local, national, and international scales. Interactions with fertility decisions are especially important at a local level, and so this handbook contains a chapter on population, poverty, and the environment.

Having noted above a way in which the scope of this handbook is narrower than that of its predecessor, we conclude by noting a way that it is broader. The *Handbook of Environmental Economics* places more emphasis on the application of economics to environmental policy issues in developing countries. Environmental economics was born and raised in universities and research institutes in rich, industrialized countries with well-developed political, legal, and market institutions. Most people in the world live in very different circumstances: poverty, restricted civil and political liberties, and traditional property rights that are backed up only weakly, if at all, by the legal system. By and large, they also live in more degraded natural surroundings – which, as economists might surmise, is no coincidence.

We believe environmental economics can play an especially important role in improving the welfare of this destitute majority. Environmental economists know more about institutional failures than do most economists, so the resources of their discipline should be especially valuable when directed toward the problems of poor countries. For this reason, we commissioned for this handbook chapters specifically on developing country issues. We also asked the authors of all the other chapters to search for examples of studies on developing countries.

The authors were helped by the fact that an increasing share of the pages of leading field journals like the *Journal of Environmental Economics and Management, Environmental and Resource Economics, Land Economics*, and *Resource and Energy Economics* are occupied by articles based on studies conducted in developing countries, and by the relatively recent launch of a new journal, *Environment and Development Economics*, that provides an outlet specifically for such research. We find it heartening that most development economics textbooks – and the latest volume of the North-Holland *Handbook of Development Economics* – now include chapters on the environment, and that most environmental economics textbooks now include chapters on the developing world. We hope the *Handbook of Environmental Economics* will accelerate this integration of development and environmental economics.

In drawing attention to the relevance and significance of environmental economics to developing countries, we are also confirming the prescience of Allen Kneese, who was one of the editors of the *Handbook of Natural Resource and Energy Economics*. More than a decade before the Brundtland Commission popularised the phrase "sustainable development", Allen published a paper with Robert Ayres titled "The sustainable economy" (*Frontiers in Social Thought – Essays in Honor of Kenneth E. Boulding*, Elsevier Science Publishers, 1976). To our knowledge, this paper was the first in the environmental economics literature to include the word "sustainable" in its title, and one of the first to examine the differences between developed and developing countries. As in so many other ways, Allen was a pioneer.

Acknowledgements

Our greatest thanks go to Christina Leijonhufvud, without whose administrative support the handbook project would have foundered. Anna Sjöström stepped in and solved multiple problems related to the figures for one of the chapters. Benjamin Vincent relieved much of the tedious work related to indexing the chapters.

Our institutions – the Beijer Institute of Ecological Economics (Mäler) and the Harvard Institute for International Development, Kennedy School of Government, and Graduate School of International Relations & Pacific Studies at the University of California, San Diego (Vincent) – provided congenial and supportive bases of operation.

Finally, a series of diligent external reviewers – Jack Caldwell, Steve Carpenter, Bill Clark, Larry Goulder, Ted Groves, Daniel Kahneman, Charles Plott, Stef Proost, Steve Schneider, Brian Walker, Jörgen Weibull – helped ensure the relevance, comprehensiveness, and accuracy of the material presented in the chapters. Any shortcomings that remain are, of course, our responsibility.

KARL-GÖRAN MÄLER and JEFFREY R. VINCENT
Stockholm, September 17, 2002

CONTENTS OF VOLUME 1

PERSPECTIVES ON ENVIRONMENTAL ECONOMICS

We invited several leading economists who did not serve as chapter authors to write short essays presenting their personal views of the development of environmental economics and the field's intellectual and practical value. The essays printed below, by Nobel Laureate Robert M. Solow, Narpat Jodha, and Hirofumi Uzawa, illustrate both the symbioses and the tensions that exist between environmental economics and the mother discipline and the enormous range of policy issues to which environmental economics theory and methods have been applied.

Robert M. Solow (Department of Economics, MIT)

Environmental economics has been a considerable success in one sense, probably the most important sense. It has advanced our understanding of environmental problems: the way they arise, and the sorts of policies that can lead to desirable improvement. Large gaps in practical knowledge remain. For example, I have the (admittedly casual) impression that damage assessment is still a fairly crude art. And no wonder: incomparably more money has been spent on climate modelling and similar research than on modelling and estimating the economic effects of climate change, although the first kind of knowledge is of limited policy use without the second kind. I do not really want to speculate about the causes of this disparity. Maybe economic research is quite generally underfunded, maybe because many people are uncomfortable with the idea of independent research into matters with high stakes, financial or ideological. Maybe there are other reasons.

In another sense, however, environmental economics has not been successful. It is not a popular specialty among the best students in the best universities, even though environmentalism is still a popular cause outside the classroom. Graduate courses in environmental economics are few and far between, with only a very small number of centers of research having achieved non-trivial scale. Here I feel more comfortable speculating about the reasons.

From one point of view, environmental economics is mainly a series of applications of the economic theory of externalities. The basic principles are fairly standard, although the variety of applications is wide, and the context can be unusual. The right way to use the basic principles and to exploit the data at hand can be hard to figure out. It may be closely tied to the particular context. One implication may be a need to work closely with engineers, climatologists, ecologists, and others. But these are not characteristics on which the academic economics profession tends to shower accolades. The highest

1

praise usually goes to generality, depth, and originality, rather than to ingenuity, care, and comprehensiveness. Researchers and teachers respond to these incentives (in accord with comparative advantage, one hopes).

Probably disciplines are not all alike in this respect, but probably there are some others that are very like economics. One might speculate that the social sciences are especially likely to undervalue those branches that share the characteristics I have ascribed to environmental economics. The reason I have in mind is that the relevant economic parameters, unlike, say, the parameters of chemical reactions, are likely to change from time to time, as the underlying circumstances evolve. (For example, shifts in technology and tastes will change relative prices, and thus also the outcome of damage assessments.) Thus results of this kind can only be temporarily valid, and that can easily dissipate part of their charm, and their prestige.

Suppose this speculation is right. (We will never know.) Does anything follow from it? I doubt that it will be possible to attract more and better students into environmental economics by convincing them that it is more valuable, more interesting, or more glamorous than they had previously thought. That never works. A more humdrum thought is to try to attract more research funding into environmental economics, on the argument that intelligent policy depends as much on economic analysis as it does on atmospheric chemistry or marine biology. That has the merit of being true.

A recent report of the U.S. National Science Board [*Environmental Science and Engineering for the 21st Century* (National Science Foundation, Washington, 2000)] urged a substantial increase in public funding for environmental analysis and public-policy research. It emphasized the importance of combining, from the planning phase on, natural-science and social-science research. Failure to do this in the past has come partly from sheer inexperience, and partly from the mutual suspicion by environmental scientists and economists that the other group lacks basic understanding and appreciation. If I knew how to bridge that gap I would say so.

Narpat S. Jodha (ICIMOD, Kathmandu)

My professional work over the last 30 years has focused on fragile environments represented by arid and semi-arid areas of India and Africa and mountain regions of the Hindu Kush-Himalayas, where nature-society interactions, particularly at a micro-community level, have been closely observed and studied. Communities in these areas almost instantly and visibly suffer in terms of a reduced range and quality of livelihood options if they damage their local environmental resources.

The first formal evidence of the above phenomenon, which stimulated my interest in the links between human well-being and the environment, was provided by a study of famine and famine relief in an arid region of Rajasthan, India in the early 1960s. The three-year study, which covered over 50 villages, revealed that villages with better managed common property resources (CPRs) – community pastures, forests, other uncultivated lands, water bodies, watershed drainage, etc. – could adjust to drought and post-drought situations much better than villages with depleted and poorly looked-after CPRs. The latter group of villages suffered a greater extent of curtailment in consumption, sale or mortgage of assets, increased indebtedness, and migration to distant places.

The insights and understanding gained through this study were further verified and sharpened through similar work during drought years of the late sixties and seventies in other parts of India and East Africa. The emphasis was on identifying factors and mechanisms that explained the differences between villages with well managed and poorly managed environmental resources. In the subsequent period, the focus of work that covered larger areas in both the dry tropics and mountain regions shifted from the drought-mitigating role of CPRs to their overall livelihood-sustaining role, as sources of physical supplies, productive employment, income, and risk mitigation, as well as various environmental services. The studies offered a unique opportunity to understand the role of group dynamics, communities' collective stake in environmental resources, social norms and their enforcement mechanisms, etc., which helped in addressing the problems of externalities, transaction costs, free riding, and upkeep of CPRs. The key factor that contributed to better management of community resources and hence more stable livelihood support, in both the dry tropics and mountain regions and even in economically/socially heterogeneous communities, was a community's crucial, at times total, dependence on the village commons for different uses. This ensured people's collective stake in the health and productivity of the local environmental resource base. This helped in evolving various mechanisms to protect and regulate the usage of community resources during good and bad crop years.

The studies and prolonged stays in villages also helped in understanding the dynamics of change leading to a gradual decline of the traditional resource management systems and a breakdown of ecosystem–social system links. This decline happened due to a number of developments, such as the introduction of external legal and administrative regulations that disregarded customary rights and obligations, the decline of a culture of group action following the penetration of market forces and population growth, increased economic and socio-political differentiation of communities, and in some cases welfare and relief-oriented public measures, which partially substituted for the functions and services of CPRs and thus made the local resources and their care less indispensable. The consequent disintegration of collective stakes led to discontinuation of various processes and practices favoring CPRs, growth of people's indifference towards local environmental resources, and finally over-extraction and depletion of those resources. The ultimate consequence included several visible/invisible changes adversely affecting the livelihood situation of village communities, especially the poorest.

Despite increased awareness and evidence about the aforementioned negative changes in nature-society interactions, not much has been possible to reverse the situation. This is partly due to the marginal status of the issues described above vis-à-vis the mainstream discourse on global environmental change and its socio-economic consequences. Consequently, while global environmental concerns have significantly moved up in the national and international policy agenda, the revival and strengthening of ecosystem–social system links at a micro-level, which is a crucial step in integrating local and global perspectives on environmental sustainability, remain an important gap in the discourse and action on the subject.

The limited and scattered local context-focused work in the field has produced some success stories of rehabilitating the said ecosystem-social system links. They do inspire some hope, but the replication and scaling-up of such experiences is a continuing challenge for both researchers and planners. Similarly, the directions for reorientation of silvicultural and related research to incorporate indigenous knowledge, to alter the composition of products and services from local environmental resources, to satisfy the changing market-oriented needs, and to induce community participation in the rehabilitation of CPRs is another yet unattended area of work. Finally, work needs to be initiated on potential coping strategies for enhanced livelihood security and environmental resource sustainability at the local level in the face of the rapid process of globalization, as the driving forces of globalization are in apparent conflict with the conditions that are conducive to community-based environmental resource management at a micro-level.

Hirofumi Uzawa (Chuo University, Japan)

During the last three decades, we have seen a significant change in the nature of social, economic, and cultural impacts on the natural environment during the process of economic development. This is symbolically illustrated by the agendas of two international conferences convened by the United Nations, the Stockholm Conference in 1972 and the Rio Conference in 1992.

The Stockholm Conference was primarily concerned with health hazards caused by intensive industrialization during the 1960s, as exemplified by the participation of patients suffering from the Minamata disease. Such degradation of the natural environment was mainly caused by the emission of chemical substances such as sulfur and nitrogen oxides, which are themselves hazardous to both human health and biological environments. In the Rio Conference, on the other hand, the main agenda concerned the degradation and destabilization of the global environment, for example global warming, which results from intensified industrialization and extended urbanization. Global warming is primarily caused by the emission of carbon dioxide and other chemical substances that by themselves are not harmful to the natural environment nor hazardous to human health, but at the global scale cause atmospheric instability and other serious environmental disequilibria.

The changing nature of the environmental impacts of economic development as indicated above has forced us to reexamine the basic premises of economic theory in general and environmental economics in particular, and to search for a theoretical framework in which the mechanisms through which the natural and social environments are interwoven with the processes of industrialization and urbanization can be analyzed closely and their social and policy implications explicitly brought out. We are particularly concerned with processes of economic development that are sustainable both with respect to the natural environment and within the market economy, and with analyzing the insti tutional arrangements and policy measures under which sustainable development may necessarily ensue. Such institutional arrangements are generally defined in terms of property right assignments to various environmental resources, with specific reference to the behavioral criteria for those social institutions and organizations that manage the resources.

One of the obvious implications is that economic incentives for individual members of society are primarily to be replied upon. Direct social control and coercion are nei- ther effective in solving global environmental problems nor desirable from social and cultural points of view.

The phenomenon of global warming is basically of anthropogenic origin, primar- ily due to the massive consumption of fossil fuels and secondarily due to the deple- tion of tropical rain forests. The predominant forces behind these human activities are economic, and any policy or institutional measures to effectively arrest the process of atmospheric disequilibrium would have to take into account economic, social, and po- litical implications.

There exist two distinct features of the phenomenon of global warming that traditional economic theory is hardly equipped to deal with. First, global warming is caused by rising concentrations of carbon dioxide and other greenhouse gases in the atmosphere. The atmosphere plays the role of social overhead capital, which is neither privately appropriated nor subject to transactions in the market. Traditional economic theory has been primarily concerned with those scarce resources that are privately appropriated and whose ownership rights are transacted on the market.

The second feature concerns the equity problem between different countries and be- tween different generations. Those who emit most of the carbon dioxide are those who benefit most from the combustion of fossil fuels, while those who suffer the most from global warming are those who benefit least from the emission of carbon dioxide.

By the same token, while the current generation enjoys a spuriously high living stan- dard from the combustion of fossil fuels, future generations will suffer from global warming and other problems related to the atmospheric concentrations of carbon diox- ide and other greenhouse gases. Again, traditional economic theory has shied away from problems involving equity and justice, restricting its realm to the efficiency aspect, with the notable exception of the classic work by Kenneth Arrow, *Social Choice and In- dividual Values* (1952), followed by those of Atkinson, Dasgupta, Sen, Williams, and others.

Thus the problem of global warming offers us a unique opportunity to reexamine the theoretical premises of traditional economic theory and to search for a theoretical framework that enables us to analyze dynamic and equity problems involving environ- mental disruption. Such a framework is provided by the theory of optimum economic growth and the theory of social overhead capital, both of which have been developed in the last three decades.

Beginning in the middle 1960s, various attempts have been made to develop full- fledged dynamic analyses for both decentralized and centralized economies. Karl-Göran

Mäler was the first to apply the techniques of optimum economic growth theory to formulate a systematic, dynamic model in which the environment was made an integral component of processes of economic development. In his classic *Environmental Economics: A Theoretical Inquiry* (1974), Mäler gave us the basic framework that can be used to analyze the economic and political circumstances under which global warming and other environmental problems occur and to find those policy and institutional arrangements that may be effectively implemented to arrest them. A large number of studies since have applied this framework to more specific cases such as forestry resources, subterranean water, coastal wetlands, common fisheries, and others.

Another distinctive feature of the atmosphere is that it is neither privately appropriated, nor is it subject to transactions in the market. Thus the atmosphere may be regarded as a component of social overhead capital, and some of the more relevant propositions in the theory of social overhead capital may be applied to examine institutional arrangements for the stabilization of atmospheric composition.

The concept of social overhead capital was originally introduced by myself in "Sur la theorie economique du capital collectif social" (*Cahiers du Seminaire d'Econometrie*, 1974), in which I explicitly brought out the mechanisms by which social overhead capital interacts with the working of market institutions and analyzed the effects social overhead capital exerts upon the distribution of real income. The concept of social overhead capital has since been extended to include the natural environment, social infrastructure, and institutional capital; to analyze explicitly the phenomena of externalities, both static and dynamic; and to examine the implications for the structure of intertemporal allocations of scarce resources that are both dynamically optimal and intergenerationally equitable.

Chapter 1

GEOPHYSICAL AND GEOCHEMICAL ASPECTS OF ENVIRONMENTAL DEGRADATION

BERT BOLIN

S. Asvagen 51, 18452 Österskär, Sweden

Contents

Handbook of Environmental Economics, Volume 1, Edited by K.-G. Mäler and J.R. Vincent

Abstract

The environmental system is characterized by an interplay of geophysical and geochemical processes that provide a setting for life. Now that human interventions are affecting the global system as a whole, it is important to distinguish between changes of natural origin and changes brought about by human activities. Major difficulties arise in doing this because of the nonlinear and chaotic nature of the interactions between the environmental and human systems. Following an initial review of basic earth science principles, this chapter focuses on five fundamental issues that are important in all quarters of the world. Two sections deal with purely atmospheric issues, air pollution near the earth's surface and depletion of ozone in the stratosphere. These sections are followed by a closer look at water pollution and water management. A specific issue, acidification of freshwaters and soils, is next dealt with in more detail. The final issue addressed in the chapter, global climate change, requires an analysis of the total environmental system. All of these environmental issues have a bearing on how humankind might be able to secure sustainable development for the future, which is touched upon in the concluding section.

Keywords

air pollution, ozone layer, water management, acidification, global climate change

JEL classification: Q15, Q23, Q24, Q25, Q4

1. Introduction

Life on Earth has developed during many hundred millions of years. This has been possible because of the favorable location of the Earth in the solar system. The planets closer to the Sun (Mercury and Venus) are much too hot to permit the existence of the kind of complex molecules that life is built around. The planets further away from the Sun, on the other hand, are cold and uninhabitable.

A so-called black body,[1] at the same distance from the Sun as the Earth and in thermodynamic balance with the radiation from the Sun, would have a temperature merely a few degrees above the freezing point. The Earth is, however, not a black body but reflects about 30% of the incoming radiation back to space – its albedo is 0.30 – while the heat radiation emitted by the Earth towards space still is about that of a black body. Therefore, if there were no atmosphere around the Earth its temperature would be merely about $-18°C$, a very harsh setting for life to thrive in. In reality the global mean surface temperature of the Earth is about $+15°C$. This is the result of the presence of an atmosphere that contains water vapor and some other so-called greenhouse gases, which in addition to creating a friendly climate provide for the possibility for a number of other requirements for life to develop.

Human activities are, however, now gradually changing the composition of the atmosphere. The concentrations of the greenhouse gases are increasing because of human emissions. The radiative balance of the Earth is being disturbed. The global average surface temperature has increased by about $0.6°C$ during the 20th century and as expressed by the IPCC (2001a) "... a significant anthropogenic contribution is required to account for surface and tropospheric trends (of temperature) over at least the last 30 years".

Continued global warming may have far-reaching environmental consequences, which, however, have not yet been conclusively established. Nor are the implications for human life on Earth and the well-being of the human race well understood. Some fundamental questions naturally arise: How sensitive is the environment with its terrestrial and marine ecosystems to human disturbance in general, be it global climate change, destruction of the stratospheric ozone layer, reduced biodiversity, acidification of precipitation fresh waters, etc.? Or is the global environment rather resilient? To what extent is it possible to predict the consequences of even more extensive exploitation of natural resources? How urgent is it to take preventive measures and to what extent is adaptation to change adequate?

A global view of environmental issues is obviously a necessity when trying to answer these kinds of questions. The transfer of energy and the motions of air and water bring about a physical interdependence of what happens in different parts of the global

---

[1] A black body absorbs all radiative energy that reaches it, and emits the maximum possible radiation at the prevailing temperature as given by Planck's radiation law. The total outgoing radiation from a black body increases proportionally to the fourth power of its temperature. If subject only to radiative forcing, the temperature of the body changes until a balance between incoming and outgoing radiation is reached.

system. (In addition, there are also biotic linkages, e.g., through migratory species and the spreading of deceases.) This very fact will be at the center of our attention. We are actually in the midst of a process of finding out more about these spatial linkages, and it is clear that there will be no easy and clear answers for a long time. Uncertainty is part of the issue. Assessments of these major environmental issues will therefore largely have to be in the form of *risk* analyses.

The environmental system has a considerable inertia. It may nevertheless occasionally be changing abruptly, if some thresholds are surpassed and these are difficult to foresee. Mostly, however, changes take place slowly, and once a change has occurred it may take decades, sometimes a century or more, to restore the original setting, if this is at all possible. Similarly, society is not able to respond and act quickly, when major issues of environmental change emerge. We are thus concerned with an analysis of the interaction between two complex, non-linear systems, the global environment and the global human society, the future development of which is only partially predictable. Some principle features of such a so-called *chaotic system* will be outlined in the next section.

The following analysis will not be a comprehensive treatment of global environmental problems, but will rather focus on a set of issues of increasing importance and complexity. Recognition of these specific issues has come gradually, and the presentation will also provide a historical perspective. A detailed analysis of the Earth system as a background for the issues that will be raised in the following can be found in Jacobson et al. (2000).

- *Local effects* of emissions of gases as well as other substances into the atmosphere and the oceans and direct physical disturbances of life on land with its fresh water systems and vegetation are usually first experienced and recognized (cf. Section 3). Preventive and protective measures in the past have therefore begun with a focus on *local* damage and *local* mitigation. High smoke stacks and filters to avoid emissions of smoke have been installed. Similarly, emissions into watercourses, lakes, and coastal waters of the sea have been reduced. Much has been done in developed countries, but new problems still emerge. The methodologies applied and the experiences gained in developed countries need be transferred more effectively to developing countries.

- The *regional* scope of environmental degradation was not widely recognized until the late 1950s.

  (i) At that time local air pollution had increased within and around industrial centers in the United States and in Europe to a degree that required organized counter measures on a *regional scale*. Sulfur emissions, primarily emanating from the burning of oil and coal that contain sulfur, acidify precipitation, lakes, rivers, and soils and thereby damage vegetation [first detected by Svante Odén in 1968; see Sweden's Case Study (1971)]. This insight meant a recognition that it was no longer sufficient to build higher chimneys; limitations of emissions would be required (cf. Sections 3 and 4).

Nature could no longer be viewed as an infinite sink, an everlasting waste-basket for human activities.

(ii) *Fresh water management* similarly requires the development of action plans for whole river basins or watershed utilization in order to come to grips with the increasing issues of water pollution and the escalating demands of water for irrigation and industrial as well as domestic use. Drainage pipes farther out into lakes or the sea would not prevent increasing damage (cf. Section 6).

- It was soon thereafter also appreciated that some substances emitted into the atmosphere might stay there for weeks, years, or even centuries, while the characteristic mixing time for the global troposphere is merely about a year or two. *Global environmental issues* were becoming increasingly important and have also caught public attention in recent decades.

    (i) It was recognized in the early 1970s that the chloro-fluoro-carbon gases (CFCs) might decrease the amount of ozone in the stratosphere, which is of fundamental importance in protecting life on earth from destructive UV radiation from the sun [Crutzen (1971), Molina and Rowland (1974); cf. Section 4]. The life times of the CFC molecules were found to be on the order of a hundred years. They therefore spread all around the globe before disappearing very slowly. The ozone hole over the Antarctic continent discovered in the 1980s [Farman et al. (1985)] was the result of emissions primarily in Europe and North America.

    (ii) At about the same time the gradual enhancement of carbon dioxide concentrations in the atmosphere and possible associated changes of the climate of the Earth were established scientifically [Manabe and Wetherald (1975)], although Arrhenius (1896) had pointed out this possible long-term effect as the result of burning fossil fuels more than hundred years ago [Ramanathan and Vogelmann (1997); see further Section 7]. It would, however, still take time until a possible human-induced climate change would became a political issue [National Academy of Sciences (1979), Bolin et al. (1986)].

Today the threat to the environment is high on the political agenda in many countries and there is every reason to believe that it will stay there for a long time to come, although local issues may still temporarily overshadow the long-term impact on the environment as a result of human activities. It is obvious that the issues referred to above have been brought to the forefront by the natural scientists, and now the societal and economic implications are emerging in full strength. Good knowledge about the issues' characteristic features is becoming increasingly important for properly addressing them. This will require a much better understanding of how the interplay between the two complex systems of the environment on one hand, and the human society on the other, might evolve in the future.

2. The environmental system

2.1. *Key characteristics*

The global environmental system comprises the atmosphere, the oceans (including marine biota), the terrestrial systems (i.e., flora, fauna and soils), the freshwater systems and the cryosphere (i.e., snow and ice). Its present state has evolved gradually since the Earth was created as a barren planet about 4500 million years ago. Early in this course of events life was born, first in the form of primitive organisms in water, which later gradually developed into the rich variety of life forms that we find today all around us on land and in the sea. On the other hand, the development of human beings and the social structures that now exist have taken place during the last perhaps million years. Humans have spread to all corners of the world only during the very last few hundred thousand years and did not become the dominant species on Earth until about the last ten thousand years.

The birth of life and its later transformations and developments have been of profound importance for the evolution of the environmental system on which the human race is so fundamentally dependent. In this process physical, chemical, and biological processes have been closely interwoven and can really not be dealt with separately. Still, this chapter will focus on the geophysics of the environment, but a number of geochemical and biological/ecological issues will necessarily be brought into this treatise. A more detailed analysis of these will, however, involve an analysis of ecosystem dynamics, which will be dealt with in the following chapter.

As already alluded to, it is important to recognize that the features of this system embrace all spatial scales from that of the Earth itself, such as the global characteristics of the circulation of the atmosphere and the oceans, to those on the micro-scale, such as the features of a rain drop or a tissue of a plant or an animal. Similarly, we will be concerned with time scales from those that are of importance when analyzing variations of the global climate during hundreds of thousand years, to those that span merely fractions of a second as in the case of the molecular exchange of gases across the air–sea interface. As a matter of fact, the simultaneous treatment of processes on these vastly different space and time scales constitutes a major difficulty in our analysis and is often the reason for present uncertainties both in trying to explain what has happened in the past and to foresee future implications of increasing human intervention.

The research efforts in environmental sciences necessarily are different compared with controlled experiments in the laboratory that are carried out in order to verify a theoretical hypothesis qualitatively or quantitatively. Such experimental set-ups can be specified and controlled. The geophysicist, geochemist, or ecologist, on the other hand, is rather analyzing in real time the behavior of natural systems of considerable extension, which are increasingly being disturbed by human activities. Hypotheses and theories about the role of various possible mechanisms are formulated and tested by comparison with observations that describe the variability and change of the systems being studied. In this way we are able to increase our understanding and reduce uncertainties only step by step. Projections of likely future changes then presumably also become more reliable, but it is often difficult to quantify the progress being made. Also, so-called abrupt changes may occur and surprise us. This feature of a so-called chaotic system therefore deserves some further consideration.

2.2. Complexity and uncertainty

The non-linear character of a chaotic system, such as the environmental system, implies that its time evolution in principle is irreversible, although it may change in a semi-periodic fashion in response to external forcing such as that due to the daily, annual, or long-term variations of solar radiation. A state may, however, occasionally be reached from which some totally different evolutions are possible, some of which may be abrupt changes from a rather smooth course that until then has characterized the time evolution of the system. The existence of such "bifurcation points" implies that the system becomes unpredictable beyond some limited period of time.

Instabilities of this kind may occur on all scales of phenomena that are possible in the system. For example, small-scale wind vortices become unpredictable within seconds, which in general is a characteristic feature of turbulent motions of gases and fluids. Instabilities at larger scales of motion can, however, be determined approximately. The behavior of a cumulus cloud may be predictable for minutes to perhaps half an hour, and a mid-latitude storm or a hurricane for a few days once it has formed. Effects on even larger scales of motion can also be grasped statistically with fair accuracy, permitting some modest skill in predicting climate change, for example, if forcing by solar radiation or human activities is prescribed. Similarly vortices that are created by major ocean currents (e.g., the Gulf stream) may prevail for weeks and their motions and development are to some degree predictable.

Abrupt changes might, for example, be associated with major structural changes of the system, e.g., sudden and major changes of the distribution of lakes and water courses as a result of the melting of continental ice sheets (which seems to have happened when the Laurentide (Canadian) ice sheet withdrew some 8000 years ago); a partial collapse of the Antarctic ice sheet and an associated substantial rise of the sea level; or the disappearance of the sea ice in the Arctic in summer time.

Because of the complexity of the Earth system as well as the human society, projections of the consequences of future changes of the environmental system, regardless if they occur naturally or are caused by human interventions, are necessarily uncertain. They should therefore be considered as possible scenarios of the future, rather than predictions, but their construction and analysis is the only tool available when trying to grasp what might happen in the future. Their most important use might be to help us to understand the sensitivity of such projections to assumptions made, and to avoid undesirable consequences of harnessing natural resources.

2.3. A few principal considerations

We should of course try to learn from past experience, particularly when concerned with local changes of the environment. For example, developing countries have obviously much to learn from achievements as well as mistakes made in developed countries during the last century. Some lessons may also be learned about regional changes, although this is more difficult because of the need for data over quite large areas for considerable periods of time, which are seldom available. Developing countries are, for example, trying to exploit the experience gained in Europe in the case of acidification, which may be difficult, however, because the ecosystems differ markedly between temperate and tropical latitudes.

Any analysis of human damage to the environment dealt with in this chapter must be based on an analysis of the probable natural setting before humans intervened significantly. However, damage of the environment is often the final stage of a long process that may not in itself have been viewed as a threat until quite late, and the word 'damage' implies a judgment of values. The present analysis will therefore be limited to a description of past and expected changes. But it is important to recognize that losses for some may imply gains for others and that even an optimal strategy may well cause major disruptions for some.

The first task must then be to analyze ongoing changes, to try to determine what might be a result of human activities and what instead are natural variations. Such analyses should not be limited to what might be viewed as destructive. Benefits as a result of human exploitation of the environment, now as well as in the future, should obviously also be accounted for. However, changes of the services that the environment provides can only be evaluated on the basis of socio-economic analyses, which are the subject matter of later chapters.

Some basic concepts that are important for the analysis of spatial and temporal variations of gases and particulate matter in the atmosphere or pollutants in water bodies, as well as for the transfer of matter between the atmosphere, terrestrial systems, and the sea, are given below [Bolin and Rodhe (1973)].

Life time, residence time, age, turn-over time, adjustment time

(Average) life time = (average) residence time is the time that a molecule of the compound being considered on average stays in a reservoir (or pool) before being chemically or radioactively transformed, or transferred into a neighboring reservoir.

Average age is the average time the molecules present in the reservoir at a given moment have spent there.

Adjustment time is the time required to establish a new (quasi)equilibrium of the partitioning of a compound between exchanging reservoirs.

These different times are usually not necessarily the same. A distribution function for the residence time of individual molecules in a reservoir as a function of the time can be defined and may differ considerably from one case to another. If the reservoir is 'well-mixed', however, and the exchange with neighboring reservoirs is random, or if radioactive decay is the means for disappearance, the distribution function is exponential and in the case of a radioactive compound equal to the radioactive decay curve. In this case average life time, average residence time, and average age are all the same. They are also equal to the *turn-over time*, the latter being defined as the total amount of the compound in the reservoir divided by the flux into (or our from) the reservoir per unit time.

Adjustment time is, on the other hand, an entirely different concept. Here the mutual exchange between reservoirs in a case of disturbances is considered and the time it will take until *net* fluxes induced between the different reservoirs because of the disturbance have declined to close to zero and a new equilibrium has been established.

The next two sections will deal with purely atmospheric issues, *air quality* and *air pollution* (Section 3) and *depletion of stratospheric ozone* (Section 4), to be followed by a closer look at *water pollution* and *the management* issues (Section 5), which concern human actions that directly affect fresh waters and soils. A specific issue is then dealt with in more detail, i.e., *acidification* (Section 6). Finally the issue of *global climate change* (Section 7) requires an analysis of the total environmental system.

All these specific environmental issues have a bearing on how humankind might be able to secure sustainable development for the future. This will be touched upon in the concluding section of this chapter (Section 8).

3. Air quality and air pollution

Air and water pollution are the results of intentional or unintentional human emissions of substances into the atmosphere and global water system. For emissions of substances that are toxic, represent health hazards, damage flora and fauna or cause other damages, e.g., on buildings or outdoor constructions, it has been common to define critical loads above which damages become serious enough to call for mitigation. Many difficulties arise, however, in such an approach. Pollution usually is a mixture of different pollutants causing different damages; their concentrations depend much on distance from the source and meteorological conditions. There is also the matter of whether the accumulated dose or peak concentrations matter, i.e., long-term and short-term exposure. The substances that are emitted may also be decomposed chemically, in the atmosphere, for example, under the influence of solar radiation, or because of attachment to cloud drops, rain drops, or snow and thereby brought back to the Earth's surface. In analogous ways water pollution may be modified while being dispersed. A closer analysis is therefore often required in order to determine an approximate strategy for response.

Methods for assessing the required reductions of emissions in order to stay within given concentration levels have been developed since the early decades of last century and are generally available [Seinfeld (1986)]. The turbulent nature of air and water in motion and the complexity of the physical, chemical, and biological processes do, however, not permit a precise determination of measures that may be required. It is obvious that dealing with air and water pollution is a problem of risk analysis.

3.1. Atmospheric composition

Air is a mixture of a large number of gases and particulate matter. Table 1 provides an inventory and important characteristics of key components. For millions of years the four most common gases of dry air have been nitrogen, oxygen, argon, and carbon dioxide. They make up more than 99.99% of dry air. Nitrogen and argon have accumulated very slowly in the atmosphere during hundreds of millions of years. Oxygen has similarly been an almost constant component of the atmosphere for a long time, but less extended, time. Its concentration is presently decreasing very slightly because of the formation of carbon dioxide when burning fossil fuels and deforestation [Keeling, Piper and Heimann (1996)]. However, this change can barely be measured because it is very small compared with its total concentration. Carbon dioxide, on the other hand, has increased from about 280 ppmv (parts per million by volume) to 370 ppmv during the last about 200 years, i.e., by about 32%.

Pre-industrial amounts of other minor components of the air as well as ongoing changes due to human activities are given in Table 1 and will be considered in more detail below (Section 3.4), but first some basic features of the atmosphere will be provided.

3.2. Structure and mixing of the atmosphere

The atmosphere is a thin shell around the Earth. 90% of the air is found below an elevation of about 16 km, and 99% below about 31 km, which is still small compared with the radius of the Earth (about 6700 km).

The global mean temperature at the Earth's surface is about $+15°C$. It decreases with elevation through the lower parts of the atmosphere (the *troposphere*) by 0.6–1.0°C per 100 m, up to the *tropopause*, at a height of about 10 km in polar regions and about 16 km at tropical latitudes (Figure 1). This decrease of temperature is the result of the heating of the Earth's surface by incoming solar radiation and vertical mixing within the troposphere.

The *stratosphere* extends from the tropopause to the stratopause at about 50 km. Here absorption of solar radiation by oxygen and ozone provides another heat source and the air temperature increases with height. The temperature stratification in the stratosphere is therefore stable, and vertical mixing is much reduced. A temperature maximum is reached at the *stratopause,* where the temperature is at about the freezing point and the air density is less than 0.1% of that at the Earth's surface.

We define the concept of *mixing time* as the time required to reduce concentration differences by a factor of about 2.7 (e). This time is dependent on the characteristic air motions and the size of the region being considered. The average vertical mixing time of

Table 1

Composition of the atmosphere in the year 2000

Gas or particulate matter	Concentration (parts per million by volume, ppmv)*	Turn-over time	Annual change (percent)	Effects[†]
Nitrogen (N$_2$O)	780,800	millions of years		
Oxygen (O$_2$)	209,400	1000s of years	about −0.001	
Argon (Ar)	9,330	millions of years		
Carbon dioxide (CO$_2$)	368	about 5 years	about +0.5	Cl (OS)
Neon (Ne)	18.2	millions of years		
Helium (He)	5.2	millions of years		
Methane (CH$_4$)	1.7	about 10 years	< +0.5	Cl, OS, OT
Krypton	1.14	millions of years		
Hydrogen (H$_2$)	0.56	about 2 years		
Nitrous oxide (N$_2$O)	0.31	about 120 years	about +0.2	Cl, OS
Carbon monoxide (CO)	0.15	about 6 months	about +1	OT, H
Ozone in troposphere	0.02	about 1 month	increasing	OT, Cl, P, H, Co
Ozone in stratosphere	0.2–1.0	variable	decreasing	OS, Cl, H
N oxides (NO, NO$_2$)	< 0.0005	a few days		OT, A, H, Co
Sulfur dioxide (SO$_2$)	< 0.0002	a few days		A, P, H, Co
Ammonia	< 0.0002	a few days		A, P
Hydrogen sulfide (H$_2$S)	< 0.0001	a few hours		A, Co
CFC-11 (CFCl$_3$)	0.00027	about 50 years	almost stable	OS, Cl
CFC-12 (CF$_2$Cl$_2$)	0.0005	about 100 years	increasing	OS, Cl
Hydrofluorocarbons		1–200 years	increasing	OS, Cl
Cl-hydrofluorocarbons:				
HCFC-22 (CHClF$_3$)	0.00013	12 years	about +5	Cl
Others	< 0.00001	1–18 years		OS
Perfluorocarbons	< 0.0001	3000–50000 years	increasing	Cl
Sulfur hexafluoride	0.00003	3200 years	increasing	Cl
Particulate matter:				
Sulfate (SO$_4$)		few days to weeks		A, H, Cl, Co
Toxic organic substances				H, P
Heavy metals, particularly lead (Pb) and cadmium (Cd)				H
Radionuclides				H

*Concentration in "pure" air, i.e., air outside population centers and industrial regions.

[†]The symbols indicate:

A: acidification;

H: health effects;

Cl: induces climate change;

Co: causes corrosion;

OS: influences the ozone concentration in the stratosphere;

OT: influences the ozone concentration in the troposphere;

P: affects plants.

Sources: IPCC (1996, 2001a), Jacobson et al. (2000).

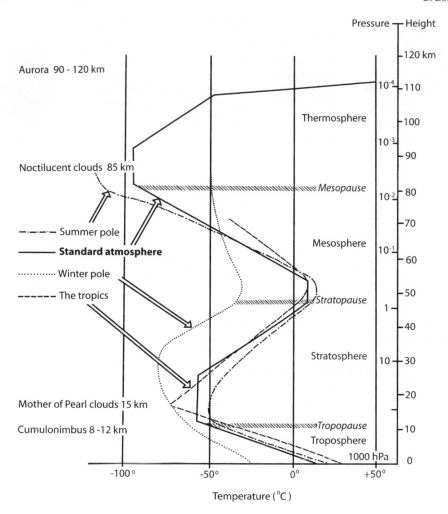

Figure 1. The vertical temperature structure of the atmosphere.

the troposphere is about a month, and cyclones and storms are the major agents for this turnover of the lowest 10–15 kilometers of the atmosphere. Horizontal mixing within the Northern or the Southern hemisphere takes several months, and for the globe as a whole a year or two. The so-called Inter-Tropical Convergence Zone (ITZ) is a rather effective barrier for N–S mixing.

Pollutants that are emitted into the atmosphere and have residence times of merely days or hours affect the composition of the atmosphere only in the immediate surroundings of the sources. Neither do they within that time spread much above the surface boundary layer (that is about a kilometer deep). In many parts of the world and particularly in the sub-tropics the stratification is often stable above this boundary layer because of a temperature inversion (the temperature increases with elevation), which effectively reduces mixing of pollutants to higher elevations.

If emissions come from rather few sources it may well be advantageous to build high chimneys and make use of the fact that horizontal mixing will have reduced the concentrations effectively in a smoke plume when it touches the ground at some distance away from the source. This way of solving the pollution problem of course becomes less effective in more densely populated areas with emissions from a number of sources.

3.3. The fundamental role of water vapor in the atmosphere

Air always contains some water vapor. The maximum amount is determined by the saturation pressure of water vapor. At most a small fraction of a gram of water may be present in the form of water vapor in a kg of dry air at temperatures far below freezing. On the other hand, the amount of water vapor may exceed 30 g/kg at a temperature of about $+30°C$, without condensation occurring. An air parcel cools when rising to higher elevations because of its expansion, and at some level condensation therefore usually occurs, which leads to cloud formation, but not necessarily to precipitation. Cloud drops are formed on tiny particles and may remain in liquid phase even if the temperature is well below freezing. Their size (radius) is usually a μm (a thousandth of a millimeter) or less and they settle only slowly through the air. Some particles initiate the formation of ice crystals, which grow more rapidly (by sublimation) than do droplets. Clouds that have formed at temperatures below freezing therefore may contain both ice crystals and super-cooled water drops. When this is the case, the cloud droplets will evaporate and provide water for the ice crystals to grow into snowflakes. These then fall more rapidly and break up into smaller fragments, which in turn serve as nuclei for the growth of new snowflakes. Precipitation thus usually begins by the formation of snow high above the Earth's surface even in summer time and at tropical latitudes. However, cloud drops may also grow quickly at temperatures above freezing at low latitudes because of the high moisture content of warm air. The droplets collide and form bigger drops that similarly lead to precipitation.

The presence of water in the form of water vapor in the atmosphere is of fundamental importance for the dynamics of atmospheric disturbances. Evaporation from water surfaces at the surface of the Earth requires energy and the water vapor carries this energy up into the atmosphere. When the air is humid, the stratification becomes unstable more easily. Towering cumulonimbus clouds can form causing heavy rain, sometimes with hail and thunder. As condensation occurs in an atmospheric disturbance the rising air is heated by the supply of this energy through the process of condensation and the disturbance intensifies. This applies to the growth of small convection clouds (fair weather clouds) into thunderstorms, middle latitude cyclones into intense storms, as well as so-called easterly waves in the tropics in the late summer and the fall being transformed into fierce hurricanes.

Clouds and precipitation are also of fundamental importance as cleaning agents for air pollution by eliminating particles (aerosols) emitted into the atmosphere as a result of human activities. If hydroscopic in nature, these may serve as condensation nuclei or are otherwise incorporated into cloud drops, raindrops and snowflakes and thereby brought back to the Earth's surface. This is called *wet deposition*. The frequency of the occurrence of precipitation largely determines the residence times for these compounds in the atmosphere, which vary between a few days in rainy climates to weeks or possibly a month or more in dry climates. Aerosols may also be directly absorbed by or attached to the vegetation or absorbed by open water surfaces, which is called *dry deposition*.

The presence of water droplets and ice crystals also influences chemical reactions in the air by providing the setting for heterogeneous chemistry, which has turned out to be of basic importance for understanding the ozone chemistry in the stratosphere (see Section 4).

3.4. Human-induced changes of atmospheric composition, air pollution

3.4.1. General features

The term air pollution usually denotes changes of atmospheric composition that have a directly disturbing and/or damaging effect on humans, outdoor human activities and constructions, as well as on flora and fauna. It is, however, of some importance to distinguish between the emissions of extraneous substances that are directly harmful to ecosystems or human health, and on the other hand emissions of compounds that are part of the natural setting, but whose concentrations are enhanced by human activities. In any case, however, the complex web of chemical reactions in the atmosphere will be modified.

Three important gases of the latter kind are of particular interest because of their fundamental role for life on Earth: carbon dioxide (CO_2), sulfur dioxide (SO_2), and nitrogen oxides (NO_x). They all are emitted primarily when burning fossil fuels, in particular, in coal- and oil-fired power plants and when being used as fuels in the transport sector. *Carbon dioxide* is chemically inert in the atmosphere and is generally harmless, but increasing concentrations influence the global climate, which will be considered further in Section 7. Human-induced *sulfur dioxide* emissions today dominate the natural circulation of sulfur (i.e., the sulfur cycle) in the industrial parts of the world, may be directly harmful to human health and are increasingly acidifying precipitation, fresh waters, and soils (see Sections 3.4.3 and 6). *Nitrogen oxide* emissions, emanating from combustion processes at high temperatures and the enhanced use of nitrogen fertilizers in agriculture, modify the natural nitrogen cycle (see Section 3.4.2). Nitrous oxide – primarily formed in the process of nutrient overturning in the soil (denitrification) – is a greenhouse gas (see Section 7.3.3). Its concentration is enhanced, particularly as a result of using artificial fertilizers. *Methane* (CH_4) is another natural component of the atmosphere, that is active as a greenhouse gas, and whose concentration has been greatly increased as a result of human activities (see Section 7.3.2). *Carbon monoxide*

(CO) is also formed naturally, but concentrations are much enhanced at middle latitudes of the northern hemisphere as a result of incomplete combustion, e.g., when burning waste from agriculture and forestry, particularly at tropical latitudes (see Section 3.4.4).

A large number of minor constituents are emitted from industrial processes, domestic activities, and increasing traffic. Most of them take part in the complex set of atmospheric photochemical reactions, which are therefore modified by these human activities. Chlorine is emitted from bleaching processes, while other organic compounds stem from the chemical industry. Unburned hydrocarbons leak into the air in the course of mining fossil fuel deposits, heavy metals leak from the manufacturing industry, etc. (see Section 3.4.5). Some of these extraneous gases are innocuous in themselves, as, for example, most *halocarbons*, but have a profound effect on the concentrations of ozone in the stratosphere and are also strong greenhouse gases. Others are directly poisonous and are harmful to flora and fauna as well as humans.

Many of these compounds spend only a short time in the atmosphere, are decomposed and disappear, but still have an impact while present. Others, which decompose slowly, travel long distances and reach any corner of the world before being transformed chemically or deposited on land, terrestrial ecosystems, into lakes, rivers, and the sea. *Heavy metals* are of course not destroyed being natural elements, are often toxic and may again be set in motion after having been hidden in deposits for long times.

We will in the following sub-sections consider in more detail these human-induced changes of atmosphere composition and their direct local impacts, while the major global threats to the environment on regional and global scales will be dealt with in later sections.

3.4.2. Nitrogen oxides (NO_2, NO, and N_2O)

Nitrogen is a fundamental element in the building of a number of organic molecules in nature but must be supplied in forms that can be used by plants. Nitrogen gas consists of two atoms that are tightly knit together, and cannot usually be directly incorporated into plant tissues. The access of nitrogen for plant growth is rather accomplished by symbioses between fungi (mycorrhiza), algae, and roots of some plants. Ammonia is formed, from which the plant can extract the nitrogen needed. Most of the nitrogen in dead organic matter is also available for plant growth and only a rather limited amount is transformed into molecular nitrogen and nitrous oxide by denitrification, and returned to the atmosphere. The circulation of nitrogen in Nature in the form of compounds that are accessible for plants has a rather small leakage back to the atmosphere. This is a fundamental feature of terrestrial ecosystems and a result of the evolution of plants and ecosystems over millions of years.

Nitrogen oxides (i.e., NO and NO_2, often denoted as NO_x) are naturally present in air in concentrations less than 0.0005 ppmv. They play a role in the formation and destruction of ozone (see Sections 3.4.6 and 4) and are poisonous and corrosive, however, only modestly so at the low concentrations that occur naturally. They are formed when nitrous oxide (N_2O) is decomposed by UV-radiation and by oxidation of nitrogen in lightning flashes. They take part in numerous natural chemical reactions and are incorporated into cloud and rain drops, ultimately ending up as nitric acid in the soil.

This natural pattern of sources and sinks has, however, changed dramatically because of human-induced emissions. NO_x concentrations have been much enhanced during the 20th century as a result of combustion in industrial regions and also when fertilizers are being used in agriculture. They are major components of industrial and urban air pollution. and contribute significantly to acidification of fresh water systems and soils (see Section 6). However, their transformation into nitrate makes them also a nutrient source for plants. Nitrogen is important for the life processes, but also poisonous and destructive when appearing as NO_x or nitric acid in enhanced concentrations. More detailed descriptions of the nitrogen cycle are available in Jacobson et al. (2000), Seinfield (1986), and IPCC (2001a, Chapter 4).

3.4.3. *Hydrogen sulfide, sulfur dioxide, and sulfuric acid (H_2S, SO_2, and H_2SO_4)*

Sulfur is also required for the formation of a number of organic compounds and is naturally circulating in nature. The gas dimethyl-sulfide is formed when organic matter is decomposed, emitted into the atmosphere and rather quickly transformed into hydrogen sulfide and sulfur dioxide by oxidation. In the presence of sun light sulfur dioxide, in turn, is further oxidized into sulfur trioxide, which is dissolved in cloud droplets as sulfuric acid and transformed into sulfate particles when the water evaporates at low humidity. The volatile sulfur compounds have long been present in the atmosphere. The small amount of sulfuric acid that is being formed naturally has similarly been part of the natural atmosphere. The lifetime of a sulfur molecule in the atmosphere is merely a few days to a few weeks.

Human activities now cause emissions of sulfur in amounts that are much larger than those that are circulating naturally. The prime source is sulfur dioxide from burning coal and oil that contain sulfur. Figure 2 shows the increase of annual emissions since the middle of last century. During the last decades of the 20th century the emissions decreased in developed countries, while a continued and rapid increase was observed in developing countries, particularly in East and Southeast Asia. Obviously human-induced emissions today dominate the circulation of sulfur in nature. The emissions as well as atmospheric concentrations of sulfur dioxide in densely populated regions, such as Western Europe, the eastern United States, and the Far East, are about ten times larger than during pre-industrial times. Because of the short lifetime of sulfur molecules in the air, enhanced concentrations of sulfur dioxide and high deposition rates of sulfuric acid and sulfate are primarily found within a few thousand kilometers of source regions. The oxidized sulfur is corrosive; it damages vegetation and thus crops. When inhaled in enhanced concentrations, the respiratory system is damaged. Major efforts have been made in developed countries to deal with this issue. Similar actions are under way in developing countries, particularly in Southeast Asia and China. More detailed information is provided in Jacobson et al. (2000) and IPCC (2001a, Chapter 5).

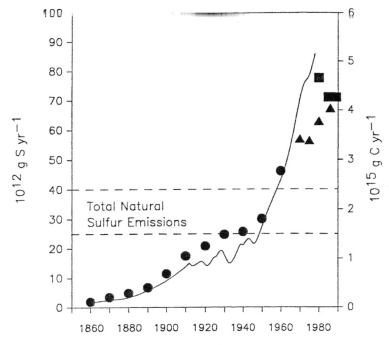

Figure 2. *Solid shapes* (left scale): annual emissions of sulfur, 1860–1990. *Solid line* (right scale): annual emissions of carbon dioxide from fossil fuel burning. Sulfur emissions stem primarily from the sulfur content of oil and coal. Source: Berresheim, Wine and Davis (1992).

Sulfur is the main pollutant that acidifies precipitation and therefore fresh water systems and soils. This environmental threat is considered in Section 6.

3.4.4. Carbon monoxide (CO)

Carbon monoxide has long been a permanent component of air and present in the atmosphere in concentrations between 0.01 and 0.2 ppmv, the global average having earlier been about 0.04 ppmv. It is formed by oxidation of hydrocarbons in the atmosphere and by decomposition of organic matter, either by bacteria or by incomplete combustion, e.g., in forest fires. It is destroyed primarily by reaction with the hydroxyl (OH) radicle whereby CO_2 is formed. Pre-industrial concentrations were already higher in the Northern than in the Southern hemisphere, a difference that has become more pronounced in the course of the 20th century.

The average residence time for CO in the atmosphere is merely about a month, thus shorter than the north–south mixing time for the atmosphere. Mixing is thus not quick enough to reduce the north–south concentration gradient that the different sources in the two hemispheres give rise to. This is the reason why pre-industrial concentrations were considerably lower in the southern than in the northern hemisphere, where more extended the terrestrial ecosystems serve as source regions.

The present enhanced CO emissions are primarily a result of incomplete combustion of fossil fuels, primarily in automobiles, and biomass burning. They are also a product of the oxidation of the enhanced amounts of hydrocarbons in the atmosphere (particularly methane) by the OH radicle. The average global concentration is therefore now about 0.07 ppmv, corresponding to about 360 Mt.[2] It may locally be very much higher in industrial areas and in heavy traffic, where at times of poor ventilation and calm weather it can reach dangerous concentration levels, i.e., exceed 10 to 20 ppmv, and constitute a health hazard. Preventive actions are being taken in developed countries, but the situation is aggravating in the rapidly growing mega-cities in the developing world. The global emissions are at present about 2500 Mt CO yr^{-1}, of which more than half is the result of human activities [see further IPCC (2001a, Chapter 4)].

3.4.5. *Volatile organic carbons (VOC) and their derivatives*

Volatile organic carbons, which include non-methane hydrocarbons (NMHC), have short residence times in the atmosphere, but still play an important role in atmospheric chemistry, and they serve as precursors for aerosol formation. A number of *hydrocarbons* are naturally present in small amounts, but concentrations have increased markedly because of human emissions. Many extraneous compounds are being added because of leakage when extracting, refining, and using oil and natural gas, and also through biomass burning to provide energy in developing countries. They are common components of urban air pollution and play an important role in the complex web of chemical reactions in polluted air, for example, in the formation of smog (the interplay of smoke and fog). See further Jacobson et al. (2000), Seinfield (1986).

Halocarbons arc simple hydrocarbons (e.g., methane, ethane) in which some or all hydrogen atoms have been replaced by chlorine, fluorine, or bromine atoms. Their presence in the atmosphere is almost exclusively the result of human activities. They are transparent, largely inert and were considered harmless when first made use of in industry in the 1930's. They have had and still have extensive industrial use, for example, as cooling agents for refrigeration and air conditioning and as cleaning agents in the micro-electronic industry.

Whole ranges of halocarbons have been emitted during the last 50 years [see further IPCC (2001a, Chapter 4)]. They have significantly modified atmospheric chemistry particularly by changing the ozone balance in the stratosphere (see Section 4), and they influence the radiative fluxes through the atmosphere and contribute thereby to the ongoing climate change (see Section 7.3).

---

[2] 1 Mt (megaton) $= 10^6$ ton.

3.4.6. Ozone (O₃)

Ozone is formed naturally in the atmosphere. Its presence is the result of the photo-dissociation of oxygen (O_2) into two oxygen atoms, one of which is energetically excited and able to initiate a series of chemical reactions involving also nitrogen oxides and hydrocarbons. Its concentration in the lower *troposphere* (often called surface ozone) was merely about 0.02 ppmv or less before the industrial revolution. It has increased since the latter part of the 19th century, particularly in industrial regions as the result of enhanced photo-oxidation in the presence of increasing amounts of NO_x in the air. Concentrations in industrial regions are now rather about 0.04 ppmv and may temporarily reach values above 0.1 ppmv in heavily polluted areas. Ozone affects both flora and fauna already at quite low concentrations. It inhibits plant growth by damaging the foliage. Respiratory organs in animals and humans may be damaged. In order to avoid increasing ozone concentrations it is necessary to reduce air pollution, particularly emissions of nitrogen oxides and hydrocarbons. This has succeeded partially in developed countries, but high ozone concentrations are now also found in densely populated regions in many developing countries.

The ozone concentration increases with elevation and reaches a maximum, about 1 ppmv, in the *stratosphere* at elevations between 20 and 25 km, but it varies considerably in space and time. It has, however, decreased very significantly during the last two to three decades because of the destructive catalytic reactions with halocarbons. The issue of stratospheric ozone depletion will be analyzed in more detail in Section 4.

3.4.7. Particulate matter

Particulate matter is present in the atmosphere because of either being emitted mechanically by the force of the wind in dry climates (mineral dust) or as a result of chemical transformation of gases emanating from natural sources (such as fires) or human activities [Seinfield (1986), IPCC (2001a, Chapter 5)].

In general, mechanically produced particles are comparatively large, several micrometers (μm) or more. They settle through the atmosphere rather quickly and their residence times are short (of the order of hours or less). However, the very smallest particles travel long distances, particularly from the major deserts of the world. For example, dust from Sahara reaches the West Indies, and dust that has been found in the ice sheets in the Antarctic and that stems from glacial times, 10,000–15,000 years ago, may have come from Asian deserts [IPCC (2001a, Chapter 5)].

Small drops form as a result of breaking ocean waves. These evaporate rather quickly if the air is not saturated and are transformed into small saline drops or solid salt particles. The haze that is often found in coastal regions, when fresh or strong winds blow from the sea, is commonly a result of the presence of small salt particles.

However, much of the particulate matter in the atmosphere stem from combustion, either directly formed in the combustion process or from the transformation of so-called precursor gases that are emitted. They are generated in combustion of fossil fuels as well as when burning biomass and organic waste. Most commonly, ash and soot particles are formed, which make the smoke visible. Filters are used successfully to reduce their escape from stationary sources. Burning of waste, particularly in developing countries, remains, however, as a major source of smoke that may contain harmful organic compounds and heavy metals, of which cadmium and lead are toxic. It is also difficult to prevent the emissions of particles from traffic exhaust. Diesel motors, using oil rather than gasoline, emit substantial amounts of particles, some of which consist of soot while others are formed from gases in the exhaust. Health hazards caused by aerosol emissions have been summarized in IPCC (2001b, Chapter 9).

Soot of course absorbs solar radiation and reduces the amount that reaches the surface of the Earth. This became obvious when a large number of oil wells were set on fire during the Gulf War in 1991. Poisonous gases and large amounts of soot were emitted. The local temperature at the Earth's surface in and around Kuwait was reduced significantly when the wells were on fire. On the other hand, the temperature rose at and above the height where the smoke stabilized because of the absorption of solar radiation.

Crutzen and Birks (1982) foresaw this effect in a study of the possible implications of extensive fires from a global nuclear war, which might lead to the emission of large amounts of soot, in turn causing extensive cooling. The heating as a result of absorption of solar radiation aloft would create a more stable stratification of the atmosphere, which would disappear slowly and which might imply serious damage because of freezing temperatures at the Earth's surface underneath a temperature inversion well into the tropics. Actually, the experience gained during the Gulf War may be viewed as a qualitative, although spatially limited, validation of the very serious environmental effects of a global nuclear war.

3.5. An integrated approach to the air pollution problem

A critical concentration of a pollutant, which must not be exceeded if damage is to be avoided, can sometimes be defined. This may be the case when concerned with the influences of poisonous substances on human health, but in most cases this is difficult. Integrated models that assess the combined effects of the many components of atmospheric pollution are still in their infancy because one and the same pollutant may have different damaging effects depending upon the subject being exposed and the standard used in defining damage. For example, sulfur dioxide is corrosive, causes acidification of soils and fresh waters, and affects human health.

It is seldom sufficient to assess the impacts of one pollutant at a time, but rather the combined effect of several pollutants may be of more concern and there may be positive or negative feedback mechanisms in the system that need careful consideration. For example, increasing amounts of halocarbons in the atmosphere mean more target molecules for destruction by OH-radicles, which will reduce their availability for oxidation of other pollutants or, e.g., methane.

Weather situations during the cold season with weak winds and lack of vertical mixing during a week or longer may lead to serious accumulation of pollutants in an atmospheric surface boundary layer of merely a few hundred meter's thickness. Young children and elderly people, particularly individuals with respiratory illnesses, may then be severely hit. Many large cities in developed countries have by now taken major steps to reduce emissions of pollutants, this being the only approach to avoid catastrophic incidents of this kind. Rapidly growing mega-cities in developing countries that have a distributed heating system with poor means for reducing emissions are today often worse off than the major cities in Europe and North America were half a century ago. The necessity of attending to the immediate and local needs may also lead to lack of interest and resources to address the long-term and global issues that also may require early attention in order to secure that preventive measures become effective early enough in order to avoid more serious damage. Solutions should be looked for that are optimal in the sense that local *and* short-term as well as global *and* long-term issues are being addressed simultaneously. This is indeed one fundamental aspect of the concept of sustainable development.

4. Depletion of the ozone layer

4.1. The natural distribution of ozone in the stratosphere

Most of the UV-radiation from the Sun of wavelengths < 240 nanometers (nm)[3] is absorbed in the stratosphere by oxygen. The excited oxygen atoms combine with other oxygen molecules to form the three-atom molecule of ozone. This production of ozone is partly balanced by decomposition by UV-radiation from the sun (primarily at wavelengths less than about 320 nm). Quasi-equilibrium is established between formation and destruction that leads to a layer with enhanced ozone concentration, about 10 km thick and with a maximum concentration of about 1 ppmv at about 15–20 km above the Earth's surface (cf. October average above Antarctica in Figure 3). The concentration and the level of the maximum vary seasonally, and are also functions of latitude because of the variations of solar radiation. The total amount of ozone in the stratosphere is equal to that of a layer merely about 3 mm thick if compressed to the air pressure at the Earth's surface. This is the so-called 'ozone layer'; see further Crutzen (1996).

The destruction of ozone by radiation may be enhanced catalytically by compounds, such as H, OH, NO, Cl, and Br. These are present naturally in the stratosphere only in small amounts and play an insignificant role for the ozone balance, except for NO. The presence of NO in the stratosphere is, however, primarily as a result of decomposition of nitrous oxide, N_2O, which contributes to the natural destruction mechanism for ozone at these levels. This was confirmed for the first time by the observation of enhanced NO concentrations and a simultaneous decrease of the ozone as a result of a major solar proton event in August 1972 [Heath, Krueger and Crutzen (1977)].

---

[3] 1 nm (nanometer) = 10^{-9} m = 10 Angströms.

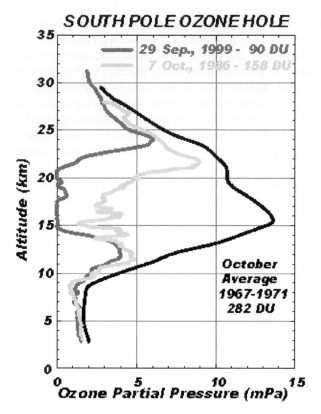

Figure 3. Vertical profiles of ozone in mPascal (= 10^{-5} mb) at McMurdo, Antarctica (latitude 78°S) in October 1986, at the time of discovery of the 'ozone hole', and in September 1999, showing complete disappearance of ozone between 15 and 20 km altitude. For comparison, the average October profile during 1967–1971 is also shown. Source: NOAA/OAR/CMDL (2002).

The absorption of UV-radiation by the ozone layer protects organic compounds and living cells from destruction and is therefore a prerequisite for the existence of life on Earth. A very small fraction was let through during pre-industrial times and has to some extent contributed to the development of skin cancer (malignant melanoma) and eye tumors. A major reduction of the ozone layer would be serious for plants, animals, and humans on Earth [Caldwell and Flint (1994), De Fabo and Noonan (1996)].

4.2. *Human impacts on the stratospheric ozone layer*

Knowledge about the mechanisms that maintain the ozone layer increased rapidly in the early 1970s, largely as a result of intensified research efforts because of fears for environmental implications of a rapid increase of supersonic transport in the stratosphere [see Crutzen (1996), Social Learning Group (2001)].

At about the same time it was also recognized that emissions of CFC gases into the atmosphere increase the chlorine concentration in the stratosphere, which in turn enhance the natural ozone destruction [Molina and Rowland (1974)]. The basic scientific principles about the proposed reactions were soon accepted, but extensive controversies arose about the magnitude of an expected reduction of the ozone content. Restrictions in the use of CFC gases were, however, agreed in some countries (e.g., in USA and Sweden), and the issue largely disappeared from the public scene, in spite of the fact that the possible implications of such a change for life on Earth might be serious. However, UNEP arranged regular meetings between countries in attempts to reach a political agreement on a Convention for the protection of the stratosphere.

In 1985 Farman and colleagues at the British Antarctic Survey discovered that the ozone layer was seriously depleted for a few months every spring (September–October) over the Antarctica and more so in recent years. This finding quickly engaged the scientific community and soon also became public knowledge. The term 'the ozone hole' was coined for the phenomenon (Figure 3). The intricate chemical interplay behind this threatening development was settled quickly [Crutzen and Arnold (1986), Toon et al. (1986)].

Under normal stratospheric conditions, the prime chemical reactions would involve NO, NO$_2$, and ClO but not ozone. Compounds would form that would not much affect the ozone concentration. It was known that nitrogen oxides by themselves would reduce the amount of ozone, being part of the 'natural' destruction mechanism, but chlorine would by itself be much more effective. What might then reduce the amounts of nitrogen oxides in the stratosphere? What is unique about the Antarctic region, the enhanced destruction of ozone essentially being limited to this distant part of our globe? The extraordinarily low temperatures in the Antarctic stratosphere in winter turned out to be the key to an answer.

At temperatures below about 200 K (i.e., $-73°$C), nitric acid crystals form and the nitrogen oxides are effectively removed, leaving the playground to chlorine. In the course of the dark southern hemisphere winter the stratosphere cools and the stage is gradually being set for a catalytic process of ozone destruction when solar radiation returns in early spring. This situation prevails until the nitric acid crystals again vaporise in the spring. Laboratory experiments and careful thermodynamic analyses have validated the correctness of this interpretation [Solomon et al. (1987)].

CFC gases have been emitted into the atmosphere during more than half a century and are present in concentrations up to about 0.0005 ppmv (CFC-12), but most of them are less abundant. When decomposed by the UV-radiation in the stratosphere, their chlorine content is set free, but only a small fraction is decomposed annually. Their residence times are 50 years or more. They are therefore quite homogeneously distributed throughout the troposphere and lower stratosphere and their stratospheric concentration is not much affected in the course of a year.

The HCFCs, on the other hand, disappear more quickly because the UV-radiation that penetrates into the troposphere is of sufficient energy to decompose them. Clouds and precipitation wash out the chlorine that is released rather quickly. Their effects as ozone-depleting substances in the stratosphere are therefore considerably less.

There is also an interesting and potentially important interplay between destruction of the ozone layer and global climate change. The expected warming at the surface of the Earth is coupled with decreasing temperature in the stratosphere (cf. Section 7), which also is partly the result of decreasing amounts of stratospheric ozone. We note then that low temperature in the stratosphere over Antarctica is the key to the formation of the ozone hole. It is not expected that equally low temperatures will develop over the Arctic, but the possibility of synergistic effects of this kind demands our continued attention. Some decline of stratospheric ozone (by up to about 10%) has in fact been observed at high latitudes of the Northern Hemisphere during the last decade.

It is finally worth noting that chlorine rather than bromine was chosen to replace the hydrogen atoms in the halocarbon molecule when the CFCs were first developed in the 1930s. The main reason for this choice was to minimize costs of manufacturing. It is fortunate that chlorine is much less effective in reducing ozone than bromine, which of course was not known when the CFC-gases were first put on the market.

4.3. Measures to protect the ozone layer against destruction

Less than two years after the discovery of the ozone hole in 1985 an agreement was reached by the countries that had signed the Vienna Convention for the Protection of the Ozone Layer. This agreement, the "Montreal Protocol", implied that the emissions of long-lived chlorine compounds would be reduced substantially within a decade. A graph of the kind shown in Figure 4 served as a basis for this agreement. The projections of Cl concentrations shown in the figure have remained about the same ever since.

Limitations of the emissions of CFC-gas emissions were agreed in 1987, but the ozone depletion became worse. The vertically integrated amount of ozone over parts of the Antarctica was at times merely 40–50% of what was observed during earlier decades. As was also foreseen, concentrations of CFC gases continued to rise because emissions were still larger than the natural annual destruction.

Since the lifetime of the CFC gases in the atmosphere is 50 years or more, the springtime decline of the ozone layer is expected to return annually for several decades into the 21st century. A recovery of the ozone layer will only come about gradually. The inertia of the system is partly associated with the time required for the CFC molecules to be dispersed up to and well into the stratosphere. It would take on the order of a decade before a drastic reduction of CFC emissions would be noticed at the level of the ozone layer.

Reductions of the emissions of CFC gases are also important in the attempts to reduce the pace of human-induced climate change (see Sections 3.4.5, 7.3 and 7.6).

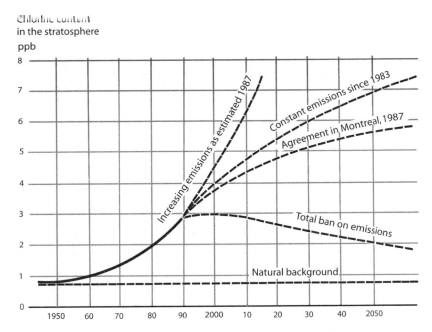

Figure 4. The observed concentration of chlorine in the stratosphere through 1987, and projected concentrations associated with different policy measures proposed in Montreal in 1987. Units: parts per billion.

5. Water pollution and water management

5.1. The global hydrological cycle

Life on Earth is crucially dependent on the availability of water and its circulation between the atmosphere, the land, and the oceans. A global overview of the magnitude of the major water reservoirs and the annual flows between them is given in Figure 5, expressed in units of 10^{12} ton and 10^{12} ton yr^{-1}, respectively.

The land surface area is $148 \cdot 10^{12}$ m^2 and that of the oceans $362 \cdot 10^{12}$ m^2. The average annual precipitation is about 0.72 and 1.14 m over land and sea, respectively. The average amount of water in the air (about $13 \cdot 10^{12}$ ton) at any one time corresponds to about 25 mm of precipitation, which, in turn, implies that a water molecule resides in the atmosphere for about ten days before being brought back to the Earth's surface in the form of precipitation. It is further noteworthy that less than half of the precipitation over land is returned to the oceans as run-off. The remainder evaporates, particularly from terrestrial ecosystems in the process of photosynthesis, which thus transfers about $59 \cdot 10^{12}$ ton yr^{-1} water back to the atmosphere. The amount of water in the soil and as ground water (on average about $18 \cdot 10^{12}$ ton) corresponds to merely about 35 mm of precipitation. Spatial variations of the availability of water are, however, huge as can be seen from the drastic differences between deserts and tropical rain forests.

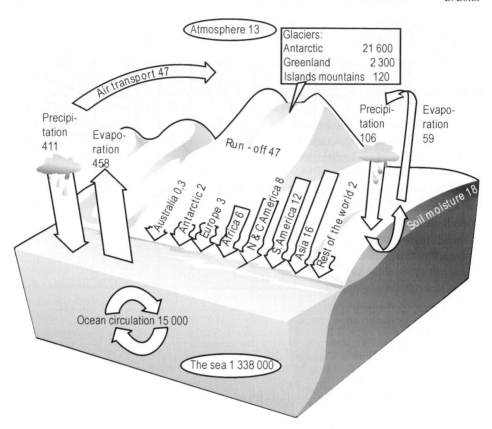

Figure 5. The global hydrological cycle. The sizes of the reservoirs are given in units of 10^{12} tons, and the fluxes between them in 10^{12} tons yr^{-1}.

As seen the terrestrial ecosystems play a fundamental role in the hydrological cycle that must be recognized in any analysis of the amount of water that might be available for human use. Water is needed for the chemical reactions in plants that transform the carbon dioxide in the air into organic matter, and water is the agent for the transport of chemical compounds within the plants. Water is lost when the stomata in the leaves open in daylight to capture the carbon dioxide molecules. It escapes because of the high humidity that is maintained at the bottom of the stomata. This creates a gradient to the outside, drier air, and water vapor diffuses out through the stomata. Up to about 1000 times more water is lost than carbon is captured in the form of carbon dioxide. The global net primary production is about $60 \cdot 10^9$ tons of carbon per year (see Section 7.3.1), while $59 \cdot 10^{12}$ water is lost globally by evapotranspiration. The presently increasing concentration of atmospheric carbon dioxide that human activities bring about is also an asset for some plants in that less water is lost for a given amount of carbon dioxide being captured. How much the net primary production is enhanced thereby is still an open question, however, since the photosynthesis also requires access to key nutrients [cf. IPCC (2001a, Chapter 3)].

A closer analysis of water on Earth, i.e., the details of the hydrological cycle, must, however, start from analyses of its regional features. The fundamental unit is the water basin or river basin defined by the runoff area associated with each river of the world, which carries rainwater to the sea or a lake with no outlet because of a dry climate. Within each such unit the terrestrial ecosystems have developed based on available water, which is dependent on topography, geology, and climate. The soils have in turn been formed during millions of years as a result of the evolving ecosystems. During the last few centuries an increasing amount of the water has been withdrawn by the people who have settled in the different regions and ever since have been dependent on its availability. Increasing difficulties to provide adequate amounts of water is now experienced in several countries, of course particularly in dry and semi-dry regions.

5.2. The human need for water

Humans need merely a few liters of water per day for survival, i.e., about a ton per year, but this presupposes that nature provides the food. The 6 billion people thus need about $0.006 \cdot 10^{12}$ ton annually, which is a very small fraction of the natural fluxes of water. In a modern society, however, the amount of household water needed is much larger to allow a decent and realistic quality of life, say 100 liters per person per day, or about 36 tons per person per year. As a matter of fact much more is actually used today in affluent societies. On the other hand, poor people in developing countries use only a fraction of this amount [see Falkenmark and Lundqvist (1998)].

But we are also dependent on the water that is needed to supply adequate amounts of food. To provide a kilogram of grain an amount of water at least several hundred times larger is needed. Requirements are similar for key industrial activities, e.g., several hundred tons of water is needed to produce one ton of paper. Each individual in an industrial country today is dependent on about 500 tons of water per year being available explicitly or implicitly for his or her needs. An increase of the world population to about 10 billion in the course of the present century implies a need for an annual supply of $5 \cdot 10^{12}$ tons of water by 2100 for the world population as a whole, i.e., about 5% of the total precipitation on land. But people do not always live where there is water, a prime reason why densely populated regions in arid climates are encountering difficulties. These facts show that more efficient use of water will be required. Recycling of water, as is practiced already today in some dry areas, may become of increasing importance. There is a need for integrated water resource management and protection of water quality (see further Section 5.4).

Such efforts must start with an analysis of available observations to determine the natural supply of precipitation through the year. Best possible spatial details are important so that extremes are not lost because of inadequate data and averaging. It is also important to record the extremes, i.e., the very highest and lowest, of observed total amounts of water in the past to be able to develop a management plan that can serve as a basis for action in case of future water shortage as well as flooding in the basin as a whole. The land cover and its change as a result of human actions will in most cases significantly change the mean and variance of the annual runoff in a river basin.

A good number of factors need be considered before increasing the withdrawal of water for new industrial activities, mining, etc. Expansion of agriculture and forestry put additional demands on water resources for irrigation, possibly in conflict with the need for an adequate water supply to natural ecosystems. Some wider safety margins might also be desirable in the future, when a changing climate may well imply marked changes of the amounts of water that will be available. Increasing populations, as is, particularly, the case in many African countries will of course deserve special attention. It is clear that limited supply of water will require careful consideration of how the available resource can best be used in the light of the societal structure and expected change in the future.

5.3. Floods and droughts

It is essential to have good knowledge of the characteristic response times for how an excess or deficiency of water in one part of a river system influences conditions in other parts, particularly downstream along the key river of the basin as well as its tributaries and flood plains. In situations when the natural water supply is well below normal conditions, activities in the lower part of the river course may be seriously harmed if an increased amount of water is withdrawn upstream. Similarly, letting excessive amounts go in case of flooding in the upper part of a river system, may just transfer a dangerous situation downstream. Many rivers flow through a number of countries and action plans need be agreed in order to optimize measures and avoid conflict. As a matter of fact, more than 200 river or lake basins in the world, populated by about 40% of the world's population and covering more than 50% of the Earth's land area, are shared by two or more countries.

Extreme precipitation events are difficult, often impossible, to foresee both with regard to timing and the regions that may be struck. Statistics of past events can provide information about the severity of phenomena that have occurred, say, merely once in 100 years, 50 years, or 10 years, etc., but data are often inadequate for more precise analyses, particularly in developing countries. It is also difficult to determine whether the frequency of extreme events is changing or not. The more extreme an event is, the longer a record is required in order to judge if the frequency of its occurrence is changing. This is the more noteworthy, when now changes of climate are expected (see Section 7).

Environmental disasters that are the result of droughts, storms, or floods, the latter two often in combination, are perhaps the most serious extreme events to be considered with the aim to secure lives and property in a future, presumably increasingly affluent society. This is becoming obvious from the fact that the values lost because of storms and flooding have increased, in some regions tenfold, even though the severity of the weather phenomena themselves may not have increased that much. Building close to the waterfront has instead become a major cause of increased risk of damage.

Changing land use may influence the occurrence of floods markedly. Flood plains have often in the past served as effective protection against severe disasters. During the last century, however, many of these often-fertile soils have been cultivated, and the construction of protective walls along the river banks has been the common precautionary measure. These may well serve their purpose under modestly extreme conditions, but there are many examples in recent years of severe catastrophes when they have been inadequate. An example is the flooding in the neighborhood of Saint Louis in USA in the mid 1990s and in Poland a few years later. It should also be recognized, that rivers carry silt towards the sea, and deposit some of it along their course. This slowly raises the riverbed and increases the vulnerability of surrounding lands. Severe damage has been recorded in the delta regions of major Chinese and Indian rivers.

5.4. Water quality

There are four major types of fresh water pollution:
- excess nutrients from sewage originating from agriculture and households;
- pathogens that spread disease, primarily also from sewage;
- silt brought into the water by excessive runoff;
- heavy metals and synthetic organic compounds from industry, mining, and agriculture, which are toxic and may accumulate in aquatic organisms.

Key nutrients, i.e., nitrogen compounds, phosphorus, and sulfur, primarily emanating from agriculture, are discharged in increasing amounts to watercourses, lakes, and the coastal zone of the sea. Ecosystems are thereby disturbed. The rate of photosynthesis increases; more biomass is formed, as can be seen in the form of plankton blooms. This may increase the amount of fish and thus be beneficial for fisheries.

On the other hand, decay of increasing amounts of organic matter requires more oxygen. Water in the surface layer is usually supplied with adequate amounts, but widespread loss of flora and fauna at greater depths may be the consequence if the amount of dissolved oxygen in the water is exhausted. This may occasionally happen in natural systems (the Black Sea is a well-known example), but it is today a frequent phenomenon, especially in lakes in densely populated regions. The nutrients in the dead organic matter end up in the bottom sediments.

Modern agriculture use large amounts of nutrients in order to enhance yields. There is often considerable leakage from the farmland that pollutes the runoff. This in turn enhances biological activities in lakes and the coastal zone (eutrophication), where most of the nutrients end up, either dissolved in the water or deposited in the bottom sediments.

Management practices are being developed in order to avoid excessive water pollution. These include conservation tillage, crop rotation, planting cover crops in winter, filter systems, and not the least fertilizer management. Such practices control erosion and may harness up to 60% of the nitrogen and phosphorus that otherwise would have been lost with the run-off from cropland.

It is of course possible to use as fertilizer the sludge that is retained from water purification, but fears of accumulation of toxic substances in agricultural products often become an obstacle for such use. Ultimately the sources of pollution will have to be reduced. The opportunities are many, and as the availability of clean water becomes scarce the costs for purification will presumably not be prohibitive. In the meantime, management must focus on canvassing the possibility for serious incidents and prepare for how to avoid them.

6. Acidification of fresh waters and soils

6.1. The basic issue

The development of terrestrial ecosystems over millions of years has established natural patterns of acidity and alkalinity, i.e., the pH,[4] of freshwater and soils. The ecosystems have optimized their structures, modes of assimilation, and growth as well as geological conditions and prevailing climate allow, and have implicitly adapted to the pH that has emerged. The composition of the soils, as well as their pH, are still changing naturally very slowly because of weathering of the bedrock and deposition of air-borne dust from neighboring regions or from far away. However, soil matter may also be lost in the form of dissolved compounds and suspended sediments, which find their way through runoff to lakes and the sea, where some end up in the bottom sediments. Natural change and adaptation of ecosystems and biomes is constantly going on.

These slow natural changes of the soils are now being modified by human induced emissions that are deposited on the lands and also because of direct human interference. The natural setting for the ecosystems is being modified. Burning fossil fuels to provide energy results in emissions of oxides that are transformed into acids when dissolved in water. The most important ones are carbon dioxide, sulfur dioxide, and nitrogen oxides, which form carbonic acid, sulfuric acid, and nitric acid, respectively.

The acidification of precipitation in turn influences the pH of freshwater and soils. Carbonic acid is a weak acid and poses no threat to terrestrial systems, but it does enhance somewhat the rate of weathering. However, the two latter acids, particularly sulfuric acid, may have significant impacts. Some soils are alkaline and thus can neutralize the acidity of the rainwater. These soils have a buffering capacity, but still their pH decreases slowly. Other soils contain comparatively small amounts of neutralizing compounds; their base saturation[5] and accordingly their pH are low, and they are more sensitive to acid precipitation.

---

[4] pH is a measure of alkalinity (pH > 7) and acidity (pH < 7). A decrease by one pH-unit implies a ten-fold increase of the number of hydrogen (H) ions in the water. Carbonic acid has a pH down to 5 6, while sulfuric acid and nitric acid may have much lower values. On the other hand, pH for seawater is about 8.2 and pH for natural soils varies between about 4.5 and 8.

[5] Base saturation is the percentage of exchangeable cations, i.e., calcium, potassium, magnesium, etc., relative to acid ions, primarily hydrogen ions.

Acidification means that the concentration of hydrogen ions is enhanced, nutrients like calcium, potassium, and other key elements are dissolved and carried away by runoff, and the base saturation is thereby lowered. This sometimes means that nutrients temporarily become more available for plants and growth may be stimulated, but when the base saturation has been brought down to below 5–10%, the fertility of the land becomes drastically reduced. The podsols of the boreal ecosystems show this typical feature and have been seriously affected by acidification in some regions. On the other hand, the nutrients that are carried away by the runoff end up in the surface layers of lakes and coastal waters of the sea, where photosynthesis may be stimulated. Excessive plankton blooms may result, which in turn enhances the flux of dead organic matter to deeper strata of the waters, where oxygen may be insufficient for its decomposition. Existing lakustrine (lake) and marine ecosystems may be destroyed. This chain of reactions is an example of the intricate interdependence of ecosystems that is a characteristic feature of our environment and which may lead to not easily foreseen consequences.

Soils being used for agriculture and forestry also change because of direct human interference. Harvesting from fields and forests implies losses of nutrients that have been extracted from the soil during growth, e.g., potassium, calcium, magnesium, ammonium. The base saturation declines. This change may be of about the same magnitude as the decrease that results from acid deposition, dissolution, and export of nutrients by runoff. Sustainable agriculture and forestry, i.e., avoiding declining yields, therefore requires that these losses be replenished. In agriculture this is done by fertilizing the fields, partly by returning organic matter in the form of straw and organic waste. Reduced tillage is also a positive action in that the decomposition and losses of organic matter in the soil are reduced. In forestry there are increasing efforts to return ashes from biomass burning to the forest soils.

As the pH decreases below 4.5–5 other elements that are present in the soil go into solution, such as aluminium, copper, and cadmium, as well as other heavy metals. Many of these pose health risks, both directly by contaminating the ground water and indirectly by the accumulation of these compounds in fish and other seafood. Some also are toxic for the flora and fauna [Seinfield (1986)].

6.2. The present global status

Today more than 90 million tons of sulfur are emitted into the atmosphere (cf. Section 3.4.3 and Figure 2), which is 3–4 times the emissions from natural sources during pre-industrial times. These human emissions are concentrated in industrialized regions; about 90% are released in the northern hemisphere. Since the average residence time for sulfur in the atmosphere is merely about a week, most of the deposition and acidification occurs within a few thousand kilometers from the source. The pattern of impact is patchy because of the variable natural buffering of soils and lakes referred to above.

In the atmosphere, the sulfur emissions form sulfate aerosols. The air becomes hazy because of the scattering of the solar radiation by these particles, as can often be seen in industrial regions. The solar radiation that reaches the Earth's surface is reduced, which reduces the global warming caused by the emission of greenhouse gases (see Section 7.3). It follows that efforts to reduce the emissions of sulfur into the atmosphere in order to reduce acidification will uncover the full warming effects of greenhouse gases, another example of the complex pattern of feedback mechanisms that character-ize the environmental system.

The seriousness of the acidification of soils and freshwater systems depends on the magnitude of the regional emissions and the soil types that are exposed [see Rodhe et al. (1988)]. The most important responses can be expected in regions of low pH, low ion exchange capacity, and high aluminium saturation. Factors such as alkaline dust, soil texture, or anion adsorption capacity can modify the expected effects. It is further important to emphasize the fact that acidification is a cumulative process in that the base-saturation is changing gradually as a result of a continuing exposure.

The boreal soils of northern Europe, northeastern USA, and Canada are especially vulnerable to acidification, but measures have been taken in the industrialized regions to ameliorate the situation. Southern parts of China have large areas of potentially sen-sitive soils, while northern China is less prone to acidification because of the abun-dance of neutralizing dust from calcareous soils. Southwestern India is also sensitive to acidification, while Australian soils generally are less acid than are soils at simi-lar latitudes in Africa and South America. Thus parts of Nigeria, southeastern Brazil, and northern Venezuela are sensitive to further exposure from industrialization. The ad-sorption of sulfate in the soils may, however, reduce the risk of serious acidification. A more detailed outline of the regions at risk requires specific consideration of the dis-tribution of soils and expected deposition from increasing emissions [see Rodhe et al. (1988)].

6.3. Preventive actions

The prime preventive action is of course to reduce emissions. To establish how this may be achieved at least cost requires the development of integrated models, which can be used to deduce the patterns of dispersion of the emissions by the winds and the resulting distribution of wet deposition (i.e., by rain) and dry deposition (i.e., by direct adsorption of sulfate particles or nitrogen oxides) under alternative assumptions about the spatial distribution of emissions. It is important to cover a sufficiently large area around a re-gion of major emissions in order to catch the total dispersion and deposition patterns. Observations of the distribution of prevailing depositions are important in order to check on the reliability of such model computations. It may also be desirable to consider how to compensate for past losses of nutrients and to enhance the base saturation of soils that have been exposed to acidification.

The outcome of such analyses might serve as a basis for international agreements on the optimal reductions of emissions. Analyses of this kind have in fact been con-ducted cooperatively by Norway, the Netherlands, and IIASA [Tuinstra (1999)] and have served as a basis for reaching agreement on how best to come to grips with acidi-fication in Europe. Similar efforts are underway in Southeast Asia and China.

7. Climate change and global warming

The threat of future human-induced climate changes is undoubtedly the most far-reaching environmental issue that has emerged until now. It will affect many sectors of society, if not all, and many measures of different kinds will have to be considered in order to reach agreement on effective mitigation and adaptation policies. A proper understanding of this issue requires the analysis of how human activities may influence the global environmental system as a whole. Burning of fossil fuels to provide the energy required for the continuing development of rich as well as poor countries is at the forefront. Major modifications of central societal and economic functions will be required.

7.1. The heat balance of the Earth

It was emphasized in the introduction that the presence of an atmosphere around the Earth that contains water vapor is fundamental since the global mean temperature thereby is enhanced to a level where life can be sustained. The presence of water is crucial for the development of the complex biochemistry that is another prerequisite for life to evolve. Human activities now have reached a stage where the global balances that have been created over millions of years are changing. First and foremost we need to understand how the heat balance of the Earth is changing [cf. IPCC (2001a, Chapter 1)].

The average incoming solar radiation is at present 342 Wm^{-2}. It varies by less than ± 0.5 Wm^{-2} on the time scale of years to a century, approximately in pace with the eleven-year sun spot cycle. About 31% is reflected back to space primarily as a result of the high reflectivity of clouds, ice, and snow. In that sense the presence of water in these forms tends to cool the Earth; see Figure 6. Thus about 235 Wm^{-2} is absorbed in the atmosphere and at the Earth's surface. That same amount of energy must leave the Earth to maintain an energy balance. Note, however, that the outgoing radiation from the Earth's *surface* at the prevailing average temperature of about $+15°$C is about 390 Wm^{-2}, i.e., about 155 Wm^{-2} more than finally escapes from the Earth. This is the result of the *natural greenhouse effect*. How does it operate?

This improvement of the heat balance of the Earth is primarily the result of the presence of water vapor in the atmosphere. The infra-red (heat) radiation that is emitted from the surface of the Earth is to a considerable extent absorbed by greenhouse gases, in particular water vapor, at quite low elevations and is re-radiated upwards as well as back towards the Earth. Not until an elevation of about 5 km is the air above sufficiently dry to permit the outgoing radiation largely to escape out towards space. The temperature at this level is merely about $-18°$C. The heat radiated out towards space at this level is considerably less than what is emitted from the surface and just about balances the incoming solar radiation. Below that level the vertical turnover and mixing of the air establishes a temperature gradient of about $0.65°$C$/100$ m. A mean surface temperature of about $+15°$C is accordingly maintained at the Earth's surface, which is some $30°$C higher than what it would be without the warming effect of the greenhouse gases. In

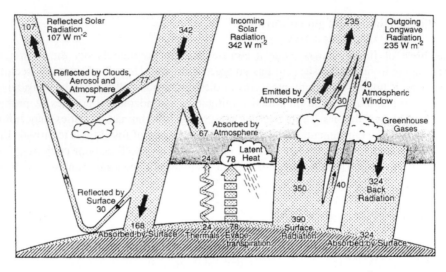

Figure 6. The Earth's global average annual energy balance. Fluxes are given in Wm^{-2}. Source: Kiehl and Trenberth (1997).

addition to water vapor, carbon dioxide, methane, nitrous oxide, and ozone also play a role. Their combined contributions are, however, merely about 5% of the total.

The heat balance is, however, not exclusively a result of radiative processes. The air is also heated directly from the surface of the Earth, primarily at low latitudes. Evaporation of water from the oceans and from vegetation on land requires energy, which is transferred to the air when water evaporates and is realized in the form of heat when condensation of the water vapor occurs in clouds and precipitation. There is generally a net transfer of heat from the Earth's surface to the air at low latitudes and in the opposite direction at high latitudes, but the net global transfer is from the Earth to the atmosphere. The human-induced increase of the concentrations of other greenhouse gases, particularly carbon dioxide, enhance this natural greenhouse effect.

7.2. Past changes of the global climate

During the last about two million years, i.e., the Quaternary period, the climate of the Earth has varied rather regularly because of variations of the incoming solar radiation resulting from small gradual changes of the Earth's orbit around the sun and the angle of the Earth's axis relative to the orbital plane, so-called Milankovič variations. The total incoming solar energy as well as its distribution with latitude have varied on time scales between 20,000 and 100,000 years. Internal feedback mechanisms have enhanced the rather small changes of the heat balance that are caused directly by these variations. For example, expansion of the areas covered by snow and ice in summertime over polar regions increases the albedo (reflectivity) of the Earth and reinforces the cooling and the formation of ice. The concentration of atmospheric carbon dioxide has varied in pace

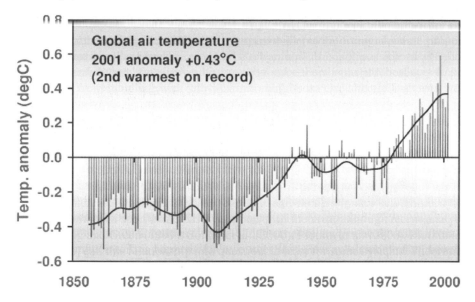

Figure 7. *Bars*: annual anomalies of combined land-surface and sea-surface air temperatures (°C) during 1856–2001, relative to 1961–1990 mean. *Solid line*: 10-year moving average. Sources: Jones et al. (1999), Palutikof (2002).

with this change and was considerably lower during the last ice age, merely about 200 ppmv, and we do not know why. This also re-enforced the cooling. Some of these past changes may sometimes have been quite abrupt on regional and local scales, but have in general been gradual over the millennia.

The extended glaciation in the North Polar Region until about 15,000 years ago was gradually replaced by the inter-glacial period that now prevails. The major continental ice sheets over large parts of Canada, Scandinavia, parts of northern Russia, and Siberia disappeared. Only the one over Greenland withstood the warming. The favorable distribution of incoming solar radiation that brought the Earth into an interglacial period is, however, now gone.

Direct measurements of temperature are available only for about the last 200 years and have really not been numerous enough to determine global changes with reasonable accuracy until about the end of the 19th century. Figure 7 shows the changes during the last 140 years [IPCC (2001a, Chapter 2)]. The variations during the latter part of the 19th century and the first decade of last century were partly caused by a series of major volcanic eruptions that occurred during this period. A rather smooth increase of the global mean temperature was observed during the thirty years 1910–1940, followed by some cooling until the 1970s. Since then a comparatively rapid increase has occurred and is still going on.

The last decade of the 20th century was undoubtedly the warmest one during the last 140 years. It is likely even to have been the warmest decade during the last millennium.

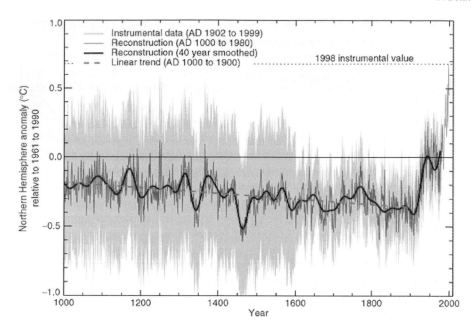

Figure 8. Northern Hemisphere temperature anomalies. The linear trend during AD 1000–1900 is shown as a dashed line, with two standard deviations shown in gray. Source: adapted from Mann, Bradley and Hughes (1999).

This conclusion is based on analyses of the temperature variations since about the year 1000 in the Northern Hemisphere, which have been determined by indirect means (see Figure 8). Variations of the width of tree rings, changes of the flora as determined by pollen analyses, the changing composition of successive layers of glacial ice, and sediments from the bottom of the deep sea that can be dated precisely tell us about past changes of the temperature.

Some interesting features stand out. The global mean temperature seems to have declined very slowly until about hundred years ago. This may be part of the slow cooling of the Earth that has occurred since the climatic optimum about 8000 years ago. The record also shows inter-annual variations of the global mean temperature of the order of ±0.2°C as well as variations on the time scale of one to a few decades. These are, however, small compared with the increase by about 0.6°C that has occurred during the last century. There are also other indicators that support the conclusion that the ongoing climate change is exceptional, as concluded by the Intergovernmental Panel on Climate Change, IPCC [see IPCC (2001a, Chapter 2)].

- Global precipitation increased on average by 5–10% during the 20th century.
- There has been an increase of extreme precipitation events in parts of the mid- and high latitudes of the Northern Hemisphere.

- Warm episodes of the El Niño/Southern Oscillation[6] (ENSO) have been more frequent, persistent, and intense since the mid-1970s.
- Mountain glaciers in many parts of the world have shrunk considerably.
- The size of the area around the North Pole that is covered by sea ice has decreased by 10–15% and the ice thickness in some regions by as much as 40% since the middle of last century.
- Sea level rise during the 20th century, 10–20 cm, was considerably greater than the average rate during the last 3000 years.

On the other hand, it has not been established that severe weather events (storms, floods and droughts) have been more frequent and more fierce than at earlier times and thus also might be a manifestation of an ongoing climate change, but this possibility cannot be excluded.

7.3. Key biogeochemical features of the climate system

In order to understand the reasons for the ongoing climate change and in particular to settle if human activities have contributed to this change, it is essential to have good knowledge about the processes that play a role in regulating the concentrations of key atmospheric components, in particular carbon dioxide, nitrous oxide, and sulfur dioxide.

7.3.1. Carbon dioxide

7.3.1.1. Circulation of carbon in nature during pre-industrial times Carbon is the fundamental element of life. Its chemistry is complex and there exist more than a million known carbon compounds, of which thousands are vital for biological processes. Atmospheric carbon dioxide serves as the carbon source for the process of photosynthesis. It dissolves in fresh water as well as seawater and forms carbonate and hydro-carbonate ions that serve as the carbon source for life in lakes and in the sea. Most of the carbon on Earth is locked up into minerals in the solid earth, but these reservoirs are of secondary interest in the present context because of their very slow exchange with other natural carbon pools.

We therefore restrict ourselves to consider the exchange of carbon between the atmosphere, the sea, and the terrestrial ecosystems (including soils), i.e., those parts of the environmental system in which significant changes occur on the time scales of decades and centuries. Figure 9 shows a schematic diagram of the exchange of carbon between these major pools, i.e., the global carbon cycle. A key issue is obviously to understand the partitioning of carbon between these major reservoirs and how this system as a

---

[6] El Niño is a semi-periodic oscillation of weather conditions in the equatorial Pacific that is characterized by a large region of higher temperatures in the atmosphere and the surface layers of the ocean that is formed and moves, every 4–5 years, from the Indonesian archipelago toward South America in the matter of about half a year.

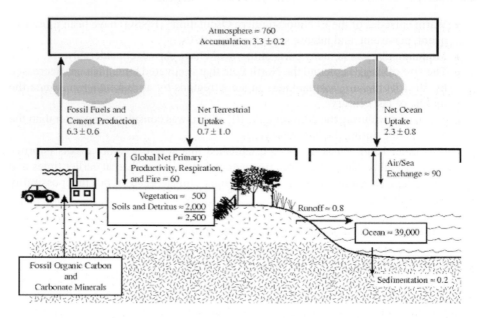

Figure 9. The perturbed global carbon cycle during the 1990s. Carbon stocks in natural reservoirs are given in gigatons (Gt; $= 10^9$ tons), and carbon flows between them are given in Gt yr^{-1}. Sources: IPCC (2000b, Chapter 1; 2001a, Chapter 3).

whole responds to the emissions of carbon dioxide that human activities now bring about [see IPCC (2001a, Chapter 3)].

Carbon dioxide *exchange between the atmosphere and the sea* is determined by the solubility of carbon dioxide in seawater and the rate of mixing of surface water with deeper strata of the oceans. The total amount of carbon dissolved in the sea is about 50 times larger than the amount present in the atmosphere. The gross flux between the atmosphere and the sea is approximately in balance, with exchange of about 90 Gt C (Carbon) yr^{-1} in both directions.[7] During pre-industrial times a remarkable overall equilibrium prevailed, and the net flux in one or the other direction was mostly less than 1 Gt C yr^{-1}.

Carbon dioxide is more easily dissolved in cold water than in warm water. Before industrialization there was therefore a net flux into the sea in polar regions and an outflow in the tropics. To maintain an approximate balance there was a net flux from the tropics towards the poles in the atmosphere. This is nowadays over-shadowed by the transfers induced as a result of emissions from fossil fuel burning primarily at middle latitudes in the Northern hemisphere.

---

[7] 1 Gt (gigaton) $= 10^9$ ton.

Terrestrial plants assimilate carbon dioxide from the atmosphere in the process of photosynthesis. The gross primary production (GPP) of the terrestrial system (i.e., uptake by the terrestrial ecosystems) is estimated to be about 120 Gt C yr^{-1}. About half of this is, however, returned to the atmosphere by nightly respiration. The global net primary production (NPP) is therefore about 60 Gt C yr^{-1} [Bolin et al. (2000)]. The residence time for a carbon dioxide molecule in the atmosphere is about four years. This is still long in comparison with the mixing time within the troposphere. The carbon dioxide concentration in the troposphere therefore varies spatially by merely five to ten ppmv over the globe, except in the boundary layer next to the Earth's surface, where the terrestrial systems cause marked variations of the concentration between day and night because of photosynthesis in the day and respiration at night.

Most of the primary terrestrial production is used for growing short-lived products, such as leaves, grass, and fine roots. When these die, most of the organic compounds decay in less than a year or two. About a quarter of the NPP is transformed into wood and is stored within forest ecosystems for decades to centuries before the trees die and decay. A small portion of the dead organic matter ends up in the soil as complex compounds that may remain there for centuries. The amount of carbon stored in the upper 0.5–1 meter of the soil is about five times larger than the amount present in above ground biomass. The carbon dioxide produced by decay is returned to the atmosphere (heterotrophic respiration) and an overall approximate carbon balance was in this way maintained in natural terrestrial systems. For more detailed analyses, see Walker et al. (1999).

7.3.1.2. Human-induced changes of the carbon cycle The carbon dioxide concentration in the atmosphere has increased markedly in the atmosphere during the last about 200 years, from about 280 ppmv to about 370 ppmv in 2001, i.e., by about 32%. This means an increase from about 595 to about 780 Gt C, i.e., by about 185 Gt C. The present average annual increase is about 1.6 ppmv (i.e., 3.3 Gt C yr^{-1}).

The atmospheric concentration of carbon dioxide varied already before human emissions became important. Figure 10 shows the variations during the last 45,000 years, obtained from analysis of the carbon dioxide content in air bubbles locked into ancient glacier ice from Greenland and the Antarctic. During the last ice age, the average concentration was merely about 200 ppmv, but about 280 ppmv during the preceding interglacial period about 120,000 years ago. The reasons for these variations are not well understood. The recent human-induced increase is, however, more than 50 times more rapid than these natural variations.

Because of combustion of fossil fuel and changing land use, the partial pressure of carbon dioxide in the atmosphere is now very much higher than during earlier times. Therefore a net flux of carbon dioxide into the sea and uptake by the terrestrial systems has been established. The partitioning between the major carbon reservoirs can be deduced rather accurately with the aid of simultaneous measurements of the annual changes of the atmospheric concentration of carbon dioxide and oxygen. Figure 11 shows the annual values during the 1990s. For the decade as a whole the carbon dioxide

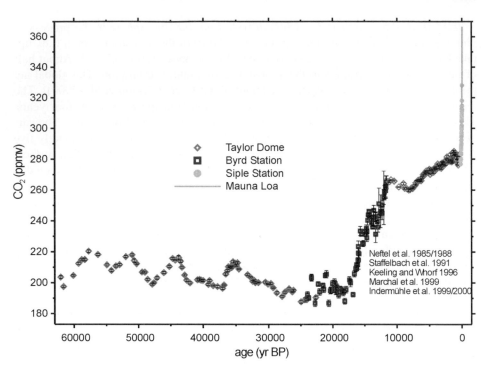

Figure 10. Variations of atmospheric carbon dioxide concentration (ppmv) during the last approximately 45,000 years. Concentrations were measured directly during the last approximately hundred years and in air bubbles preserved in glacier ice in Antarctica and Greenland for earlier time periods. Source: Indermühle et al. (1999).

concentration increased from 352.0 to 367.2 ppmv, i.e., by 15.2 ppmv, which means on average by 3.3 Gt C yr^{-1}.[8] The total emissions during this period amounted to 6.3 Gt C yr^{-1}.

When fossil fuels are burned we know how much oxygen is consumed, and if all the emissions would stay in the atmosphere we would see changes of carbon dioxide and oxygen concentrations in the atmosphere that would follow the straight line "fossil fuel burning". We also know how much oxygen is released when photosynthesis extracts carbon dioxide from the atmosphere. Net uptake would change these concentrations as shown by the slanting curve in the range denoted "land uptake". On the other hand, carbon dioxide uptake by the oceans does not affect the oxygen concentrations. As can be seen, there is only one way to reach the observed concentrations in 2000 from the lower right end point, which corresponds to all emissions staying in the atmosphere.

---

[8] 1 ppmv carbon dioxide = 2.12 Gt carbon.

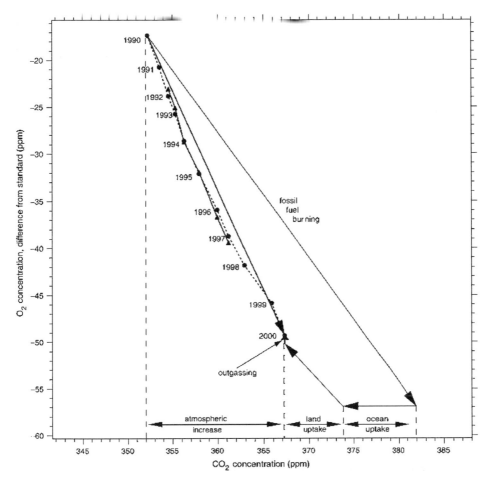

Figure 11. Changes in atmospheric concentrations of carbon dioxide and oxygen during the 1990s. Units: parts per million by volume (ppmv). Sources: Keeling, Piper and Heimann (1996), IPCC (2001a, Chapter 3).

This yields an ocean uptake of about 1.7 Gt C yr^{-1} and a terrestrial uptake of about 1.4 Gt C yr^{-1}.

This rather large net terrestrial uptake is the more noteworthy as there at the same time have been net emissions to the atmosphere because of deforestation and changing land use, which are estimated to have amounted to 1.7 ± 1.0 Gt C yr^{-1} during the 1980s [Houghton (1999), Houghton et al. (2000)]. If the latter remained about the same during the 1990s, 3.1 ± 1.5 Gt C yr^{-1} must have been absorbed elsewhere in the terrestrial system during that decade.

A number of different processes might contribute to this important response of the terrestrial ecosystems:

- An increased rate of photosynthesis because of higher atmospheric concentrations of carbon dioxide.
- Enhanced growth in the terrestrial systems because of deposition of airborne nutrients, particularly fixed nitrogen due as a result of human activities.
- Improved land use practices, particularly in forestry, resulting in an increasing amount of carbon in above ground biomass, e.g., timber, particularly, at middle and high latitudes in the Northern Hemisphere.
- Reduced tillage in agriculture to preserve the pool of organic carbon in soils.
- Changing climate, although we do not know how an increasing global temperature might change the uptake.

In summary, at present human activities result in total emissions of carbon dioxide of about 8.0 Gt C yr^{-1}. About 40% stays in the atmosphere, 20% goes into the oceans, and 40% into the terrestrial ecosystems.

A crucial question arises: will the terrestrial uptake continue in the future? In a long-term perspective there is necessarily a limit to the uptake by vegetation. Because of human activities there is a continuing decrease of the area covered by forests on earth. We also know from laboratory experiments that the increase of the rate of photosynthesis as a result of enhanced carbon dioxide concentrations will gradually decrease and cease as atmospheric concentrations approach 500–600 ppmv. Soil respiration increases, as does the flux of carbon dioxide to the atmosphere, when climate gets warmer. We need to understand these feedback mechanisms better in order to be able to assess how best to limit the increase of atmospheric carbon dioxide.

7.3.1.3. Stabilization of atmospheric carbon dioxide concentrations   Total fossil fuel resources that can be extracted at reasonable prices are large. Oil and gas resources have been estimated to at least 500 Gt C and coal > 3000 Gt C [Nakicenovic et al. (1996)]. Total reserves, i.e., identified resources, are well above 1000 Gt C to be compared with total emissions so far of about 300 Gt C. Abundant fossil fuel resources, particularly of coal, permit an increasing use of fossil fuels for a long time to come, although the availability of oil and natural gas will start to decline in the next half-century.

Emission scenarios that stabilize concentrations at alternative concentration levels have been deduced on the basis of our present knowledge of the carbon cycle [IPCC (2001a, Technical Summary), Bolin and Kheshgi (2001)]; see Figure 12. For example, under the assumption of a modest net uptake by terrestrial ecosystems, to achieve a stabilization level of 550 ppmv, annual emissions due to fossil fuel burning must not exceed 9–13.5 Gt C yr^{-1} and must decrease to well below the present level during the latter part of the 21st century. In order not to exceed 450 ppmv, emissions must not exceed 7–11 Gt C yr^{-1}, and the decline would have to occur more quickly and begin within few decades.

To understand the implications of the very unequal use of fossil fuels amongst the countries of the world also requires knowledge about the carbon cycle. Figure 13 shows *per capita* emissions of carbon (as carbon dioxide) in nine regions of the world in 1999, as well as the percentage of their respective contributions to the total [Bolin and Kheshgi

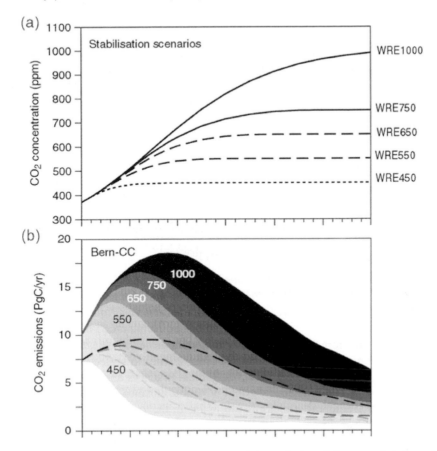

Figure 12. (a) Stabilization scenarios for carbon dioxide and (b) required restrictions on emissions to achieve them. The ranges of permissible emission curves from 1990 forward are shown for different stabilization levels, i.e., 450, 550, 650, 750, and 1000 ppmv. Source: IPCC (2001a, Chapter 3 and Technical Summary).

(2001)]. The average *per capita* emissions from developed countries, including countries in economic transition, was ca 3.1 C ton yr^{-1} in 1999, from developing countries ca 0.6 ton C yr^{-1}, and for the world as a whole ca 1.1 C ton yr^{-1}. The emission scenarios for stabilization at or below 550 ppmv show that, even if developed countries reduced their emissions by more than 50% during the next half-century and did not increase their populations, developing countries would not be able on average to emit more than ca 1.3 ton C yr^{-1} per capita, because of a likely increase of their populations from almost 5 billion to about 8 billion. This is only about 40% of per capita emissions in developed countries at present. Stabilization at 450 ppmv is hardly possible, since it would require emission reductions in developed countries by 50–70% within the next about 30 years, even if developing countries did not increase their average per capita emission beyond about 0.7 ton C yr^{-1}.

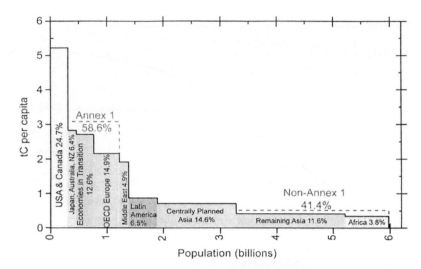

Figure 13. Per capita emissions (ton C yr^{-1}) of carbon dioxide from fossil fuel burning and cement production in 1999 in nine regions of the world. Percentages of global emissions are also shown. Source: Bolin and Kheshgi (2001).

A question is therefore being asked: can the terrestrial sinks be used in order to relax these stringent requirements for emission reductions? The most effective measure would of course be to introduce sustainable management practices and to reduce the clear-cutting of forests. Forest plantations would of course enhance storage in the terrestrial system but would require that measures to maintain these planted forests were put in place. The additional storage would, however, be modest and not provide a long-term solution. Reduced use of fossil fuels, as the primary source for future energy needs, is a necessity.

Other technical options for recovery and storage of carbon dioxide are conceivable. Efforts are being made to use abandoned oil or gas fields for such purposes. There are also geological formations that might be used [see, e.g., Parson (1998)]. Technical feasibility and costs need further analysis. In any case, it would take decades to build the infrastructure that would be required.

7.3.2. Methane

The amount of methane in the atmosphere represents a balance between the emissions from various human sources and decomposition, primarily through reactions with the OH-radicle that is formed by the decomposition of water vapor by UV-radiation. Natural emissions of methane to the atmosphere, about 200 Mt yr^{-1}, had by the end of last century increased to about 550 Mt yr^{-1}. The following human sources contribute about one quarter each: (1) the exploitation, transportation, and use of fossil fuels, (2) the rearing of ruminants, (3) the decomposition of waste and burning of biomass, and (4) wet rice cultivation. Modest reductions in these emissions can stop the rate of increase of methane concentrations. About a ten percent reduction would stabilize the atmospheric concentrations at the present level within about a decade. This fast response is due to the fact that the lifetime of methane in the atmosphere is short.

7.3.3. Nitrous oxide

The nitrous oxide molecule is largely inert and harmless. It is not part of any chemical reactions in the atmosphere, but it is simply decomposed by UV-radiation in the stratosphere. Because of its rather weak absorption lines and slow transfer up to these levels its residence time in the atmosphere is about 150 years. It is therefore quite evenly distributed over the globe.

Nitrous oxide is formed primarily in the soils by natural denitrification of nitrogen compounds and is thence emitted into the atmosphere. Some nitrous oxide is also formed when burning fossil fuels, primarily in combustion engines. The increasing nitrous oxide concentration during the last century is primarily a result of the increasing use of nitrogen fertilizers in agriculture, whereby the total turnover of nitrogen in the soil has increased and the rate of denitrification similarly so. This is a slow process, however, and so is the leakage of nitrous oxide from the soil back to the atmosphere. Nitrogen in the terrestrial ecosystems will therefore probably increase further and so therefore will the emissions to the atmosphere for decades to come.

7.3.4. Other greenhouse gases subject to restricted use

Hydrofluorocarbons (HFCs), perfluorocarbons (PFCs), and sulfur hexafluoride (SF_6) are also effective greenhouse gases and have therefore been included in the Kyoto Protocol of the Climate Convention for emission reductions. Their concentrations are still small but increasing. Most of the HFCs have atmospheric residence times of a few years, few more than 50 years, and emission reductions are required to stabilize their concentrations. PFCs and SF_6 are very long-lived and may stay in the atmosphere for thousands of years. Emissions must be prohibited in order to stabilize their concentrations. They are becoming semi-permanent constituents of the atmosphere [see further IPCC (2001a, Chapters 4 and 6)].

7.4. Has the recent climate change been caused by human activities?

The average global concentrations of the natural greenhouse gases have increased during the last 200 years: carbon dioxide by 32%, methane by about 150%, nitrous oxide by about 16%. Ozone concentrations have increased in the lower parts of the atmosphere in and around regions of widespread industrial activity. On the other hand, stratospheric ozone has decreased significantly in polar regions, particularly in Antarctica. Aerosol concentrations, particularly, sulfur aerosols from the combustion of sulfur rich oil and coal (cf. Sections 3.4.3 and 6), have increased and enhanced the reflection of incoming

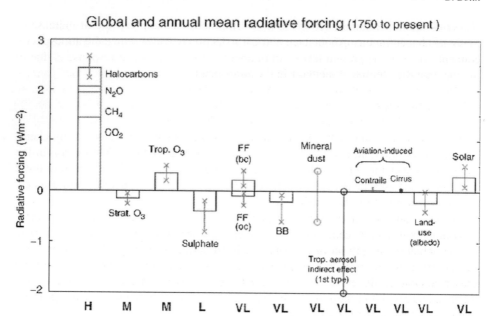

Figure 14. Radiative forcing (Wm^{-2}) of the atmosphere as a result of its changing composition through 1999. The effects of aerosols and changing solar radiation are included. FF = fossil fuels (bc = black carbon, oc = organic carbon), BB = biomass burning. H = high, M = medium, L = low, VL = very low. Source: IPCC (2001a, Chapter 6).

solar radiation and accordingly reduced the warming due to enhanced greenhouse gas concentrations.

Have these well-established changes been the cause of the ongoing climate change? Figure 14 summarizes the changes in radiative forcing of the atmosphere that have oc-curred during the last about 200 years due to these changes in atmospheric composition, as well as the estimated uncertainties in these determinations [see IPCC (2001a, Chap-ter 6)]. The estimates are based on well-established radiative transfer models.

The total enhancement of the greenhouse effect amounts to 2.6 ± 0.4 Wm^{-2}, includ-ing also the net effect of the changes of ozone in the troposphere and the stratosphere. A little more than half of this amount is the result of the increase of carbon dioxide. Ex-cept for the contribution by ozone, this total change is rather homogeneously distributed over the globe, but generally larger at low latitudes than in polar regions.

The reductions or increases of the radiative fluxes as the result of aerosol emissions are also shown in Figure 14, but their values are much more uncertain. Their spatial distribution over the Earth's surface is also patchy, because of the short residence times of particulate matter. The role of sulfate aerosols is best understood. The globally aver-aged reduction in radiative forcing resulting from their presence in cloud-free air (the direct effect) is estimated to be 0.8 ± 0.6 Wm⁻², while the indirect effect, which is the result of changes in the optical properties of clouds in the presence of aerosols, is poorly known.

The intensity of solar radiation has increased somewhat during the 20th century, but this change is small compared to the changes caused by greenhouse gas emissions and will probably decline during the present century. Past volcanic eruptions have also had an effect if emissions reach the stratosphere, but still only temporarily because the aerosols emitted in an eruption disappear within a few years.

These changes of radiative fluxes can be translated into a change of the global mean temperature with the aid of an advanced climate model. Such a model is simply a global weather forecasting model that describes the whole environmental system and its dynamical behavior in some detail and also includes considerations of slowly varying components such as the oceans and terrestrial ecosystems. First a model run is made with fixed atmospheric composition and then with gradually increasing concentrations of greenhouse gases and aerosols as observed until now. So far, only the direct effects of sulfate aerosols have been included in such models. Water vapor concentrations are not prescribed but are determined by the model. The difference between two such experiments is interpreted as the climate change that might have been caused by the observed change of atmospheric composition.

Model simulations for the last century have been compared to observed changes. The following conclusions can be drawn [IPCC (2001a)]:

- The warming during the past 100 years is very unlikely[9] to be the result of internal natural variability of the climate system.
- Simulations of the climatic response to natural and anthropogenic forcing have yielded clear evidence of a human influence on the climate during the last 35–50 years. They also indicate that natural forcing may have contributed significantly to the observed warming in the first half of the 20th century.
- The best agreement between model simulations and observations during the last 140 years is obtained when the anthropogenic and natural forcing factors are combined.
- The sensitivity of the climate system to an imposed change of the radiative forcing has been reasonably well established: a doubling of the carbon dioxide concentration, or an equivalent change of a mix of greenhouse gases will lead to an average global warming of 1.5–4.5°C. The uncertainty of this estimate primarily depends on the difficulty in accounting properly for a number of feedback mechanisms that are at work. The most important one is the likely increase of the amount of water vapor in the atmosphere that will occur if the temperature increases. Without this positive feedback the temperature increase resulting from doubling carbon dioxide

[9] The IPCC uses the following terminology: *virtually certain* (greater than 99% chance that the statement is true), *very likely* (90–99% chance), *likely* (66–90% chance), *medium likelihood* (33–66% chance), *unlikely* (10–33% chance), *very unlikely* (<10% chance).

concentrations would be merely about 1.2°C. Other feedback mechanisms, e.g., those resulting from changes in the extension of snow and ice and vegetative cover, also play an important role.

- The inertia of the climate system is considerable. In particular, the warming of the surface layers of the oceans takes time because of the slow heat transfer to deeper layers. Only about 70–80% of the expected global warming as a result of the changing composition of the atmosphere has probably as yet been realized.
- Global warming will decline only slowly because of the long residence time of key greenhouse gases in the atmosphere, while the negative forcing due to aerosols will disappear in pace with the reduction of emissions, since their residence time in the atmosphere is merely a few weeks.

7.5. Expected future changes of the global climate

The IPCC has generated a number of scenarios that describe alternative futures of the global society as dependent on changes of the world population, differences between developed and developing countries, availability of different means for providing primary energy, technical innovation, economic development, land-use change, etc. [IPCC (2000a)]. Future emissions of greenhouse gases and aerosols have been projected, and this in turn has provided an input for derivation of scenarios of atmospheric composition and changes of the global mean temperature (Figure 15).

It is to be noted that the scenario that shows the least changes has been based on an increase of the world population to about 9 billion people by the middle of the 21st century and subsequently a decline to merely 7 billion at the end of the century, and also an early gradual transition to other sources of primary energy than fossil fuels. On the other hand, large changes occur in scenarios with a more rapidly increasing world population, rising above 10 billion at the end of the century, and continuing use of abundant fossil fuel resources as well as more rapid economic development [IPCC (2000a)].

The analysis can be summarized as follows [see IPCC (2001, Summary)]:

- The global mean surface temperature is projected to increase by the end of this century by 2.0 to 6.5°C above pre-industrial conditions, if no concerted actions for mitigation are taken.
- It is very likely that nearly all land areas on average will warm more rapidly than the globe as a whole, particularly those at northern high latitudes in the cold season. Most notable of these are the northern regions of North America and northern and central Asia, where the global mean warming may exceed the global mean by more than 40%.
- It is likely that precipitation will increase over northern middle to high latitudes and Antarctica, which also implies increased frequency of heavy precipitation and flooding. Larger year-to-year variations in precipitation are likely over most areas at low latitudes.

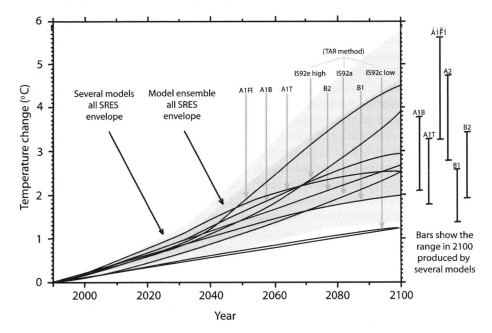

Figure 15. Global mean surface temperature projections for six IPCC illustrative emission scenarios (SRES) and three earlier IPCC scenarios: IS92e (high, ending at 3.9°C), IS92a (medium, ending at 2.5°C), IS92c (low, ending at 1.2°C). SRES scenario A2 nearly coincides with the earlier scenario IS92e and thus is not drawn as a separate line. The darker shading is the envelope of all 42 SRES scenarios, but only using the results from models with average climate sensitivity; the lighter shading includes results for the whole range of model sensitivities. The vertical bars to the right show the uncertainty ranges for the indicated scenarios under different assumptions about the sensitivity of the climate system to greenhouse gas forcing. Sources: earlier scenarios, IPCC (1992), SRES scenarios, IPCC (2000a, Technical Summary).

- It is very likely that the heat index[10] will increase over land areas.
- It is likely that global warming will increase the following:
    - (i)   summer continental drying and associated risk of drought;
    - (ii) tropical cyclone mean and peak wind and precipitation intensities;
    - (iii) Asian summer monsoon precipitation variability.
- Global mean sea level is projected to rise by 10 to 90 cm during the 21st century, primarily due to seawater expansion and loss of mass from glaciers and ice caps.
- A major climate change is expected to persist for several centuries.
- It is also possible to obtain a rough idea about expected regional changes of the daily temperature cycle, probability of heat waves or frosts, likeliness of periods of excessive precipitation and floods, changes of the length of the growing season, statistics for possible occurrence of storms and gales, etc. So far results are limited [cf. IPCC (2001a) and also IPCC (1992)].

---

[10] Heat index refers to a combination of temperature and humidity that measures effects on human comfort.

It should, however, be recalled that the climate system is chaotic in nature and that instabilities might exist that are not predictable. Reference is often made to the fact that the temperature records from Greenland during the last interglacial period show large variations on time scales from decades to centuries, but that this is not a feature of the present interglacial. The reason for this difference is not understood. It has also been pointed out that the temporary disappearance of the Arctic sea ice in summertime might initiate unexpected changes elsewhere. A change of the thermohaline ocean circulation,[11] which in turn would affect major ocean currents, is another threat about which we cannot as yet make trustworthy projections.

The IPCC conclusions, on which the present summary has been based, are supported by a large majority of the scientists active in the field, but there are still some few that challenge them. Their objections should not be ignored but rather analyzed further [IPCC (2001a)]. There are still uncertainties, but the risk of a substantial change of climate is considerable and must be taken seriously. The main issues are now, how serious would the impacts of global climate change be, and how urgent is the need to take early action [see IPCC (2001b)]?

A geophysical/geochemical background as sketched here is essential to address these issues properly, but the scope of such analyses should be widened: in particular, what does the risk of human-induced changes of climate imply for our search for sustainable development? It is becoming increasingly clear that a closer integration of our knowledge about the climate change issue as seen from the perspective of natural sciences with the analysis of the socio-economic issues that arise in this context is a necessity.[12]

8. Environmental stresses and sustainability

The Earth's environmental system is clearly disturbed by our dependence on natural resources. The future will be shaped by the interplay of human activities and natural processes. The term anthropocene has been coined for this emerging era of a changing Earth that now is succeeding the holocene. Sustainable development requires creating the conditions that are supportive of a sustained improvement of human well being that simultaneously maintains the planet's life support systems and biodiversity. We need to unravel the determinants of vulnerability and resilience of the nature-society systems as they are shaped by multiple interactions on all time and space scales [Gunderson and Holling (2002)].

Sustainable development must necessarily take into account cultural and other differences between people and countries. There is not one way to a sustainable society but many, amongst which choices will have to be made. These will differ dependent on

---

[11] Thermohaline circulation is the vertical overturning of the sea as a result of the distribution of temperature and salinity.

[12] For an analysis of the economics of climate change, see the Handbook chapter by Charles Kolstad and Michael Toman.

human values and traditions. In fact, it might even not be desirable to aim for a common concept of sustainable development, because it would not recognize the fundamental value of a heterogeneous world. The issue should rather be addressed in an inverse way, i.e., to try to reach agreement on what should be avoided in order to assure that there will be attractive options for sustainable development, even if a more specific definition of the concept is not agreed upon [Fisher (2000)]. This is what Schellnhuber et al. (1997) call "to avoid degradation syndromes".

There is a need to recognize fully the non-linear features of the combined socio-economic/natural system. As yet our knowledge is limited, but there is much to learn from the possible instabilities and chaotic behavior that are associated with the geophysical-geochemical interactions that have been the subject of this chapter.

References

Arrhenius, S. (1896), "On the influence of carbonic acid in the air upon the temperature of the ground", Phil. Magazine 42:237–276.

Berresheim, H., P.H. Wine and D.D. Davis (1992), Sulfur in the atmosphere, in: H.B. Singh, ed., Composition, Chemistry, and Climate of the Atmosphere (Van Nostrand/Reinhold, New York).

Bolin, B., and H. Rodhe (1973), "A note on the concept of age distribution and transit time in natural reservoirs", Tellus 25:58–62.

Bolin, B., B.R. Döös, J. Jäger and R.A. Warrick (1986), The Greenhouse Effect, Climate Change, and Ecosystems, SCOPE, Vol. 29 (Wiley, Chichester, UK) 541 pp.

Bolin, B., and H. Kheshgi (2001), "On strategies for reducing greenhouse gas emissions", PNAS 98:4850–4853.

Bolin, B., R. Sukumar, P. Ciais, W. Cramer, P. Jarvis, H. Kheshgi, C. Nobre, S. Semenov and W. Steffen (2000), "Global perspective", in: R. Watson, I. Noble, B. Bolin, N.H. Ravindranath, D. Verardo and D. Dokken, eds., Land Use, Land-Use Change, and Forestry (Cambridge University Press, Cambridge) 27–51.

Caldwell, M.M., and S.D. Flint (1994), "Stratospheric ozone reduction, solar UV-B radiation and terrestrial ecosystems", Climate Change 28:375–394.

Crutzen, P.J. (1971), "Ozone production rates in an oxygen, hydrogen, nitrogen oxide atmosphere", J. Geophys. Res. 76:7311–7327.

Crutzen, P.J., and J.W. Birks (1982), "The atmosphere after a nuclear war: Twilight at noon", Ambio 11:114–125.

Crutzen, P.J., and F. Arnold (1986), "Nitric acid cloud formation in the cold Antarctic stratosphere: a major cause for spring time 'ozone hole' ", Nature 324:651.

Crutzen, P.J. (1996), "My life with O_3, NO_x and other YZO_xs", in: Les prix Nobel, 1995 (The Nobel Foundation, Stockholm) 116–176.

De Fabo, E.C., and F.P. Noonan (1996), "UV radiation and human health effects", Internat. J. Environmental Studies 51:257–268.

Falkenmark, M., and J. Lundqvist (1998), "Towards water security: political determination and human adaptation crucial", Natural Resources Forum 21(1):37–51.

Farman, J.C., B.G. Gardiner and J.D. Shouklin (1985), "Large losses of total ozone in the Antarctica reveal seasonal ClO_x/NO_x interaction", Nature 315:207–210.

Fisher, A.C. (2000), "Introduction to a special issue on irreversibility", Resource and Energy Economics 22:189–196.

Gunderson, L.H., and C.S. Holling (2002), Panarchy. Understanding Transformations in Human and Natural Systems (Island Press, Washington, DC) 507 pp.

Heath, D.F., A.J. Krueger and P. Crutzen (1977), "Solar proton event: Influence on stratospheric ozone", Science 197:886.

Houghton, R.A. (1999), "The annual net flux of carbon from changes in land use, 1850–1990", Tellus 50B:298–313.

Houghton, R.A., D.L. Skole, C.A. Nobre, J.L. Hakler, K.T. Lawrence and W.H. Chomentowski (2000), "Annual fluxes of carbon from deforestation and regrowth in the Brazilian Amazon", Nature 403:301–304.

Indermühle, A., T.F. Stocker, F. Joos, H. Fischer, H.J. Smith, M. Wahlen, B. Deck, D. Mastrolanni, J. Tschumi, T. Blunier, R. Meyer and B. Stauffer (1999), "Holocene carbon-cycle dynamics based on CO_2 trapped in ice at Taylor Dome, Antarctica", Nature 398:121–126.

IPCC (1992), Climate Change 1992 (Cambridge University Press, Cambridge, UK) 200 pp.

IPCC (2000a), Special Report on Emissions Scenarios (Cambridge University Press, Cambridge, UK) 595 pp.

IPCC (2000b), Land Use, Land-Use Change, and Forestry, A Special Report of the IPCC (Cambridge University Press, Cambridge, UK) 377 pp.

IPCC (2001a), Climate Change 2001. The Scientific Basis, Working Group I, Third Assessment Report to the Intergovernmental Panel on Climate Change (Cambridge University Press, Cambridge, UK) 881 pp.

IPCC (2001b), Climate Change 2001: Impacts, Adaptation and Vulnerability (Cambridge University Press, Cambridge, UK) 1032 pp.

Jacobson, M., R. Charlson, H. Rodhe and G. Orians (2000), Earth System Science; From Biogeochemical Cycles to Global Change (Academic Press, New York) 523 pp.

Jones, P.D., M. New, D.E. Parker, S. Martin and I.G. Rigot (1999), "Surface air temperature and its changes over the past 150 years", Reviews of Geophysics 37:173–199.

Keeling, R.F., S.C. Piper and M. Heimann (1996), "Global and hemispheric sinks deduced from changes in atmospheric O_2 concentration", Nature 381:218–221.

Kiehl, J.T., and K.E. Trenberth (1997), "Earth's annual global mean energy budget", Bull. Am. Met. Soc. 78:197–208.

Manabe, S., and R.T. Wetherald (1975), "The effect of doubling of CO_2 concentration on the climate of a general circulation model", J. Geoph. Res. 85:5529–5554.

Mann, M.E., R.S. Bradley and M.K. Hughes (1999), "Global-scale temperatures during the past millennium: Inferences, uncertainties and limitations", Geophys. Res. Lett. 26:759–762.

Molina, M.J., and F.S. Rowland (1974), "Stratospheric sink for chloro-fluoro-methanes: chlorine-catalysed destruction of ozone", Nature 249:810–812.

Nakicenovic, N., A. Gruebler, H. Ishitani, T. Johansson, G. Marland, J. Moreira and H.-H. Rogner (1996), Energy Primer, in: Climate Change 1995, Impact, Adaptation and Mitigation, IPCC (Cambridge University Press, Cambridge, UK) 75–92.

National Academy of Sciences (1979), "Carbon dioxide and climate", A Scientific Assessment (US Climate Research Board, Washington, DC) 22 pp.

NOAA/OAR/CMDL (2002), South Pole Ozone Program, available at http://cmdl.noaa.gov/ozvv/ozsondes/spo/index.html.

Palutikof, J. (2002), "Global temperature record", Climate Research Unit, University of East Anglia, available at www//uea.ac.uk/kyndal.centre.

Parson, E.A. (1998), "Fossil fuel without carbon dioxide emissions", Science 282:1053–1054.

Ramanathan, V., and A.M. Vogelmann (1997), "Greenhouse effect, atmospheric solar absorption and the earth's radiation budget: From the Arrhenius–Langley era to the 1990s", Ambio 26:38–46.

Rodhe, H., E. Cowling, J. Galbally and R. Herrera (1988), "Acidification and regional air pollution in the Tropics", in: H. Rodhe and R. Herrera, eds., Acidification in Tropical Countries SCOPE, Vol. 36 (Wiley, Chichester, UK) 3–39.

Schellnhuber, H.-J., A. Block, M. Cassel-Gintz, J. Kropp, G. Lammel, W. Lass, R. Lienenkamp, C. Loose, M. Lüdeke, O. Moldenhauer, G. Petschell-Held, M. Plöchl and F. Reusswig (1997), "Syndromes of global change", GAIA 6:19–31.

Seinfield, J.H. (1986), Atmospheric Chemistry and Physics of Air Pollution (Wiley, New York) 280 pp.

Social Learning Group (2001), Learning to Manage Global Environmental Risks: A Comparative History of Social Response to Climate Change, Ozone Depletion and Acid Rain (MIT Press, Cambridge, MA).

Solomon, S., G.H. Mount, R.W. Sanders and A.L. Schmeltekopf (1987), "Visible spectroscopy at McMurdo Station, Antarctica: Observations of OClO", J. Geophys. Res. 93:8329.

Sweden's Case Study for the UN Conference on the Human Environment (1971), "Air pollution across national boundaries", The Impact on the Environment of Sulfur in Air and Precipitation (Royal Ministry for Foreign Affairs, Stockholm, Sweden) 96 pp.

Toon, O.B., P. Hamill, R.P. Turco and J. Pinto (1986), "Condensation of HNO_3 and HCl in the winter polar stratosphere", Geophys. Res. Lett. 13:1284.

Tuinstra, W. (1999), "Using computer models in international negotiation", The Environment 41(9):32–43.

Walker, B., W. Steffen, J. Canadell and J. Ingram (1999), The Terrestrial Biosphere and Global Change (Cambridge University Press, Cambridge, UK) 439 pp.

Chapter 2

ECOSYSTEM DYNAMICS

SIMON A. LEVIN and STEPHEN W. PACALA

Department of Ecology and Evolutionary Biology, Princeton University, Princeton, NJ 08544-1003, USA

Contents

Handbook of Environmental Economics, Volume 1, Edited by K.-G. Mäler and J.R. Vincent

Abstract

From ecosystems we derive food and fiber, fuel and pharmaceuticals. Ecosystems mediate local and regional climates, stabilize soils, purify water, and in general provide a nearly endless list of services essential to life as we know it. To understand how to manage these services it is essential to understand how ecological communities are organized and how to measure the biological diversity they contain. Ecological communities are comprised of many species, which are in turn made up of large numbers of individuals, each with their own separate ecological and evolutionary agendas. Not all species are equal as regards their role in maintaining the functioning of ecosystems or their resiliency in the face of stress. This chapter explains how ecosystems evolve and function as complex adaptive systems. It examines ecological systems at scales from the small to the large, from the individual to the collective to the community, from the leaf to the plant to the biosphere (including the global carbon cycle). It reviews theoretical and empirical models of ecosystem dynamics, which are highly nonlinear and contain the potential for qualitative and irreversible shifts. It considers applications to forests, fisheries, grasslands, and freshwater lakes.

Keywords

ecosystems, communities, biodiversity, global carbon, evolution

JEL classification: Q22, Q23, Q25

1. Introduction

Ecosystems are the meeting grounds on which species interact, the integrated networks of biotic and abiotic elements through which materials and information flow, and that support our continued existence on the planet. From ecosystems we derive food and fiber, fuel and pharmaceuticals. Ecosystems mediate local and regional climates, stabilize soils, purify water and in general provide a nearly endless list of services essential to life as we know it [Daily (1997)]. The case for the preservation of ecosystems and these services is manifestly clear; the essential challenges are in the details of how to do it.

Ecosystem services come in a tremendous diversity of forms, some realized and many potential. This, of course, makes the problem of valuation immensely difficult. We derive many direct benefits from ecosystems, most obviously things like food, fuel, fiber and timber. These are typically part of the economic marketplace, so we have some good ideas about how to assess their worth. Furthermore, it is generally estimated that about 30 percent of the pharmaceuticals currently available were derived from natural sources; what is more difficult to evaluate is the potential for finding new pharmaceuticals, perhaps from species not yet identified. We use biodiversity further for recreation and spiritual enhancement. Travel methods provide a beginning to approximate how humans value these services, but do not account for the sense of well-being people derive simply from knowing pieces of Nature exist that they will never visit, or the ethical or cultural dimensions.

It is the indirect benefits that make the problem even more difficult. As discussed elsewhere in this paper, ecosystems provide pollination services, sustain critical biogeochemical cycles, mediate climate, provide habitat for a vast diversity of species, protect against droughts and floods and maintain the quality of our air and water. We are just beginning to understand the relationships between biodiversity and ecosystem functioning, what aspects of that functioning are most crucial to sustaining our way of life, and the degree to which we could substitute (and at what cost) for those services were particular natural systems lost.

The goal of managing and preserving ecosystems is confounded by immense problems of scale. For the purpose of this article, ecosystems may refer to landscapes ranging from lakes or watersheds to the biosphere. Typically, the boundaries of ecosystems are not well delineated, but are defined operationally for the convenience of the investigator or manager. Hence, ecosystems are fractal-like entities, with structure and organization on multiple nested scales, and that are not sharply distinguished from their neighboring ecosystems.

The problem of scale [Levin (1992)] has a number of implications for management. First of all, it introduces problems of externalities. The activities of humans, especially regarding the utilization of biodiversity and the discharge of materials into the environment, typically are such that costs and benefits are not realized on the same scales. For example, extraction of resources from the ocean, or release of effluents into the air or water, creates costs for society that are not normally reflected in the ways the market prices things. We live in a global commons [Hardin (1968)], in which there is not adequate financial or moral incentive for people to behave in the common good. We can expect individuals to care about pollution in their own neighborhoods and localities, but less so as the scale of impact increases; this mismatch of scales is, indeed, the source of many of the environmental problems we face today [Levin (1999)].

More generally, ecological communities – the biotic essence of ecosystems – are comprised of many species, which are in turn made up of large numbers of individuals, each with their own separate ecological and evolutionary agendas. The dynamics of ecosystems emerge from the collective dynamics of huge numbers of individual parts, and in turn feed back to influence those parts. To understand how to preserve the services that ecosystems provide it is essential to understand how communities are organized, and which are the most relevant ways to measure biodiversity. Not all species are equal as regards their role in maintaining functioning of ecosystems, or their resiliency in the face of stress. Thus it is essential to develop ways to relate processes at the level of individual organisms to the populations of which they are members, and to the communities and ecosystems in which they reside. We must learn to scale from the small to the large, from the individual to the collective to the community, from the leaf to the plant to the biosphere. We need, in effect, to build a statistical mechanics of ecological communities, founded upon a combination of observation, controlled experimentation and simulation, and mathematical theory.

The problems we face will be familiar to economists, who well recognize the need to integrate micro- and macro-perspectives, and to relate the dynamics of societies to the way individuals make decisions. They will also recognize the context dependence of decision making, and that in consequence the dynamics of systems are highly nonlinear, hence constrained by the accidents of history. It is these issues, and how to deal with them, that will form the core of this paper.

2. The nature of communities and ecosystems

Ecological communities are not entities constructed de novo, by a developmental process akin to the ontogeny of an individual organism. They have not been shaped through natural selection for their macroscopic properties, as have been organisms, because they are not reproduced as units that faithfully replicate genetic material. Rather, they are complex adaptive systems [Levin (1998)], whose collective properties emerge from interactions and a process of selection operating at myriad levels of organization below the whole system. This is why concepts of system health, as applied to ecosystems, can be so misleading. Ecosystems are loosely defined assemblages of interacting elements, exhibiting structure and functioning at almost every scale of organization, and interchanging genetic material so freely with other ecosystems that they cannot represent evolutionary units. Ecosystems do evolve, in concert with processes of succession, immigration and emigration; and certainly ecosystems provide the context for broad evolutionary responses to environment. But that evolution is more the result of the evolution of their component species, weakly interacting across multiple scales of space and time, rather than the evolution of coherent entities in the classical Darwinian sense.

Though this view is widely accepted among ecologists, and certainly among evolutionists, that has not always been the case. In the early part of the Twentieth Century, two views of ecological communities defined the poles of a spectrum of perspectives. Henry Gleason argued that species were individualistically distributed along environmental gradients (such as altitude), thereby defining communities that intergraded continuously with one another, without clear boundaries demarcating where one community ended and another began. In contrast, however, the great synecologist Frederick Clements viewed communities as superorganisms, comprised of species that occurred together, and which hence formed evolutionary units. The view was not so different from that familiar today among Gaia theorists, who derive inspiration from Hutton's eighteenth century characterization of the Earth as a superorganism, "whose proper study is by physiology" [Hutton (1788)].

Though Clements's perspective was the dominant one for a time, today we know that the truth is much closer to that of Gleason. The proof came largely from detailed studies of the distributions of species along gradients, and sophisticated mathematical analyses [Curtis (1956), Whittaker (1975)]. Species are, to a first approximation, individualistically distributed along gradients, in accordance with the conceptualization of Gleason. That is not to say that coevolution of species does not occur. Certainly host–parasite pairs provide the kind of close associations that tie together the evolutionary fates of species, and so too do mutualistic pairs such as lichens and plant-mycorrhizal associations. But these examples are for the most part restricted to pairs of species, in which interactions are tight [Ehrlich and Raven (1964)]; it is hard to cite convincing examples of strong coevolution involving three or more species. Indeed, it is clear that coevolutionary forces also operate to shape the dynamics of much larger assemblages, for example, plants and their herbivores. Plants develop defensive chemicals to repel herbivory, and herbivores in turn develop detoxification mechanisms to overcome these defenses. But these interactions are typically highly generalized, involving what is termed "diffuse coevolution", in which evolutionary changes in one species of plant or herbivore have only weak influence over other particular species [Ehrlich and Raven (1964), Futuyma and Slatkin (1983)]. The human immune system is a parallel example, involving a diffuse (and adaptive) response to a suite of potential enemies, rather than a focused response to a particular one.

Despite the demise of the superorganism concept, Clements's influence remains justifiably strong in our understanding of the dynamics of ecological communities. It was Clements (1916), and Shelford (1911a, 1911b), who helped shape our understanding of succession (Section 5), the description of how communities become formed over ecological time, and the notion of the "climax community", the endpoint of the successional progression. Clements made clear how the nature of the regional climax community depended upon local climatic factors, in particular temperature and moisture regimes, in addition to other factors such as soil characteristics and fire history. In the next section we turn to the patterns that emerge in the distribution of vegetation types.

3. Terrestrial ecosystem patterns

Climate and vegetation: Laminar flow in a turbulent ecological matrix

Although many aspects of ecosystems vary unpredictably in space and time, a few of the most important are orderly and predictable everywhere on the globe, indicating universally applicable regulation and relatively simple underlying causes. For example, it is possible to glance at the vegetation anywhere on the earth and deduce details about the climate, the amount of solar energy captured each year by plants, and the amount of water transpired by vegetation from the soil into the atmosphere (evaporation controlled by plants is labeled *transpiration*). This section describes some of the most important such patterns in terrestrial systems and outlines the emerging mechanistic understanding of their causes. Although three-quarters of the earth is covered by water, humans live on land; hence there has been much more work on fine tuning our understanding of terrestrial vegetation while taking a broader brush approach to the marine. Still, especially when considering the global biogeochemical cycles, and in particular the flux of carbon, it is vital to consider the role of the oceans. We shall return to this theme in later sections.

Climate and biomes

At the largest scales, vegetation is grouped into classes called biomes. These include:
1. Tundra – treeless regions underlain by permafrost.
2. Taiga – a mixture of arctic shrubs and short (<10 m) needle-leaf trees.
3. Boreal forest – cold-adapted conifer forest with early successional broad-leaf trees such as birches (*Betula*) and aspens (*Populus*).
4. Temperate deciduous forest – dominated by broad-leaf trees that lose leaves in winter.
5. Temperate rain forest – dominated by immense needle-leaf trees like those in Northwest North America.
6. Grassland – dominated by perennial grasses and usually with periodic fires.
7. Shrub-steppe – dominated by drought-adapted shrubs.
8. Savanna – grassland with intermingled small trees.
9. Tropical deciduous forest – broad-leaf trees that lose leaves in the dry season(s).
10. Tropical evergreen rain forest.
11. Desert.

The global distribution of these biomes shows remarkable association with climate. Figure 1 plots annual average rainfall on one axis and annual mean temperature on the

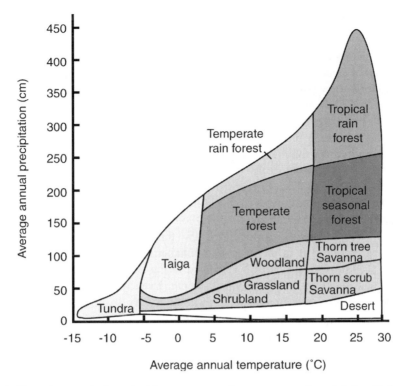

Figure 1. Vegetation type is determined by average annual temperature and precipitation. Redrawn from Gould and Keeton (1996, page 1183).

other [from Whittaker (1975)]. The envelope in the diagram is triangular because there are no cold regions of the earth with very wet climates (cold air cannot hold as much water as warm air). Note that the biomes separate neatly into regions of different climate, and that biomes dominated by plants with similar size are arranged in diagonal bands, with the largest plants at the top and the smallest at the bottom. The significance of the diagonal banding probably relates to water availability in the soil. Soil moisture depends on the mix of loss due to evapotranspiration (evaporation from the soil and transpiration by plants) and gains due to precipitation. Thus, one could hold soil moisture constant by increasing rainfall and evapotranspiration simultaneously. Because the latter increases with temperature, diagonal bands in the figure correspond roughly to equally available soil water. Figure 1 is one of many similar schemes relating vegetation structure to climate, which show the over-riding importance of water availability at the largest scales [MacArthur and Connell (1966), Holdridge (1967), Whittaker (1975)].

Primary production and climate

Virtually all energy used by living organisms is solar energy captured by green plants and used to make carbon–carbon bonds in organic molecules during photosynthesis. The carbon for these chemical bonds usually is taken either directly or indirectly from carbon dioxide. The reverse process, called respiration, provides energy by breaking the bonds and reconverting organic carbon back to carbon dioxide. The amount of energy per unit time that green plants have left from photosynthesis after deduction of the respiration required for maintenance metabolism is called net primary production or NPP. NPP is an important quantity because it is the rate at which energy is provided to an ecosystem. NPP is usually reported in units of carbon mass per year, with the understanding that it actually refers to the bond energy in the mass.

The global NPP is approximately 70–90 billion metric tons of carbon per year [Whittaker and Likens (1973), Field et al. (1998)]. Two thirds of this is produced on land, and the remaining one-third is in the oceans [Whittaker and Likens (1973), Bunt (1975), Field et al. (1998)]. Oceanic NPP is restricted to the surface because light necessary for photosynthesis attenuates quickly with depth. Even in clear water, photosynthesis provides a surplus over maintenance respiration only within the top 100 meters. Oceanic NPP is also strongly limited by nutrients because of a phenomenon known as the biological pump. The microscopic plants that dominate oceanic photosynthesis (phytoplankton) have very short lifetimes, primarily because the majority are killed by predators. To grow, phytoplankton must absorb from the surrounding water the carbon, nitrogen, phosphorus and other materials necessary to produce living tissue. When phytoplankton die, they sink, and so these materials are perpetually moved from the surface to the bottom. This creates a gradient, with the nutrients in short supply at the surface and abundant at depth. By separating the sunlight from the nutrients, the biological pump strongly limits oceanic NPP. Open ocean areas are only as productive as terrestrial deserts (e.g., <100 g carbon per m^2 per year) [Whittaker and Likens (1973), Bunt (1975), Field et al. (1998)]. The abundant fish that support commercial fisheries are concentrated along continental margins, because the shallow water there brings nutrients close to available sunlight and allows high NPP.

Like the distribution of biomes, terrestrial NPP is strongly determined by climate. Figure 2 shows the "so-called" MIAMI model of Lieth, Boy and Wolover [Whittaker and Likens (1973)] that relates terrestrial NPP to mean annual temperature and rainfall. The NPP predicted by the model is the minimum of the values given by the two regression functions. NPP averages 1000 g C m^{-2} y^{-1} in warm wet tropical forests, one-third to two-thirds this value in temperate forests, one-third or less in boreal forests, one-tenth to one-third in grasslands, and less than one-tenth in tundra and deserts.

Like oceanic ecosystems, terrestrial areas may also be limited by nutrients – this is, after all, the reason for the use of chemical fertilizers in agricultural systems – but the significance of nutrient limitation is diminished in terrestrial systems by the absence of a mechanism like the biological pump. In general, nitrogen is the most important limiting

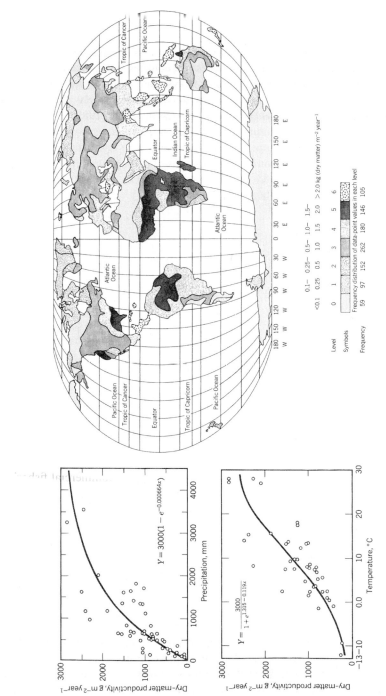

Figure 2. Lieth's Miami model. The net primary production for a location is given as the minimum of the two values predicted by the curves at the left.

nutrient in temperate terrestrial ecosystems, while phosphorus is most important in the tropics [Schlesinger (1997)].

Eventually, all of the organic carbon produced as NPP is respired by animals and microorganisms; but over time scales of decades to centuries, most ecosystems accumulate C, in a long transient phase. In terrestrial systems, a small fraction ($<10\%$) is consumed by herbivores and so respiration by decomposers largely balances NPP [Elton (1927), Whittaker and Likens (1973), McNaughton et al. (1989)]. When NPP is greater than total respiration by animals and microorganisms, carbon accumulates in the ecosystem. Although this cannot occur indefinitely, there is substantial evidence that NPP currently exceeds respiration by 2 billion tons of carbon annually in terrestrial ecosystems in the northern hemisphere [Tans, Fung and Takahashi (1990), Battle et al. (2000), Bousquet et al. (2000), Pacala et al. (2001)].

Models of ecosystem physiology

During the past decade, mechanistic models have been developed that predict patterns like those described above. These models work to large extent because the biochemistry of photosynthesis is conserved everywhere. Most are based on the work of Farquhar and Sharkey (1982) [see also Collatz et al. (1991, 1992), Foley et al. (1996)], who developed models of photosynthesis and leaf respiration that are predictive and consistent with the known biochemistry, and yet remain algebraically simple. The model for the most common form of photosynthesis predicts the instantaneous rate of photosynthesis minus leaf respiration (A_N) as a function of absorbed photosynthetically active radiation (Q_{sol}), leaf temperature (T_L), and the concentration of CO_2 within the air spaces inside the leaf (C_i). The functional form for A_N describes the two primary steps in photosynthesis: light capture by chlorophyll and the use of this energy to fix CO_2 with the enzyme rubisco (ribulose bisphosphate carboxylase) [Farquhar and Sharkey (1982), Collatz et al. (1991, 1992), Foley et al. (1996)].

For simplicity, let us assume that T_L is equal to air temperature. Carbon dioxide moves between the atmosphere and the inside of the leaf through small valves called stomates. If we assume that the leaf is roughly in equilibrium, the rate of CO_2 movement into the leaf must balance the rate of consumption of carbon dioxide by the leaf, and the dynamics can be captured in a simple diffusion model. The rate of consumption is simply A_N because CO_2 is consumed to make the organic carbon. Thus $A_N = (g_s/1.6)(C_A - C_i)$, where the stomatal conductance, g_s, determines how open the stomates are and C_A is the concentration of CO_2 in air (now approximately 350 ppm). Values of stomatal conductance are usually defined for water vapor; the factor of 1.6 simply converts these into values for CO_2.

To complete the model we require an equation for g_s that captures the regulation of stomatal opening and closure. Stomates control both the rate of CO_2 gain and the rate of water loss by transpiration (water vapor escapes almost entirely by diffusing through open stomates). The algorithm that controls stomatal openness balances the costs of water loss against the benefits of carbon gain and the regulatory mechanisms, and is still the subject of active research [Farquhar and Sharkey (1982), Collatz et al. (1991, 1992), Leuning (1995), Grace (1997)]. Nonetheless, over 15 years ago Ball, Woodrow and Berry (1986) produced a simple phenomenological model that has stood the test of time. This model gives g_s as a linear (affine) function of A_N, with slope an increasing function of atmospheric humidity (H) and a decreasing function of leaf temperature: $g_s = m(H, T_L)A_N + b$. In total, we now have three equations (equations for net photosynthesis, CO_2 diffusion and stomatal conductance) and three unknowns (A_N, C_i, and g_s), with three environmental inputs (T_A, Q_{sol}, and H; assuming $T_L = T_A$).

The rate of net photosynthesis in these models is in units of mass of CO_2 per unit leaf area per unit time. To use the models to predict ecosystem-wide NPP, one also needs to know the number of leaf layers and how light attenuates from the top layer down. The number and temporal dynamics of leaf layers can be obtained from satellite measurements, which are available for the entire globe [Potter et al. (1993), Field et al. (1998)]. Alternatively, one can predict these quantities with a marginally more complicated model [VEMAP and Members (1995), Field et al. (1998)]. Light attenuation through the canopy is simply exponential; this is known as Beer's Law.

Thus, if one knows air temperature, humidity, solar radiation and leaf layers present throughout the year, then one can sum the NPP predicted by our simple model throughout the year for each layer, and compute an annual ecosystem NPP with considerable accuracy [Bonan (1995), Foley et al. (1996), Haxeltine and Prentice (1996), Friend et al. (1997)]. In addition, because water vapor escapes from leaves through the stomates, the rate of evaporative water loss (E) can be computed from simple one-dimensional diffusion as: $E = g_s(H_L(T_L) - H)$, where $H_L(T_L)$ is the humidity of the saturated air inside the leaf, which in turn depends only on leaf temperature [Aber and Melillo (1982), Leuning (1995)] . This model predicts water loss as accurately as the NPP model predicts carbon gain.

One must exert caution, however. Using aggregated (big leaf) models to describe vegetation effectively ignores competitive interactions within the vegetation. Over time scales of a few decades, this can lead to substantial errors [Bolker et al. (1995), Walker, Kinzig and Langridge (1999)].

The current crop of global ecosystem models used to predict climate, weather and the response of the Earth system to human impacts are generally underpinned by close relatives, the ecophysiological models described above [Sellers et al. (1986), Potter et al. (1993), VEMAP and Members (1995), Foley et al. (1996), Haxeltine and Prentice (1996), Haxeltine, Prentice and Creswell (1996), Friend et al. (1997), Potter et al. (1998), Moorcroft, Hurtt and Pacala (2001)]. By coupling an ecophysiological model to a model of below-ground decomposition, nitrogen cycling and soil moisture, and a knowledge of the way in which plants compete and allocate carbon, these models increasingly capture the causes of predictable large-scale patterns, like those in Figures 1 and 2.

4. Ecosystem assembly

The climax community discussed earlier describes a steady state mosaic achieved only after a long progression of states in the development of a community, and involving not a single homogeneous state but a patchwork of successional stages. The actual assembly process of a community involves some clearly distinguishable stages, each characterized by a characteristic set of species. Volcanic eruptions, massive blowdowns and ocean upwellings all disrupt the progression towards a maintenance of equilibrium, conferring added importance to transient phenomena.

Succession

In 1883, the Pacific island of Krakatoa (Krakatau) blew up, after a series of volcanic eruptions of lesser magnitude. Following the 1883 eruption, what remained of the island was covered with ash and pumice, wiping out whatever biotic life there was. What ensued was a long period of recolonization of the island, a process called *primary succession*. Mosses and lichens arrived early, followed by flowering plants and other more complex species. Such successional development follows well established patterns wherever it occurs; other recent examples are the colonization of the island of Surtsey, which formed off Iceland as a result of volcanic eruptions there, and the post-eruption recolonization of mountains such as Mt. St. Helens, in the State of Washington.

Volcanic eruptions are particularly dramatic disturbances of ecosystems, and occur relatively infrequently. More common are disturbances, such as those caused by treefalls, windthrows, lumbering, or even fire, which remove the dominant plants, but leave the basic physical characteristics of the substrate unchanged. Following such events, a successional process takes place similar to primary succession, but with a head start. This sequence, called *secondary succession*, is a ubiquitous feature of most forests, due to the frequent patterns of disturbance caused by the deaths of individual trees, and broader forms of disruption. Furthermore, analogous successional patterns are to found among grassland plants, intertidal algae and invertebrates, and marine plankton [Reynolds (1997, 2000)]. Indeed, localized disturbance followed by some form of succession is a common feature of almost any ecosystem.

Why is succession important to us in thinking about economic valuation, and about the sustainability of ecosystem services? Succession is a form of sharing of resources, in which some species are opportunistic (exploiting new but short-lived opportunities), while others appear later but enjoy competitive dominance over the early colonists. In between, there is a virtual continuum of types, along an axis of increasing competitive ability and decreasing colonization ability. It is not hard to think of parallels in the economic marketplace. The proliferation of roles within the successional hierarchy becomes reinforced, over broader spatial and longer temporal scales, through evolutionary diversification; and it is in that diversification that the system's resiliency is encoded. Recognition of this has led to more enlightened fire management strategies, in which small, controlled burns are recognized as being vital for the local renewal of the sub strate, and for the maintenance of overall system resiliency.

Spatial scale and biodiversity

The importance of localized disturbance, and the patterns of recovery that ensue, make clear that our understanding of the dynamics of any ecosystem will depend on the scale at which we study it. On small scales, ecosystems are open, highly variable systems, exhibiting dynamical patterns that are predictable only in a statistical sense. As the scale enlarges, the system becomes relatively more closed, more heterogeneous, and more predictable. Not surprisingly, measures of diversity – in particular, the number of species observed – increase with spatial scale. The classical models reflect self-similarity, in which the number of species increases as a power of the area, although recent analyses [Harte, Kinzig and Green (1999), Condit et al. (2000), Plotkin et al. (2000), Plotkin and Levin (2001)] call this relationship into question, at least for some systems. Theoretical analyses [Durrett and Levin (1996), Hubbell (2001)] also support the notion that the power law relation cannot hold over all scales.

What accounts for the observed relationships between number of species and area? Such questions led Robert MacArthur and Edward Wilson, in the 1960s, to turn their attention to islands in the ocean, and to examine the processes controlling species abundances there. MacArthur and Wilson (1967) argued that the species assemblages on islands reflects a balance between processes of colonization and extinction (Figure 3),

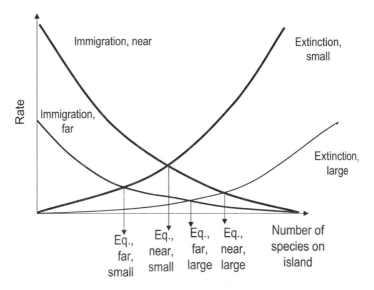

Figure 3. Immigration and extinction rates depend on the number of species on an island, the size of the island, and the distance from the mainland [after MacArthur and Wilson (1967)].

just as species diversity over broader and longer scales reflects a balance between speciation and extinction. They reasoned that the biota of islands would be constantly in flux, as new species arrived at islands, and as others went extinct. Colonization rates should depend primarily upon how far an island was from a mainland source (though island size can also play a role), whereas extinction rates depend primarily upon island size (and the number of species on the island). Based on these generalizations, they formulated an "equilibrium theory of island biogeography", as described in their influential 1967 book. In the equilibrium theory, the transient events of colonization dynamics are ignored, and the focus is restricted to the asymptotic equilibrium number of species. Quantitative aspects of the theory were then tested by Daniel Simberloff, who fumigated mangrove islands and studied their recolonization by various insect species [Simberloff and Wilson (1969)]. Subsequently, the theory has been tested and applied in a number of other settings, most notably in providing a theoretical basis for conservation practice [but see Simberloff and Abele (1976)].

Islands in the ocean provide a relatively simple system for formulating and testing theory, and it was natural that MacArthur and Wilson started there. Their own interests, however, extended much beyond such systems, especially to include islands of habitat in terrestrial environments; and it is the applicability of the theory to such situations that accounts for the great influence and popularity it has enjoyed. Natural landscapes are patchworks of habitats of differing characteristics, and the activities of humans have exacerbated patterns of fragmentation through land-clearing for agriculture, housing and other activities. What formerly were vast virgin stands of forest, for example, have been replaced by patchworks of suitable habitat for particular species, and reduced species diversity. Each patch of suitable habitat is, in a very real sense, an island in a sea of unsuitable habitat; species persist as metapopulations [Levins (1969), Gilpin and Hanski (1991)], organized into small local populations that interact only occasionally with other local populations. To some extent, this is the natural state of affairs; but human activities clearly have changed the nature of the island landscape. Such a perspective has informed conservation strategies, for example for the spotted owl in the old-growth forests of the Northwestern United States; but it can be applied more generally to ensembles of species that occupy these terrestrial archipelagoes. It can also be overused [Simberloff and Abele (1976)].

As we have seen earlier, the equilibrium theory of island biogeography is inadequate for capturing the dynamics of many ecosystems, most notably temperate and tropical forests. In these, recurrent disturbances are too frequent to allow the neglect of transients. Small islands form through the deaths of individual trees, larger islands through windthrows, and still larger islands through forest fires. Each source of disturbance leaves its own signature on the local environment, and each creates a spectrum of opportunities for those species adapted to the various stages of secondary succession. Local unpredictability hence becomes, globally, the most predictable aspect of these systems, and this is reflected in ecological dynamics, as well as in evolutionary scenarios [Levin and Paine (1974)]. Local disturbance, hence, becomes the key mechanism maintaining diversity, and maintaining system resilience [Paine and Vadas (1969), Connell (1971) Levin and Paine (1974)].

Given such a dynamic, it became essential to expand the theory of island biogeography to address non-equilibrium situations. Levin and Paine did this for the intertidal regions, complementing transient models of the dynamics of island biota with dynamic models of island creation and generation, operating over longer time scales and broader spatial scales. Since their paper, and with the advent of high-speed computation, these models have been joined by spatially explicit descriptions of ecosystems, built either on interacting particle models (stochastic cellular automata), coupled map lattices or individual-based models that track every individual. Remarkably, one of the great successes of the modern theory [Hurtt et al. (1998), Moorcroft, Hurtt and Pacala (2001)] has been to show that detailed forest simulation models can be collapsed into the Levin–Paine framework, providing more robust prediction than could be achieved in other ways.

Forest simulation models

Forest simulation models are an essential complement to the understanding of what maintains diversity, and for management purposes. Because the basic information about climate effects, for example, is manifest at the level of the individual, that is the natural place to begin for modeling. Ultimately, one needs more robust models however, in which macroscopic descriptions are derived from microscopic rules through judicious aggregation. We treat both here. One particularly successful kind of individual-based models is the forest "gap" model developed originally by Botkin, Janak and Wallis (1972) nearly thirty years ago. A forest gap model is a stochastic process that predicts the birth, dispersal, growth, reproduction and death of every individual plant throughout its life [see Shugart (1984)]. The plants are usually located in a lattice of tree-sized cells, but space may also be continuous [i.e., Pacala and Deutschman (1996)]. The individual plants compete with one another by shading and by depleting local soil water and nutrients. Modern versions increasingly rely on ecophysiological formulations, such as those in Section 3, to predict individual growth and resource uptake [Friend et al. (1997), Moorcroft, Hurtt and Pacala (2001)].

Forest gap models have a demonstrated capacity to reproduce observed ecological succession, how tree species composition varies in response to past changes in climate and across current spatial gradients in soils and topography, and how forests recover from management and harvesting [Shugart (1984), Pacala and Deutschman (1996), Friend et al. (1997)]. They are widely used to manage timber and public lands. Forest gap models are successful largely because the scale at which they are formulated is appropriate both to represent the critical processes and to gather the data necessary to measure their components in the field. In particular, the explicit inclusion of tree-sized spatial areas is critical, because regeneration of the forest canopy requires the occurrence of canopy gaps [Pacala and Deutschman (1996)].

However, the benefits of the individual-based formulation are balanced by the limited spatial area that can be modeled because of the need to simulate each and every individual. One cannot predict the biosphere by simulating every plant on Earth. For this reason, considerable effort has been expended to discover macroscopic equations that govern a large ensemble of linked tree-sized areas. These would be analogous to physical models that predict the macroscopic behavior of a fluid from the underlying stochastic rules of molecular motion. Formally, the necessary equations would describe the region-specific first moments of the stochastic individual-based model.

In the past five years, a number of efforts have successfully scaled-up individual-based models, describing interactions ranging from host–parasite interactions, to evolutionary games and plant competition [Durrett and Levin (1994a, 1994b), Durrett and Neuhauser (1994), Rand, Keeling and Wilson (1995), Keeling, Rand and Morris (1997), Neuhauser and Pacala (1999), Bolker, Pacala and Levin (2000)]. Most of these rely on so-called second-order (first two moments) closure schemes [Levin and Pacala (1997)]. Recently, as discussed earlier, however, [Hurtt et al. (1998), Moorcroft, Hurtt and Pacala (2001)] have shown how to collapse the full equations into a set of biogeographic equations of the form of the Levin and Paine (1974) model; this simplification provides a close approximation to the first moment of forest gap models. This formulation overcomes the spatial limitations of gap models; Moorcroft, Hurtt and Pacala (2001) use it successfully to model the entire Amazon basin and the rest of the Neotropics.

5. Ecosystems as self-organizing systems

Ecosystems are prototypical examples of self-organizing systems, assembled from parts that are molded by evolution over broader spatial scales and longer temporal scales. As such, the process of self-assembly is governed in part by chance events of colonization and seasonality, which influence a nonlinear dynamic in ways that can lead to multiple endpoints. When a gap forms in a forest, the trees that will first recolonize these islands of opportunity are those that happen to be waiting in the understory, or those whose seeds become established first. At broad enough scales relative to the scale of disturbance, the system will maintain its essential character as a spatial–temporal mosaic. But at small scales, or if disturbance is an event (such as volcanic eruption) that affects large areas, the forest could be shifted into one with a very different appearance.

In the face of small, localized disturbances, the source of new seeds to populate gaps is a fairly predictable one; and indeed, the whole successional dynamic of the forest often can be well-described as a crazy-quilt made up of patches that change states probabilistically, according to simple Markovian rules [Horn (1976)]. In the face of large-scale disturbances, however, nonlinear models are essential, and include the possibilities of complex dynamics and multiple stable states.

Population dynamics

The simplest nonlinear model of population growth, now well-studied within the context of chaos theory, is the one-dimensional map

$$x(t+1) = f\big(x(t)\big), \tag{1}$$

in which $x(t)$ is the population size at time t, and f is some functional form such as the logistic,

$$f(z) = z + rz\left(1 - \frac{z}{K}\right). \tag{2}$$

Models of this sort allow the projection of population size forward, and have proved especially useful in the management of fisheries, where the logistic form is often replaced either by the Beverton–Holt relation

$$f(z) = \frac{Krz}{K + rz} \tag{3}$$

or the Ricker equation

$$f(z) = rz \exp\big(-b(z - K)\big). \tag{4}$$

Beverton, Holt and Ricker were three of the main figures in building a quantitative theory of fisheries management [Ricker (1954), Beverton and Holt (1957)]. Ricker's 1954 paper was a classic, in which he explored the complicated periodic, quasiperiodic and chaotic dynamics that some fisheries models could produce. May (1974) put these sorts of observations into a theory of complex dynamics, providing a link to similar investigations in other branches of science [Lorenz (1963, 1964)]. The key result, as developed by May, is that as the parameter r (the so-called intrinsic rate of increase of the population) is increased in (4) or (6), the dynamics of the system (3) will transition from a stable equilibrium (at K) to a cycle of length 2 generations, and from there through a characteristic sequence of bifurcations to cycles of powers of 2. Finally, for sufficiently high values of r, a region is entered in which cycles have no fixed period, but exhibit chaotic dynamics, with great sensitivity to initial conditions.

These simple models provide only a starting point for investigation. They can be and have been extended in a number of directions, to include for example aspects of the structure of the population, the variation in the external environment and interactions with other species. Not surprisingly, the potential for complex events goes up rapidly as the dimensionality of the system expands. Most populations, including of course fish populations, are not composed of individuals all of the same age or size, and demographic extensions of standard models are needed. Levin and Goodyear (1980), for example, explore age-structured versions of (1), and show that the already fascinating story told by Ricker and May becomes even more interesting, as the interaction of competing periods produces quasiperiodic dynamics, more regular than the chaotic patterns but without a single dominant period. Models of multispecies communities similarly increase the potential for complex dynamics. Although continuous-time versions of (1) do not show chaotic dynamics, it is easy to generate such dynamics in continuous-time multi-species systems of 3 species or more.

There is a long history of investigation of models of interacting species, of the general form

$$\frac{dx_i}{dt} = F(x_1, x_2, \ldots), \quad i = 1, 2, \ldots, n, \tag{5}$$

where x_i is the density of species i. These models differ from those already discussed in that time is continuous, rather than discrete; but complicated dynamics can still result. Chaos can only appear for $n \geqslant 3$, but periodic dynamics are quite easily obtained for $n = 2$.

The earliest such models in ecology were put forward by the great mathematician, Vito Volterra (1926), and by Alfred Lotka (1925). Volterra was attracted to these problems at the behest of his son-in-law, the fisheries biologist Umberto d'Ancona, who was interested in the fluctuations of the Adriatic fisheries. Volterra's equations of predator and prey exhibited a natural oscillatory tendency, and more elaborate versions of these models indeed exhibit stable limit cycle behavior. For species in competition, multiple stable states could emerge. More complicated models of the dynamics of trophic networks of species, in which there are multiple chains of interaction among species of all persuasions, can exhibit any dynamical behavior imaginable.

System "flips"

The potential for multiple stable states and path dependence is a matter of particular concern, because of the possibility that anthropogenic or other activities could cause a system to flip precipitously from one configuration to another. Ecological and related examples abound. For example, the reintroduction of the sea otter into large areas of California has the potential to flip the fishery form one dominated by shellfish to one dominated by finfish, because of the keystone role of otters in the system [Estes and Palmisano (1974), Levin (2000)]; the possible economic consequences of this are obvious. Land-use patterns across broad regions of the Middle East, the Sahel and elsewhere have changed, and can change, systems from rich and fertile areas to deserts, as the loss of vegetation leads to erosion, changes in hydrologic cycles, and permanent alterations in climate regimes. Walker and his colleagues [Walker, Langridge and McFarlane (1997)] have demonstrated, in a series of important papers, the potential for system flips in grazing systems.

At broader scales, there is evidence that major global climatic changes have occurred over relatively short periods of time, and that such dramatic changes could occur again in association with qualitative shifts in patterns of circulation [Broecker, Peteet and Rind (1985), Manabe and Stouffer (1988)]. Such concerns have directed attention to the resiliency of ecological systems [Holling (1986), Levin et al. (1998)], and to their ability to sustain the services humans derive from them. They have also emphasized the need for precautionary and adaptive approaches for dealing with uncertainty [Arrow et al. (2000)].

Two cases in point involve the dynamics of grasslands [Walker (1995)] and of temperate lakes under nutrient inputs [Carpenter and Kitchell (1993), Scheffer (1998), Carpenter (2001)]. Such lakes can exist either in eutrophic (nutrient-rich) states or oligotrophic (nutrient-poor) ones, or somewhere in-between. Oligotrophic lakes have high water clarity, which disappears under eutrophic conditions, in which high algal levels choke the system.

Under certain conditions, it appears that the transition from oligotrophy to eutrophy may be sudden, as a result of the input of phosphorus and other nutrients from agriculture and other human activities. There are dramatic differences in the services to be derived from these two situations; eutrophic lakes, besot with algae, are unattractive and unsuitable habitats for fish and fishermen. From an economic point of view, the differences are essentially those between night and day.

In the border areas between the United States and Canada, both shallow and stratified lakes are naturally limited in phosphorus levels, but have been altered by anthropogenic inputs of phosphorus. Carpenter, Ludwig and Brock (1999) assume that the store of phosphorous (x) suspended in algae changes according to a simple law of the form

$$\frac{dx}{dt} = a - bx + f(x). \tag{6}$$

Here a denotes loading (phosphorous inputs from the watershed), b is the purification rate, and $f(x)$ is internal loading. $f(x)$ is assumed to be "S-shaped". Thus, for low phosphorus levels, most additions will be stored in the lakebed; that is, for low stocks of phosphorous, additions tend to be stored in the lakebed; as x is increased, there is a larger return to the water.

In the absence of external loading, this system may exist either in a eutrophic state (if purification rates are low), or in a clear oligotrophic one (if purification rates are high). As loading levels are increased, even in the presence of high purification, the system will flip to a eutrophic one. At intermediate loading rates, there are two stable states, eutrophy and oligotrophy; which is achieved will depend upon the lake's history. Eutrophy is reversible, at economic cost, if loading levels are reduced; indeed, still for high purification, reducing loading from intermediate levels and then increasing them back to those levels could shift a eutrophic system into the more desirable oligotrophic state. At intermediate purification rates (b), even without external loading, the system has two stable states – an oligotrophic one and a eutrophic one, since purification may not be rapid enough to clear natural loading once it gets above a threshold. Hence, if the lake is in its oligotrophic state but loading levels are increased sufficiently, it will flip to the eutrophic state. Once this occurs, even reducing loading levels to zero would not restore the eutrophic condition. Although it is possible that the lake could be purified by expensive technological intervention, this is hardly a desirable turn of events. It furthermore illustrates that some changes might be in effect irreversible.

The example of lake flips is by way of example only, but illustrates why so much recent attention has been directed to the potential for multiple stable states, and for issues of resiliency. To explore this further, one needs a clear understanding about the connections between system structure and function, especially diversity and resiliency. We turn to these issues in the next section.

6. Biodiversity and ecosystem functioning, and relations to ecosystem services

Biodiversity and ecosystem functioning

In the middle of this century Elton (1958) argued that:

"Ecosystem productivity should increase with biodiversity because a mixture of plant species that specialize on exploiting different resources (or the same resources under different conditions) should outperform any single-species".

He further argued

"The stability of NPP and other ecosystem-level measures should increase with biodiversity because a diverse portfolio is more stable than a narrow one".

This view that complexity necessarily begets stability was undercut by theoretical studies by May (1974), who argued that stability indeed usually decreases with increased system complexity, and by simple experiments of two-species competition that usually showed higher productivity in monoculture (single species stand) than in mixture [Trenbath (1974)]. Care must be exercised in interpretation of such results, however. Indeed, more complex systems are less stable in the sense of return to an equilibrium point of species abundances; but that very variability can increase the potential for the system to maintain the ecosystem level stability to which Elton referred [see, for example, Holling (1973), Levin (1999)].

McNaughton (1977) published a contrary view based on experiments in African grasslands that involved mixtures of many species. McNaughton's study was not repeated until a recent explosion of activity led by Naeem et al. (1995), Tilman and coworkers (1996), and a massive cooperative study in several European counties [Kinzig, Pacala and Tilman (2002)]. This body of work overwhelmingly endorses the earlier view of Elton. Net primary production increases with biodiversity in the experiments and asymptotes at approximately 20 plant species, although there are reasons to suspect that the asymptote might occur at substantially higher levels of diversity in a fluctuating environment [Kinzig, Pacala and Tilman (2002)].

There are two reasons for the positive association between production and diversity. In the first year or two, the effect is caused largely by a sampling phenomenon. That is, monocultures differ markedly in production because some species are better adapted than others to the experimental conditions. Because the best performing species tend to dominate multispecies stands and because multispecies stands are likely to contain the best performing species, high diversity mixtures tend to perform as well as the best monoculture and thus better than the average monoculture. After the first year or two, this sampling effect is replaced by a more powerful effect in which the multispecies stands substantially outperform even the best performing low diversity stands. The reason for this result is not yet clear, but the original explanation of specialization by the species on different resources is a likely candidate. In addition, recent theoretical studies and empirical work by David Tilman (2001) also support the earlier view that the stability of production increases with plant diversity, and show that stability asymptotes only at very high levels of diversity.

It is important to understand that stability of production is conferred by high diversity in part because the individual species abundances are less stable at high diversity than at low diversity. This is because high diversity stands tend to be dominated by the species that is best adapted to current conditions. But the identity of this species changes as conditions change from year to year (i.e., the dominant species tends to be drought-adapted in drought years, and wet-adapted species in wet years [Tilman, Wedin and Knops (1996)], thereby maintaining high productivity in all years. Thus, there is inevitable tension between management for the stability of particular species and management for the stability of production. Norberg et al. (2001) have laid the foundations for a quantitative theory of diversity/productivity relationships in fluctuating environments.

Measurement of biodiversity

The measurement of biological diversity has been a fascination of ecologists ever since the subject of ecology began. The most straightforward way to measure diversity is just to count the number of species found, but this measure has a number of drawbacks. Most important among these is that a simple species count gives no indication of how species are distributed within a community. It is intuitive that a community with 10,000 individuals and 10 species, but in which 99% of the individuals are from a single species, is less diverse than one in which the 10 species are more or less equally abundant; and indeed, such differences in diversity have fundamental implications for the functioning of the system, and for its resiliency. Secondly, simple species counts necessarily ignore the differences among species in terms of their importance to ecosystem processes, or to the resiliency of the ecological community.

The first problem has been addressed in a number of ways. One method is to introduce as a measure of diversity (or *evenness*) the probability that two individuals drawn at

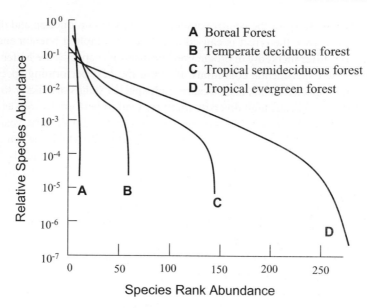

Figure 4. Dominance–diversity curves for diverse ecosystems. Relative species-abundance in relation to rank. A – boreal forest; B – temperate deciduous forest; C – tropical semideciduous forest; D – tropical evergreen forest. Redrawn from Hubbell (1979).

random from the community are different. This measure, known as Simpson's index, is given by

$$D = \frac{1}{\sum_{i=1}^{S} P_i^2},$$
(7)

where P_i is the proportion of individuals that belong to species i, and S in the number of species.

Alternatively, other measures have been introduced, with the most common being the Shannon–Wiener index,

$$H = -\sum_{i=1}^{S} P_i \ln P_i$$
(8)

derived from information theory [Shannon and Weaver (1949)]. Any single measure of diversity is flawed, in that it collapses a multidimensional concept into a single dimension, so it is usual to report several diversity measures, and to complement these with dominance–diversity curves such as those shown in Figure 4. The horizontal axis refers to species, ranked according to their abundance from the most abundant to the least; the vertical axis is relative abundance, scaled so that the maximum is unity.

Valuing biodiversity

A basic problem with diversity measures, in addition to those already addressed, is their lack of connection to the functional roles of species within an ecosystem. There is a growing recognition that biodiversity is important to ecosystem services, and that not all aspects of biodiversity are equally important. Naive measures of biodiversity treat friend and foe alike, according equal merit to pollinator species and pests, for example. It is well-appreciated that some species (such as the otters already mentioned) are keystones in ecosystems, and that their removal would engender profound consequences for the persistence of the functioning of those systems; whereas other species, like the chestnut in the forests of the northeastern United States, can disappear without noticeable effect. Some efforts have been made to take this into account [Weitzman (1992)], e.g., by weighting species according to their perceived importance. But any such effort is flawed not only by the lack of agreed-upon measures of the worth of a species, but more importantly by the fact that ecosystems are dynamic assemblages in which the functional role of a species is integrally dependent upon the context. Hence, any linear measure of biodiversity, which seeks to assess the status of a system by adding together the values of its parts, is bound to be inadequate at best, and misleading at worst. Indeed, in many cases, it might be more natural to focus on functional groups of organisms, perhaps even cutting across species, rather than on preserving individual species [Walker, Kinzig and Langridge (1999)]. Static measures oversimplify the situation, and complicate the decision-making. A preferred approach, therefore, is through scenario development, in which dynamical models of ecosystems are used to assess the ecosystem consequences of perturbations.

Similar objections, stated more emphatically, apply to efforts to value the biodiversity of all of the world's ecosystems [Costanza et al. (1997)]. Recognizing that the biodiversity of the world is essential for maintaining not just the fluxes of gases through our atmosphere, but all aspects of our life support systems, Costanza et al. set out to estimate the total value of the services biodiversity provides. Unfortunately, for essentially the same reasons measures of biodiversity fail to tell us much about system functioning, total valuation of biodiversity is impossible, and is not a feasible tool for decision-making.

Costanza et al. began by attempting to estimate the market values of the services ecosystems provide, and thereby to derive per-unit measures of what those systems are worth in today's markets. The methods used necessarily were diverse, combining (as was to be expected) revealed preferences with estimates derived from contingent valuation. The diversity of approaches used, and inconsistency of many of the estimates, is a problem in itself. More fundamental, however, is the context dependence of the measures obtained. Because one is not considering alternative scenarios, but rather simply attempting to extrapolate from marginal valuations to derive global measures, those measures become meaningless for making management decisions.

The true merits of economic valuation are to be seen when it is used for distinguishing between two alternatives, involving the total functioning of an ecosystem. An excellent example [Chichilnisky and Heal (1998)] is given by the demonstration of huge economic benefits by New York City in restoring the Catskill watershed as a natural water filtration system, rather than replacing it with a filtration plant that would have cost billions to construct, and 300 million dollars a year to operate.

The most promising approach to valuation is to recognize the ecosystem as a dynamic interacting system, not as the sum of independent entities. It does make sense to attempt to value the services an ecosystem provides, or a part of an ecosystem, because one can consider (as for the Catskill watershed) the costs and benefits of maintaining the system, or of replacing it with something else. Such options, of course, do not apply to the biosphere *in toto*, because there are no alternatives to weigh against it. In the valuation of an ecosystem, however, it is essential to study it as an interconnected network, with all the complex dynamics that are typical of nonlinear systems. For that purpose, the most reliable approach to guiding decision making is through the development of robust simulation models, which can be the basis for informed scenario development. This has been the approach followed with success in predictions of global climate change (see Chapter 1 by Bolin), and has formed the basis for economic models aimed at estimating the costs and benefits of action to reduce inputs into the environment [Nordhaus and Boyer (2000)].

7. Linkages to global biogeochemical cycling: The global carbon cycle

Because humans currently release annually into the atmosphere approximately seven billion metric tons of carbon from fossil fuel and cement production, the atmospheric concentration of CO_2 is increasing, and brings with it enhanced greenhouse forcing of climate. Four things are interesting about the pattern increase in Figure 5. First, the large oscillation in the Northern Hemisphere is caused by the seasonal cycle of land plants. Photosynthesis begins following leaf-out in the Spring, and the consumption of CO_2 by plants drives down the atmospheric concentration at a rate considerably quicker than the long-term anthropogenic rate of increase. Following leaf drop in the Fall, respiration releases CO_2 from the land surface and drives the concentration up, again with slope steeper than the long-term anthropogenic signal. This shows graphically that the terrestrial biosphere annually takes up and releases considerably more CO_2 than do humans (terrestrial NPP is 50–70 billion tons of carbon vs. 7 billion tons from humans). Obviously, a relatively small proportional imbalance in the biosphere could have a very large effect on atmospheric carbon.

The oscillation in Figure 5 is reduced in the tropics simply because the tropics are largely aseasonal. If you look closely, you can see that the oscillation in the Southern Hemisphere is opposite in phase to the signal in the Northern Hemisphere, because of the opposite seasonality there. The amplitude of the oscillation is smaller in the Southern Hemisphere because it contains a much smaller land mass than the Northern Hemisphere. Oceanic ecosystems have very little impact on the seasonal cycle of CO_2 in the atmosphere because oceanic plants obtain their CO_2 from the water, and superabundant CO_2 is stored in the oceans in the form of dissolved carbonate ($CO_2 + H_2O$

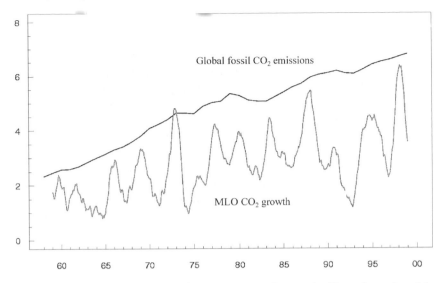

Figure 5. Productivity in relation to precipitation, temperature and geography. The surface carbon sink: solid line – global CO_2 emissions; gray line – increase of CO_2 in the atmosphere. The difference (black minus gray) is taken up by the land and ocean.

gives H_2CO_3; there are 37,000 billion tons of carbon stored in the oceans in this form, compared to only approximately 700 in the atmosphere).

The second curious thing about the pattern in Figure 5 is that the rate of increase is too small to be explained if the terrestrial biosphere is in carbon balance (that is if it is not steadily gaining carbon). Of the seven billion metric tons released by humans, only three billion on average make it into the atmosphere (Figure 6). Studies of oceanic carbon show definitively that the oceans take up approximately two billion metric tons annually [Sarmiento et al. (1999), Battle et al. (2000), Bousquet et al. (2000)]. This leaves approximately two billion metric tons that must be taken up by the terrestrial biosphere. Two billion metric tons is a lot of carbon – more than the amount above ground in 100,000 km^2 of tropical rain forest. This number is even more surprising when one considers that humans are in the process of deforesting the tropics. The IPCC (Intergovernmental Panel on Climate Change) reports that between one and two billion metric tons of carbon are released into the atmosphere by tropical deforestation. The estimate of two billion metric tons taken up by the biosphere is over and above the uptake necessary to cancel any deforestation losses.

The third curious thing about the pattern in Figures 5 and 6 is that the annual rate of increase of atmospheric CO_2 varies enormously from year to year. For example, in the 1990s the largest annual average rate was 5 billion metric tons higher than the smallest rate. A variety of sources of evidence indicate that most of this interannual variation is caused by year-to-year fluctuations in the amount of carbon taken up by the terrestrial biosphere [Battle et al. (2000), Bousquet et al. (2000)]. Moreover, because most of the

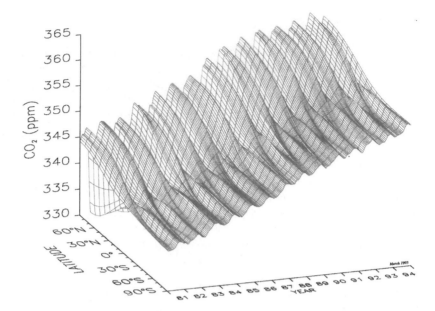

Figure 6. Three-dimensional representation of the global distribution of atmospheric carbon dioxide in the marine boundary layer assuming no variation with longitude. Data from NOAA/CMDL Global Cooperative Air Sampling Network, P. Tans and T. Conway, NOAA/CMDL Carbon Cycle Group, Boulder, CO.

variation in the atmospheric growth rate of CO_2 is associated with El Niño events, it appears that normal fluctuations in weather are enough to turn on and off imbalances between respiration and photosynthesis on land that are approximately as large as the annual release of fossil fuel CO_2. This is to be expected in retrospect, given the strong dependence of terrestrial vegetation on climate discussed in Section 3. It also implies that effects of climate change on vegetation could have a larger effect on atmospheric CO_2 than the direct anthropogenic release, and thus makes an understanding of the response of the terrestrial biosphere to climate an essential component of forecasting climate change caused by greenhouse gases.

The final curious thing about the pattern in Figure 5 is that the latitudinal gradient in atmospheric CO_2 is too small to be explained by the latitudinal gradient in fossil fuel consumption and the rate of inter-hemispheric air exchange [Tans, Fung and Takahashi (1990), Fan et al. (1998)]. The implication is that the ecological sequestration of two billion metric tons of carbon occurs in the Northern hemisphere, and probably in north temperate latitudes. This conclusion is now supported by a variety of separate lines of evidence [Tans, Fung and Takahashi (1990), Myneni et al. (1997), Randerson et al. (1997), Pacala et al. (2001)].

A model of carbon dynamics

To understand the dynamics of carbon uptake by the terrestrial biosphere, we now consider a simple model. It is a distillation of the more complex models used in global change studies, but retains the essential features and some quantitative accuracy. The model has been checked against the extensive U.S. Forest Service Inventory, and works well for forests in the temperate zone. Much of the north-temperate ecological sequestration probably occurs in forests because most of the living carbon on the planet is in forest trees.

Let $C(t)$ be the amount of living carbon (kg m^{-2}) in a location t years after a major disturbance like clear cutting, abandonment from agriculture, or fire. The rate of increase of $C(t)$ as it recovers from disturbance is governed by its NPP, which we label $g(t)$, and losses due to leaf litter and death of fine roots $(L(t))$ and whole tree death $(\mu C(t))$. The function for NPP generally rises quickly as carbon accumulates on the site, until the total leaf area of the site stabilizes, and then reaches an asymptote. A function that works well is $g(t) = G(1 - e^{-kC(t)})$, where G and k are constants. Leaf and root litter are approximately proportional to leaf area, and thus $L(t)$ is approximately proportional to $g(t)$. Reasonable parameter values for G from Section 3 are 1.0 (kg m^{-2} y^{-1}) in tropical forest, 0.6 in temperate deciduous forest, and 0.4 in boreal forest, with k approximately equal to 1.0 m^2 kg^{-1} (canopy closure when there is 20 tons of living carbon per hectare on the site) and $L(t)$ about two thirds of $g(t)$. The mortality rate of trees, μ, ranges from 0.1 y^{-1} in fire-prone sites such as pine barrens on sand plains to 0.02 y^{-1} in wet tropical forest and 0.01 y^{-1} in temperate deciduous forest [Caspersen et al. (2000)].

For the purposes of illustration, we take k to be very large, and so both $g(t)$ and $L(t)$ are constant. Using $3L(t) = 2g(t)$ we obtain $dC(t)/dt = G/3 - \mu C(t)$, which has the solution $C(t) = (G/(3\mu))(1 - e^{-\mu t}) + C(0)$. This equation is interesting because it predicts the amount of living carbon in old growth ($G/(3\mu) = 15$ kg m^{-2} in wet tropical forest and 20 kg m^{-2} in temperate forest), and shows that stands continue to take up large amounts of carbon well into succession. Both the amount of carbon eventually stored and the time scale of successional carbon uptake are set by the longevities of individual trees, which vary considerably among species. The long time scale of carbon uptake is a conservative result because $g(t)$ actually reaches its asymptotic value after a delay from five (wet tropics) to twenty (deciduous temperate) or more years. In cases where the soil is severely depleted by human land use, the delay until $g(t)$ reaches an asymptote may take considerably longer [Aber and Melillo (1982)]. Also, the mortality rate μ tends to decrease during succession as short-lived early successional species are replaced by longer-lived late-successional species. Finally, carbon also accumulates in the leaf litter and soil, and this accumulation lags behind the accumulation of living biomass.

A variety of excellent and predictive models of below-ground carbon accumulation are now available. Most follow the CENTURY model [Parton, Stewart and Cole (1988)] in separating dead organic carbon into pools with different decay rates. Bolker, Pacala and Parton (1998) show that at most two pools are necessary to obtain a good quantitative answer. Let S_f be the amount of organic matter below ground that decomposes rapidly. The decay rate for this pool is typical of leaf or fine root litter: $\lambda_f = 2.0$ y^{-1}. Similarly, let S_s be organic matter that decomposes slowly with a time scale typical of wood: $\lambda_f = 0.04$ y^{-1}. The exact value of these decay rates will depend to some extent on the chemical composition of the species present and strongly on temperature and soil moisture. The rates of decomposition increase from zero at freezing to the above values at 10–20°C, and then decline precipitously above 40–45°C [Parton, Stewart and Cole (1988)]. Similarly, decomposition stops in water-logged or very dry soils, or if there is no available soil nitrogen [Parton, Stewart and Cole (1988)]. But in most cases during the growing season, the above values give reasonable results. In most models of decomposition there is also a passive pool, but this has little impact at successional time scales because fluxes in and out of it are very small (i.e., with decay time scales of centuries to millennia).

The majority of leaf and fine root litter enters the fast pool, while virtually all the wood in a tree enters the slow pool. Thus a reasonable approximation is:

$$\frac{dS_f(t)}{dt} = \frac{2G}{3} - \lambda_f S_f(t),$$

$$\frac{dS_s(t)}{dt} = \mu C(t) - \lambda_s S_s(t). \tag{9}$$

The fast pool equilibrates within a very few years at the small value of $2G/(3\lambda_f)$, whereas the slow pool tracks the slow growth in living carbon. The slow pool lags only marginally behind the quasi-equilibrium value of $\mu C(t)/\lambda_s$ because λ_s is typically several fold larger than μ. Thus, a good rule of thumb is that the total uptake of carbon during successional recovery is approximately 25 percent larger than the rate of increase of living biomass. This accords well with the many ongoing measurements of carbon uptake. At the AMERIFLUX and EUROFLUX measurement sites [Baldocchi et al. (1996)], carbon uptake is typically 2/3 above ground, as predicted by the model, because trees have approximately 4/5 of their biomass above ground. Note also that our simple model yields estimates of total carbon uptake within the proper range of 0.1–0.3 kg m^{-2} y^{-1} for stands last clear-cut in the first half of the 20th century in the temperate zone [Baldocchi et al. (1996), Caspersen et al. (2000)].

8. The evolution of interactions and ecosystems, and the maintenance of ecosystem services: From Darwin to Gaia

Ecosystems provide humans not only with the resources we extract directly – fish and fowl, fiber and pharmaceuticals, fuel and other products that lead us to value individual species; they also provide a suite of indirect services, such as the stabilization of climate, the availability of pollination services and the sequestration of toxic materials, that allow the very survival of life as we know it. It is these ecosystem services that make the challenge of managing natural resources so great. Unless we can identify and value what might be lost, how can we factor the social costs into the decision-making framework? How can we create a market that properly incorporates these externalities?

The valuation of ecosystem services is perhaps the greatest challenge of environmental management. Any effort to make a list of services is bound to be incomplete, since many of the benefits are potential, unrealized, perhaps unknown or even unknowable. And for those indirect benefits that are most compelling – climate mediation, the provision of clean air and water, the maintenance of pollination, the mitigation of floods and drought, the sustenance of fertile soil [Daily (1997)] – there is no generally accepted formula for quantifying benefits. Indeed, any such quantification must be context dependent, and highly sensitive to the scale of an action. Ethical issues must play an important role, especially in trying to resolve the fact that different people will value possible benefits in different ways. Contingent valuation is a notoriously problematic mechanism for getting at such issues, but other methods are little better, given the paucity of information. What are the rights of future generations, and how shall we discount the future? Is there a workable notion of sustainability to guide our decision-making. These are, of course, familiar problems to economists, but familiarity has not yet led to much progress in untying the multiple Gordian knots.

When attention is focused on a small area, the problem of ecosystem services is difficult enough. As we change the landscape on broader and broader scales, however, we are confronted with externalities that affect landscapes and, as with global change, the biosphere itself. To address the social costs of such activities, we must be able not only to assess the value of ecosystem services, but how robust they are in the face of our activities. For this we can turn to a limited extent to the sorts of modeling exercises addressed earlier, but there always will remain a core of ineluctable unknowability, a limitation to our ability to predict consequences of our actions. Holling (1986) has argued eloquently that the one thing that should not surprise us is that there will be surprises, and that we must manage systems accordingly. This means minimizing the tendency to use inflexible strategies, maintaining the character of ecosystems as complex adaptive systems, with the ability to respond to change.

But we should not be misled; ecosystems are limited in their capacity to respond to perturbations in ways that maintain their basic functioning. Evolution operates on the genomes of organisms to confer some robustness in the way they respond to changing environments – behavioral and physiological responses are largely reversible ways in which organisms adjust to environmental variation. Under more extreme situations, possibly irreversible developmental choices may allow adjustments to be made. Gaia theorists would wish that similar principles apply to ecosystems and the biosphere; but evolution does not apply in the same way at those higher levels. Selection operates within ecosystems, not among them; extant ecosystems are not the offspring of parent ecosystems that had higher fitnesses than other ecosystems. This is not to say that processes at the level of ecosystems and the biosphere are not stable in the face of change; we know that there is remarkable constancy in many of the macroscopic features, most notably in the cycles of elements that sustain system functioning. The challenge, however, is to understand the limits to that resiliency, and how it arises in the face of Darwinian challenges at lower levels of organization [Levin (1999)].

Ultimately, it is essential that, as we stress ecosystems, we recognize that there are limits to our capacity to exploit them, and the potential for qualitative and irreversible shifts. We must endeavor to reduce uncertainty to the greatest extent possible, but heed Holling's caveat that there will always remain surprises. Managing our natural resources is management under uncertainty, and a balancing of risks and putative benefits.

Acknowledgements

We are pleased to acknowledge the very helpful comments of Steve Carpenter, Karl-Göran Mäler, Jeff Vincent and Brian Walker, and the support of the David and Lucile Packard Foundation, award 8910-48190, the Andrew W. Mellon Foundation, and the National Science Foundation, award DEB-0083566.

References

Aber, J., and J. Melillo (1982), "Fortnite: a computer model of organic matter and nitrogen dynamics in forest ecosystems", University of Wisconsin Research Bulletin R3130 (Madison, WI, USA).

Arrow, K., G. Daily, P. Dasgupta, S. Levin, K.-G. Mäler, E. Maskin, D. Starrett, T. Sterner and T. Tietenberg (2000), "Managing ecosystem resources", Environmental Science & Technology 34:1401–1406.

Baldocchi, D., R. Valentini, S. Running, W. Oechel and R. Dahlman (1996), "Strategies for measuring and modeling carbon dioxide and water vapor fluxes over terrestrial ecosystems", Global Change Biology 2:159–169.

Ball, J., I. Woodrow and J. Berry (1986), A model predicting stomatal conductance and its contribution to the control of photosynthesis under different environmental conditions (Martinus–Nijhoff, Dordrecht, The Netherlands).

Battle, M., M.L. Bender, P.P. Tans, J.W.C. White, J.T. Ellis, T. Conway and R.J. Francey (2000), "Global carbon sinks and their variability inferred from atmospheric O_2 and $\delta^{13}C$", Science 287:2467–2470.

Beverton, R.J.H., and S.J. Holt (1957), "On the dynamics of exploited fish populations", Fisheries Investment (Series 2).

Bolker, B.M., S.W. Pacala, F.A. Bazzaz, C.D. Canham and S.A. Levin (1995), "Species diversity and ecosystem response carbon dioxide fertilization: Conclusions from a temperate forest model", Global Change Biology 1:373–381.

Bolker, B.M., and S.W. Pacala (1999), "Spatial moment equations for plant competition: Understanding spatial strategies and the advantages of short dispersal", American Naturalist 153:575–602.

Bolker, B.M., S.W. Pacala and S.A. Levin (2000), "Moment methods for stochastic processes in continuous space and time", in: U. Dieckmann, R. Law and J.A.J. Metz, eds., The Geometry of Ecological Interactions: Simplifying Spatial Complexity (Cambridge Univ. Press, Cambridge) 388–411.

Bolker, B.M., S.W. Pacala and W.J. Parton, Jr. (1998), "Linear analysis of soil decomposition: Insights from CENTURY model", Ecological Applications 8:425–439.

Bonan, G.B. (1995), "Land-atmosphere CO_2 exchange simulated by a land surface process model coupled to an atmospheric general circulation model", Journal of Geophysical Research 100(D2):2817–2831.

Botkin, D.B., J.F. Janak and J.R. Wallis (1972), "Some ecological consequences of a computer model of forest growth", Journal of Ecology 60:849–873.

Bousquet, P., P. Peylin, P. Ciais, C. Le Quere, P. Friedlingstin and P.P. Tans (2000), "Regional changes in carbon dioxide fluxes of land and oceans since 1980", Science 290:1342–1346.

Broecker, W.S., D.M. Peteet and D. Rind (1985), "Does the ocean–atmosphere system have more than one stable mode of operation?", Nature 315:21–26.

Bunt, J.S. (1975), "Primary productivity of marine ecosystems", in: H. Leith and R.H. Whittaker, eds., Primary Productivity of the Biosphere (Springer, New York) 169–215.

Carpenter, S.R. (2001), "Alternate states of ecosystems: Evidence and some implications", in: M.C. Press, N.J. Huntley and S.A. Levin, eds., Ecology: Achievement and Challenge (Blackwell Science, Oxford, UK) 357–383.

Carpenter, S.R., and J.F. Kitchell (eds.) (1993), The Trophic Cascade in Lakes (Cambridge Univ. Press, Cambridge).

Carpenter, S.R., D. Ludwig and W.A. Brock (1999), "Management of eutrophication for lakes subject to potentially irreversible change", Ecological Applications 9:751–771.

Caspersen, J.P., S.W. Pacala, J.C. Jenkins, G.C. Hurtt, P.R. Moorcroft and R.A. Birdsey (2000), "Contributions of land-use history to carbon accumulation in U.S. forests", Science 290:1148–1156.

Chichilnisky, G., and G. Heal (1998), "Economic returns from the biosphere", Nature 391:629–630.

Clements, F.E. (1916), "Plant succession: an analysis of the development of vegetation", Publ 242 (Carnegie Institute of Washington, Washington, DC).

Collatz, G., J. Ball, C. Grivet and J. Berry (1991), "Physiological and environmental regulation of stomatal conductance, photosynthesis and transpiration: a model that includes a laminar boundary layer", Agricultural and Forest Meterology 54:107–136.

Collatz, G., M. Ribas-Carbo and J. Berry (1992), "Coupled photosynthesis-stomatal conductance model for leaves of c_4 plants", Australian Journal of Plant Physiology 19:519–538.

Condit, R., P.S. Ashton, P. Baker, S. Bunyavejchewin, S. Gunatilleke, N. Gunatilleke, S.P. Hubbell, R.B. Foster, A. Itoh, J.V. LaFrankie, H.S. Lee, E. Losos, N. Manokara, R. Sukumar and T. Yamakura (2000), "Spatial patterns in the distribution of tropical tree species", Science 288:1414–1418.

Connell, J.H. (1971), "On the roles of natural enemies in preventing competitive exclusion in some marine animals in rain forest trees", in: B.J. den Boer and G.R. Gradwell, eds., Dynamics of Populations: Proceedings of the Advanced Study of Institute in Dynamics of Numbers in Populations, Oosterbeck, The Netherlands, 7–18 September 1970 (Centre for Agricultural Publishing and Documentation, Wageningen) 298–310.

Costanza, R.R., R. d'Arge, R. de Groot, S. Farber, M. Grasso, B. Hannon, K. Limburg, S. Naeem, R.V. O'Neill, J. Paruelo, R.G. Raskin, P. Sutton and M. van den Belt (1997), "The value of the world's ecosystem services and natural capital", Nature 387:253–260.

Curtis, J.T. (1956), "The modification of mid-latitude grasslands and forests by man", in: W.L. Thomas, ed., Man's Role in Changing the Face of the Earth (Univ. of Chicago Press, Chicago, IL) 721–736.

Daily, G.C. (ed.) (1997), Nature's Services: Societal Dependence on Natural Ecosystems (Island Press, Washington, DC).

Durrett, R., and S.A. Levin (1994a), "The importance of being discrete and (spatial)", Theoretical Population Biology 46:363–394.

Durrett, R., and S.A. Levin (1994b), "Stochastic spatial models: A user's guide to ecological applications", Philosophical Transactions of the Royal Society of London, Series B 343:329–350.

Durrett, R., and S.A. Levin (1996), "Spatial models for species area curves", Journal of Theoretical Biology 179:119–127.

Durrett, R., and C. Neuhauser (1994), "Particle systems and reaction–diffusion equations", Annals of Probability 22:289–333.

Durrett, R., and C. Neuhauser (1997), "Coexistence results for some competition models", Annals of Probability 7:10–45.

Ehrlich, P.R., and P.H. Raven (1964), "Butterflies and plants: a study in coevolution", Evolution 18:586–608.

Elton, C.S. (1927), Animal Ecology (Sidgwick & Jackson, London).

Elton, C.S. (1958), The Ecology of Invasions by Animals and Plants (Methuen, London).

Estes, J.A., and J.F. Palmisano (1974), "Sea otters: Their role in structuring nearshore communities", Science 185:1058–1060.

Fan, S., M. Gloor, J. Mahlman, S. Pacala, J. Sarmiento, T. Takashi and P. Tans (1998), "A large terrestrial carbon sink in North America implied by atmospheric and oceanic carbon dioxide data and models", Science 282:442–446.

Farquhar, G.D., and T.D. Sharkey (1982), "Stomatal conductance and photosynthesis", Annual Review of Plant Physiology 33:317–345.

Field, C.B., M.J. Behrenfeld, J.T. Randerson and P. Falkowski (1998), "Primary production of the biosphere: Integrating terrestrial and oceanic components", Science 281:237–240.

Foley, J.A., I.C. Prentice, N. Ramankutty, S. Levis, D. Pollard, S. Sitch and A. Haxeltine (1996), "An integrated biosphere model of land surface processes, terrestrial carbon balance, and vegetation dynamics", Global Biogeochemical Cycles 10:603–628.

Friend, A., A. Stevens, R. Knox and M. Cannell (1997), "A process-based, terrestrial biosphere model of ecosystem dynamics", Ecological Modeling 95:249–287.

Futuyma, D., and M. Slatkin (eds.) (1983), Coevolution (Sinauer, Sunderland, MA).

Gilpin, M., and I. Hanski (eds.) (1991), Metapopulation Dynamics: Empirical and Theoretical Investigations (Academic Press, London).

Gould, J.L., and W.T. Keeton (1996), Biological Science, 5th edn. (Norton, New York).

Grace, J. (1997), "Plant water relations", in: M.J. Crawley, ed., Plant Ecology (Blackwell Science, Oxford, UK) 28–50.

Hardin, G. (1968), "The tragedy of the commons", Science 162:1243–1248.

Harte, J., A. Kinzig and J. Green (1999), "Self-similarity in the distribution and abundance of species", Science 284:334–336.

Haxeltine, A., and I.C. Prentice (1996), "Biome 3: An equilibrium terrestrial biosphere model based on ecophysiological constraints, resource availability, and competition among plant functional types", Global Biogeochemical Cycles 10:693–709.

Haxeltine, A., I.C. Prentice and I.D. Creswell (1996), "A coupled carbon and water flux model to predict vegetation structure", Journal of Vegetation Science 7:651–666.

Holdridge, L. (1967), Life Zone Ecology (Tropical Science Center, San Jose, Costa Rica).

Holling, C.S. (1973), "Resilience and stability of ecological systems", Annual Review of Ecology and Systematics 4:1–23.

Holling, C.S. (1986), "The resilience of terrestrial ecosystems; local surprise and global change", in: W.C. Clark and R.E. Munn, eds., Sustainable Development of the Biosphere (Cambridge Univ. Press, Cambridge) 292–317.

Horn, H.S. (1976), "Forest succession", Scientific American 232:90–98.

Hubbell, S.P. (1979), "Tree dispersion, abundance, and diversity in a tropical dry forest", Science 203:1299–1309.

Hubbell, S.P. (2001), The Unified Neutral Theory of Biodiversity and Biogeography (Princeton Univ. Press, Princeton, NJ).

Hurtt, G.C., P.R. Moorcroft, S.W. Pacala and S.A. Levin (1998), "Terrestrial models and global change: challenges for the future", Global Change Biology 4:581–590.

Hutton, J. (1788), "Theory of the Earth; or an investigation of the laws observable in the composition, dissolution, and restoration of land upon the globe", Royal Society of Edinburgh, Transactions 1:209–304.

Keeling, M.J., D.A. Rand and A.J. Morris (1997), "Correlation models for childhood diseases", Proceedings of the Royal Society of London, Series B 264:1149–1156.

Kinzig, A.P., S.W. Pacala and G.D. Tilman (eds.) (2002), The Functional Consequences of Biodiversity: Empirical Progress and Theoretical Extensions (Princeton Univ. Press, Princeton, NJ).

Leuning, R. (1995), "A critical appraisal of a combined stomatal-photosynthesis model for c_3 plants", Plant Cell and Environment 18:339–355.

Levin, S.A. (1992), "The problem of pattern and scale in ecology", Ecology 73:1943–1967.

Levin, S.A. (1998), "Ecosystems and the biosphere as complex adaptive systems", Ecosystems 1:431–436.

Levin, S.A. (1999), Fragile Dominion: Complexity and the Commons (Perseus Books, Reading, MA).

Levin, S.A. (2000), "Multiple scales and the maintenance of biodiversity", Ecosystems 3:498–506.

Levin, S.A., S. Barrett, S. Aniyar, W. Baumol, C. Bliss, B. Bolin, P. Dasgupta, P. Ehrlich, C. Folke, I.-M. Gren, C.S. Holling, A. Jansson, B.-O. Jansson, D. Martin, K.-G. Maler, C. Perrings and E. Sheshinsky (1998), "Resilience in natural and socioeconomic systems", Environment and Developmental Economics 3:225–236.

Levin, S.A., and C.P. Goodyear (1980), "Analysis of an age-structured fishery models", Journal of Mathematical Biology 9:245–274.

Levin, S.A., and S.W. Pacala (1997), "Theories of simplification and scaling of spatially distributed processes", in: D. Tilman and P. Kareiva, eds., Spatial Ecology: The Role of Space in Population Dynamics and Interspecific Interactions (Princeton Univ. Press, Princeton, NJ) 271–296.

Levin, S.A., and R.T. Paine (1974), "Disturbance, patch formation, and community structure", Proceedings of the National Academy of Science, USA 71:2744–2747.

Levins, R. (1969), "Some demographic and genetic consequences of environmental heterogeneity for biological control", Bulletin of the Entomological Society of America 15:237–240.

Lorenz, E. (1963), "Deterministic nonperiodic flow", J. Atmospheric Sciences 20:448–464.

Lorenz, E. (1964), "The problem of deducing the climate from the governing equations", Tellus 16:1–11.

Lotka, A.J. (1925), Elements of Physical Biology, reprinted 1956 with corrections and bibliography as Elements of Mathematical Biology (Dover, New York) edition (Williams and Wilkins, Baltimore, MD).

MacArthur, R.H., and J.H. Connell (1966), The Biology of Populations (Wiley, New York).

MacArthur, R.H., and E.O. Wilson (1967), The Theory of Island Biogeography (Princeton Univ. Press, Princeton, NJ).

Manabe, S., and R.J. Stouffer (1988), "Two stable equilibria of a coupled ocean–atmosphere model", Journal of Climate 1:841–866.

May, R.M. (1974), "Biological populations with non-overlapping generations: stable points, stable cycles, and chaos", Journal of Theoretical Biology 49:511–524.

McNaughton, S.J. (1977), "Diversity and stability of ecological communities", American Naturalist 111:515–525.

McNaughton, S.J., M. Osterheld, D.A. Frank and K.J. Williams (1989), "Ecosystem-level patterns of primary productivity and herbivory in terrestrial habitats", Nature 341:142–144.

Moorcroft, P.R., G.C. Hurtt and S.W. Pacala (2001), "A method for scaling vegetation dynamics: The ecosystem demography model (ED)", Ecological Monographs 71:557–586.

Myneni, R.B., C.D. Keeling, C.J. Tucker, G. Asrar and R.R. Nemani (1997), "Increased plant growth in the northern high latitudes from 1981 to 1991", Nature 386:698–702.

Naeem, S., L.J. Thompson, S.P. Lawler, J.H. Lawton and R.M. Woodfin (1995), "Biodiversity and ecosystem functioning: empirical evidence from experimental microcosms", Philosophical Transactions of the Royal Society of London, Series B 347:249–262.

Neuhauser, C., and S.W. Pacala (1999), "An explicitly spatial version of the Lotka–Volterra model with interspecific competition", Annals of Applied Probability 9:1226–1259.

Norberg, J., D.P. Swaney, J. Dushoff, J. Lin, R. Casagrandi and S.A. Levin (2001), "Phenotypic diversity and ecosystem functioning in changing environments: A theoretical framework", Proceedings of the National Academy of Sciences, USA 98:11376–11381.

Nordhaus, W.D., and J. Boyer (2000), Warming the World: Economic Models of Global Warming (MIT Press, Cambridge, MA).

Pacala, S.W., C.D. Canham, J. Saponara, J.A. Silander, R.K. Kobe and E. Ribbens (1996), "Forest models defined by field measurements: II. Estimation, error analysis and dynamics", Ecological Monographs 66:1–43.

Pacala, S.W., and D.J. Deutschman (1996), "Details that matter: The spatial distribution of individual trees maintains forest ecosystem function", Oikos 74:357–365.

Pacala, S.W., G.C. Hurtt, R.A. Houghton, R.A. Birdsey, L. Heath, E.T. Sundquist, E.T. Stallard, D. Baker, P. Peylin, P. Moorcroft, J. Caspersen, E. Shevliakova, M.E. Harmon, S.-M. Fan, J.L. Sarmiento,

C. Goodale, C.B. Field, M. Gloor and D. Schimel (2001), "Consistent land- and atmosphere-based U.S. carbon sinks", Science 292:2316–2320.

Paine, R.T., and R.L. Vadas (1969), "The effects of grazing by sea urchins, *Strongylocentrotus spp.*, on benthic algal population", Limnology and Oceanography 14:710–719.

Parton, R.T., J.W.B. Stewart and C.V. Cole (1988), "Dynamics of C.N.P, and S in grassland soils: a model", Biogeochemistry 5:109–131.

Plotkin, J.B., and S.A. Levin (2001), "The spatial distribution and abundances of species: Lessons from tropical forests", Comments on Theoretical Biology 6:251–278.

Plotkin, J.B., M. Potts, D. Yu, S. Bunyavejchewin, R. Condit, R. Foster, S. Hubbell, J. LaFrankie, N. Manokaran, L. Seng, R. Sukumar, M. Nowak and P.S. Ashton (2000), "Predicting species diversity in tropical forests", Proceedings of the National Academy of Sciences, U.S.A. 97:10850–10854.

Potter, C.S., J.T. Davidson, S. Klooster and D. Nepstad (1998), "Regional application of an ecosystem production model for studies of biogeochemistry in a Brazilian Amazonia", Global Change Biology 4:315–333.

Potter, C.S., J.T. Randerson, C.B. Field, P.A. Matson, P.M. Vitousek, H.A. Mooney and S.A. Klooster (1993), "Terrestrial ecosystem production: A process model based on global satellite and surface data", Global Biogeochemical Cycles 7:811–841.

Rand, D.A., M. Keeling and H.B. Wilson (1995), "Invasions, stability, and evolution to criticality in spatially extended, artificial host–pathogen ecologies", Proceedings of the Royal Society of London, Series B 259:55–63.

Randerson, J.T., M.V. Thompson, T.J. Conway, I.Y. Fung and C.B. Field (1997), "The contribution of terrestrial sources and sinks to trends in the seasonal cycle of atmospheric carbon dioxide", Global Biogeochemical Cycles 11:535–560.

Reynolds, C.S. (1997), Vegetation Processes in the Pelagic: A Model for Ecosystem Theory (Ecology Institute, Oldendorf/Luhe, Germany).

Reynolds, C.S. (2000), "Status and role of plankton", in: S.A. Levin, ed., Encyclopedia of Biodiversity (Academic Press, San Francisco, CA) 569–599.

Ricker, W.E. (1954), "Stock and recruitment", Journal of Fisheries Research Board Canada 11:559–623.

Sarmiento, J.L., S.C. Wofsy, A.S. Denning, W. Easterling, C. Field, I. Fung, R. Keeling, J. McCarthy, S.W. Pacala, W.M. Post, D. Schimel, E. Sundquist, P. Tans, R. Weiss, J. Yoder and E. Shea (1999), A U.S. Carbon Cycle Science Plan: A Report of the Carbon and Climate Working Group (U.S. Global Change Research Program, Washington, DC).

Scheffer, M. (1998), The Ecology of Shallow Lake (Chapman & Hall, London).

Schlesinger, W.H. (1997), Biogeochemistry: An Analysis of Global Change (Academic Press, San Diego, CA).

Sellers, P.J., Y. Mintz, Y.C. Sud and A. Dalcher (1986), "A simple biosphere model (SiB) for use with general circulation models", Journal of the Atmospheric Sciences 43:505–531.

Shannon, C.E., and W. Weaver (1949), The Mathematical Theory of Communication (Univ. of Illinois Press, Urbana, IL).

Shelford, V.E. (1911a), "Ecological succession: stream fishes and the method of physiographic analysis", Biological Bulletin 21:9–34.

Shelford, V.E. (1911b), "Ecological succession: pond fishes", Biological Bulletin 21:127–151.

Shugart, H.H. (1984), A Theory of Forest Dynamics (Springer, New York).

Simberloff, D., and E.O. Wilson (1969), "Experimental zoogeography of islands: The colonization of empty islands", Ecology 50:278–296.

Simberloff, D.S., and L.G. Abele (1976), "Island biogeography and conservation practice", Science 191:285–286.

Tans, P.P., I.Y. Fung and T. Takahashi (1990), "Observational constraints on the global atmospheric CO_2 budget", Science 247:1431–1438.

Tilman, D. (2001), "Effects of diversity and composition on grassland stability and productivity", in: M.C. Press, N.J. Huntley and S.A. Levin, eds., Ecology: Achievement and Challenge (Blackwell Science, London) 183–207.

Tilman, D., D. Wedin and J. Knops (1996), "Productivity and sustainability influenced by biodiversity in grassland ecosystems", Nature 379:718–720.

Trenbath, B.R. (1974), "Neighbour effects in the genus *Avena II*. Comparison of weed species", Journal of Applied Ecology 12:189–200.

VEMAP Members (1995), "Vegetation/ecosystem modeling and analysis project: Comparing biogeography and biogeochemistry models in a continental-scale study of terrestrial ecosystem responses to climate change and CO_2 doubling", Global Biogeochemical Cycles 9:407–434.

Volterra, V. (1926), "Variazioni e fluttuazioni del numero d'individui in specie animale conviventi", Mem. Rediconti Accad. Nazionale dei Lincei (Ser. 6) 2:31–113.

Walker, B., A. Kinzig and J. Langridge (1999), "Plant attribute diversity, resilience, and ecosystem function: The nature and significance of dominant and minor species", Ecosystems 2(2):95–113.

Walker, B.H. (1995), "Rangeland ecology: Managing change in biodiversity", in: C. Perrings, K.-G. Mäler, C. Folke, C.S. Holling and B.-O. Jansson, eds., Biodiversity Conservation (Kluwer, Dordrecht, The Netherlands) 69–85.

Walker, B.H., J.L. Langridge and F. McFarlane (1997), "Resilience of an Australian savanna grassland to selective and non-selective perturbations", Australian Journal of Ecology 22:125–135.

Weitzman, M.L. (1992), "On diversity", The Quarterly Journal of Economics 107:363–405.

Whittaker, R.H. (1975), Communities and Ecosystems, 2nd edn. (Macmillan Co., New York).

Whittaker, R.H., and G.E. Likens (1973), "Carbon in the biota", in: G.M. Woodwell and E.V. Pecan, eds., Carbon and the Biosphere (U.S. AEC, Washington, DC).

Chapter 3

PROPERTY RIGHTS, PUBLIC GOODS AND THE ENVIRONMENT

DAVID A. STARRETT

Stanford University, CA, USA

Contents

Handbook of Environmental Economics, Volume 1, Edited by K.-G. Mäler and J.R. Vincent

Abstract

We delineate the various ways in which rights to environmental and other resources can be assigned to individuals or groups. We then examine models of individual and group interactions, drawing out their implications for the ways in which resources will be utilized and managed under various rights assignments. Resources are classified into various groups (such as "collective" and "private") depending on the type of rights assignment that is most appropriate, and we critically examine situations in which it is claimed that certain combinations of rights and rules of behavior will lead to an "ideal" allocation of the associated resources. We argue that in all but a very limited set of circumstances, efficient allocations will require at the least some form of social intervention, and we discuss both formal and informal models of social organization toward this end. Various distortions are identified that may arise when incorrect assignments of rights are utilized. We discuss various practical ways of correcting for these distortions using instruments such as taxes, quotas, and markets for pollution permits.

Keywords

property rights, public goods, Coase theorem, open access, self-organizing systems, externalities

JEL classification: C71, C72, C78, D60, D61, D62, D71, H4

1. Introduction

In discussions of environmental management, the view is sometimes taken that if we could get the assignment of property rights correct, the desired conservation policies could be achieved by parties exercising those rights. We discuss here the various ways in which rights to environmental resources could be assigned and the variety of social institutions that have been or could be created to enforce these rights. We will see that the appropriate rights scheme varies depending on the nature of the resource involved, and we will try to elucidate the factors that determine whether or not the position suggested in our opening sentence is justified.

At the outset, it should be clear that the assignment and enforcement of property rights is essential to facilitate any allocation of resources by private parties. Indeed, unless there is a proscription against theft, ownership has no real meaning and no one would pay anything for any valuable asset that could not be nailed down.

The assignment and enforcement of property rights is a way of institutionalizing ownership of resources. In capitalist societies it is implicitly assumed that the assignment of private rights is a good thing and further that the costs of enforcement (through a system of laws, police to monitor them and courts to settle disputes) are negligible compared to the benefits so derived. Socialist societies are not willing to go as far in assigning rights to individuals but rather seek to assign some of them to collectives (see below). However, economic analysis of either type of system still has generally ignored the costs of enforcement. We will see later that there are times when these costs should not be ignored, but defer such discussion until we have laid the appropriate framework.

2. Taxonomy of property rights[1]

There are at least two distinct dimensions on which property rights regimes may differ: (1) the scope of the exercising group and (2) the degree of control granted to the exercising group. In category (1) we will distinguish four levels: *private, collective, government, open*. A *private property right* is one that is exercised by a single individual. The right to one's own labor time is an example of this type. A *collective property right* is one exercised by some specific group (the collective). Examples of collectives are traditional fishing or herding cooperatives and homeowners' associations. When the collective is a political entity, we refer to the associated right as a *government property right*, where the entity could be anything from a county to a nation. Here, examples would be regional and national parks. When the collective is "all comers" we refer to the right as an *open right*. Examples here would be unregulated fisheries and open range.

Rights can involve varying degrees of control over the associated resource. We distinguish here between rights to *use* and rights to *regulate*. Use rights include *access* and *withdrawal*.

---

[1] The taxonomy adopted here is roughly that of Schlager and Ostrom (1992). Also, see that paper for references to the earlier literature.

Access: The right to enjoy or experience the resource but without changing it quantitatively or qualitatively.

Withdrawal: The right to diminish the resource in some specified quantitative or qualitative way.

As examples, a person who enters a national park has access but not withdrawal rights. By contrast, a person who enters a national forest with a cutting permit has both access and withdrawal rights (to a specified amount of firewood or timber).

Regulation rights include *management, exclusion,* and *alienation*.

Management: The right to transform the resource by making improvements or otherwise altering the nature of the resource and to determine how any associated benefits or costs are to be distributed.

Exclusion: The right to determine who will have what access or withdrawal rights, on what terms these rights will be granted, and how these rights may be transferred.

Alienation: The right to sell or lease either of the other regulation rights.

A complete rights regime for managing some resource must assign each of the five control rights to some individual or collective. In many practical cases, different control rights will be assigned to different collectives. For example, in the case of a condominium homeowners' association, access rights and that part of exclusion which deals with access are assigned to individual owners, whereas the remaining rights are exercised by the collective of owners (although in some cases, individual owners retain some of the regulation rights as well). We can complete any system of rights by using the natural convention that if some control right is not assigned, then it is automatically an open right. For example, in national forests access is treated as an open right whereas the other rights typically are exercised by the government. Of course, if a particular right is open, exclusion and alienation become meaningless with respect to that right.

In much of the sequel we will be concerned with the normative question of how to assign and exercise rights to environmental resources in an "optimal" way. We will see that the answer will differ from resource to resource and indeed that for some resources, there may be no optimal design. We begin with the natural benchmark of private property.

3. Scope and limitations of private property

A resource is private property if all of the rights with respect to that resource are assigned to an individual. For example, if I own my own home with no liens, I have the right to exclude (decide who may enter), I may make improvements at my own expense and I can sell whenever I want.[2] It has long been a tenet of capitalist economics that

---

[2] In reality, even here some rights are assigned to the state – the right to enter with a search warrant and the right of eminent domain. But we will ignore these exceptions in our discussions of private property.

for a large class of resources, private property is a good thing in that private property regimes facilitate an efficient allocation of resources through the use of markets. Since this is well trodden ground, we sketch only briefly the main argument.[3]

Establishment of private property rights is a necessary precursor to the use of markets and indeed is usually directly associated with the presence of market institutions. There are strong reasons why this should be so, as once the rights are in place there are incentives for individuals to create markets and under certain circumstances we know that markets are an efficient way of allocating resources. Let us see how this might work. Suppose that there are no enforced regulations on the disposal of household garbage. Then, ignoring the possibility of altruistic behavior, we would expect everyone to dump their garbage on someone else. And this outcome will likely be inefficient in that some people will have isolated sites where dumps would not have large disutility, whereas others will not. Now, once a right is established whereby I cannot dump on you without your permission, persons with isolated sites have incentives to offer dumping services for compensation from those with comparatively high disutility. Both parties are winners as long as the price is set between their relative disutilities, so economic efficiency is improved. As long as none of the parties involved have enough power to influence the market clearing price, the outcome will be an efficient allocation of garbage. In the parlance of economic theory, markets succeed in internalizing the *externality* created when I dump garbage on you without your permission. Not only has the private property right promoted efficient allocation but it has done so automatically, without the need for interference except for the enforcement of the right. This is the major virtue of the "invisible hand".

3.1. The problem of open access

The preceding example suggests more generally that whenever private property rights are not assigned, the associated resource necessarily must take on a "public" character, by which we mean that any individual's decision to use or degrade the item necessarily has repercussions on others. In such situations, the social benefit from individual consumption is necessarily different from the private benefit and we may expect that unregulated private decision making will mis-allocate the resource. The classic paradigm in the property rights literature arises from the use of open access rights to some natural resource. This situation and the difficulties it entails frequently are referred to as "the problem of the commons". We illustrate with the case of cattle grazing on a piece of open access land.

We assume a production function for beef: $y = f(a, K)$, where a represents the number of head of cattle and K for the acreage of the rangeland (other inputs are suppressed

---

[3] There is no attempt here to give a complete treatment of the "first theorem of welfare economics" as that would take us too far afield and there are many excellent treatments available. For a relatively nontechnical textbook treatment, see Varian (1978). The classic technical exposition is Debreu (1959). For a more complete exposition of the example used here, see Starrett (1988).

for simplicity). Ranchers will be indexed by i and we make the "common pool" assumption that all cattle put on the land will mingle in such a way that each gets its share of the fodder. Thus, if rancher i puts a_i cattle on the land and others put on a_{-i} (so $a = a_i + a_{-i}$), output to rancher i will be

$$\left(\frac{a_i}{a_i + a_{-i}}\right) f(a_i + a_{-i}, K).$$

We treat this range as small relative to the total cattle market so that prices can be treated as given, p_y for meat and p_c for cattle (alternatively we can think of this as a partial equilibrium analysis). Then, the first best use of the range will be determined by choosing the number of cattle to maximize profits: $p_y f(a, K) - p_c a$, so the optimal choice of a must satisfy:

$$p_y \frac{\partial f}{\partial a} = p_c,$$

that is, cattle should be chosen so that the price equals the value of the marginal product (VMP) in producing beef.

However, if there are many users of the common, rancher i will choose the size of his herd to maximize:

$$p_y \left(\frac{a_i}{a_i + a_{-i}}\right) f(a_i + a_{-i}, K) - p_c a_i$$

which generates the following first order condition for choice of a_i:

$$\left[\frac{a_{-i}}{a}\right] \frac{p_y f(a, K)}{a} + \left[\frac{a_i}{a}\right] \frac{p_y \partial f(a, K)}{\partial a} = p_c.$$

Thus, we see that the rancher will choose his herd so that the price of cattle is equal to a *weighted average* of the marginal and average product of the extra cow. For added simplicity let us assume that each rancher is small relative to the whole so that the first term in square brackets above is approximately equal to one and the second approximately equal to zero. (Note that the same analysis will apply if the rancher ignores the effect that the last cow he adds will have on the grazing opportunities of his intra-marginal herd.) Then our rancher will add cows until the value of the *average product* (VAP) of the extra cow is equal to its price.[4]

---

[4] More generally, we can show that if all ranchers are identical then the expected equilibrium outcome will be one in which all ranchers graze the same number of cows and that the degree of overuse on the common will be increasing in the number of users. For more on the concept of equilibrium involved here and the general presumption of inefficient outcomes, see Section 5.

So we see that open access will lead to distortions in behavior to the extent that the average and marginal products of cows differ on the common. If there were no crowding out effect so that the marginal product was independent of the number of cows, then of course marginal and average products would be the same and there would be no distortion. However, once the range land starts to fill up, marginal product will begin to fall and therefore will be below the average product. At that point open access will lead to overgrazing as each rancher adds cows beyond the point where VMP equals price, to the point where VAP equals price. In fact the VMP might actually be negative at the equilibrium point.

The distortion can be explained in terms of externalities; when one rancher adds a cow, there is less fodder available to others' cows so that their profits are marginally reduced. Since the extra cow earns its owner VAP but only contributes VMP to the total, this external cost is measured as VAP minus VMP. (As an exercise, the reader might derive this formula using calculus.) Because this extra social cost is ignored, the rancher adds cows beyond the socially optimal holding capacity and the common is overgrazed.

As with the case of garbage, the introduction of private property rights and associated markets can be used to internalize this externality. Here the land is being treated as a free good under open access, whereas it has scarcity value (due to the crowding out). If the land is treated as private property[5] and traded on markets this scarcity value will be reflected in land rent and the rancher will either have to pay this rent to expand his herd or suffer himself the loss in marginal product of adding cattle to fixed land. Without doing a full analysis, we can argue that the rent will exactly internalize our externality. Assuming that there are no other fixed factors to producing beef,[6] we expect the production function to be constant returns to scale; that is doubling the land and doubling the cows should serve to double the beef. For such functions it is well known that the competitive value of the factors of production exactly exhausts the value of final product. Here this means that the value of the land as input to producing beef plus the value of the cows in producing beef should equal the total value of the beef. It follows that the value of the land per cow employed is equal to VAP minus VMP. Consequently if a rancher is willing to rent the land needed for an extra cow, he must be willing to pay the externality cost, now reflected in the scarcity rent on land. Alternatively, if he adds an extra cow to a fixed piece of land he absorbs the externality cost. On the margin, he will be indifferent between these two options and an efficient use of the land will result.

3.2. Potential conflict with equity

Of course, we know there are limitations to market efficiency, and these translate naturally to shortcomings of private property rights. We take up a philosophical objection

---

[5] In this case, there will be some costs of exclusion (e.g., building fences) associated with the enforcement of private property rights. These are ignored here but we will have more to say about this in the sequel.

[6] If there are other factors of production and they are priced correctly the same analysis will apply.

first and turn to intrinsic difficulties in the next section. The exclusive use of private property rights has implications for the distribution of income. Indeed, the most common argument in favor of socialism (in which some subset of resources is not assigned as private property, but rather owned collectively) claims that capitalism generates an allocation of resources that is inequitable, in that some agents wind up commanding a disproportionate share of resources. One might think that it would be possible to achieve any desired distribution simply by rearranging property rights. In principle that is true, but implementation would involve some degree of slavery as we might have to assign a talented person's labor time to someone with less talent. Assuming we rule out slavery, as most societies now do, we can make the income distribution more even only by use of taxes and transfers. There is a large literature on the design of tax/transfer systems with the aim of creating ones that do not distort economic incentives.[7] (Note that even if we allowed slavery, there would be incentive problems in eliciting effort once we take into account the costs of monitoring.) Most economists believe that such 'incentive compatible' schemes (if possible at all) are impractical so that any attempt to redistribute income must entail some loss of economic efficiency.[8]

This conflict between egalitarian distribution and efficient allocation through the market system is an old problem without a satisfactory solution.[9] The presence of this conflict is the justification for the socialist position that some resources should be treated as collectively owned even when it would be possible to assign and enforce private property rights. But even without the socialist's view, there are more fundamental impediments to the use of private property rights, as we will see in the next section.

4. Publicness and the need for collective rights

As we saw in the previous section, the private assignment of property rights can serve to internalize what would otherwise be damaging externalities. Unfortunately, for many goods and services and especially for many environmental resources it is difficult to make such assignments effective. The issues here are generally well understood and there are many excellent textbook expositions.[10] Therefore, we will confine ourselves here to a brief summary together with references to that literature.

---

[7] For a discussion of various types of taxes and their distortions, see chapters in Handbook of Public Economics, Vol. 1 (1987), or a text such as Boadway and Wildasin (1984), Atkinson and Stiglitz (1980), or Laffont (1989).

[8] However, we will see later that it may sometimes be possible in the context of environmental resources to use the assignment of rights in such a way as to affect the distribution of income without incurring distorting incentive effects.

[9] There is, however, a school of thought which I will refer to as the "entitlements" school that has it that people are entitled to what they start with and therefore, that the resulting market distribution is in fact equitable. For an exposition of this view, see Nozick (1974).

[10] See, for example, Baumol and Oates (1988), Boadway and Wildasin (1984) or Oakland (1987).

The biggest impediment to use of markets to allocate environmental resources is *non-appropriability* – namely, the difficulty or impossibility of enforcing a private property right. A pure example of such a resource is "clean air". It is not possible to assign an individual the right to clean air over his property since there is no practical way to prevent that air from mingling with "dirty" air coming from elsewhere. Or alternatively, we might say that the costs of enforcing a private right (by erecting barriers) is prohibitive. Other environmental resources with similar character include fish in the ocean, greenhouse gas concentrations and lake water quality. Even when a property right can be enforced it still might not be desirable if the costs of enforcement (which as we have said are usually ignored) are too high. For example, as indicated earlier, grazing land can be treated as private property only at the cost of building fences or walls. When the density of use is sufficiently small, the benefits of efficiency may not be worth this cost so we may prefer "open access".

Even when exclusion is costlessly possible, it may not always be desirable from an efficiency standpoint. This happens for resources that have an element of *nonrivalry*. A resource possesses some degree of nonrivalry if my use of it does not completely preclude your use. As an example of pure nonrivalry, consider radio or television signals.[11] My use of the signal to obtain reception does not in any way preclude your using the same signal. In this case, we can in principle exclude some users at a finite cost (through the use of scramblers) but it is inefficient to do so; once the signal is sent (and the associated costs sunk) the greatest benefit will be derived from allowing all potential users free access. Another commodity with nearly the same character is information. Once it is produced the costs of dissemination are minimal so that efficient management would dictate free access. Notice that if we do choose not to exclude then the associated commodity takes on the same public character as we identify with the inherently nonappropriable goods. Goods, services and resources that are either nonappropriable or nonrival we label as *collective* – items in this class cannot be efficiently allocated through the use of unregulated private property rights. The extreme examples in this class possess both properties. "National defense" and "ozone protection" are cases in point. In both cases, it is not possible to exclude citizens from the benefits nor is their a cost of allowing additional users (or enjoyers). Our concept of collectiveness is intentionally broad so as to encompass the various types of pure, impure and local public goods introduced in the economics literature.

Although we will treat them relatively symmetrically here, there are some important distinctions between the two types of collectiveness. This is because while exclusion is a binary concept (either you do or you do not, the decision generally depending on the costs of exclusion compared to potential benefits), rivalry is a matter of degree. The lanes of a highway can be either completely nonrival (when few cars are present) or completely rival (when there is queuing to get on) or anything in between. These

---

[11] The two-way dichotomy involving appropriability and rivalry is widely attributed to Richard Musgrave. See Musgrave (1959).

distinctions play important roles in determining the optimal allocation of associated resources, and are central to the theory of local public goods, but further discussion is beyond the scope of this chapter.[12]

"Collectiveness" does not entirely preclude the use of markets but does imply that if employed they will work inefficiently at best. There are two ways in which markets might be used in this context – for nonrival goods, we can exclude and force agents to pay for use or, for nonexcludable goods, we can allow agents to contract for use but not exclude. As an example of the first type we may use patents to exclude potential users from free access to information, as a way of providing incentives for the production of such information. But we would be even better off if we could provide those incentives in another way, since the patent royalty will deter agents whose potential benefit, though positive, falls short of the royalty. As an example of the second type, suppose we sell community safety through the market. That is we allow citizens individually to purchase police time for patrolling the town streets. It is possible that there would be some purchases on this market but we argue that these will be lower than desirable for the overall social good; when any particular citizen purchases police time, most of the benefits go to others (referred to as "free riders" in the literature) who will benefit equally from the police presence. Here (in contrast to the case of privately divisible goods such as bread or steel) the private benefit from purchase is lower than the social benefit. Since market price can only reflect the private benefit, police protection will be underprovided in this case compared to the first best. The problem here again can be viewed in terms of externalities: when one agent purchases services he confers external benefits (or costs) on others who will also be affected. For further discussion of the voluntary (private) provision of public goods, see Chapter 4 (by Baland and Platteau).[13]

Thus, we see that when the resource in question is collective in nature we will have to assign at least some of the rights collectively if we seek to achieve optimal management. Our problem becomes one of determining the appropriate collective group for each type of control rights and to design procedures whereby these groups will be induced to make the right decisions.[14]

5. Outcomes under decentralized decision making

When private property rights are appropriate and are properly defined and enforced, we have seen that decentralized decision making through the use of markets can generate

---

[12] For surveys of the theory of local public goods, see Rubinfeld (1987), Starrett (1988, Chapters 5 and 11), or Cornes and Sandler (1996).

[13] See also Oakland (1987), and Inman (1987) for additional perspectives on market provision of collective goods.

[14] For general discussions of the appropriate assignment of rights see Barzel (1989), Bromley (1991), and North (1990).

an efficient allocation of resources. Decentralization has the desirable features that implementation requires little or no communication and coordination among the agents. Here we ask what we should expect from decentralized behavior in more general situations where private property rights are either undesirable or unenforceable. Throughout the sequel we will assume that agents have *complete information* about the allocation situation at hand. In particular they know the payoff relevant outcomes for them as a function of the actions taken by all participants, and they observe those actions, at least after the fact.[15] Given this context, we must determine how agents will act when they know their actions will have observed effects on third parties. And the answer is very likely to depend on context: for example, if the agents interact with each other more than once, each will surely realize that actions taken today will be observed by others and consequently may well influence subsequent behavior. But before discussing such complications let us consider the case of one-time interaction.

5.1. Static interaction

Assume first that we are in a static (timeless) world where agents come together only once and all relevant outcomes are determined by their simultaneous actions. Even here, there is no single behavior that will be convincing in all situations, especially when there are small numbers of players (as, for example, in a product duopoly) who will be acutely aware of their strategic interaction. We start with an example which is fairly representative of situations with "free rider" incentives, and where the strategic issues are easily resolved. Suppose there are two firms (I, II) both of whom use a common lake. Each uses lake water as an input and also possibly as a repository for waste. For simplicity we assume the firms are identical and that the only strategic choice they have is whether to dump effluent into the lake or treat their waste. Assume the following per firm costs and benefits:

Cost of treatment:	6,
Benefits from clean water:	8,
Benefits if one firm dumps:	5,
Benefits if both firms dump:	0.

We can represent this strategic situation in a two by two matrix "game form":

		Firm II	
		treat	dump
Firm I	treat	(2, 2)	(-1, 5)
	dump	(5, -1)	(0, 0)

---

[15] Note that without this assumption it would be impossible to predict an outcome without specifying exactly what each agent knows and what she believes about things she does not know.

There are four possible combinations of strategies (each firm can treat or dump) and the table numbers indicate the net payoffs to (firm I, firm II) of the associated strategies. For example, if both firms treat, they each get benefits of 8 and pay treatment costs of 6 for net return of 2. Or if firm I treats but firm II dumps, they each get benefits of 5 but firm I pays costs 6, so the net returns are $(-1, 5)$, etc.[16]

Note that regardless of what firm I is expected to do firm II wants to dump (and vice versa): if firm I treats, firm II will get 5 rather than 2 by dumping and if firm I dumps he gets 0 rather than -1 by dumping. The incentive to "free ride" is dominant and in the absence of communication we should certainly expect both firms to dump. As a consequence they will reach an inefficient solution since they would be better off coordinating on a joint treatment strategy. Unfortunately, many environmental interaction situations have this property that free riding is a dominant strategy so we should not be surprised when we observe excessive pollution in decentralized, unregulated situations.

Of course, not all games have this "dominant strategy" structure. For example, in the grazing model presented earlier, one rancher's decision on how many cows to graze surely will depend on the numbers grazed by others. But in such somewhat more complicated situations it still may be reasonable to assume that all agents take as given the behavior of others and make their own choices to optimize against those expected behaviors. The corresponding outcome of decentralized decision making is a set of behaviors such that each agent's choices are optimal for him, taking as given the corresponding choices of others. In game theory we refer to these behaviors as *Nash behavior* and the corresponding outcome as *Nash equilibrium*. This view of behavior and the associated concept of Nash equilibrium seems the most plausible outcome in many static situations where there are relatively large numbers of players and communication is difficult or impossible.[17] Recall that we already employed it informally in discussion the problem of open access where we showed that it led to an inefficient outcome. Now, we claim a general presumption of inefficiency; namely, we argue that in any situation where agents' choices affect each other's payoffs in significant ways, the Nash outcome is almost certain to be inefficient from the point of view of the group as a whole.

Suppose that we are in a general situation in which payoffs of the various agents depend on actions they all take. Let a^i stand for the (vector) of decision variables available to agent i. To the extent that agents face constraints, we assume that they can be solved for a dependent set as functions of some independent subset, and that a^i represents the independent subset. The matrix of all actions will be simply denoted a (without a superscript). Further, the notation (a^i, b^{-i}) will mean the configuration in which agent i plays from configuration a whereas everyone else plays from b. Let $P^i(a)$ stand for the

---

[16] The reader will note that the payoff structure here is the same as that of the famous "prisoner's dilemma" game wherein two suspects would be best off if they kept their mouths shut, but each finds it a dominant strategy to implicate the other. Many games involving economic externalities have this same structure.

[17] However, the Nash assumption can be criticized on a number of grounds. For example, we can see that it is always disconfirmed out of equilibrium. For further discussion of this assumption and possible alternatives, see Fudenberg and Tirole (1991, Chapters 1 and 2).

objective function of agent i. Now a^* will be an equilibrium outcome for the group if each finds it best to use her a^* decision as long as she expects everyone else to do so as well; that is:

$$\text{for all } i, \quad P^i\left(a^*\right) \geqslant P^i\left(a^i, a^{*-i}\right), \text{ for all feasible choices } a^i.$$

We now argue that equilibrium in this context will generically be nonoptimal from the point of view of the group as a whole. This conclusion will hold no matter how we choose to weight individual payoffs in defining the group objective. Suppose we assign weights w_j and consider the social objective $W(a) = \sum_j w_j P^j(a)$. Thinking of a^i as one dimensional, we can define its marginal social benefit and marginal private benefit as

$$MSB^i(a) = \sum_j w_j \frac{\partial P^j}{\partial a^i},$$

$$MPB^i(a) = \frac{\partial P^i}{\partial a^i}.$$

Now, by the definition of Nash equilibrium (a^*), $MPB^i(a^*) = 0$ so

$$MSB^i\left(a^*\right) = \sum_{j \neq i} w_j \frac{\partial P^j}{\partial a^i}.$$

Thus, as long as the interdependences are generic, we expect to find the MSB's not equal to zero at equilibrium so the group can be made better off through marginal changes in private choice variables. Further, we can measure the marginal external benefit of choice a^i as $MSB^i(a) - MPB^i(a)$.

5.2. Repeated interactions through time

Let us now generalize to contexts where agents interact with each other on an ongoing basis. This is typical of many environmental situations where groups of people use the same grazing land, forest resources, water sources, and the like. It seems likely that behaviors might be different in this situation than with static interaction. In particular, agents might be deterred from "antisocial" behavior by the fact that it will be observed (at least after the fact) by others and might lead to retaliation.[18] Here, we examine the possibilities in the special case where the same "static interaction" is repeated by the

---

[18] The intuition here is quite old and not easily attributable. Indeed, the formalizations are frequently referred to as "folk theorems". See Friedman (1971) and Axlerod (1984) for good expositions.

same group of players a specified number of times (which might be finite or infinite).[19] In the game theory literature from which we draw, the static interaction is referred to as the *stage game*, the strategic situation defined by an infinite number of repetitions as the *supergame*, and the remaining opportunities for interaction after a certain date is reached as the *continuation game* from that date. To simplify further we assume that there is a unique Nash equilibrium (a^*) in the stage game. Note that this is true in both the examples we gave above.

In what follows it is critical what agents are able to observe and how they use that information. Here we assume that agent actions are observable to all after the fact and that agents will consider conditioning future choices on what they have observed. Note that these assumptions are conducive to generating the modified behavior suggested above, and indeed that if actions are completely unobservable it is hard to see how the fact of repetition alone would make any difference to behavior.[20] Let h_t represent the history of play up to date t. For example, in the water resource game above, the history would be a recording for each player as to whether or not he treated his waste in each preceding period, and in the open access range, the number of cows each rancher grazed in each preceding period. Then, a strategy for player i in period t is a function that maps each potentially observed history into a current action. We represent this function as $a_t^i = \sigma_t^i(h_t)$. An example of a strategy in the water resource game is "tit for tat": namely play today exactly as your opponent played yesterday – so if the other firm treated last period, you treat now, but if he dumped, you dump now. Note that for that particular strategy, the history before yesterday is ignored and indeed we do not require that all potential information be used in determining strategy.

We assume that agents value each continuation game as the discounted sum of payoffs from the associated series of stage games using a constant discount factor δ, that is,

$$V_t^i = \sum_{\tau=t}^{T} \delta^{\tau-t} P^i\left(\sigma_\tau(h_\tau)\right),$$

where V represents the continuation value and T stands for the time horizon (date at which the last stage game is played). It seems reasonable to require that agents act in any continuation game just as they would in a static game; therefore we require of an equilibrium sequence of actions that it constitute a Nash equilibrium in every continuation game entered. This is a special case of the *subgame perfect equilibrium*

---

[19] For a discussion of bargaining models with other structures of ongoing interaction, see Fudenberg and Tirole (1991) or Moulin (1986).

[20] Actually, even if individual actions are unobservable, agents may be able to infer something about such actions from observation of aggregates. For example in our water resource model, each agent can tell what his opponent did simply by observing the quality of lake water. When commonly observed variables can be used to identify private behavior, the results reported here will carry over with only slight modification [see Fudenberg and Tirole (1991, Chapter 5)]. Note that such identification is likely to be easiest with a small number of players.

for general sequential games. We are interested in determining what outcomes could be observed that are subgame perfect equilibria.

5.2.1. Finitely repeated games

Suppose first that T is finite; that is there is a date where it is known that the strategic interaction will end. For such games, subgame perfect equilibria can be determined by backward induction. Namely we can determine how the last stage game will be played. Knowing this, we can back up one period and determine how the continuation payoffs will depend on actions in that period, solve for the Nash equilibrium in that game and continue backward to the present. From this, we can show the somewhat surprising outcome that there is little scope for generating cooperation in finitely repeated games. To see this, observe that since the stage game is played in the last period continuation, equilibrium requires that all agents play according to a^*. But then, in the preceding period, continuation payoffs are just the discounted constant value of a^* plus the values of the current stage game. Consequently, subgame perfection requires that a^* be played in the penultimate period as well. Continuing the backward induction, we find that a^* must be played in all previous periods as well and there is no scope for cooperation at all.[21]

Many people find this conclusion counterintuitive. Namely it seems that if there are many periods to go, agents would want to play in a way that encourages cooperation, at least for a while. Further, when games of this type are simulated experimentally, players generally are observed to cooperate in the early stages when there are still many periods remaining. These considerations have led some to question subgame perfection and the concept of individual rationality that lies behind it. The interested reader is referred to game theory texts such as Fudenberg and Tirole (1991), as further discussion is beyond the scope of this chapter.

5.2.2. Infinitely repeated games

Part of the problem with the finitely repeated structure may be that in most real life situations there is no obvious last period of interaction, even though all agents know that the relationships will end at some unspecified date in the future. Consequently, use of an infinite horizon ($T = \infty$) may be a better approximation to the strategic situation that agents feel they face. With an infinite horizon, backward induction is no longer

---

[21] This rather severe conclusion does depend critically on our restriction to subgame perfect equilibria and the assumption that our stage game has a unique Nash equilibrium. If there are ways of punishing agents and thereby holding them to utilities below what they would get at a^*, these can be used to induce cooperative behavior in a Nash equilibrium of the finitely repeated game. However, this outcome cannot be subgame perfect [see Benoit and Krishna (1987)]. Also if there are multiple static Nash equilibria, the opportunity to switch among them can induce some degree of cooperation even in subgame perfect equilibria [see Benoit and Krishna (1985)].

available and we must find other ways of solving the game. And as we shall see, the outcomes can be quite different. Indeed, we will show informally that any stage game outcome that gives agents at least as much as they get from a^* can be supported as a subgame perfect outcome in the supergame if the discount factor is sufficiently high (that is the discount rate is sufficiently low, indicating that agents are relatively patient).

It is important to note first that even here, subgame perfection does not rule out the noncooperative outcome. Namely, the strategy: "play according to a^* in every period no matter what you observe" is always subgame perfect. The logic here is just as it was in the finite horizon case. If I enter a continuation game with the belief that everybody will play according to a^* in the future regardless of what I do now, then I will want to act just as if I were in a static stage game and will want to play according to a^* now.

What is different now is that many other outcomes are possible as well, at least when the discount factor is relatively high. To see this, let us see what would be required to support an outcome in the stage game with payoffs $v = P(a^v) \geqslant P(a^*)$ where a^v is the action that generates payoffs v. The idea is to employ the following punishment strategies: in any continuation game all agents will play according to actions a^v as long as everyone has played that way in all previous stage games, but will "punish" by playing according to a^* if any agent has played anything other than a^v in any previous stage game. Clearly if these strategies are consistently played, the outcome will be that payoffs v are realized in every period, so we want to know when these "trigger" strategies constitute a subgame perfect equilibrium.

With the trigger strategies, there are only two possible continuation games that can be entered, depending on whether or not someone has previously "defected". In the case where someone has defected, all expect play according to a^* subsequently and we have already seen that "a^* forever" is subgame perfect in that situation. Consider now the continuation wherein no one has defected heretofore. Let us see whether agent i will want to deviate from the cooperative strategy. If she chooses to deviate, she will play her best response to the cooperative strategy a^{v-i} from others, yielding for her a payoff w^i. For example, in our water resource game, the best response when your opponent treats is to dump and the corresponding payoff will be 5. Consequently, since she knows that after deviation, a^* will be played, we find:[22]

$$\text{value of deviation} = w^i + \sum_{t=1}^{\infty} \delta^t p^{*i} = w^i + \left[\frac{\delta}{1-\delta}\right] p^{*i}.$$

Therefore, the trigger strategies will be subgame perfect as long as the value of "cooperation" which is $v^i/(1-\delta)$ is at least as large as the value of deviation, namely if

$$v^i \geqslant (1-\delta)w^i + \delta p^{*i}, \quad \text{or} \quad \delta \geqslant \frac{w^i - v^i}{w^i - p^{*i}}.$$

[22] Since this game has a "time stationary" structure (continuation payoffs depend only on past history and not on calendar date), the choice of current date is arbitrary and we start from date zero.

So we see that if the discount factor is sufficiently large, the trigger strategies are sub game perfect and the stage game payoff v is supported as a subgame perfect equilibrium. For example, in the water resources game, we can support the cooperative play of always treating waste (yielding $V = (2, 2)$) as long as the discount factor is at least as large as $(5 - 2)/5 = 0.6$. Equivalently the corresponding discount rate would have to be less than or equal to 66%, a weak requirement!

Thus, we see that situations with repeated interactions give groups of agents an opportunity to foster cooperative behavior without explicitly entering into binding contracts. However, it would be wrong to say that efficient outcomes are being generated here by strictly decentralized behavior. Before these methods will work, there must be some common agreement on what it is we are trying to achieve (which V) and a common understanding of how sanctions will be used. In our simplified water resources model there was only one efficient choice, but generally we expect an entire "Pareto frontier" to choose from and resolving on a particular element requires reconciling preference differences. For example, in the grazing example, the surplus can be distributed in many ways determined by the numbers of cows assigned to a particular herdsman, and this allocation must be agreed to before any cooperative behavior will be enforceable. Further, the agreement on when and how to punish must be part of the social norms mutually accepted by the group. Thus, it is hard to see how cooperative behavior can be generated without substantial communication and bargaining. We turn next to an examination of what can be achieved when such interaction is allowed.

6. The Coase theorem and limitations

There is an old argument in the literature that as long as (1) property rights are assigned and enforced in an exhaustive (complete) way, (2) there is free and costless communication among agents, and (3) the control rules allow for bargaining among the collective of all affected parties, this collective will always reach an outcome through bargaining that is Pareto efficient. A stronger version of this "theorem" would have it that the allocation outcome reached is independent of how the rights are assigned. This second version generally is not true unless income effects are negligible since changing the assignment of property rights will change the distribution of income and, if preferences differ among consumers, also change the demands for goods and services. However, the first version may still be true. If it is, we could say that there is something like the invisible hand for collective property rights.

The simplest argument in support of the Coase theorem goes as follows.[23] Suppose an allocation has been proposed that is Pareto inefficient; that is, there is a change in

---

[23] This argument was first espoused by Coase (1960). However, even then he was aware of the fact that its force would be mitigated by the presence of transactions costs (see below). There are many general discussions of the Coase theorem and limitations. See, for example, Alchian and Demsetz (1973), Cooter (1987), Hoffman and Spitzer (1982) and Hurwicz (1995).

the allocation that will make at least one member of the collective better off and no one worse off. Then, if the parties who would be made better off proposed making this change, all rational parties should accept. This argument is independent of who has rights. If I have the right to pollute but am polluting to such a high level that you would pay more than it matters to me to get the level reduced, then you pay, but we are both better off. Alternatively, if you had the right to clean air, I wind up paying to pollute up to the level where my cost is worth marginally more than the extra pollution damage. In either case, we would continue to bargain until an efficient point is reached where the marginal benefit of extra pollution to the polluter is exactly offset by the marginal costs to pollutee. We already made this argument for the case of physical garbage but it should work just as well for collective goods as long as all affected parties are actively involved.

Although this argument sounds reasonable it involves many pitfalls and considerable care is required to specify precisely the conditions under which it is correct. Indeed, we believe that it is quite difficult to state and prove a "Coase theorem" precisely. Here we will confine ourselves to a discussion of necessary conditions for its validity.

6.1. Costless communications and implementation

We have already seen that unrestricted communication among the agents is essential to an efficient outcome. In situations with static interaction we argued that uncoordinated behavior would almost certainly lead to inefficient outcomes. And even in situations of repeated interaction some coordination on the desired outcome and punishment strategies would be required in order to achieve efficiency.

But it is not enough that agents can communicate freely – this communication and any associated rules for enforcement and implementation must also be costless. When there are small numbers, this assumption may be pretty reasonable, but it becomes less so as the numbers increase. The presence of nonnegligible transactions costs has independent negative implications for the Coase theorem. To illustrate, consider the following example involving a flood control project. A dam can be built for $10M which will provide $9M in benefits to a small group of people. The dam is to be paid for by taxing 10M people $1 each. This project clearly is Pareto inefficient since each of the taxpayers would be better off canceling the tax and paying $.91 to the benefitting group, an offer that group should accept since it provides them with $9.1M in benefits. But will this bargain be struck? Can the amorphous group of taxpayers coordinate? And even if they could would it be worth their effort given that each only avoids a $1 liability? It seems clear that if the costs imposed are sufficiently small and the numbers involved sufficiently large we will not see such outcomes due to the fact that transactions costs, though small, are not negligible. Indeed, projects of this type (sometimes referred to as "porkbarrel" projects) wherein benefits go to a relatively small group with costs being paid out of general revenue are a staple of government budgeting in much of the world. Note that the nature of property rights does matter now and will influence the outcome:

if taxpayers have the right to refuse the dam, and the developer is thus forced to make his case, the project will fail.

Similar reasoning suggests that bargaining alone will not "solve" the problem of open access. Suppose that there are many ranchers that graze their cattle on an unfenced piece of land. Then in the absence of bargaining, we saw that each will graze too many cattle since the social cost of an extra cow is imposed mostly on other ranchers (whose cows have less feed available). However, if all the ranchers get together, they should agree to a bargain wherein each reduces his herd by a small amount to the point where the marginal social cost imposed equals the marginal private benefit. But if there are large numbers involved, coordinating on this strategy and monitoring to enforce it may be more trouble than it is worth. Indeed, there is empirical evidence from studies of traditional societies that efficient management of common property is generally achieved only when the numbers are relatively small, and not always even then.[24]

6.2. Commitment requirement

Further, agreements must be enforceable in the sense already articulated in section 1. In the context of bargaining, this means that agreements must be observable to an enforcing party that can guarantee compliance. Without such assurance, I could take your money, agreeing to cut my pollution, and then renege; and you, anticipating this, would not pay.

There are differences of opinion in the literature as to what is allowable as part of a Coasian bargain. One view would have it that any arrangements for compliance and enforcement must be thought of as "outside" the bargain. Under this view, free riding on agreements is relatively easy and it must be left to mechanism design to make agreements incentive compatible. This is roughly the view taken by Baliga and Maskin in Chapter 7 (Mechanism Design for the Environment).

Here we take the view that Coasian bargains take place in the context of a social contract and can take advantage of its rules and regulations. However, we still need to take account of the limitations of these rules. Generally society imposes some limits on the type and character of agreements that are enforceable. For example, if a steel mill signs an agreement not to emit smoke from its stacks, there is no practical way to ensure that smoke is not emitted – the best that can be done is to impose some kind of sanction if and when smoke is emitted.

Once we recognize these restrictions we must require that agreements reached be "self enforcing" in that participants would prefer to remain in the agreement rather than revert to some fallback wherein they accept whatever sanctions are available. Such requirements are formalized in cooperative game theory by requiring that bargaining outcomes be *core allocations*. Informally, an allocation is in the core only if it gives

---

[24] For a discussion of institutional arrangements for handling the problem of open access, and examples of success and failure, see Schotter (1981), Martin (1989), Eggertsson (1990), Ostrom (1990), North (1990), and Hanna, Folke and Mäler (1996). Also for a report on outcomes from experimental design, see Ostrom (1999).

every *subcoalition* (subset of the whole group) of the society at least as much as they could guarantee themselves by "going it alone" – that is by withdrawing from the larger group, accepting whatever sanctions the rest of society can impose, and optimizing internally subject to those restrictions.

It should be clear that the limits to bargains imposed by restricting to the core will depend on what kind of sanctions are allowed – if we could throw all deviants in jail for life and strip them of all resources the restrictions would not bind at all. And while such draconian measures are unreasonable, we saw that something similar could be achieved in situations where the bargainers deal with one another on an ongoing basis. However, when sanctions are effectively limited, we will argue that restrictions to core allocations may preclude efficient bargains and further that the way in which property rights are assigned can be crucial to enforcing the efficient bargain.

These facts were first elucidated in the "garbage game" of Shapley and Shubik.[25] Here we modify this game somewhat to give it richer structure. There are three players in the game (a, b, c) endowed respectively with $(1, 2, 3)$ units of garbage. All players have the same cost of absorbing garbage on their property; if b units are absorbed the cost in numeraire dollars is b^2. The players are free to exchange garbage and money in any way they like and each agent's net cost $C(\cdot)$ will be the sum of net payments to others and the value of garbage ultimately absorbed. Thus, cost is measured in a common unit and this is a game of *transferable utility* in the parlance of game theory.

Given the convex cost of absorption, the efficient outcome here will be for agent a to absorb one unit of garbage from agent c (therefore generating equal absorption) in exchange for some compensation (the amount of which is irrelevant to efficiency) and we want to know whether free bargaining will lead to this outcome. We will examine this question under two different property rights regimes. Under the first which we label *exclusion rights*, no one can dump garbage unless rights are granted by the dumpee, whereas under the second (*possession rights*) the holder of garbage has the right to dispose of it as he likes, although he may be deterred by compensation. (In the air pollution analog, exclusion corresponds to the right to clean air, whereas possession corresponds to the right to pollute.)

6.2.1. Exclusion rights

Under the exclusion rule, any individual or subcoalition that chooses to go it alone must absorb their own garbage, but can prevent others from dumping on them. Using this rule, we can compute the cost to each coalition of going it alone. There are seven coalitions to consider: the three singletons, three pairs and the one grand coalition. We find:

$$C(a) = 1, \quad C(b) = 4, \quad C(c) = 9,$$
$$C(a, b) = 4.5, \quad C(a, c) = 8, \quad C(b, c) = 12.5,$$
$$C(a, b, c) = 12.$$

[25] See Shapley and Shubik (1969), the further discussion in Starrett (1973) and Aivazian and Callen (1981).

Note that in each case we have computed the total cost to the group so the cost is independent of money transfers among them and indicates the cheapest absorption cost they can jointly manage. So for coalition (a, b), they must absorb 3 units and the cheapest way is for each to absorb $3/2$ at total cost $2(9/4) = 4.5$, and other costs are computed similarly.

A proposed allocation (*imputation* in the game theory literature) will assign a net dollar cost (u_i) to agent i. A necessary condition for this to be a core allocation is that for every coalition (Γ) the total costs imposed upon it be no greater than its value, that is:

$$\sum_{i \in \Gamma} u_i \leqslant C(\Gamma), \quad \text{for all } \Gamma. \tag{1}$$

When condition (1) fails we say that the allocation U is *blocked* by coalition Γ. In addition an allocation must be feasible for the group as a whole. This means that the grand allocation must absorb at least its value in cost. In conjunction with the corresponding constraint in (1), the grand coalition must absorb exactly its value, that is:

$$u_a + u_b + u_c = C(a, b, c) = 12. \tag{2}$$

Conditions (1) and (2) together characterize the core allocations. Any allocation satisfying these conditions cannot be blocked by any subcoalition and thus is stable against any potential defection. Note first that any allocation in the core must be Pareto efficient; otherwise it is blocked by the coalition of the whole. Thus as long as the core exists, bargaining under its rules will lead to an efficient outcome.

In the current situation we can verify existence by exhibiting a core allocation: $u = (0, 4, 8)$. Note that this corresponds to an efficient arrangement whereby agent c transfers one unit of garbage to agent a and pays him \$4 to accept it, while agent b simply absorbs his own garbage and is not involved in side payments. To verify that this is a core allocation we merely need check that no agent or pair of agents can do any better by defecting. For example, the pair (a, b) would incur costs \$4.5 by themselves and only incur \$4 under this proposal. The reader might note that the core is not unique here – there is some leeway in side payments that is consistent with the core side constraints. We would need to know more in order to specify a unique outcome. One possible conclusion would be that a market develops as in Section 2, and if there were enough players so that no one could influence the resulting price, that price would specify the equilibrium transfers.

6.2.2. Possession rights

When property rights are changed, we know that an income redistribution occurs and here that means that coalition values change. The question for us here is what effect if any this has on core allocations. Now, there is some ambiguity in what subcoalitions can expect if they defect. In particular, if agent a decides to go it alone (and dump his garbage elsewhere) what should he assume about the amount of garbage he will be forced to absorb? Here we will take the (fairly standard) position that he expects the worst – namely that all outside garbage will end up in his property. Note in particular, that this assumption makes core existence "most likely" since it makes defection "most costly". With this convention, coalition values are as follows:

$$C(a) = 25, \quad C(b) = 16, \quad C(c) = 9,$$
$$C(a, b) = 9, \quad C(a, c) = 4, \quad C(b, c) = 1,$$
$$C(a, b, c) = 12.$$

Now in searching for a core allocation we must at least satisfy the following pairwise coalition constraints:

$$u_a + u_b \leqslant 9, \quad u_a + u_c \leqslant 4, \quad u_b + u_c \leqslant 1.$$

Adding these constraints together and dividing by 2 yields

$$u_a + u_b + u_c \leqslant 7.$$

But there is no feasible allocation that satisfies this last inequality, so the core is empty.

The problem here is that with possession rights, it is tempting for a "large" subgroup (here of size 2) to gang up and dump everything on the lone outsider and there is no efficient way to absorb all the garbage without leaving some such subgroup an incentive to defect. It is not clear what we should expect to happen in this circumstance. Perhaps a powerful subcoalition will impose its will, but in principle the isolated party could always bribe them out of the corresponding inefficient dumping arrangement. More likely when negotiations break down, everybody will be forced to go it alone and again the outcome will be inefficient.

Thus, we see that in the absence of firm rules guaranteeing compliance and commitment, the arrangement of property rights can have a significant impact on the outcome, in particular determining whether or not bargaining will always generate an efficient outcome. Intuitively, this happens because the way in which rights are allocated helps determine what sanctions can be imposed on defectors, and these matter.

6.3. Perfect information requirement

Another necessary condition for validity of the Coase theorem is that there be perfect information – everyone must know what everybody else knows. In particular, each agent must know the preferences and characteristics of others. We illustrate with an example from insurance against personal injury. For simplicity, suppose the risk is binary – either you are hurt a fixed amount or not at all. Assume further that everyone agrees that the cost of the accident is a fixed number C. There are three kinds of agent. Two of these types are risk averse but differ in the probability of an accident (one being inherently more cautious than the other). The third type is risk neutral so will not demand insurance at fair odds, but will be willing to provide insurance to either of the other types at fair odds. In this world it is well known that the first best (Pareto efficient) outcome is for each risk averse type to receive full insurance at fair odds. However, if the insurer cannot tell the types apart, this outcome cannot be achieved since the bad (high risk) type will always want to portray himself as a good type. Consequently, the insurer would find herself writing all contracts at the good odds rate, would consequently lose money in expectation, and therefore would prefer not to do business.

The deterrent to efficient bargaining in the example emerges in a wide range of contexts. Whenever there is private information, then (ignoring the possibility of altruistic behavior) an agent has private incentive not to reveal information that will harm his bargaining position. But without this information it is not possible for the collective to know the efficient outcome, much less enforce it. Of course, in these situations, there will be incentives for the "good" types to signal their good information. However, the signaling itself must be costly in order to have credible information content, so that generally the resulting outcome still is not first best efficient.[26]

Thus, the validity of the Coase theorem in the presence of private information becomes one of optimal mechanism design.[27] Is it possible for an arbiter to design a system of messages and decisions/rewards based on messages so that the agents reveal a sufficient amount of their private information to specify an outcome that is Pareto efficient. Although there are some success stories here, there are serious limitations to what can be achieved. See Chapter 7 (by Baliga and Maskin) for discussion of mechanism design.

We conclude that bargaining alone is not likely to be a practical or efficient rule for managing the collective, especially when the required collective is large. From our perspective, this is unfortunate since many environmental collectives must be large in order to include all affected parties. For example, air pollution collectives must be at least national in scope. Worse yet, the ocean fisheries and greenhouse gasses collective must be global. Therefore in many environmental management situations we are left with a design problem of how to organize institutions and rules for exercising collective property rights in a way that best achieves goals of the relevant collectives.

7. Methods and rules for managing collective property rights

The mechanism design approach to management gives us a way of stating and analyzing the management problem in a precise mathematical language. However, there is a rich

---

[26] The classic reference in signalling is Spence (1975). See also discussion in Laffont (1987).

[27] For a survey of incentive issues as they relate to the allocation of public goods, see Laffont (1987). Various information issues are addressed in Schulze and d'Arge (1974), Dasgupta, Hammond and Maskin (1980) and Farrell (1987).

literature (both theoretical and empirical) that looks at the same range of issues in a much less formal way, with the general aim of assigning rights, identifying institutions and framing rules that will improve the allocation of collective resources even if full efficiency cannot be achieved.[28] Many structures have been studied involving ways of assigning and enforcing rights in such a way as to best elicit information and internalize externalities.

7.1. Self-organizing systems

Systems wherein collectives set up governance structures and enforcement procedures are sometimes referred to as "self-organizing" systems. Among the features of these systems that seem to generate the most desirable outcomes are (1) hierarchical structures whereby decisions with respect to a given collective good are made by an agency with the same collective constituency, (2) access rules and arrangements whereby users are well known to one another, (3) intertemporal structures whereby agents deal with each other on a repeated ongoing basis.[29] For each of these features there are sound reasons why they should be effective and some formal theoretical results.

Hierarchical structures have been studied formally primarily in the context of industrial organization.[30] There it has been shown that such structures often are an economical way of passing information within an organization and further that by designing the collectives of different sizes in the various layers of the hierarchy and decentralizing decision-making, it is possible to match the decision-making group to the relevant collective. For example, in the context of political collectives, the use of several layers of government (federal, state, country, city) enables decisions involving the national collective (such as national defense) to be made by the federal government, whereas those involving a local collective (for example, a city park) to be made by a more local (city) collective. Unfortunately, it is rarely possible to match the decision unit with appropriate collective exactly, in which case we must deal with spillover externalities whereby some of the benefits go to agents not in the decision collective. For example, in the case of the city park, benefits will generally accrue to visitors passing through from elsewhere. We discuss below the use of financial instruments to correct for these spillovers.

Arrangements whereby the decision-making agents are well known to each other and deal with each other on a regular ongoing basis foster cooperation in a number of ways. The desire to maintain a good reputation with peers provides incentives for agents to be truthful with each other and to follow collective rules. Further, when relationships are repeated over time, the collective can institute rules that serve to punish those who

---

[28] See, for example, surveys in Eggertsson (1990) and Bromley (1991).

[29] For further discussion of these issues, see, for example, Riker and Ordershook (1973) or Schotter (1981).

[30] See discussion and references in Demsetz (1988) and Williamson (1975).

do not cooperate as we saw in our discussion of repeated games, further bolstering the incentives to play by the cooperative rules.[31]

7.2. Correction for externality[32]

In situations where the decision collective is not exactly commensurate with the affected group, we expect externalities to occur – namely, some of the costs or benefits of action will fall on outsiders (as in our example where a city park is visited by outsiders). Whenever this happens, we expect incentives to be distorted in the same kinds of ways as we saw when private rights are assigned to collective resources. However, in these situations we may be able to use "market like" instruments to restore correct incentives. The standard method of correction for externality is to impose a tax per unit at the rate of external costs (or subsidy for external benefits). For example, since the burning of coal contributes to air pollution and global warming, a tax should be added to the price paid by users to producers of coal, the tax being equal to marginal social damages from these effects. In this way the price paid by users will reflect both the private costs of production and the social externality costs; then each user will equate the marginal benefit from use with the full marginal social cost, thereby generating a socially optimal use level.

This reasoning can be used to justify a variety of "green" taxes in situations where externalities damaging to the environment (from private decisions) can be identified. Of course, to reach the first best, we must be able to identify and measure the marginal external damage and monitor and enforce the volume of emissions so as to set appropriate tax rates and collect corresponding tax revenues. And agents have the same kind of incentives not to reveal their true preferences as we identified in connection with Coasian bargaining.

Even in situations where free rider externalities make it impossible to determine and/or enforce first best conservation principles, there are a variety of "second best" policies available that will be better than doing nothing. For example, a method widely used in the United States involves the setting of environmental standards – that is in the case of air pollution from the burning of coal, each polluter can be assigned a quota. As long as we are certain that the unregulated level of pollution is too high, a social improvement can be achieved by assigning quotas in such a way that the total emissions are reduced.

---

[31] We saw earlier how these ideas have been formalized in the theory of repeated games. In the empirical literature there is evidence that punishment strategies are used though not always in quite the way predicted by theory [see Ostrom (1990, 1999) and references therein].

[32] In the sequel of this section we will discuss remedies for externalities in the abstract. See Ashby and Anderson (1981), Kneese and Bower (1968) and Chapter 9 (by Stavins) for much more detail on practical matters of implementation. More detailed textbook treatments can be found in Hanley, Shogren and White (1997) and Kolstad (2000).

However, it has long been recognized that we can do even better than this by assigning initial rights to pollute (that correspond to the quotas above) and then allowing trade. An example of this method is the use of emission permits to regulate the atmospheric concentration of SO_2. Unless we can correctly measure people's disutility from smog we may not be able to determine the optimal concentration. However, whatever concentration is chosen, we can use tradeable permits as a vehicle for achieving that level in a least cost manner. Further, by varying the ways in which permits are assigned, we may be able to alter the distribution of income in desirable ways.

Let us see how this might work for the case of SO_2. First we would need to identify the region to be regulated and all the sources of emissions in this region. Further we must have a monitoring system in place that enables the regulator to verify levels of emissions. Next, we assign property rights by giving an initial allocation of permits (rights to emit a pound (or ton) of SO_2 into the atmosphere) to each emitter. Then they can be allowed to buy additional permits or sell some of their allocation on a permit market. Just as in the case of the garbage example discussed at the outset, if different emitters have different opportunity costs of emissions, there will be trades on this market (with high cost emitters purchasing from those with lower costs) and in equilibrium total emissions will be achieved at least opportunity cost. And if we observe voluntary trades taking place, we can be sure that the emissions market is Pareto superior to the simple setting of standards. Further, note that the information requirements are the same. In both cases, we must be able to monitor the levels of emissions, but nothing else.

Assuming that the emitters are all firms, the distribution of income will be affected by the rights assignment only indirectly through the ownership shares in these firms. Alternatively, we could affect the distribution directly by assigning rights to consumers (or consumer groups) and requiring firms to purchase rights to pollute from them. This would be tantamount to giving those consumers rights to clean air and requiring polluters to compensate them for degradation. Returning to our earlier discussions of efficiency versus equity, we see that there is some scope in the assignment of these kinds of property rights to affect the distribution of income in a way that does not distort incentives. Indeed, it has been suggested in the "north/south" debate that rights to global pollution should be assigned in such a way as to transfer wealth from the "have" (north) nations to the "have not" (south). Unfortunately the scope for such transfers (even if the political will is there) are probably too small to eliminate the efficiency–equity conflict.[33]

It is worth pointing out that aside from distributional considerations the outcome achieved by a permit market can also be achieved through the use of an externality tax.[34] Any equilibrium permit price could have been imposed as a tax rate and thereby achieve roughly the same total emissions level with the same degree of efficiency. However, now

---

[33] For discussion of the special problems that are present when appropriate collectives cross national boundaries, see Dasgupta and Mäler (1992).

[34] For a theoretical analysis of different pollution control instruments, including emissions trading, emissions taxes, and regulatory standards, see Chapter 6 (by Helfand, Berck and Maull).

the informational requirements for the two schemes are different. In a permit scheme, the total quantity is specified and the marginal valuation is revealed by the equilibrium permit price, whereas in a tax scheme, the price is specified and total emissions revealed through choice.

This distinction takes on extra significance in a world of uncertainty. When there is uncertainty in economic production relationships, if quantity is specified, this uncertainty will show up in random variation in the associated price, whereas if price is specified there will be random variation in the corresponding quantity. Thus, in this situation, there may be a preference between these two methods depending on which uncertainty is more costly.[35]

8. Conclusions

We have discussed in this chapter the various ways in which property rights can be assigned to environmental resources and indicated which type of rights are appropriate depending on the characteristics of the associated resource. While indicating that there is no magic bullet that will solve all collective resource problems, we have identified ways of improving collective allocations and cited evidence that motivated groups sometimes do a better job of management when collective rights are properly identified than might have been predicted by theory. Thus, while recognizing that environmental problems are acute, we believe that careful management of collective property rights has considerable potential for generating improvements.

References[36]

Aivazian, V., and J. Callen (1981), "The Coase theorem and the empty core", Journal of Law and Economics 24:175–181.

Alchian, A., and H. Demsetz (1973), "The property rights paradigm", Journal of Economic History 33(1):16–27.

Ashby, E., and M. Anderson (1981), The Politics of Clean Air (Oxford Univ. Press, Oxford).

Atkinson, A., and J. Stiglitz (1980), Lectures on Public Economics (McGraw-Hill, New York).

Auerbach, A., and M. Feldstein (eds.) (1987), Handbook of Public Economics, Vols. 1, 2 (North-Holland, Amsterdam).

Axlerod, R. (1984), The Evolution of Cooperation (Basic Books, New York).

Barzel, Y. (1989), Economic Analysis of Property Rights (Cambridge Univ. Press, New York).

Baumol, W., and W. Oates (1988), The Theory of Environmental Policy (Cambridge Univ. Press, New York).

Benoit, M., and V. Krishna (1985), "Finitely repeated games", Econometrica 53:890–904.

Benoit, M., and V. Krishna (1987), "Nash equilibria of finitely repeated games", International Journal of Game Theory 16:163–185.

[35] See Weitzman (1974) and Dasgupta, Hammond and Maskin (1980).

[36] Rather than attempt to cite all relevant secondary sources, we have chosen to list a few seminal pieces and the major tertiary (textbook, handbook and monograph) sources.

Boadway, R., and D. Wildasin (1984), Public Sector Economics (Little Brown, Boston, MA).

Bromley, D. (1991), Environment and Economy: Property Rights and Public Policy (Blackwell, Oxford).

Coase, R. (1960), "The problem of social cost", Journal of Law and Economics 3:1–44.

Cooter, R. (1987), "Coase theorem", in: Eatwell et al., eds., The New Palfrey, a Dictionary of Economics, (Macmillan & Co, London).

Cornes, R., and T. Sandler (1996), The Theory of Externalities, Public Goods and Club Goods, 2nd edn. (Cambridge Univ. Press, Cambridge).

Dasgupta, P., P. Hammond and E. Maskin (1980), "On imperfect information and optimal pollution control", Review of Economic Studies 47:857–860.

Dasgupta, P., and K.-G. Mäler (1992), The Economics of Transnational Commons (Clarendon Press, Oxford).

Debreu, G. (1959), Theory of Value (Wiley, New York).

Demsetz, H. (1988), The Organization of Economic Activity (Blackwell, Oxford).

Eggertsson, T. (1990), Economic Behavior and Institutions (Cambridge Univ. Press, New York).

Farrell, J. (1987), "Information and the Coase theorem", Journal of Economic Perspectives 113–129.

Friedman J. (1971), "A noncooperative equilibrium for supergames", Review of Economic Studies 38:1–12.

Fudenberg, D., and J. Tirole (1991), Game Theory (MIT Press, Cambridge, MA).

Hanna, S., C. Folke and K.-G. Mäler (eds.) (1996), Rights to Nature (Island Press, Washington, DC).

Hanley, N., J. Shogren and B. White (1997), Environmental Economics in Theory and Practice (Oxford University Press, Oxford).

Hoffman, E., and M. Spitzer (1982), "The Coase theorem: some empirical tests", Journal of Law and Economics 25:73–98.

Hurwicz, L. (1995), "What is the Coase theorem?", in: Japan and the World Economy 47–74.

Inman, R. (1987), "Markets, Government, and the 'new' political economy", in: Handbook for Public Economics, Vol. 2 (Chapter 12) Op.Cit.

Kneese, A., and B. Bower (1968), Managing Water Quality: Economics, Technology, Institutions (Johns Hopkins Press, Baltimore, MD).

Kolstad, C. (2000), Environmental Economics (Oxford Univ. Press, Oxford).

Laffont, J. (1989), Fundamentals of Public Economics (MIT Press, Cambridge, MA).

Laffont, J. (1987) "Incentives and the allocation of public goods", in: Handbook of Public Economics, Vol. 1, Op.Cit.

Martin, F. (1989/1992), Common-Pool Resources and Collective Action: A Bibliography (Indiana University Press, Bloomington, IN).

Moulin, H. (1986), Game Theory for the Social Sciences (New York Univ. Press, New York).

Musgrave R. (1959), Theory of Public Finance (McGraw-Hill, New York).

North, D. (1990), Institutions, Institutional Change and Economic Performance (Cambridge Univ. Press, New York).

Nozick, R. (1974), Anarchy, State and Utopia (Basic Books, New York).

Oakland, W. (1987), "Theory of public goods", in: Handbook of Public Economics, Vol. 2, Op.Cit.

Ostrom, E. (1990), Governing the Commons: The Evolution of Institutions for Collective Action (Cambridge Univ. Press, New York).

Ostrom, E. (1999), "Coping with tragedies of the commons", Annual Review of Political Science 2:493–535.

Riker, W., and P. Ordershook (1973), An Introduction to Positive Political Theory (Prentice-Hall, Englewood Cliffs, NJ).

Schlager, E., and D. Ostrom (1992), "Property-rights regimes and natural resources: a conceptual analysis", Land Economics 68(3):249–262.

Schotter, A. (1981), The Economic Theory of Social Institutions (Cambridge Univ. Press, New York).

Schulze, W., and R. d'Arge (1974), "The Coase proposition, information constraints and long run equilibrium", American Economic Review 64:763–772.

Shapley, L., and M. Shubik (1969), "On the core of an economic system with externalities", American Economic Review 59:678–684.

Spence, M. (1975), Market Signaling (Harvard Univ. Press, Cambridge, MA).

Starrett, D. (1973), "A note on externalities and the core", Econometrica 11.175–103.

Starrett, D. (1988), Foundations of Public Economics (Cambridge Univ. Press, New York).

Varian, H. (1978), Microeconomic Theory (Norton, New York).

Weitzman, M. (1974), "Prices versus quantities", The Review of Economic Studies 41:477–491.

Williamson, O. (1975), Markets and Hierarchies (Free Press, New York).

Chapter 4

ECONOMICS OF COMMON PROPERTY MANAGEMENT REGIMES

JEAN-MARIE BALAND and JEAN-PHILIPPE PLATTEAU

Centre de Recherche en Economie du Développement (CRED), Department of Economics, Faculty of Economics, Business, and Social Sciences, University of Namur, Belgium

Contents

Handbook of Environmental Economics, Volume 1, Edited by K.-G. Mäler and J.R. Vincent

Abstract

The purpose of this chapter is to identify the reasons for collective action failures and successes in natural resource management, and to understand, in the light of economic theory, the mode of operation of the factors involved whenever possible. In the first section, we clarify the notion of a common property management regime and provide cautionary remarks about estimation methodologies commonly used. In Section 2, we focus on the general case where common property regulation is feasible yet only if governance costs are kept to a reasonable level. Emphasis is placed on such factors as the size of the user group, income or wealth inequality, and availability of exit opportunities. Special attention is paid to the aspect of inequality since this has remained a rather confused issue in much of the empirical literature. Economic theory can contribute significantly to improving our understanding of the manner in which it bears upon collective action. In Section 3, we discuss cognitive problems as an important impediment to the design and implementation of efficient common property management systems. We also present evidence of the deleterious effects resulting from the absence or inappropriateness of state interventions, particularly where they are motivated by private interests. In Section 4, the importance, under a co-management approach, of appropriate incentive systems at both the village and state levels is underlined and illustrated.

Keywords

commons, regulation, inequality, co-management

JEL classification: H41, O13, Q20, Q30

1. Introduction

1.1. Motivation and outline of the chapter

During the last decades we have witnessed an impressive upsurge of empirical literature dealing with common-property resources (CPR). Responsible for most of these writings are social scientists of different brands, particularly sociologists, anthropologists, geographers, human ecologists, and political scientists. So far, the contribution of economists to the accumulation of empirical knowledge about such resources has been rather modest unlike their theoretical efforts, which have conspicuously multiplied during the same period. The available empirical literature is quite disparate in the sense that it relies on different sorts of evidence and methods of investigation. There are in-depth case studies of one or two resource or village-level management systems – as illustrated by the works of Ensminger (1990) on Orma pastoralists (Kenya), and of Alexander (1982) on Sri Lankan beach seine fisheries; comparative, essentially qualitative assessments of various field situations located in a rather restricted area – as exemplified by the works of Wade (1988a, 1988b) on irrigation and grazing resources in a southern Indian state, and of Peters (1994) on grazing resources in Botswana; cross-sectional quantitative analyses of resource management systems belonging to different environments – see, e.g., the works of Tang (1991, 1992), Lam (1998), Bardhan (2000) and Fujita, Hayami and Kikuchi (1999) on irrigation systems, or that of Gaspart and Platteau (2001) on fishermen's regulatory schemes in Senegal; and broad-sweeping overviews of observations made in different regions and countries, possibly involving different types of resources, such as the works of Singh (1994) and Sengupta (1991) on India and the Philippines.

A striking feature of most of these studies lies in the fact that their authors are generally convinced that, given the glaring failure of state ownership experiences in developing countries, collective, community-based regulation holds out the best prospects for efficient management of village-level natural resources. Yet, since they recognize at the same time that the balance sheet of actual experiences of common property management is mixed, the central aim of their inquiries is typically to understand the reasons that can account for these varying levels of performance of user-managed resource systems. Moreover, such a mixed record prompts many of them to believe that a realistic solution to the problem of village resource conservation will necessarily entail a certain level of co-management between direct users, on the one hand, and state authorities or specialized agencies, on the other hand.

The purpose of this chapter is to identify the reasons for collective action failures and successes in natural resource management as they emerge, whether explicitly or implicitly, from the burgeoning literature of the last decades. It is also to try, in the light of economic theory, to have the best analytical understanding of the mode of operation of the factors involved whenever this is possible. Our analysis is presented in four successive sections. In Section 1, of which these introductory remarks form a part, we clarify the notion of a common property management regime, provide cautionary remarks about estimation methodologies commonly used (a point that is elaborated further in the Appendix) and analyze two set of polar circumstances, one under which common property regulation is very unlikely to succeed, and another where it is the only solution available. In Section 2, we focus on the general case where common property regulation is feasible yet only if governance costs are kept to a reasonable level. Here, emphasis is shifted to factors on which recent theoretical endeavors can shed a new light, namely the size of the user group, income or wealth inequality, and availability of exit opportunities. The role of these factors is assessed within the framework of three theoretical models corresponding to different characteristics of the common property situation contemplated. Special attention is paid to the aspect of inequality since this has remained a rather confused issue in much of the empirical literature and economic theory can contribute significantly to improving our understanding of the manner it bears upon collective action. In Section 3, we discuss two important impediments to the design and implementation of efficient common property management systems. Cognitive problems and state actions are analyzed successively under this heading. In Section 4, before we summarize our main points, the importance of appropriate incentive systems under a co-management approach to village-level resources is underlined and illustrated.

1.2. Common property management regime

Before embarking upon this agenda, it is necessary to be precise about what we mean by a common property management regime. A common property management regime implies that various restrictions are imposed on members of a well-defined group of people regarding the manner in which they may use local-level resources. In other words, common property management implies collective regulations regarding both membership and the way to use the resource, and the existence of monitoring and sanctioning procedures so that those rules can be effectively enforced. Below, we offer three detailed illustrations of what collective regulation can mean in the specific context of three different natural resources: regulation of village forestry and pastures in Tokugawa rural Japan, regulation of villages pastures in Rajasthan (India), and regulation of access to the sea in coastal communities of Southern Sri Lanka.

1.2.1. Management of village commons in Japan (1600–1867)

McKean (1986) collected materials on three Japanese villages – Hirano, Nagaike, and Yamanaka – with a view to assessing the way local commons were regulated on a micro-basis during the Tokugawa period (1600–1867). At that time in Japan, about half of the surface of forests and uncultivated mountain meadows were held and managed in common by rural villages, the other half being under imperial or private property.

Japanese villages actually succeeded in developing "management techniques to protect their common lands for centuries without experiencing the tragedy of the commons" [McKean (1986, p. 534)]. They used extremely detailed rules of access and conservation procedures. For example, in order to prevent the *kaya* – a grass grown to produce thatch for roofs – from being cut at an immature stage for horse fodder, villagers usually designated an area with *kaya* as 'closed' during the growing season. On the other hand, to ensure that daily cutting of fresh fodder for draught animals and pack-horses did not deplete the supply available for winter, villagers in Hirano designated one open area for daily cutting of fresh grass and another closed area as a source of grass to be dried into fodder for the winter [McKean (1986, pp. 553, 554)].

Village forests were essentially divided into two zones: open patches of forest and closed reserves. Villagers could enter the first zone at any time "as long as they obeyed rules about taking fallen wood first, cutting only certain kinds of trees and then only those that were smaller than a certain diameter, and only with cutting tools of limited strength (to guarantee that no tree of really substantial size could be cut)"; or "about leaving so much height on a cut plant so that it could regenerate, or taking only a certain portion of a cluster of similar plants to make sure the parent plant could propagate itself, or collecting a certain species only after flowering and fruiting, and so on". Also, to limit the quantity of plants collected, village authorities could prescribe the size of the sack or container used for that purpose. To control access to the first zone in a tighter way, the same authorities could also issue entry permits "carved on a little wooden ticket and marked "entrance permit for one person" " [McKean (1986, pp. 554, 555)].

As for the closed reserves, they were set aside "for items that had to be left undisturbed until maturity and harvested all at once at just the right time, or that the commons supplied in only adequate, not abundant, amounts". The time for collection and the rules to be followed by each collector were decided by the village headman. For example, if the supply of a given natural product was limited, "the reserve might be declared open for a brief period (two or three days) and households allowed to send in only one able-bodied adult to collect only what could be cut in that time". Precise rules for harvesting varied from product to product and from village to village, yet, as a matter of principle, they "appeared to be a judicious combination that rewarded strength and hard work but also severely limited the circumstances in which cutting was allowed, which ensured that the total supply was not threatened and no extreme inequality appeared among households in a given year or among *kumi* (groups of households) over time" [McKean (1986, pp. 555, 556)]. The latter requirement could drive the local authorities to devise fixed rotational sequences so that each household or group of households had access to patches of varying quality.

It is important to note that there were written rules about the obligation of each household to contribute a share to collective work intended for maintaining the commons, such as systematic programmes for harvesting and weeding certain plants in a particular sequence to increase the natural production of the plants they wanted; or the burning of the common meadow lands which was conducted each year to burn off hard and woody grasses and thorny plants (and kill pests), and which involved "cutting nine-foot firebreaks ahead of time, carefully monitoring the blaze, and occasional fire-fighting when the flames jumped the firebreak" [McKean (1986, pp. 558, 559)].

Apparently, Japanese villages widely resorted to selective inducements and punishments in order to ensure due respect of the written codes which most of them had to govern their CPRs. Regarding inducements, we are thus told that "there was an intrinsic pride in the importance of doing one's duty by the commons and in preserving the vil lage's well-being; a young man brought credit to his family and future by doing the job properly . . ." [McKean (1986, p. 564)]. Regarding punishments, the evidence is that "violating rules that protected the commons was viewed as one of the most terrible offenses a villager could commit against his peers, and the penalties were very serious" [McKean (1986, p. 561)]. In order to detect rule infractions, purposeful monitoring was practiced in the form of groups of detectives destined to constantly patrolling the commons: "the detectives would patrol the commons on horseback every day looking for intruders, in effect enforcing exclusionary rules". Their job was considered "one of the most prestigious and responsible available to a young man" [McKean (1986, pp. 560, 561)]. According to villages, these positions changed hands more or less frequently and, in some of them, all eligible males had to take a turn, so that no family was without its full labor supply for very long. In the poorest villages, specialized but rotating detectives did not exist (probably because people were too poor to spare the required labor), yet anyone could report violations [McKean (1986, p. 561)].

When necessary, Japanese villages did not hesitate to threaten to use their more powerful sanctions: "ostracism in increasingly severe stages, followed by banishment". Ostracism – which implies that the village community "cuts off all contact with the offender except for assistance at funerals and fire-fighting" – was thus resorted to in gradual stages, "starting with social contact and only escalating to economic relations if the offender did not express remorse and modify his behavior". Moreover, "to ensure that the villagers would remember to shun contact with someone subjected to ostracism, that person might be expected to wear distinctive clothing (a flashy red belt or pair of unmatched socks)" [McKean (1986, p. 562)].

1.2.2. Management of village pastures in Rajasthan (India)

In the arid areas of Western Rajasthan, before the modern state was formed by conglomerating the princely states of Rajputana, communal grazing lands used to be under the effective control of big landlords known as *jagirdars* who took upon themselves the task of deciding and implementing "conservation measures which ensured considerable stability to these resources" [Shanmugaratnam (1996, p. 172)]. Thus they charged grazing taxes, organized rotational grazing around evenly scattered water-points, decreed the periodical closure of parts of the commons and periodical restrictions on entry of certain animal species, appointed watchmen to monitor compliance with the grazing regulations, imposed penalties on herd owners found guilty of violating them, and used their authority to extract regular labor contributions for maintenance works from poorer users. Such measures had the effect of conserving perennial grass species and trees and of allowing effective rotational grazing thanks to proper maintenance of water points [Jodha (1987, 1989)].

1.2.3. Regulating access to the sea in a coastal fishing community in Sri Lanka

The sea tenure system discovered by Alexander (1980, 1982) and by Amarasinghe (1989) in the beach-seine fishing communities of southern Sri Lanka involves the rotational assignment of fishing spots in such a manner that all fishermen get equal access not only to fish but also to fish in the best spots.

Beach-seining is a fishing technique that requires a rather large water space (since it is intended for catching a whole school of fish) located close to the shore and with a smooth sea bottom. Furthermore, the laying of a beach-seine and its subsequent hauling takes only a few hours so that, on a particular day, only a maximum of 4 nets can be operated on each suitable location. Moreover, incomes from beach-seining are significantly affected by the timing of fishing operations, within a daytime as well as across seasons.

From Alexander's account [Alexander (1980, pp. 97–102), (1982, Chapter 7)], the collective management scheme is defined by the following rules:

(1) Membership in the village community, whether hereditary or acquired in a lifetime, involves a right of access to the community-controlled sea area (local fishermen belong to the same kinship group).

(2) Ownership of a net carries the obligation to work it when required.

(3) There is no labor market and, since a beach-seine normally requires eight fishermen to operate, joint ownership of nets is the rule and the usual ownership share in a net is $1/8$. In fact, each net is divided into eight sections or shares but "once the net is in operation individuals have no particular rights to the sections they have contributed" [Alexander (1982, p. 142)].

(4) Net shares, and the accompanying access right to the sea, can be transferred only through inheritance [Alexander (1982, p. 203)].

Equal access is guaranteed through a turnover system that determines turns in a sequence of net-hauling rights. Thus, in the village studied by Alexander, the fishing area is divided into two stations: the harbor side (from which most big catches come) and the rock side. The net cycle begins on the harbor side and, after a net has had the dawn turn on that side, it is entitled to the dawn turn on the rock side on the next day. Subsequently, it may be used on the rock side each day once the net immediately following it in the sequence has been used. The sequence of net use over a period of five days, where each of the eight existing nets is denoted by its number, is shown in Table 1 [adapted from Alexander (1982, p. 145)]. Over the full net cycle, each net is thus operated eight times.

1.3. Methodological considerations

At this stage, we must make an important methodological caveat. Indeed, identifying the factors underlying variations in performance levels for common property management is a much more arduous task than may appear at first sight. A major difficulty lies in the fact that the estimation procedure usually applied in cross-sectional studies

Table 1
Fishing station

Day	Harbor				Rock			
	Dawn		Night		Dawn		Night	
One	5	6	7	8	4	3	2	1
Two	6	7	8	9	5	4	3	2
Three	7	8	9	10	6	5	4	3
Four	8	9	10	11	7	6	5	4
Five	9	10	11	12	8	7	6	5

based on large samples of user groups aimed at such identification is fraught with se-rious endogeneity effects that compound the classical problems raised by this type of methodology.[1] In actual fact, many studies overlook the endogeneity relationships be-tween the organizational form, the expected gains, and the user characteristics. Thus, when concluding that small and homogeneous groups are more conducive to collective action than groups with opposite characteristics, authors tend to forget that group size and composition may themselves be the product of a decision made by users in the light of their specific environmental conditions (see the Appendix for a detailed exposition of this point).

To highlight the reasons behind successes and failures of collective action in village-level resource management, in-depth case studies of particular user communities and resource systems are therefore extremely useful as a complement to cross-section sta-tistical analyses. They allow researchers to have a solid understanding of the details of the common property management mechanism and to use individual users or house-holds as observation units. Moreover, such data can be subject to quantitative analysis in order to identify the determinants of individual participation to the collective mech-anism [see White and Runge (1995), Gaspart et al. (1998)]. From there, it should be possible to shed light on important questions, such as whether participation is more in-tense and more widespread among rich than among poor users, among village leaders than among common people, etc. In addition, by reconstructing chains of past events, it is possible to uncover dynamic processes at work in the user group and the village society as a whole, in particular the processes of group formation and rule setting. Dis-tinguishing user characteristics that are exogenous from those that are endogenous and understanding how the processes of endogeneization take place then become a feasible task. Finally, careful recording of the characteristics of the user community, the resource system, the harvesting technology, and the role of state agencies as well as a detailed assessment of the initial conditions and the outcomes of endogenous processes, should allow one to draw useful lessons about community- and village-level characteristics

---

[1] Panel data are less subject to the aforementioned problem. Unfortunately, they are not yet available for the type of enquiry considered here.

conducive to effective common property management. This, however, necessitates that a comparative analysis can be carried out on the basis of the available case studies.

It must be emphasized that, to be really insightful, especially in a comparative analytical perspective, it is important that in-depth case studies are guided by well-defined research questions and grounded in solid theory whenever it exists. Unfortunately, this requirement is too rarely met in reality, perhaps because many village studies are still inspired by a holistic approach that lays stress on the opposite need to avoid precise questions and hypotheses considered as so many blinkers prone to bias observations and results.

1.4. Two polar cases: Collective regulation is doomed or indispensable

1.4.1. The rationale for resource division

As we know from standard economic theory, the problem with resources under open access or under community ownership limited to membership rules – what Baland and Platteau (1996) call 'unregulated common property' – is the fact that externalities are not properly internalized. If there were no transaction costs, 'regulated common property' (i.e., collective ownership with both membership rules and rules of use) would be equivalent to private (and state) ownership. Yet, the presence of pervasive transaction costs in real world situations tilts the balance in favor of private ownership: as a matter of fact, being under the control of a single person, private ownership avoids all kinds of negotiation costs necessary to reach a collective agreement as well as all the governance costs that have to be incurred with a view to monitoring and enforcing such agreements. There is no escape from the fact that regulation often remains imperfect as it is difficult to eliminate all the inefficiencies arising from a collective mode of exploitation. The remaining inefficiencies must therefore be considered as genuine costs of maintaining the commons. In other words, regulation is necessarily imperfect, and a fully efficient outcome cannot be expected to result from the joint exploitation of a natural resource.

On the other side of the balance sheet, there are several advantages of (regulated) common property that may possibly compensate for the above shortcoming of imperfect internalization of externalities. The first advantage lies in scale economies. These exist almost always on the side of costs, if the alternative is to divide the resource into several private portions that have to be enclosed and protected. Indeed, the costs of negotiating, defining and enforcing private property rights are increasing with the physical base of the resource: the more spread the resource base (or the less concentrated the resource) the higher the costs of delimiting and defending the resource 'territory'.[2] Scale economies may also exist on the side of benefits, being present either in the resource

---

[2] Note that the indivisibility of natural resources needs not always arise from their spatial spreading. Thus, for example, a water source is extremely difficult to apportion to several users even though it is a highly concentrated resource.

itself or in complementary factors. Resources offering multiple products because they form part of an overall ecosystem are a good example of the former kind of situation (think of forests or mangrove areas). On the other hand, the obvious advantage of co-ordinating the herding of animals so as to economize on shepherd labor in extensive grazing activities is probably the best illustration of the way scale economies in a complementary factor may prevent the division of a resource domain [see, e.g., Dahlman (1980), Netting (1981), Binswanger, McIntire and Udry (1989), Bromley (1989, p. 16), Nugent and Sanchez (1993)].

Risk-pooling considerations constitute another important and well-known advantage of common property. When a resource has a low predictability (that is, when the variance in its value per unit of time per unit area is high), the need to insure against the variability of its returns across time and space militate against resource division [McCloskey (1976), Dahlman (1980), Runge (1986), Dasgupta (1993, pp. 288, 289), Nugent and Sanchez (1993, p. 107), Singleton (1999)].[3] Indeed, users are generally reluctant to divide it into smaller portions because they would thereby lose the risk-pooling benefits provided by the resource kept whole. Such loss is especially noticeable in the case of extensive herding and fishing (whether in inland or marine fisheries). Indeed, herders (fishermen) may need to have access to a wide portfolio of pasture lands (fishing spots) insofar as, at any given time, wide spatial variations in yields result from climatic or other environmental factors. Note carefully that, to be valid, the risk-pooling argument presupposes the existence of positive enforcement costs of private property, otherwise scattered privately-owned parcels could well provide the required insurance to resource users [see, however, Baland and Francois (2001)].

1.4.2. Where resource division is inevitable

There are thus arguments *pro* and *contra* the division of a resource and the issue cannot be settled on a priori grounds but only in a contextual manner [Baland and Platteau (1998b), Platteau (2000, Chapter 3)]. This said, polar situations arise where the choice is evident. One such situation occurs when the profitability of resource division and privatization is so high that this option is incontrovertible. This is likely to happen when there are no scale economies (on the side of benefits) nor risk-sharing advantages in keeping the resource domain whole. However, low are the inefficiency losses resulting from common property, the corresponding costs are still too high compared with the savings that can be realized on enforcement costs.

The most relevant case here is that which arises when the value of a resource is high owing to population growth and/or market integration and delimitation or enforcement

---

[3] High yield variance is also one of the five conditions favorable to common property listed by Netting (1976). The four other conditions are a low value of resource productivity (see supra), few possibilities of intensification, a large area needed for effective use (see supra), and a large labor- and capital-investing group [Netting (1976, p. 144)].

costs of private property are moderate. In such circumstances, indeed, even a rather effective system of collective regulation can easily result in comparatively large amounts of rents foregone by not dividing the resource. Hence the frequent emphasis in the literature on the unit value of natural resources as one of the main determinants of its privatization (division) [Dyson-Hudson and Smith (1978), Cashdan (1983), Levine (1984), Wade (1988a, p. 215), Libecap (1989, p. 21), Dasgupta (1993, pp. 288, 289), Noronha (1997, p. 49), Baland and Platteau (1999), Singleton (1999), Platteau (2000, Chapter 3)]. For example, in his classical study of the Swiss Alps, Netting contrasts the lowlands of the valley which are fertile and therefore tend to be privately appropriated with the more arid highlands which are used as communal (summer) pastures under the authority of the village council [Netting (1972, 1976, 1981, 1982)]. Generally, when a resource requires substantial investments and regular maintenance to be conserved and improved while the option of privatization is available (enforcement costs of private property are not too high), that option generally proves to be irresistible and attempts to keep the resource under common property are bound to fail despite the best intentions and efforts of users or community organizers. The natural transformation of the ownership system over agricultural lands in areas subject to rapid population growth and agricultural intensification offers a vivid illustration of this principle [see Boserup (1965, pp. 79–81)].

1.4.3. Where resource division is prohibitively costly: The evolutionary view

Another polar case arises when, unlike in the above kind of situations, the option of private ownership is not available because enforcement of private property rights is prohibitively costly or indivisibilities are pervasive. One can think of resources that are highly mobile over large expanses of territory (open-sea fishing and hunting, in particular), or of those, such as irrigation water, which require a collective infrastructure to be exploited. In these peculiar circumstances, collective regulation under the common property regime is the only way to avoid the inefficient management and/or the degradation of the resource under conditions of open access.

The question is then how can one be assured that the required mode of resource governance will be established and maintained? It is at this juncture that the evolutionary doctrine must be brought into the picture since it stresses the considerable ability of rural communities to adapt to changing circumstances. In conformity with the induced institutional innovation hypothesis [Kikuchi and Hayami (1980), Hayami and Ruttan (1985), Binswanger and McIntire (1987)], its proponents argue that collective regulation of a resource may evolve under (moderate) population growth when privatization remains prohibitively costly.

Thus, Hayami and Kikuchi point out that, under these conditions, "the social structure becomes tighter and more cohesive in response to a greater need to co-ordinate and control the use of resources as they become increasingly more scarce". As scarcity increases and competition is intensified, rules are defined more clearly and enforced more rigorously, whether they serve to define rights and obligations among people on the use of the resource or to settle possible conflicts. Inasmuch as the environment determines which resources are scarce, which are difficult to privatize, and which are relatively easy to handle at the village level, environmental conditions are "a critical variable in the formation of village structure" [Hayami and Kikuchi (1981, pp. 21, 22), Hayami (1997, p. 92)].[4]

Close to Hayami and Kikuchi is the position taken by Wade on the basis of detailed village studies in South India [Wade (1988a)]. According to him, indeed, "villagers will deliberately concert their actions only to achieve intensely felt needs which could not be met by individual responses", that is, "they will straightforwardly come together to follow corporate arrangements" whenever "the net material benefits provided to all or most cultivators are high" [Wade (1988a, pp. 185–188, 211)]. Wade is actually confident that when no other, more decentralized alternative is available, villagers will somehow succeed in overcoming the incentive problems associated with collective action. Lawry reaches a similar conclusion: "collective action is more likely to result where the common resource is critical to local incomes and is scarce", and where privatization appears unfeasible or too costly [Lawry (1989b, pp. 7–9)]. Boserup herself did not think differently: "when grazing opportunities become scarce, rules are likely to be laid down concerning the number of animals a cultivator family is allowed to have and the amount of straw the cultivator is allowed to remove with the harvest" [Boserup (1965, p. 85)]. Such a view, it must be noted, is shared by those political scientists for whom leadership and rules arise within social groups in response to the need to regulate the allocation of resources under conditions of scarcity [see, e.g., Tyler (1990, pp. 66, 67)].

1.4.4. Where resource division is prohibitively costly: The unregulated common property

The evolutionary view sounds too optimistic, however. Collective management is not the only possible outcome in the presence of prohibitively high privatization costs and the value of the resource justifies its regulation. Indeed, one cannot exclude the possibility that the resource is going to be overexploited and depleted under either an open

---

[4]   As an example, Hayami and Kikuchi mention the differential evolution of rural social organizations (as measured by the tightness in community structure) in Thailand's Central Plain, on the one hand, and in Japan and the mountainous areas of the Philippines (such as in the Ilocos region), on the other hand. In their account, the role of population density and topography (mountainous terrain is more congenial to local-level infrastructure) is quite predominant: while rice farming in the major part of Thailand's Central Plain depends on the annual flooding of a major river delta, which is beyond the control of peasants either individually or in local cooperative groups, in Japan and the Ilocos region of the Philippines it initially developed in fan-shaped terraces in the valley bottoms and inter-mountain basins, a topography which makes local cooperation effective in controlling a water supply based on small streams [Hayami and Kikuchi (1981, pp. 22, 23)]. Note that when public goods such as irrigation infrastructure require a spatial basis exceeding the rather confined limits of a village community, state intervention is called for. A complete, but daring, induced institutional innovation theory would argue that the right kind of state would emerge in such circumstances to solve the public good problem that conditions people's survival. See Wittfogel (1957), Hunt (1989), and Allen (1997) for attempts in that direction. For a critique, see Weiss and Hobson (1995, pp. 85–87).

access or an unregulated common property regime. Under open access, there are no rules guiding access to the resource, nor rules governing its use. In such circumstances, the Tragedy of the Commons occurs: productive inputs are wastefully used and the scarcity rents associated with the resource are totally dissipated [Hardin (1968)].

Under unregulated common property, membership of user groups is fixed, yet there are no rules regarding the use of the resource that rights-holders are supposed to follow. The absence of rules of use may be due to various reasons: external factors, such as the low value (or the abundance) of the common property resource, or adverse state policies and actions that have the effect of emasculating community-based organizations, and internal factors, by which we mean the inability of a group of users to collectively organize with a view to manage local resources better.

Because collective regulation is costly and, as we have just seen, cannot be assumed to arise exactly when it is needed, it is important to enquire into the outcome of unregulated common property [the open access situation is too well known to warrant a new discussion here; see Chapter 3 (by David Starrett)]. Moreover, even a collective regulation regime requires that people act in a non-regulated manner (in a non-cooperative way) to bring it about. Indeed, if resource users, or a sufficient number of them, do not decide to contribute in a significant way to collective actions leading to the establishment of a resource management scheme, such a scheme will never come to light. Lastly, for people to have an incentive to participate in collective regulation, what they earn under the collective arrangement must exceed what they obtain in the absence of regulation. This condition constitutes their so-called participation constraint. In other words, the equilibrium outcome under unregulated common property provides a benchmark against which resource users evaluate the usefulness of a coordinated management scheme.

For all these reasons, attention in Section 2 will be focused on the outcomes of games representing decentralized patterns of interactions among resource users. These games are essentially static, thus mirroring the existing state of economic theory on the subject. In Section 3, the implications of the analysis in Section 2 for the design of collective regulation mechanisms will be investigated.

2. Simple models of non-cooperative behavior and some implications for cooperative behavior

Since in reality there is a rich variety of common property situations, it is impossible to account for all of them in terms of a single analytical model. Three main models are considered in detail below. The first model examines a situation where users share the benefits from joint exploitation of a common property resource in direct proportion to the relative amounts of their appropriation effort, which they freely decide. The second model depicts the classical problem of voluntary contributions to a pure public good. The third model is based on the assumption that users benefit from the common good produced (or the common bad avoided) with the help of their aggregate contribution in proportion of their predetermined share of 'interest' in this good. I.e., it is not a pure public good. The impacts of income inequality and group size will be analyzed with respect to each model. The presence of exit opportunities will be considered only in terms of the third model, which is particularly suitable for that purpose.

The core versions of the three models are discussed in three successive subsections while a number of interesting variants around the third model are examined in a separate subsection following the main presentation. Before Section 4 is concluded, we leave the framework of completely decentralized decisions to address an oft-neglected issue, that of the impact of inequality on collective regulation. As in the previous sections, references to the relevant empirical literature are provided whenever appropriate.

2.1. The first model: The appropriation problem

2.1.1. The model

Consider a situation in which agents jointly exploit a common property resource by individually choosing their individual level of harvesting. Villagers thus decide the number of hours they spend in the forest gathering fuelwood, fishermen decide the number of boats they operate in a common fishery, or herders decide on the number of animals to let graze on the common pasture, etc. In all these situations, the level of harvesting effort decided by an individual agent has an impact not only on the collective level of exploitation of the resource, but also on his share of the collective harvest, which is usually directly proportional to his effort level. In contrast with the analysis proposed in the next sections where individual shares in the collective output are pre-determined, we consider here the case where individual shares are endogenous.

Assume that n agents jointly exploit a common property resource and share their benefits in direct proportion to the relative amount of appropriation efforts they have chosen to put in. Let g_i stand for the appropriation effort of agent i. Total output can then be written as:

$$G = G\left(\sum_j g_j\right),$$

where $G'' < 0$ and G' can be positive or negative. For the sake of notational simplicity, we assume that costs are nil, so that the profit accruing to agent i is simply:

$$\Pi_i = \frac{g_i}{\sum_{j=1}^n g_j} G\left(\sum_j g_j\right).$$

In a Nash equilibrium, each agent maximizes his profit by choosing his own level of effort, taking the level of effort provided by the others as given. Raising the level of effort has two separate effects on profits: it may increase (or reduce) the aggregate output to be shared by all users, and it increases the individual's share in aggregate output. This is expressed in the first-order condition:

$$\frac{\partial \Pi_i}{\partial g_i} = \frac{\sum_{k \neq i} g_k}{(\sum_{j=1}^n g_j)^2} G\left(\sum_j g_j\right) + \frac{g_i}{\sum_j g_j} G'\left(\sum_j g_j\right) = 0.$$

In the above equation, the first term is always positive. In equilibrium, the second term must therefore be negative, which implies that at the Nash equilibrium agents set the total amount of effort in such a way that its marginal productivity is negative (or, if costs were positive, below marginal cost). Inefficiency arises because, at the efficient point where marginal productivity is nil, agents have an incentive to increase their effort level since it increases their share in aggregate output.[5]

2.1.2. Impact of inequality

As the objective functions of the individual users are profit (and not utility) functions, wealth or income inequality has no direct impact on individual or aggregate level of effort. When utility functions instead of profit functions are used to describe the objectives of resource users, as will be done in the next two sections, it is no longer immaterial how wealth or income is distributed among these users as effort is then subject to income effects. This being said, the simple analytical framework presented above becomes useful to determine the impact of inequality when indirect effects operating through constraints on individual efforts are taken into consideration. More precisely, we contemplate the possibility that the distribution of wealth translates into a distribution of the constraints on the amount of effort that an agent can choose [see Baland and Platteau (1999) for numerical examples]. For example, to decide the number of boats to buy, a fisherman must have enough initial wealth either to directly finance the purchases of these boats or to secure access to the credit market. Absent this constraint, all individuals contribute the same amount of effort regardless of their wealth. On the contrary, if the constraint binds, poorer fishermen are constrained while wealthier users freely decide their level of effort. Consider a disequalizing transfer from (i) an agent who was previously unconstrained and is now constrained, or (ii) from an agent who was previously constrained to an unconstrained (wealthier) agent. Such a transfer has the effect of reducing the aggregate level of effort, thereby making the use of the common resource more efficient.

To show this result, let us consider the first-order condition above. Suppose that the transfers benefitted an unconstrained agent.[6] (We leave to the reader the discussion of

---

[5] Note that, if the users on the commons are simultaneously Cournot competitors on the output market, there exists an optimal number of users such that the commons is efficiently used [see Cornes, Mason and Sandler (1986)].

[6] If more than one agent benefits from the transfer, one can always decompose the transfer as the sum of successive transfers to a single agent. Similarly, if the transfer originates from more than one agent, one can always decompose such a transfer as the sum of transfers from each agent.

the case where such transfer benefits a constrained agent.) The constrained user, who lost from the transfer, reduces his own level of effort by an amount equal to the change in his constraint. If after the transfer, the benefitting user, who is unconstrained, increases his effort level so that the aggregate level remains unchanged, the derivative of the profit function given above would be negative: the first term on the right-hand side would be smaller, and, bearing in mind that G' is negative at the Nash equilibrium, the second term would be more negative.[7] Hence, the post-transfer level of effort chosen by this agent will never be such that the aggregate level (increases or) remains constant. As can be checked from the first-order condition above, he does increase his level of effort (i.e., efforts by the agents are strategic substitutes in this model), but this increase is smaller than the reduction in the effort levels of the losing agent. It should be noted that, in some circumstances, a disequalizing change in the distribution of wealth may have such an impact on the aggregate level of effort that the welfare of all users is increased. Further discussion and details can be found in Baland and Platteau (1997b).

In a recent contribution, Bardhan, Ghatak and Karaianov (2000) propose an alternative approach for the analysis of wealth inequality in an appropriation problem. In their model, the production function includes two complementary inputs, the amount of appropriable resources from the commons (as in the traditional approach to appropriation problems described above) and a private input. To analyze the effects of wealth differences, they assume that the amount of private input owned is equal to own wealth and that the market for the private input is largely imperfect. They then show that greater inequality, in general, tended to reduce the efficiency of the decentralized outcome.

Another alternative approach would simply consider that the cost of the input for use of the common property differs across users, with lower costs for presumably wealthier users. This is the approach followed by Aggarwal and Narayan (1999). These two authors propose an interesting model of groundwater appropriation in a dynamic two stage game, where, in the first stage, players determine their extraction capacity (which represents the maximal depth of extraction), while in the second stage, they decide their extraction path over an infinite horizon (that is the rate of utilization of this capacity). Assuming that the cost of capacity varies across agents (for instance, as a function of initial wealth under imperfect capital markets), they show that inefficiency in extraction is a U-shaped function of a mean-preserving spread in the costs of capacity.[8] While that

---

[7] Since g_i increases while $\sum_{j=1}^{n} g_j$ is assumed to be constant, $\sum_{k \neq i} g_k$ is smaller. As a result, $\sum_{k \neq i} g_k / (\sum_{j=1}^{n} g_j)^2$ is also smaller. On the other hand, $g_i / \sum_j g_j$ increases, causing the second term to be more negative. The first-order condition is therefore violated.

[8] As such capacity can be more generally interpreted as a fixed cost, poorer agents may even decide not to invest at all. The analysis is partly motivated by evidence on groundwater extraction in India, where poorer farmers did not generally invest in wells, or invested in wells of low depth. They were quickly driven out by the decline in the water tables [see Bhatia (1992) and Aggarwal (2000)]. Similarly in fishing, poor artisanal fishermen may prove unable to purchase the type of high-powered vessels used by richer investors. As a consequence of this technological race, their access to the resource is reduced to the point of crowding them out of the sector.

paper can be commended for its genuine effort to explicitly bring dynamic considerations into the analysis of common property resource, the approach to inequality directly equates a mean-preserving spread in wealth (which makes sense as a comparative statics experiment) to mean-preserving spreads in input costs, which are less closely related to inequality. It is not clear that these two concepts are strictly monotonic to one another. To see this, consider the case where, below a certain wealth level, agents do not have access to formal capital markets and must turn to the moneylender for the purchase of the input. Clearly, then, any reduction in their wealth cannot change their input cost. In this situation, a mean-preserving spread in wealth reduces the input cost of better endowed agents (as they have more collateral to offer on the formal credit market) while leaving this cost unchanged for poorer agents. Such a disequalizing change increases aggregate harvesting. These effects are totally different from those of a mean-preserving spread in input costs, which would imply that the input cost rises for the lesser endowed agents. Such effects are also at variance with those obtained under a different scenario, where rich users have an unlimited access to the formal capital market. In these conditions, an increase in the wealth of the rich does not change their input price, while poorer agents are driven out of the formal market by the reduction in their wealth. Clearly, caution is required when interpreting results based on mean-preserving spreads in costs rather than in wealth.

2.1.3. *Impact of group size*

In the above model, the impact of an increase in the number of users can easily be established. Indeed, given a Nash equilibrium among n agents, as described by the above first-order condition, a new agent entering into the resource will always decide to exert a positive amount of effort, since the average return to effort is positive. The other agents will react to this increase in the use of the commons by reducing their own level of effort, but never to such an extent that the total level of effort diminishes. This can be shown by totally differentiating the first-order condition for agent i with respect to an increase in effort by others. As a result, an increase in the number of users unambiguously raises the Nash equilibrium level of exploitation of the resource, thereby increasing the inefficiencies of its joint use [see also the discussion in Dasgupta and Heal (1979, pp. 55–73)].

2.2. *The second model: Voluntary contributions to a pure public good*

2.2.1. *The model*

The management of common property resources often involves the production of public goods, such as water control infrastructure, drains in watersheds, or a fishing pier. Moreover, as will be argued later, participation in tasks of collective organization such as maintenance of resource-related infrastructures, monitoring abidance of management rules, protection of the resource against outsiders, etc. can also be analyzed as a problem of voluntary contributions to a public good.

Consider a group of n consumers, with utility functions:

$$U_i = U_i(x_i, G),$$

where x_i represents the amount of private good consumed by consumer i, and G is the amount of public good in the economy. The public good, G, is equal to the sum of the gifts, g_i, made by each consumer i, so that:

$$G = \sum_i g_i.$$

Each consumer i is endowed with a level of wealth w_i, which he allocates between his consumption of the private good, x_i, and his donation, g_i. Letting G_{-i} represents the sum of all gifts made by all the consumers other than i, we can define a Nash equilibrium as a vector of gifts (g_i^*) such that, for each i, (x_i^*, g_i^*) is the solution to the following maximization problem:

$$\max_{x_i, g_i} U_i\left(x_i, g_i + G_{-i}^*\right) \quad \text{s.t.} \quad x_i + g_i = w_i, \quad g_i \geq 0,$$

in which the last constraint requires gifts to be non-negative. In other words, a consumer cannot resell the public good provided by the others in order to increase his own consumption of the private good.

Bergstrom, Blume and Varian (1986) propose an interesting reformulation of the problem which considerably simplifies the analysis and provides the key intuition for the comparative statics results. Let us concentrate on the choice made by one consumer. When he decides the amount of his gift, consumer i chooses simultaneously the equilibrium level of G itself: indeed if he chooses to make zero gift, he chooses $G = G_{-i}$, and if he chooses to make a positive gift, he chooses $G > G_{-i}$. Thus, the maximization problem above can also be written as:

$$\max_{x_i, G} U_i(x_i, G) \quad \text{s.t.} \quad x_i + G = w_i + G_{-i}^*, \quad G \geq G_{-i}^*.$$

In other words, we can choose to make the total amount of public good provided by the other agents of the economy appear explicitly in the budget constraint of consumer i. The above problem can be easily represented with the help of a standard consumer theory diagram. This is done in Figure 1, which has been drawn for a consumer who chooses to make a positive contribution in equilibrium.

In Figure 1, the bold segment of the downward-sloping 45°-line represents the feasible points on the budget line of consumer i, under the restriction that voluntary gifts, g_i, are non-negative. At the equilibrium, consumer i chooses to consume x_i^* and to contribute g_i^*, so that the total amount of public good in the economy is equal to $G_{-i}^* + g_i^*$.

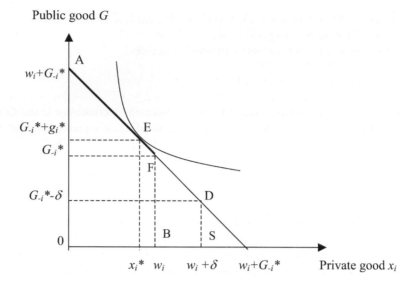

Figure 1. Consumer choice with voluntary provision to a public good.

In the equilibrium, if we consider the set of all agents, while some of them choose to contribute positive gifts, others may decide not to contribute at all and to select the corner solution $(w_i, 0)$ in which they spend all their individual income on the private good.

Bergstrom, Blume and Varian (1986) demonstrate that there exists a unique Nash equilibrium for this game. This equilibrium is typically inefficient. It can also be easily characterized if we assume that all consumers have identical preferences and that the public good is a (strictly) normal good. In such circumstances, there is a critical level of wealth, w^*, such that, below that level, all consumers choose to spend all their income on the private good. All consumers with a higher level of wealth choose to contribute a positive amount which is exactly equal to $w_i - w^*$. Indeed, as they all consume the same amount of public good, those agents must also consume the same amount of private good. (Suppose that this is not the case: consider an agent who chooses to consume more private good. Since G is a normal good, this agent would also choose to consume more public good, hence a contradiction.)

2.2.2. Impact of income distribution

Equipped with the above simple analytical framework, we can now examine the effects of changes in income distribution on the equilibrium amount of the public good.[9] By contrast with the preceding section, the use of utility functions allows for income effects

---

[9] For a related discussion in the sociological literature, see Oliver, Marwell and Tuxera (1985), Oliver and Marwell (1988) and Heckatorn (1993).

to play a role in the determination of individual contributions. While preferences can differ across agents, we still assume that G is a (strictly) normal good. We first analyze redistributions of income among contributing agents such that no consumer looses more income than his original contribution.

Consider agent i who has received δ more units of income. Suppose that, after the redistribution, all other agents contribute exactly their original contribution minus δ. In other words, they, as a whole, reduce their contribution by exactly the amount of net wealth that has been transferred to agent i. In such a situation, the budget line of agent i is exactly identical to the original one, since it is now equal to $(w_i + \delta) + (G^*_{-i} - \delta)$. His budget set is now enlarged from AFB0 to ADS0 in Figure 1. Agent i thus chooses exactly the same equilibrium bundle E than before: he consumes x_i^* units of the private good and contributes $g_i^* + \delta$ to the public good. The extra amount consumer i decides to contribute to the public good is equal to his income change, provided that all other consumers do likewise, which implies here that they reduce their aggregate contribution by δ. Since this is true for all agents between whom income has been redistributed, the equilibrium level of the public good is unaffected by non-drastic transfers between contributing agents. In other words, income redistribution among contributors has no effect on the equilibrium amount of the public good as long as the set of contributors is left unchanged, which is guaranteed here by the fact that no one loses more than his original contribution. This result is known as the 'neutrality' theorem, and it was originally discovered by Warr (1983) [see also Becker (1974, 1981) and Cornes and Sandler (1985)]: "When a single public good is provided at positive levels by private individuals, its provision is unaffected by a redistribution of income. This holds ... despite differences in marginal propensities to contribute to the public good" [Warr (1983, p. 207)].

Before turning to other types of income transfers, it is worth stressing three important conditions that must be met for this result to hold: (1) no contributor loses more than his original contribution, (2) G is a pure public good, and (3) the consumer's utility does not depend on his own gift, but only on the aggregate amount of public good provided.

As we have assumed that the public good is a normal good, the intuition suggests that changes in the distribution of income that increase the aggregate wealth of the contributing agents increases the equilibrium amount of public good provided [see Proposition 4 in Bergstrom, Blume and Varian (1986)]. This holds true for simple transfers of wealth from a non-contributor to a contributor, but it also applies to more complicated redistributions of income, which, for instance, involve a former contributor having become too poor to continue contributing. The critical factor in this result is that the aggregate wealth of the set of contributors is increased. As the preceding result has made clear, the way this increase is distributed among them is of no consequence.

However, as consumers can differ in preferences, the redistribution of income discussed here cannot be directly related to inequality. Indeed, nothing so far prohibits a situation in which poor consumers, who have a 'strong' preference for the public good, contribute while the rich consumers, with other preferences, do not have a high enough income to be induced to contribute. In such a situation, redistributions of income that would increase the aggregate provision of the public good are equalizing. That would not be the case, however, if all consumers have the same preferences. Since contributions are then increasing with income, it immediately follows that disequalizing transfers increase the aggregate provision of the public good, provided that the transfers increase the aggregate wealth of the contributing agents. In particular, if a contributor loses more than his original contribution to the benefit of other contributors, aggregate contributions rise. This is because, as explained in the above section, there is a critical level of wealth below which an agent does not contribute. Transferring an amount equal to his contribution to other contributors does not change the aggregate level of the public good, but the extra amounts transferred increase it as they are equivalent to a transfer from a non-contributing to a contributing agent. It then follows that the smaller the set of contributors with a constant level of aggregate wealth, the larger the aggregate provision of the public good. The highest level of public good will be provided if the whole income is concentrated in the hands of a single consumer.

Some of these predictions were tested in a laboratory experiment by Chan et al. (1996). An important result is that redistributing income from non-contributing to contributing individuals increases the aggregate provision of the public good. The experiment also shows that poorer individuals tend to contribute more and richer individuals less than what could be predicted on the basis of the model. For surveys of related experimental economics evidence, see Ostrom (2000) and Ledyard (1995).

2.2.3. Impact of group size

The effects of group size on the collective provision of a public good can easily be derived from the above discussion.[10] Indeed, as long as the public good is strictly normal, an increase in the number of agents in the economy such that the wealth of the original members is unchanged can have two effects. The new agents can be non-contributors. In this case, the aggregate wealth of the contributors is unchanged and the equilibrium level of public good provided is left unaffected. Alternatively, the new agents can be contributors, in which case the equilibrium level of the public good also increases, in a way similar to that resulting from an increase in the aggregate wealth of the contributors. It is possible that, individually, former contributors reduce their contributions, yet the aggregate level of G must be higher in the new equilibrium. As a result, an increase in the number of agents that leaves the wealth of the original agents unchanged does not reduce the aggregate provision of the public good.

If the marginal propensity to consume the public good is close to zero, contributors will reduce their contributions by almost the same amount as the voluntary contribution of the new agent. In equilibrium, therefore, the level of provision is hardly modified. In this situation, the contribution of the new agent almost completely crowds out the

---

[10] Formal proofs of what follows can be derived by adequately reinterpreting the discussion on state provision of public goods and Theorem 6 in Bergstrom, Blume and Varian (1986, p. 42).

Public good *G*

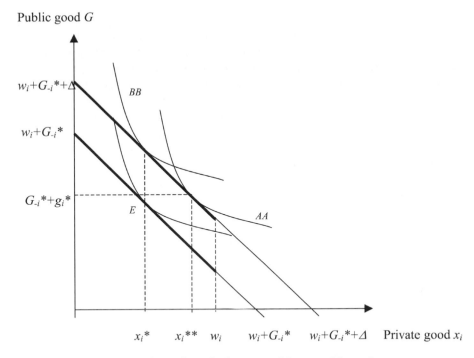

Figure 2. Group size and voluntary provision to a public good.

original voluntary gifts. In contrast, if the marginal propensity to consume the public good is close to one, the income effect of the new contribution is almost entirely spent on the public good. As a result, the equilibrium level of the latter is increased by almost the same amount as the contribution of the new agent: the crowding-out effect on former contributions is negligible.[11] These two cases are illustrated in Figure 2, where the contribution of the new agent is indicated by Δ. The indifference curve AA depicts the case of strong crowding out, while the curve BB depicts the case of weak crowding out.

As we have emphasized above, the increases in group size discussed so far are not supposed to change the wealth of the original members of the society. Suppose now that the new members do not bring any wealth of their own but receive transfers from the original members of the society. In this case, four different situations can obtain. First, the new members' wealth comes from former non-contributors and the new members

---

[11] One would expect that, when the public good is an inferior good, the new contributing agent will cause a more than proportional decline in the provision by the others, which would reduce the aggregate provision of the public good. Conversely, when the private good is an inferior good, the aggregate provision will increase more than proportionately to the contribution of the new agent. Unfortunately, since it is not clear that the Nash equilibrium is unique under those conditions, the comparative statics become less meaningful [see also Chamberlin (1974)].

decide not to contribute: the equilibrium level of the public good is then unchanged. Second, the new members' wealth also comes from non-contributors but the new members decide to contribute to the public good. The aggregate level of the public good is increased. Third, the new members' wealth comes from contributors but the new members decide not to contribute: such a change decreases the aggregate provision of the public good. Lastly, the new members' wealth comes from contributors and the new members contribute. The equilibrium level of the aggregate contributions will increase, stay constant, or decrease depending on whether the set of contributors is larger, equal, or smaller than the original one, because the aggregate wealth of the contributors is left unchanged. The main conclusion is therefore that, even if the new members do not bring any new resources to the society, so that, on average, original members of the society are poorer, the equilibrium level of the public good provided does not necessarily fall.[12]

2.3. The third model: Voluntary contributions to a common good

2.3.1. The model

Think of the use of large mesh-sized nets instead of small mesh nets to avoid capturing immature fishes in a common fishery, or the building and maintenance of anti-erosive barriers in a hilly area, or else the collective maintenance of irrigation channels. In all these situations, the benefits of the 'public good', which is called a *common good* in what follows, are not enjoyed by all agents in the same proportion. Clearly, it is the fisherman with the largest fleet, and therefore the largest share in total fish catches, who benefits most from the protection of juveniles through the adoption of appropriate mesh sizes. In these circumstances, the agents benefit from the common good produced (or the common 'bad' avoided) in proportion to their share or their 'interest' in the good. This share is often directly related to their ownership of the relevant factors of production. Thus, in the case of a fishery composed of n fishermen, the share of fisherman i, s_i, can be thought of as being equal to the number of boats he owns, B_i, in proportion to the total number of boats in the fishery, if we assume that only one type of boat technology is available. Similarly, the share of farmer i in the collective irrigation system s_i is, at least for the sake of many of the relevant issues, equal to the ratio of his landholdings to the total service area operated under this system. His share in the benefits of the common good can thus be written as:[13]

$$s_i = \frac{B_i}{\sum_{j=1}^{n} B_j}.$$

---

[12] Isaac and Walker (1988) propose experimental evidence on the impact of group size on public good provision.

[13] We use this representation for simplicity. All the results discussed in this section would also hold with a more general definition of s_i, such as $s_i = f(B_i, \sum_j B_j)$, with $f_1' > 0$, $f_2' < 0$. If different boat types exist, B_i can be measured in terms of engine horsepower, for example.

We consider a set of agents who voluntarily contribute an amount g_i to the creation of a common good, G, of which they draw benefits in proportion of their interest, s_i, which is given. In the model examined in Section 2, interests were endogenous. In a Nash equilibrium, each agent i maximizes:

$$\max_{x_i, g_i} U_i\left(x_i, s_i\left(g_i + G^*_{-i}\right)\right) \quad \text{s.t.} \quad x_i + g_i = w_i, \quad g_i \geqslant 0.$$

2.3.2. Impact of income distribution

As one can immediately see in the above equation, two dimensions of wealth distribution are worth discussing: the distribution of 'income', w_i, and the distribution of 'shares', s_i, which we consider in turn.

Assume as in the second section that the common good is a strictly normal good; i.e., the marginal propensity to consume the good is strictly positive. In these circumstances, the model is identical to the pure public good model. The difference of shares between two agents can indeed be simply reinterpreted as a difference in preferences, which the model of public good analyzed in Section 2 allowed for. The effects of changes in the distribution of income, w_i, are then identical to the ones reported there. In particular, any change in the distribution that increases the aggregate income of the contributing agents increases the provision of the public good.

We can now turn to the distribution of shares, s_i. Assume that agents have the same preferences and face the same constraints. In the Nash equilibrium, there is a critical share such that all agents with larger shares provide positive contributions, while those with smaller shares do not contribute. Also, contributions are increasing with the share of the contributing agents. As a result, any disequalizing transfer of shares from a non-contributor to a contributor increases the overall provision of the common good.[14] The impact of transfers between contributors is ambiguous, if we do not make additional assumptions on the utility function. It will actually depend on whether the increased contribution by the winning agent outweighs the reduction decided by the losing agent. However, it is clear in this framework that the largest and most efficient voluntary provision of the common good occurs when a single agent concentrates all the shares, in conformity with Olson's (1965) well-known contribution.

2.3.3. Impact of group size

The impact of an increase in group size will depend on whether or not the existing shares have to be redistributed or not. Consider the case of an irrigation network, where the common good under study is the maintenance of the main canal. The benefits of such maintenance to a particular farmer are proportional to the irrigated area he cultivates.

---

[14] The above results also hold true if the distribution of income closely follows the distribution of shares, so agents with higher shares have a higher income as well.

Suppose that some farmers, in the proximity of the irrigation scheme, decide to convert their lands into irrigated fields. This decision does not increase the maintenance costs of the main canal, nor does it alter the shares of the former users, but it increases the number of beneficiaries in the scheme, as it adds new 'shares'. In such a situation, if the new users voluntarily contribute a positive amount to the scheme, the new equilibrium level of aggregate contributions will be higher. If the new users decide not to contribute, it will be left unchanged.

The situation will, however, be different if the total amount of shares is left unchanged. With an increase in group size, the existing shares have to be redistributed among a larger number of users. The impact of such a redistribution on the aggregate provision of the public good will depend on the precise pattern of redistribution that takes place. For instance, suppose that the new users contribute in the equilibrium, and have obtained their shares from small agents, who previously did not contribute. This unambiguously increases the aggregate supply of the common good. Suppose conversely that the new users do not contribute in equilibrium and have obtained their shares from a large contributor, who, after the transfer, decides not to contribute any more. This transfer unambiguously reduces the aggregate provision of the common good. The effects here are thus closely related to our discussion on the impact of a redistribution of shares. But the essential point is that, even when shares have to be redistributed towards new users, an increase in group size does not necessarily reduce the level of the common good in the economy.

2.4. Variants of the third model

2.4.1. Linear objective function

Many analyses of cooperation on common property resources adopt a model which is, in essence, very close to the one presented in the above section. They may nevertheless propose different assumptions regarding the objective function of the agents, or the technology, which we now discuss.

In many contexts, the use of a general utility function is not adequate and, as in the model of Section 1, one may prefer the use of a profit function of the following form:

$$\Pi_i = -g_i + s_i G\left(\sum_j g_j\right),$$

where G is a concave function of aggregate contributions by all agents (non-linearity is needed to avoid unbounded solutions) while, as above, g_i represents the contribution by agent i, and s_i, his share in the benefits of the common good.[15] Consider that

---

[15] One may argue that such an approach makes sense when the product of the resource can easily be sold on external markets. If the users cannot sell the resource output but uses it for self-consumption purposes, then the utility approach of Section 2.1 is more appropriate.

agents choose the level of their individual contributions to maximize profits. In a Nash equilibrium, the first-order condition for an interior solution is:

$$-1 + s_i G'\left(\sum_j g_j\right) = 0,$$

which we obtained by deriving the profit function above with respect to g_i. Note that, for this condition to hold, G' must be positive. In equilibrium, as $\sum_j g_j$ is the same for all agents, the above equation cannot be satisfied for more than one agent: if it holds for agent i, with share s_i, it cannot hold for an agent with another share s_j, with $s_j \neq s_i$. In equilibrium, the agent with the largest share is then alone to contribute. All others will choose to free ride (the corner solution) and set the level of their contribution equal to zero. If there is more than one agent with the highest share, then there is a continuum of Nash equilibria such that all the agents with the highest share collectively contribute according to the first-order condition above.

Comparative statics on the Nash equilibrium follow easily. Any change in the distribution of shares such that the level of the highest share is increased does raise the level of contribution to the common good. Small transfers between non-contributing agents – small enough so that none of them gets a larger share than the highest one in the original situation – and increases in group size that do not affect the share of the contributing agent have no impact on the common good.

2.4.2. Constraints on contributions

As in the case of models of appropriation, one may argue that people are not always in a position to contribute the amount they want. They may thus be subject to constraints on their feasible contributions. (The utility framework used above included this effect by allowing a varying marginal rate of substitution between the common good and individual gifts.) For instance, fishermen may contribute to the conservation of the fishery by releasing immature fishes caught in their nets, but the amount they can contribute is clearly limited by the number of nets they operate. At least, this is true unless they make contracts with other fishermen, yet such contracts are fraught with enforcement problems. Or, a farmer can usually contribute to anti-erosive works only on the fields he cultivates. Doing so on neighboring fields would also imply complicated, and perhaps infeasible, contracting.

Starting with a situation in which every agent faces the same constraint, the Nash equilibrium will be such that agents with the highest share, s_i, contribute up to the level of their constraints – the left-hand side of the first-order condition above is positive for them – while agents with smaller shares do not. Consider now the effects of transfers in the capacity to contribute. Once again, disequalizing transfers of capacities from the non-contributing agents to the constrained contributors do increase aggregate contributions. Moreover, since the maximal contribution that an agent is willing to make is an increasing function of his share, transfers of capacities to the agent with the highest share also increase the provision of the common good. A more thorough elaboration of this argument can be found in [Baland and Platteau (1997b, pp. 458–472)].

Dayton-Johnson and Bardhan (1999) propose a more sophisticated two-period model designed to highlight the effects of asset inequality and constrained capacity to contribute on cooperation in a fishery.[16] Consider a group of two fishermen endowed with a fishing capacity, c_i, which is expressed in terms of the amount of catchable fish. Fishermen live for two periods. If total capacity exceeds the amount of available fish, each fisherman gets a share of it that is equal to his share of total capacity, s_i, defined as $s_i = c_i/(c_1 + c_2)$. Each fisherman has to choose how much to fish in each period, up to his own capacity. In period 1, the stock of fish is equal to F_1. Once fishing has taken place in period 1, what is left of the stock grows at a gross rate r to constitute the stock in period 2, F_2. To keep the discussion simple, we assume that $c_i < F_1$ and that $c_1 + c_2 > rF_1$.

We proceed by backward induction. In period 2, the two fishermen use their whole capacity – there is nothing to be gained from self-restraint in that period – so that the benefit to agent i is just equal to $s_i F_2$. In equilibrium, fishermen thus share the fish stock in period 2 in proportion of their capacities. Now turn to period 1. Obviously, since r is positive, efficiency requires catches to be nil in period 1: for each unit of fish caught in period 1, indeed, r units of fish are foregone in period 2. Consider the problem faced by fisherman 1. Let us begin by assuming that the other fisherman decides not to fish in period 1. Then, fisherman 1 will decide also not to fish a given amount of fish, say g_1 (that is, in the terminology used before, his contribution to the future common good), provided his future benefits of doing so outweigh the current cost, that is if $s_1 r g_1 > g_1 \Leftrightarrow s_1 > 1/r$. If this holds true for both fishermen, then conservation is a Nash equilibrium. In other words, if everyone's share in the future common good (harvest) is large enough, there is an equilibrium under which everyone makes his highest possible contribution to conserving the resource till period 2 is reached. Contributions are naturally constrained since negative catches are not allowed in the first period.

Consider now a situation in which fisherman 2 operates in period 1. If fisherman 2 decides to take some fish in period 1, he will always fish up to his capacity, since $c_2 < F_1$ and payoffs are linear in catches. Fisherman 1 will then decide to fish as much as he can in period 1 if:

$$s_1 r(F_1 - c_2) < s_1 F_1 \quad \Longleftrightarrow \quad c_2 > \frac{F_1(r-1)}{r}. \tag{1}$$

In other words, if fisherman 2 is large enough, he will catch enough fish in period 1 to deter fisherman 1 from leaving fish in the water and sharing them in period 2 with fisherman 2. If a similar expression also holds for fisherman 2, then resource depletion in period 1 is also a Nash equilibrium.

[16] See also Bardhan, Bowles and Gintis (1999).

It may also be the case that a fisherman, say fisherman 2, unilaterally decides to fish in period 1, even though fisherman 1 decides to preserve the resource (i.e., the condition above is not satisfied). He then free rides on the conservation effort of the other fisherman, which occurs if:

$$s_2 r F_1 < c_2 + s_2 r (F_1 - c_2) \quad \Longleftrightarrow \quad s_2 r < 1. \tag{2}$$

That is, the share of fisherman 2 is too small to induce him to contribute to conservation of the fish stock.

Given these conditions, one can easily distinguish three types of situations. In the first situation, the shares of the two fishermen are not too different. The game then exhibits two Nash equilibria, one in which both fishermen preserve the resource and the other one where they exhaust the whole stock in period 1. In the second situation, the shares of the two fishermen are different enough to make both conditions (1) and (2) simultaneously satisfied. Resource depletion in the first period is the unique Nash equilibrium. Finally, in a third situation, the shares of the two fishermen are very different, and it is a dominant strategy for the fisherman with the small share to fish in periods 1 and 2, while the larger one fishes only in period 2. Partial conservation is then the unique Nash equilibrium. As more shares are given to the larger fisherman (i.e., to the only contributor), the stock of the resource is better managed. As in the other models, efficiency obtains if one fisherman concentrates all the shares. As the authors conclude, "the relationship between inequality and conservation can be U-shaped: at very low and very high levels of inequality, conservation is possible, while for middle range of inequality, it is not" [Dayton-Johnson and Bardhan (1999, p. 26)].

In this model, the constraints placed on the capacity to contribute allow for situations of 'partial cooperation', in which conservation of the resource is only partly achieved, with some agents contributing and others free-riding, because the agents with the larger shares are willing to contribute more but cannot do so. Simultaneously, the model realistically incorporates a particular technology with effects that are very similar to those of increasing returns: one's incentive to contribute is increasing in the contributions made by others. This yields different levels of 'cooperation' that could be sustained under a Nash equilibrium for the same parameter values. In fact, such a multiplicity of equilibria is a common feature of all models of common good with increasing returns in contributions[17] and constrained capacities [see, e.g., Gaspart et al. (1998) examined below]. It tends to appear when agents have similar shares in the common good. The inefficient situation in which no agent contributes vanishes when one agent concentrates enough shares and capacities to prompt him to contribute, even if he is alone to do so. However, as in the model of Dayton-Johnson and Bardhan (1999), an 'intermediate' level

---

[17] If the production function is concave in the aggregate contributions, contributions by others reduce one agent's incentives to contribute. Those incentives are thus highest when contributions are nil (the inefficient situation). Such multiplicity cannot arise with a concave technology.

of inequality may distort incentives in such a way that the unique Nash equilibrium is characterized by the absence of any conservation effort, while, under a more equal distribution, the efficient outcome is an equilibrium. For similar results in the context of an anti-erosive management scheme, see Baland and Platteau (1997a).

2.4.3. Non-convexities

So far, we have typically assumed that the technology was either concave or linear. [As discussed in Bergstrom, Blume and Varian (1986, p. 31), all the results that have been obtained with a linear technology also hold if the public good is a concave function of aggregate contributions.] However, there exist a number of situations related to common property resources where technology displays non-convexities and threshold phenomena,[18] for instance because of set-up costs in the building of a common infrastructure, or because of a minimum threshold level beyond which the resource cannot reproduce itself and disappears. The above-discussed model of Bardhan and Dayton-Johnson (1999) actually incorporated such threshold phenomena.

We can illustrate the main impact of non-convexities on the voluntary provision to a common good with the help of a simple model proposed by Gaspart et al. (1998). The profit function is the same as the one given above. Assume that agents can decide to contribute an amount g_i to the building of a common infrastructure, such that $G(\sum_j g_j) = 1$ if $\sum_j g_j$ is greater or equal to a constant C, and $G = 0$ otherwise. In other words, aggregate contributions must reach a critical level for the public good to yield any benefit. The discontinuous character of the production function can easily be justified for a number of collective infrastructures such as the building of a drain in watershed management (adequate drainage can hardly be achieved by a half-completed drain), the erection of contour bunds to prevent erosion, or the digging of a well. While such a model is arguably very specific, it captures in a simple way the main effects of a non-convexity. For a more general approach, see Baland and Platteau (1997b). We also assume that no agent would have an incentive to produce alone, so that all shares are such that $s_i < C$.

First note that there is a Nash equilibrium under which no agent contributes. It corresponds to a coordination failure so that no investment occurs, even though there are situations under which everyone, even those who contribute, would benefit from it. Clearly, even though one must not exclude a priori such an equilibrium situation, it remains unlikely in the type of closed and small communities we have in mind here (see below for a related discussion).

Let us now look at the other Nash equilibria. Clearly, if the other agents contribute enough so that agent i's contribution is both profitable for agent i and needed for the production of the public good, agent i will contribute just the amount necessary for the

---

[18] See, in particular, Baland and Platteau (1997a) for a discussion of non-convexities in the realm of common property resources.

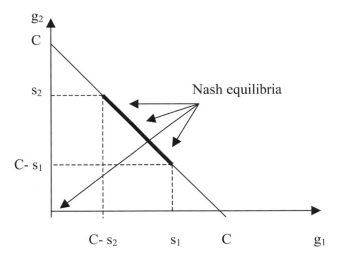

Figure 3. Reaction functions in a fixed-cost model of a common good.

completion of the public good. In other words, in a Nash equilibrium, agent i's best response is:

$$g_i = \begin{cases} C - \sum_{j \neq i} g_j & \text{if } C > \sum_{j \neq i} g_j \geqslant C - s_i, \\ 0 & \text{otherwise.} \end{cases}$$

The reaction functions and the various equilibria of a two-player version of this game are illustrated in Figure 3. The situation in which aggregate contributions cover the fixed cost of the public good (the bold segment in Figure 3) corresponds to a continuum of Nash equilibria which can be characterized by two conditions: (i) no agent contributes more than his own share in the common good, and (ii) the sum of all contributions equal the fixed cost C. While it is true that agents with larger shares will tend to appear more frequently in the possible equilibria, and that their equilibrium contributions will on average be more important, no further precise prediction can be inferred from this model. In particular, one can easily construct examples of equilibria in which only the smallest agents contribute to the public good.[19] Also, given the multiplicity of equilibria, it is hard to get meaningful comparative statics results. However, it should be noted that when the distribution of shares is such that one agent has enough shares to undertake the project alone ($s_i > C$), the Nash equilibrium under which no agent contributes disappears. In other words, the coordination failure associated with the inefficient Nash equilibrium can be trivially solved by an appropriate transfer of shares.

---

[19] For instance, suppose that you have n small agents, with share s_S, and 1 large agent with share s_L. If $n \cdot s_S > G$, there exist equilibria where m small agents contribute s_S, with $m = G/s_S$, while the others, including the large agent, contribute zero.

The main lesson to be drawn from the above model is that the presence of non-convexities, which are particularly likely in the context of common property resources, imply multiple equilibria and the possibility of coordination failures. Hence, when non-convexities are likely, empirical approaches must be tailored specifically to address these problems in order to deliver meaningful and relevant results.

2.4.4. *Other technological assumptions*

There are two alternative technological assumptions that are worth investigating. First, it is not clear that contributions are always continuous. In some instances, contributions are lump sum, or at least come in discrete amounts. Typically, participation in collective organizations requires a minimum of physical presence at meetings and activities.[20] To illustrate the types of questions that can be raised here, assume indeed that contributions are lump sum, that their cost is identical across all agents, and that benefits are linear in aggregate contributions. Two cases can then arise. First, interests, if equally divided, may not be important enough to motivate participation, say because the resource is too poor relative to the number of potential users. In this case, the highest aggregate level of contributions will be reached by concentrating the distribution of interests on a subgroup of agents. By contrast, if equally divided interests are important enough, then an increase in inequality (or an increase in group size) that makes participation non-profitable for some agents reduces the aggregate level of contributions. The latter case may be related to the results obtained by Easter and Palanisami (1986) as cited in Bardhan, Ghatak and Karaianov (2000) in their study of water organizations in India and Thailand. They found that a higher variance in farm size at village level is negatively correlated with the formation of water organizations. Such formation may well be properly described by the above technology.

Second, we have so far assumed that individual contributions are perfect substitutes: they can simply be added up to get the aggregate amount of contributions. The fungibility of money is the major argument behind the assumption of perfect substitutability of individual contributions. Since, in many collective undertakings, contributions tend to be made in kind (if only by the physical presence of the agents concerned) rather than in cash, this assumption may not be adequate.[21] This is evident when people are of different skills or talents, or make their contributions by different ways that have to

---

[20] In a recent paper, La Ferrara analyzed the determinants of participation to groups, who could exclude non-members from benefitting the collective good they provide to their members. While her theoretical analysis pointed to the ambiguous impact of inequality on participation, her empirical estimates on participation to informal groups in rural Tanzania supported the view that higher inequality at the village level was detrimental to the average level of participation [La Ferrara (2001)]. See also Alesina and La Ferrara (2000) and Di Pasquale and Glaeser (1998) for related empirical approaches in the United States.

[21] Cornes and Sandler (1996, pp. 184–190) propose an example of voluntary provision of a public good, where contributions are not perfect substitutes, and show that the 'neutrality theorem' fails to hold in this case [see also Cornes and Sandler (1994)].

be combined together. The nature of the common good provided may also be such that contributions are not perfect substitutes. This applies when nothing short of universal participation may lead to its production.

Consider also the other polar case, in which contributions are perfect complements. The level of the common good is then limited by the contribution of the agent with the smallest incentive to contribute or the smallest interest. An equal distribution of interests obviously maximizes the aggregate contribution. Examples can be constructed where this holds true even when the elasticity of substitution is substantially high (greater than 1).

2.4.5. *Inequality and exit opportunities*

The importance of existing exit opportunities for the possibility of 'cooperation' on common property resources has often been emphasized in the literature. Agents who can easily substitute other income-earning activities for their working on the commons have less interest and less commitment towards its preservation. Alternatively, agents with better access to alternative income opportunities can be seen as users with a comparatively high opportunity cost of labor, which has the effect of reducing their participation in CPR management activities if these activities are in the form of labor contributions [Baland and Platteau (1998a)]. These situations can usually be analyzed with the help of the model of provision to a common good discussed in this section, where agents with lower interests can be thought as those who enjoy better exit opportunities. For a more explicit discussion, see Dayton-Johnson and Bardhan (1999).

The relationship between wealth and exit opportunities, however, depends on the specific context under analysis. For instance, while in some cases, agents with exit opportunities are poor migrants who can easily move to other places, in other cases, they are wealthy villagers who have access to outside employment or technologies which allow them to easily relocate their activities (such as owners of industrial vessels in the case of fisheries).[22] For instance, in a Ugandan forest participatory management scheme which we visited, villagers are regularly confronted by those two types of agents: there are poor charcoal-makers, who continuously shift from one place to another and entertain no genuine relation with local villagers, and there are rich businessmen from the capital city, who hire workers and own trucks in order to cut timber at night. Those two activities are legally prohibited, but enforcement remains problematic.

The varying presence of exit opportunities may not only account for different rates of individual participation in the production of a local common good, but also for varying degrees of success in organizing and maintaining common goods. As an example of the former, one can cite the well-known situations where absent herd owners show

---

[22] Note, however, that if labor markets are risky, poor migrants remain eager to retain their assets in the native village and to keep them in good conditions. In this way, indeed, they maintain access to a reliable fallback option.

much less restraint in their use of common village or peri-urban pastures than local residents whose incomes heavily depend on pastoral activities and have few possibilities of moving their herds elsewhere [see, e.g., Shanmugaratnam et al. (1992, pp. 20–26)]. Analogous examples from the fishing sector are easily forthcoming where the opposition is between owners of industrial fishing vessels (say, bottom trawlers) that are essentially mobile and small-scale artisanal boats that are attached to local waters [see, e.g., Baland and Platteau (1996, pp. 303, 304)]. As an example of the latter case, consider a cross-sectional study of irrigators' associations in 25 irrigation systems in six provinces of the Philippines, all under the command of the National Irrigation Administration (NIA). In their study, Fujita, Hayami and Kikuchi (1999) have shown that there is a negative relationship between the effectiveness of collective action by water users and the availability of exit options from farm to non-farm economic activities.

So far, exit opportunities appear to have only negative effects on the ability and willingness of resource users to contribute to the common good. Yet, positive effects may also be present that have to be balanced out against these negative effects before a final judgement is pronounced about the impact of exit opportunities [see, e.g., Tang (1992, p. 21)]. More precisely, because of the highly imperfect credit markets that prevail in most rural areas of developing countries, resource users with access to other sources of income may be better able to participate in collective undertakings that demand substantial capital investment or sacrifice before producing benefits. In particular, a significant reduction in the rate of harvesting may be required to rehabilitate a degraded resource, to replenish a deteriorating water basin, to restore a village forest or grazing space, or to bring back vanishing fish species into the sea. We are here in a case where contributions to CPR conservation take the form of income or cash income rather than labor.

Absent outside income opportunities, an external intervention is needed to provide users with proper incentives to conserve their endangered resource. This explains the success of village-based reforestation schemes, such as the Arabari experiment in West Bengal (India) or the Guesselbodi forest reserve project in Niger, which have been attentive to the need of poor resource users to be duly compensated for temporary losses of income [Cernea (1989)]. Clearly, the amount of assistance needed will depend on the bio-physical characteristics of the resource and, more particularly, on its level of productivity at the beginning of the incentive scheme. It is a priori possible that users require only a small trigger in order to move from a shutdown to a conservation path. At the other extreme, the possibility also exists that bio-physical conditions are initially so bad and/or conservation practices so ineffective that no conservation strategy is going to be profitable in the long run: production will never be sustainable, whatever the conservation efforts undertaken. In this case, there is no way out of creating alternative income-earning opportunities for the rural poor and, if the natural resources on which their subsistence presently depends have a value for the society, the sooner these new opportunities are created, the better it is (since the resources may be completely degraded if they come too late). Yet, if conservation is a feasible strategy, it bears emphasis that a policy of assistance to poor resource users is likely to be all the more effective as they have in general great incentives to seek to conserve their resource base precisely because they have limited alternative income sources [Baland and Platteau (1996, p. 294)].

Assuming conservation to be feasible and socially desirable (say, on account of external effects), it would nevertheless be wrong to believe, as hinted at above, that only poor users require being compensated during a critical resource recovery period. Indeed, users with attractive alternative income opportunities may have a strong incentive to follow a shutdown path of resource exploitation: if the rate of return on the resource conservation investment falls below the return achievable by allocating production factors to alternative uses, they will draw down the resource to the point where production cannot be continued since by so doing they can reap the high returns from resource over-exploitation and, thereafter, shift to an alternative activity [Pagiola (1993)]. Obviously, the amount of incentives required to make these fortunate users shift to a conservation path may be much higher than in the case of poor users. In this regard, it is revealing that, in northern India, farm forestry schemes have been much more successful in districts with poor laterite soils than in better-endowed districts [Baland and Platteau, (1996, Chapter 11)].

To sum up, assuming perfect rural credit markets, the availability of alternative income opportunities is a factor adverse to effective collective action with respect to CPRs when it has the effect of raising the opportunity cost of labor or of reducing the users' interest in the resource. When real-world credit market imperfections and liquidity constraints are taken into account, however, access to such opportunities might prove useful, especially if poor CPR users enjoy the benefits of such access and conservation efforts are needed to restore the state of the resource.

2.5. Some lessons of the non-cooperative framework for collective regulation

By viewing the organizational tasks related to collective regulation as a common or a public good, one can draw a first set of implications regarding the setting up of a collective management scheme. First, the richer an agent is (in income, wealth, or 'interest'), the more he will tend to contribute. Voluntary contributions are increasing with wealth. This general result partially confirms Olson's intuition: "the greater the interest in the collective good of any single member, the greater the likelihood that member will get such a significant proportion of the total benefit from the collective good that he will gain from seeing that the good is provided, even if he has to pay all of the cost himself" [Olson (1965, p. 34)].

However, that contributions are positively related to wealth does not imply that regressive redistributions of wealth necessarily increase the aggregate provision of the public good or the intensity of collective regulation. It is therefore wrong to conclude that: "In smaller groups marked by considerable degrees of inequality ... there is the greatest likelihood that a collective good will be provided" [Olson (1965, p. 34)]. More precisely, when some poor agents do not contribute, redistributing income from those agents to contributing agents increases provision of the collective good. In contrast, redistribution between contributing agents has ambiguous effects. Still, it remains true that public good provision is highest when a single agent concentrates all the wealth, as Olson correctly hypothesized.

There exist a large number of case studies dealing with this issue which tend in their vast majority to confirm the hypothesis that agents with the highest interest contribute a larger share of the collective good. For example, in his in-depth study of irrigation systems in South India, Wade demonstrates that the effectiveness of a local irrigation council "depends on its councillors all having substantial private interests in seeing that it works, and that interest is greater a larger a person's landholding" [Wade (1987, p. 230)]. The claims that large landowners can make "are sufficiently large for some of them to be motivated to pay a major share of the organizational costs" [Wade (1988a, p. 190)].[23] A similar observation has been made by Gaspart et al. (1998) in their detailed analysis of voluntary participation by villagers in a watershed management scheme in Ginshi (Ethiopia). They indeed find a positive relationship between the size of one's potential interest and the amount of effort spent on the building site. As the authors have pointed out, despite the indeterminacy of their theoretical results, the equilibrium selection process in the village studied has apparently been based on a norm of proportionality between contributions and benefits. In another study, Gaspart and Platteau (2001) have shown econometrically that in the Senegalese fishing community where effort-regulation schemes have been most successful, they have been strongly supported by an elite of wealthy and comparatively old fishermen who have played a major role in initiating and enforcing them. Similar evidence can be found for watershed management in Haiti [White and Runge (1995)], rural cooperatives in the Netherlands [Braverman et al. (1991)], grazing schemes in Lesotho [Swallow and Bromley (1995)] and Rajasthan [Shanmugaratnam (1996)], erosion control in Mexico [Garcia-Barrios and Garcia-Barrios (1990)], forest management in China [Menzies (1994)], and irrigation systems in Nepal [Laitos (1986), Ostrom and Gardner (1993)].

However, a number of empirical studies of irrigation schemes in developing countries conclude that higher inequality in landholdings (or farm income) tends to reduce the overall level of maintenance, even though it simultaneously induces larger agents to support a bigger share of the collective costs [see Tang (1991), Dayton-Johnson (1998) and Bardhan (2000)]. This is in conformity with our above-stated proviso. Relatedly, in their analysis of sugar cooperatives in Maharashtra, Banerjee et al. (2001) show how the weight of wealthy and influential users in collective decision-making tends to distort collective regulation towards their interest, at the cost of efficiency. Their empirical estimates show that distortions (and inefficiency) in collective regulation tend to increase when inequality is larger among users.

Inequality affects collective regulation not only through voluntary participation in the organizational tasks involved, but also through the regulatory possibilities that the available instruments allow for. From this second standpoint, inequality appears as much less

---

[23] In the same study, Wade (1988a) also argues that small size might be detrimental to the success of collective action.

favorable to collective regulation. There are indeed a number of arguments to support the view that wealth or skills inequality between users makes regulation less efficient. First, in the presence of inequality, regulation is more difficult to design and implement as regulatory instruments are imperfect[24] and are often limited to uniform quotas, or constant tax rates. Baland and Platteau (1998a), following Kanbur (1992), propose a number of examples aimed at highlighting this difficulty. In particular, as resource users are more different, the regulated solution tends to be less efficient. It is also more likely that some of the users will be hurt by the regulation proposed, in the absence of compensatory transfer schemes. As a result, if we require the regulated solution to Pareto-dominate the ex ante unregulated situation, as inequality rises, the Pareto-dominating regulation, if it exists, tends to be less efficient.

That regulation tends to be more difficult to implement in the presence of inequality is supported by the well-known analysis of shrimp fishery in Texas by Johnson and Libecap (1982, pp. 1006, 1010):

> Contracting costs are high among heterogeneous fishermen, who vary principally with regard to fishing skill. The differential yields that result from heterogeneity affect the willingness to organize with others for specific regulations. ... regulations that pose disproportionate constraints on certain classes of fishermen will be opposed by those adversely affected. (...) Indeed, if fishermen had equal abilities and yields, the net gains from effort controls would be evenly spread, and given the large estimates of rent dissipation in many fisheries, rules governing effort or catch would be quickly adopted. (...) For example, total effort could be restricted through uniform quotas for eligible fishermen. But if fishermen are heterogeneous, uniform quotas will be costly to assign and enforce because of opposition from more productive fishermen. Without side payments (which are difficult to administer), uniform quotas leave more productive fishermen worse off.

The literature dealing with CPR management in village societies abounds with examples of uniform quotas and taxes, or transfer payments destined for equalizing individual incomes. As illustrated by the case of Japan, uniform quotas can be observed even in rural communities characterized by strong economic differentiation. According to McKean (1986), indeed, in preindustrial Japanese villages, a relatively egalitarian treatment of all villagers with respect to use of local CPRs went hand in hand with inequality in private landholdings and political power. For descriptions of similar systems in India and Nepal, see Guha (1985, p. 1940) and Arnold and Campbell (1986, p. 436).

Given the inherent defects of uniform quotas (or taxes), one may wonder why more differentiated instruments are not put into practice. A first explanation is the information problem. Indeed, when information about the performance of individual users is imperfect, the latter have an incentive to lie about their real endowments or their true use of

---

[24] For a thorough discussion of the limitations in the use of such instruments, and in particular the importance of equal treatment of community members and the prohibition of monetary compensations, in the case of common property resources, see Baland and Platteau (1999, pp. 782–784).

the resource. To avoid endless arguments and conflicts, communities tend to have re-course to systems of uniform treatment of all users, irrespective of their type. Evidence from Senegal artisanal fisheries confirms that fishermen are reluctant to differentiate fishing quotas according to individual skill levels or performance. As noted by Gaspart and Platteau (2001), many fishermen actually denied that skill differentials exist in their community and they "actually took pains to explain that better performances on the part of some fishermen are only transient phenomena likely to be reversed as soon as luck turns its back on them to favor other fishing units" (p. 14).

Another powerful motive against unequal treatment of different users rests upon the traditional ethics of village communities. Typically, indeed, access to communal re-sources is mediated through membership in a social group. The relation is reciprocal: on the one hand, group membership is the basis of social rights, and, on the other hand, maintaining access to a share of the corporate productive assets serves to validate mem-bership in the group. In these conditions, an unequal treatment of users would be consid-ered to introduce or reflect a hierarchy of social status. In the same logic, communities tend to avoid monetary payments to a fraction of their members as such payments will be viewed by the people concerned as a manoeuvre aimed at buying their exclusion from customary entitlements [Berry (1984), Bourdieu (1977, 1980)].

That the regulated outcome may not be efficient is an important conclusion that should prompt us to critically assess field experiences with resource management schemes. This is all the more so as there is a general tendency in the empirical lit-erature to confuse the means with the ends by inferring from the simple existence of regulatory instruments that the resource concerned is properly managed or conserved. Field enquiries typically focus on the question as to whether rules have been laid out and whether they are effectively enforced (e.g., what are the detection and monitor-ing methods used, what is the incidence of rule violation, etc.). For example, studies dealing with forestry or irrigation schemes have a tendency to describe in considerable detail the various rules established by a user community to regulate access to the for-est or water as well as the monitoring and sanction systems created to enforce them [see Ostrom (1990, 1992) and Baland and Platteau (1996, Chapter 12) for references]. An effort is then generally undertaken to identify the characteristics of those user com-munities that have shown their ability to devise and apply membership or use rules as though these rules were necessarily conducive to efficient management of local-level re-sources. Typically, the possibility that rules do not support an efficient outcome or that they are infringed because they are considered to be inefficient or hurting the interests of violators is rarely contemplated.

Three points remain to be made. The first one has to do with the impact of group size on collective regulation. This impact is actually ambiguous, and it depends critically on the following factors: (i) whether initial wealth or interest has to be divided with the new agents or not, and (ii) whether the cost of providing the collective good increases with the number of users. Clearly, when new agents do not reduce the wealth of the former contributors, and when the cost of the collective good does not depend on the size of the user group, increases in group size have a positive impact on collective provision. See Aggarwal and Goyal (1999) for an analysis of the case of scale economies in monitoring costs. By contrast, many empirical studies conclude that successful management schemes tend to be run by small user groups or communities [see, e.g., Ostrom (1990, 2000), Tang (1992), Wilson and Thompson (1993)].

Poverty is another important dimension that has not been touched upon. It is generally argued that poverty may drive people to contemplate short-term strategies, with heavy consequences for the future state of the resource [Baland and Platteau (1996), Pagiola (1993), Perrings (1996)].[25] Typically, poor people do not have access to the capital market. They also tend to be more prone to adverse income shocks, with little ability to self-insure. When their income is low, they would be willing to dissave, that is to use credit markets to transfer income from future periods to the present, but they cannot. They will therefore use alternative and inefficient ways to dissave, as a substitute to their access to the capital market. One such means is to over-exploit the commons. Poverty, when it implies poor access to credit and insurance, is an additional factor of inefficiency in the use of the commons.[26] However, insofar as poor people have few alternative income opportunities available to them, they also tend to have more stakes, and thus more incentives, to take measures to protect common property resources.

Finally, for most of our analysis, we focussed on Nash equilibria in games of finite duration. While this may adequately represent a large number of field situations, there also exist cases where the game played by users on the commons is more realistically depicted as an infinitely repeated game. In such games, as is well known, there is a plethora of equilibrium strategies. In particular, the efficient outcome can possibly be sustained in equilibrium [see, e.g., Abreu (1986, 1988)]. In such games, however, the problem becomes one of equilibrium selection. Such selection can be based on evolutionary processes [see, in particular, Sethi and Somanathan (1996)], social norms and customs, or other mechanisms. Such a study lies beyond the scope of the present review. For some insights, see Dasgupta (1993), Baland and Platteau (1996, Chapter 12), Wade (1988a), and Bardhan and Udry (1999, pp. 173–177). It is probably in the context of infinitely repeated games that the literature emphasizes most the advantages of small size for 'cooperation'. It is thus argued that in small and closed communities, people know each other well and can communicate easily (to coordinate on the 'good' equilibrium, for instance), reputation can play a role, and actions taken by others are easily observed. Also, people tend to be related through more dense and multiplex relationships, which makes defection in one sphere of social or economic life punishable in many other spheres, such as through social ostracism.

---

[25] In a recent contribution, Ternstrom (2001) examines the implications for cooperation on the commons of Dasgupta's hypothesis according to which, at low levels of income, an agent's utility is an S-shaped function of consumption. Under different settings, she shows that the prospects for cooperation are highest when agents' expected income is located around the inflexion point, where the marginal benefits of cooperation, and the costs of defection, are highest. Cooperation is less easily sustained when agents are poorer (or richer).

[26] For a parallel discussion relating to child labor and poverty, see Baland and Robinson (2000).

3. Impediments to the design and implementation of efficient common property management systems

3.1. Information on resource characteristics

3.1.1. Cognitive problems

People appear to be naturally inclined to believe that resources are abundant till the proof of the contrary is driven into their eyes and minds. Even when resources have been degraded, users may deny that they are responsible for the damage. This is typical of fishing and hunting societies, which deal with resources that often move over vast territories and are part of complex ecological systems about whose functioning even specialists can strongly disagree. In particular, these systems are characterized by numerous chains of interdependencies among plant and animal species that make for unpredictable behaviors of harvestable elements. In these conditions, it would be surprising if users had a clear and correct perception of the consequences of their actions.

Especially when resource systems have the kind of physical characteristics described above, users tend to view the flow product of a resource system as given rather than as an outcome which they may themselves influence through their own harvesting behavior. In other words, they do not perceive the relationship that exists between the stock and the flow of a resource nor the causal link between their own actions and the level of this stock. Or, in technical terms, they are not aware of the existence of a sustainable yield curve in so far as they do not have a clear grasp of the fact that today's choices may constrain the set of future choice possibilities. As a result, they do not perceive themselves as actors in a strategic game resembling the Prisoner's Dilemma, as is generally assumed by economists. Or, in more general terms, they misperceive the game that they are playing.

The anthropological literature provides us with interesting examples of hunting and fishing societies where the agents of ecological destruction have a poor understanding of their role in this process. Particularly illuminating is the case of the *Ponams* fishermen of Papua New Guinea studied by Carrier (1987). They opposed a government plan aimed at the conservation of fish and other marine resources, which played an extremely important role in their life, because they refused to ascribe declining fish catches to a decrease in fish population. Instead, they held that catches fell because fish became wary: "fish themselves are the agents of ecological change and the cause of decreased catches" [Carrier (1987, pp. 153–155)]. Unlike the Western view, which tends to see human action as the principal agency of ecological disruption, their conception was therefore based on the belief that external agencies are at the heart of environmental phenomena.

In a study of boreal forest Algonquians, Brightman (1987) found that although Indian tribes could recognize their proximate role as agents in the destruction of animals (most notably, the beaver), they would never admit final responsibility; the ultimate cause of any ecological change was necessarily located in the decisions of supernatural beings.

The idea that hunting pressure could reduce species populations in the long term and on a large scale was all the more absent in traditional Algonquian culture. Game animals killed by hunters were thought to spontaneously regenerate after death or reincarnate as fetal animals. Thus, *Cree* trappers could well imagine that "an adult, trapped animal was 'the same one' that had been killed the previous winter", thus reflecting their profound belief that their environment was one of primordial abundance. This ecological optimism was actually reinforced by the feeling that "game could not be destroyed but only temporarily displaced and that "animals were 'given' to hunters when they were needed" [Brightman (1987, pp. 131–133)].

Interestingly, in another study describing the present-day life of the *Cree* Indian fishermen in the James Bay, Berkes has reached the same conclusion as Brightman and Carrier: at least up until recently, these fishermen believed that "fish is an inexhaustible resource, and that the numbers available are independent of the size of the previous harvest" [Berkes (1987, p. 84)]. In the words of Berkes [Berkes (1987, pp. 85, 86); see also Martin (1979, p. 285)]:

> *Cree* practices violate nearly every conservation-oriented, indirect-effort control measure in the repertory of contemporary scientific fisheries management ... In many of these [Inuit or Eskimo] groups, as with the *Cree*, it is the animals who are considered to be making the decisions; hunters are passive. Any management system claiming to maximize productivity by manipulating the animals is considered arrogant.

That the problem persists even to this day in both developed and developing countries is confirmed by recent observations, for example in marine fisheries. Thus, in Toyama Bay (Japan), Platteau and Seki (2000) have found that boat skippers have attempted to form groups, sometimes with durable success, in order to regulate fishing effort and limit fish landings. Yet, the stated motive behind these collective actions is not the conservation of the resource but the increase of fishermen's market power *vis-à-vis* merchants. Interviews with fishermen revealed their total skepticism regarding the idea that the stock of shrimps can be influenced by the total amount of fishing effort. For them, fish come in the bay from elsewhere, e.g., the open ocean where it spawns and breeds, and the quantity available for the current season is fixed and determined by ecological factors out of human control. If there is less fish in local waters this year as compared to last year, it is because, for some natural reason, fish has moved in greater quantities to an adjacent fishing space.

In another recent study, Gaspart and Platteau (2001) have shown that fishermen's statements about the need to conserve the fish stock may be delusive. In Senegalese villages where collective regulation of fish landings or sea trips has been carried out for at least several years – quite an extraordinary achievement in itself – the basic preoccupation of the fishermen has been with increasing producer prices rather than with managing the resource with a view to conserving it. When they mention biological effects, most of the time they do it in a perfunctory manner. They do not seriously consider the possibility of their being partly responsible for overfishing; therefore, the idea that they could combat environmental degradation by restricting their own fishing effort seems alien to most of them.[27] Revealingly, there is a clear tendency among them to externalize the problem by blaming industrial fishing vessels, or migrant fishermen who operate other fishing techniques, for the destruction of fish resources.[28]

The tendency to 'blame the other' for stock depletion is typical of almost all artisanal fishing communities, including those in Europe. It is thus generally acknowledged that enforcement of fishing quotas laid out in accordance with the European Community's 'blue' policy is difficult as fishermen have various ways of evading them by under-reporting their catches. A central problem is the fishermen's belief that limiting their fishing effort cannot improve the state of the stock because the main cause of stock depletion does not lie with them but with their neighbors (the French and the Spaniards for British fishermen, the Spaniards and the British for French fishermen) or with external sea-roaming operators.

Poor understanding of interactions between human behavior and the environment may sometimes characterize users of other resources than fish and game. Thus, for example, in the Kgatleng district of southeastern Botswana, privatization of grazing lands through private ownership of boreholes did not have the effect of preventing overgrazing. Tswana people were eager to maintain or expand their cattle herds and "they continued to consider that the cause of overgrazing was the lack of rain, which forced concentration of cattle around scarce water" [Peters (1994, p. 82)]. The solution for them lay in the digging of new boreholes and not in the enforcement of stock limits such as could be achieved through the spacing of water points. It is therefore not surprising that "most ranches have had no better record in herd and range management than the typical cattle-post system – and sometimes the record has been worse" [Peters (1994, p. 220)]. To take yet another example, present overgrazing in Mexican pastoral *ejidos* has been partly blamed on the *ejidatorios'* limited technical understanding "regarding the complex interdependence of individual grazing decisions and the impact these choices have on the range resource and, ultimately, on livestock productivity and human welfare". In particular, "local understanding of elementary soil–water–plant–animal relationships is rudimentary" [Wilson and Thompson (1993, p. 314), see also Cernea (1989, p. 61), for

---

[27] Another finding of Gaspart and Platteau's study is that fishermen who are relatively educated (they have more than six years of French or Coranic school) tend to mention biological effects, whether in conjunction with economic effects or not, more often than the other fishermen. The fact that environmental problems are nowadays a widely publicized issue, both at school and in the media, probably explains why many relatively educated fishermen refer to the biological effect of output regulation.

[28] There is no denying that industrial fishing can wreak havoc in maritime fisheries as the history of recent decades amply testifies across the world. Yet, small-scale fishermen often take too much comfort from this fact to conceal from themselves the painful truth that they can also have their share of the blame owing to the rapid expansion of the artisanal fishing fleet and the tremendous improvements in the artisanal fishing technology. The same can be said of the accusations by Kayar's fishermen that migrant operators from Saint-Louis must bear serious responsibility for destruction of the fish stock under the dubious pretext that the dead fish trapped in their bottom-set nets tend to frighten the living fish out of the area, a statement that does not stand scientific scrutiny.

forestry].[29] The same problem is sometimes noted also for village forests [Ribot (1999, p. 32)].

A final remark is in order. The absence of any measure or scheme intended for the conservation of the resource does not imply that no collective action is undertaken to regulate its use. In point of fact, rules solving assignment problems are often adopted whenever an excessive number of claims are laid to the flow product of a resource. This is especially evident in the case of fishing where rotation of fishermen around various fishing spots is frequently practiced to reduce the opportunities for conflicts that inevitably arise when the most productive resource sites are congested. See the illustration from Sri Lanka in Section 1.2.3, and also Berkes (1986), Cordell and McKean (1986), Levieil (1987), Hannesson (1988), Baland and Platteau (1996, p. 208), and Platteau and Seki (2000). In a study based on thirty case studies of fisheries, Schlager (1990, 1994) has found that the existence of assignment externalities is significantly related to whether or not fishermen have adopted systematic procedures for fishing operations.

Because assignment schemes are designed to solve pressing and incontrovertible problems of congestion – if nothing is done, high transaction costs have to be incurred and grave conflicts can erupt – and because their effects are generally clear and predictable, they are often adopted by resource users. See Guha (1985), Arnold and Campbell (1986), and McKean (1986) for forests; Lawry (1989a) for grazing activities; and Messerschmidt (1986), Ostrom (1992), Tang (1992), and Mahdi (1986) for irrigation water. Hence, a group of users able to regulate access to a resource not only by laying down membership rules defining rights-holders but also by setting rules governing access of members to various portions of the resource domain may well be unwilling, or unaware of the necessity, to take up management measures designed for a better conservation of the resource.

As we have explained, fishing is an activity that fits in especially well with the above configuration of assignment rules but no management rules. Canal irrigation also falls in the same category, yet for another reason: if access to water involves pervasive congestion problems, then water is generally not liable to be depleted or overexploited as a result of irrigators' actions so there is no need to take up conservation measures (the total quantity of water available to a group of irrigators is exogenously fixed for the season depending on rainfall and behavior of the global irrigation system). From that point of view, canal irrigation water narrowly resembles beach-seine fishing, a technique that has no destructive potential because the nets are operated from the beach and for which the existence of fishing turns is often observed [Alexander (1980, 1982), Amarasinghe (1989)].

---

[29] This said, one should be wary of inferring actual beliefs of people from their explicit statements. In point of fact, we cannot rule out the possibility that respondents strategically conceal their true beliefs about the resource stock and the impact of their behavior on it. By feigning to believe that the resource is abundant, they self-justify their opportunistic behavior and their reluctance to give it up. Unfortunately, it is extremely difficult empirically to uncover the true beliefs of people since they may not be easily revealed by observation of actual behavior and the recording of professed opinions.

3.1.2. The evolutionary view and learning processes

Even if we adhere to the evolutionary view that the ability to organize collectively develops gradually as a response to emerging needs, a transition problem clearly remains. Indeed, nothing ensures that the resource will not be degraded during the time span required for the users to realize the main causes of the disaster and to organize themselves. As pointed out by the authors of a World Bank study [Gregersen, Draper and Elz (1989, pp. 9, 144)]:

> Rural people deplete their forest and soil capital, often unaware that they are destroying their future source of fuel, fodder, and soil protection. The people do not realize the danger until the local forest is nearly gone and they must go farther into the countryside to find fuelwood. When the stock is eventually used up, ... the extent of the crisis becomes evident ... There is no 'fast fix' when this happens ... Deforestation by local people using wood for local uses can be a slow and largely unnoticed process; realization of the damage may come too late for them to do anything about it without significant outside intervention.

If the resource can nevertheless be replenished, then the evolutionary argument may just be reformulated by saying that a society needs to be confronted by a natural disaster before being able not only to realize the extent of the problem but also to organize itself so as to avoid its repetition in the future. Thus, a recent survey of more than 70 empirical studies has concluded that the normal pattern is for erosion and grazing degradation at first to worsen as population numbers and cropping frequencies rise [Templeton and Scherr (1999)]. It is only at a later stage that new forms of land management designed to offset these trends and to raise land productivity are eventually induced. In the meantime, population densities have increased quite significantly, from 25 to 100 inhabitants per square kilometer on average.

According to McKean (1986) also, it is only under the most acute pressure that Japanese villagers were driven in the Tokugawa period to adopt management measures to protect their natural resources from destruction [see Section 1.2.1, and, for more details, McKean (1986, p. 558)]. In the 17th century, this pressure took the form of a genuine ecological crisis manifested in considerable deforestation following a sudden surge in the demand for timber caused by the rapid construction of cities and castles after the return of peace conditions. During the 16th century the country had been devastated by a widespread civil war. In the author's own words [McKean (1986, p. 549)]:

> For our purposes, the significance of this episode of deforestation during the 17th century is threefold: visible deforestation seems to have made villagers aware of the very real risks of overuse and enabled them to develop and enforce stricter rules for conservation on their own initiative to save their forests and commons from the same fate. Rather than destroying the commons, deforestation resulted in increased institutionalization of village rights to common land. And it promoted the development of literally thousands of highly codified sets of regulations for the conservation of forests and the use of all commons.

It is particularly interesting to note that, in response to this visible experience of deforestation, Japanese villages have chosen not only to adopt strict conservation measures
but also to regulate access to common lands in such a way as to discourage population
growth. Equal rights of access to communal resources were allotted on a household
basis so as to avoid giving advantages to large families and thus discourage population growth. To make this system effective, the formation of a branch household from
the main family was subject to approval by village authorities. The latter "recognized
that creating an additional household would enlarge the number of claimants on the
commons without enlarging the commons" and were therefore reluctant to grant such a
permission. In some villages, "no new household were permitted unless an old one died
out for lack of heirs" [McKean (1986, pp. 551–553)].

The same idea that learning is a time-consuming and hazardous process that may need
a disastrous experience to be triggered comes out of the study by Brightman (1987) on
the ecological attitudes and practices among boreal forest Algonquians. In that study,
we are told that, following intensified predation, limited-access land tenure and conservation eventually appeared as adjustments to depleted environments. Conservation
was actually "a postcontact innovation that did not develop on any scale prior to game
depletions in the early 1800s ..." [Brightman (1987, p. 12)]. Among the *Crees* and
other groups, continuous and visible experiences of game shortages slowly led Algonquians to question their traditional beliefs and, after a certain point, to reinterpret them
in a way more consistent with the changed circumstances. In the words of Brightman,
"*Crees* encountered in the game shortages a contradiction of cosmic proportions: despite conventional ritual treatment, animals were not renewing themselves but were
disappearing". Game shortages therefore "motivated a reinterpretation of indiscrimi-
nate or 'wasteful' hunting and trapping as offensive to animals and to the spirit entities
regulating each species". These spirit entities were now imagined to interfere with the
harvests of hunters who trapped unselectively, and conservation was redefined as a religious obligation, the violation of which caused severe punishment in the form of game
shortages [Brightman (1987, pp. 136–139)].

The same lack of, or delayed, awareness accounts for the oft-noted difficulty in establishing village woodlots in rural communities whose tradition has been long centered
on the priority of (communal) pasturage. In some of these communities woodlots are
established in the face of considerable opposition which can occasionally lead to the
purposeful destruction of fencing and of young trees, but will most commonly manifest
itself in the current damages caused by "individual stock owners and herdboys seeking
grazing for their animals and unimpressed by the need to protect the woodlot" [Bruce
(1986, p. 116)]. In the same vein, see Arnold and Campbell (1986, pp. 429, 430) and
Cernea (1989, p. 30); see also Cernea (1985).

Probably the main lesson from the above illustrations is that villagers' awareness
about the real causes behind the degradation of their environmental resources needs to
be propped up especially when environmental change is rapid owing to fast population growth and accelerated processes of market integration. This can be done through

the work of external agents who help villagers to articulate their traditional knowledge about their local resources with the changes that are occurring.

3.2. The state as a major actor

There are two major ways in which the state can impede village-level management of common property resources. The first is by abstaining from providing services, such as technical expertise, conflict-resolution mechanisms, and legal support, which villagers need to be effective managers of their local resources. The second is by the way of interventions, deliberate or not, that have the effect of undermining the ability or willingness of villagers to cooperate towards that purpose. In many notable cases, these two aspects cannot be easily disentangled because they are just different manifestations of a policy that is actually opposed to local management of natural resources. Let us now look in more detail at the ways through which the state has contributed to stifle local initiatives in matters of resource management, by citing a number of illustrative examples. The evidence is presented in two separate subsections, one devoted to the top-down approach to village-level resource management and the other to state interventions and policies geared toward supporting private business interests.

3.2.1. A top-down approach to village-level resource management

States can obviously have different views about the role of user communities in owning and managing local natural resources. At one extreme, a state can decide to give maximum empowerment to user communities by passing legislation specifically designed to support common-property systems. For example, in Japan the legal sanctioning and constitutional guaranteeing of the property rights of fishing communities organized as co-operative associations over inshore waters has been an important factor underlying the success of a highly original system of decentralized management of coastal fisheries [Ruddle (1987), Asada, Hirasawa and Nagasaki (1983), Platteau and Seki (2000)]. In South Korea, Nepal, and the Philippines, the law entitles village communities to generate the necessary rules, regulations, and operational measures required to enforce local-level collective management of forestry and other village resources. This has implied that the state relinquishes power and responsibility to manage forests, irrigation water, and fishing areas through village programmes and committees [Arnold (2000), Pomeroy, Katon and Harkes (1999)].

In Indonesia, the government supported decentralized management of coastal fisheries by granting fishing communities exclusive tenure rights over a delimited portion of the sea. In this case, however, the motive behind the legal support of user groups was purely political: to placate mounting Islamist and nationalist opposition forces by earmarking local waters for (Muslim) Indonesian small-scale fishermen at the expense of Chinese-owned industrial vessels [Mathew (1990), Baland and Platteau (1996, pp. 258–260)]. Given the heavy involvement of former President Suharto in Chinese business ventures, it is not surprising that enforcement of the exclusive fishing rights of small-scale coastal fishermen was much less rigorous than in the case of Japan. The class-orientation of the Indonesian regime determined a strategy that, despite appearances to the contrary, has not resulted in the devolution of *de facto* property rights to artisanal fishermen communities.

In other, much more frequent instances, state interventions have had the intended or unintended effect of destroying local capacities for collective regulation. In the case of Sahelian forests, for instance, when scarcity became apparent in the late 1960s, state-imposed rules "emasculating local organization" have hindered resource management efforts at local level. "As it happened, most villages had lost their power of independent activity as the result of efforts of both the colonial and independent regimes to establish controls over major forms of organization in rural areas. Villages (or quarters within them) had no authority to enforce sanctions against violators of locally devised use rules" [Thomson, Feeny and Oakerson (1986, p. 399), see also Ribot (1999)]. A common pattern initiated during colonial times was to integrate village chiefs into the state as an administrative extension. This formula gave rise to ambiguity and tension due to the dual allegiances of chiefs downward to their people and upward to the central state. Chiefs in the now-independent states of Africa have continued to be regarded as tools in the hands of the administration, which they actually are most of the time [Ribot (1999, pp. 13, 14)].

Rural councils, which are often in charge of various management tasks regarding local resources, are not necessarily more accountable to the people. In many countries, particularly in West Africa, they just do not represent the villagers but political parties and cooperatives initiated from above and are effectively controlled by political and administrative machineries. In the words of Ribot: "even if rural councils were openly elected, they are not independent decision making bodies", their official role is "merely to advise and assist the sous-préfet on political and administrative matters ... they are administrative links to the central government" behaving exactly like colonial village and canton chiefs [Ribot (1999, p. 18)]. Under these conditions, it is not surprising that village bodies officially recognized by the state are frequently politicized with all the expected corroding effects on people's ability to organize and to manage their resources. In Senegal, for example, rural councils are at times "nothing more than sections of the Socialist Party".

The above problems are characteristic not only of Africa but also of many countries in Asia and Latin America. In India, an expert scholar in irrigation issues writes, "the authority imposes a set of programmes upon the farmers which, instead of promoting their participation, actually restricts the expression of cooperation by narrowing its scope. In course of time this erodes the cooperative spirit that existed earlier" [Sengupta (1991, p. 251)]. It is therefore not surprising that, when the state decides on water allocation and distribution, frequent rule violations are reported [Bardhan (2000, p. 15); see also Lam (1998, pp. 175–178), regarding Nepal].[30] The Indian experience is in stark

---

[30] Thus, Wai Fung Lam writes that in Nepal, unfortunately, "enhancing farmers" participation is frequently interpreted as an exercise of tutelage by irrigation officials to tell farmers what to do and how to fit their

contrast with the aforementioned experience of the Philippines where the irrigation administration systematically encourages and actually requires the formation of irrigation
associations. Experimentation with diverse forms of such associative organizations is
promoted by the state, which disseminates knowledge about the most successful experiences. This much more interactive approach is apparently rooted in a well-established
tradition that dates back to the Spanish colonial era [Sengupta (1991, pp. 38–54)]. In the
words of Sengupta again: "There is no difference between irrigators in the Philippines
and those in India or elsewhere as far as their readiness to cooperate is concerned. It is
the technocratic distrust for people's capabilities which prevents effective intervention"
[Sengupta (1991, p. 80)].

Note carefully that there are at least two distinct forms taken by demoralization or
lack of motivation among villagers when the state or political authorities intrude too
much in their own internal affairs instead of supporting their own initiatives with technical expertise and legal backing. First, villagers do not consider rural councils manipulated from above as legitimate bodies that represent them. As a result, they oppose the
decisions and choices made with every means at their disposal. Moreover, politicization
of these councils may easily increase tensions between various village factions if they
are allied with different political parties. As a result, common property management
will be more difficult [Singh (1994, pp. 215, 216)].

Second, villagers form an expectation that the state will in any event perform the
management tasks required or pay for failures if they arise. Consequently, they tend to
shun participating in village schemes and programmes. In Nepal, for example, entrepreneurial energy in the irrigators' communities tends to be directed toward getting money
or construction contracts from the specialized government agency instead of organizing operation and maintenance among the irrigators themselves. Indeed, when external
government funding for construction and maintenance is made available to a community, the members soon develop an expectation that repairs and maintenance jobs will
be done by the state [Tang (1992, p. 135); see also Arnold and Campbell (1986, p. 430),
Azhar (1993, pp. 117, 118), Bromley and Chapagain (1984, p. 872), Agarwal and Narain
(1989, pp. 13, 27), Lam (1998, pp. 181, 193)].

3.2.2. *Active collusion with private business interests*

Collusion with private business interests at the expense of commoners may manifest itself under the form of macro-economic policies biased in favor of the former. This happens, for example, when the government heavily subsidizes the intrusion into traditional
resource territories of private companies. Thus, in Brazil, the exemption from taxation
of virtually all agricultural income combined with the rule that logging is regarded as

---

effort in the Operations and Maintenance plan laid down by the officials" [Lam (1998, pp. 186, 187)]. It is
therefore not surprising that farmers have difficulties in perceiving water committees set up at the initiative
of the Department of Irrigation as 'their' organizations. Rather, they tend to view them as the 'administrative
arm' of that department [Lam (1998, p. 208)].

proof of land occupancy has strongly encouraged rich people to acquire forest lands for the purposes of maximum exploitation [Mahar (1988), Binswanger (1991), Barraclough and Ghimire (1995, Chapter 3)]. Not only deforestation but also dispossession or disfranchisement of traditional user communities have resulted from this ill-conceived policy. Subsidization of imported trawler boats in many Asian countries constitutes another striking example of a government-supported thoughtless acceleration of the rate of exploitation of a commons which ended up disfranchising small-scale fishermen and crew laborers [Kurien (1978), for Kerala in South India].

Active support of business ventures by governmental agencies generally placed under the thumb of high-level political authorities may be motivated by the latter's direct interests in these ventures or by their desire to oblige political supporters and friends following a logic of patronage politics. In Botswana, both motivations are at work behind the state's support given to the wealthy cattle elite for the *de facto* privatization of grazing lands and the concomitant erosion of a traditional institution (the *kgotla*) charged with their management. In the words of Peters [Peters (1994, p. 22)]:

> There is no doubt that some of the highly placed members of the government and party who promote the policy benefit directly as wealthy cattle and borehole owners. In addition, the government's apparent blindness to criticism in calling for expansion of the policy is partially driven by political considerations. Much of the strongest support for the ruling Botswana Democratic Party comes from the wealthier cattle owners. The resistance of government to consider either increasing taxation of cattle owners (a point of contention with the World Bank) or the abolition of dual rights seems attributable to fear of jeopardizing this support.

Governmental support of powerful business interests intermingled with state agencies can take on much more direct and brutal forms as illustrated by the following examples. In the Philippines, the now defunct government agency Panamin (Presidential Assistance to National Minorities), which was officially set up for the purpose of protecting the indigenous people's rights and interests, actually played a disastrous role amounting to sheer betrayal of its mission: "far from preventing the pillage of indigenous lands by mining companies, loggers and hydropower projects, Panamin collaborated with the armed forces in depriving the peoples of their ancestral lands" [Colchester (1994, p. 74)]. The majority of this agency's board members "came from wealthy industrialist families, many of whom had direct financial interests in companies encroaching on indigenous lands". This was certainly the case for Manuel Elizalde, a relative of President Marcos, who played the key role in Panamin. His political base and personal wealth, indeed, lay in extractive concerns such as mining, logging and agribusiness. Moreover, he maintained his own private army in Cotobato in Mindanao to fight the insurgent indigenous peoples who, in despair, took up arms against the government by joining the communist insurgency group [Colchester (1994, p. 75)].

In forests in East Kalimantan, Indonesia, customary rights-holders have been unfairly deprived of access to village commons as a result of licensing rights awarded to private companies. In the early 1970s, the Indonesian government granted timber concessions to a large number of foreign and national companies, and, according to Jessup and Peluso, this had several detrimental effects on local communities. In particular, "despite their legal right to collect minor forest products within timber concessions, villagers have at times been denied entry to those areas, and timber company personnel have otherwise infringed on the rights of local residents". For instance, there is evidence of timber company guards confiscating rattan from collectors, loggers raiding caves and selling the stolen birds' nests to unauthorized buyers, and timber companies illegally cutting Borneo ironwood, a species reserved for local use. These acts sometimes led to violent confrontations between local inhabitants and company guards or loggers [Jessup and Peluso (1986, pp. 520, 521)].

In Sarawak, Malaysia, forest-dwellers feel equally helpless even though the process of dispossession has been less brutal and less overt than in Mindanao. We are thus told that "the corrupting influence of the timber trade has promoted the domination of the economy by nepotistic, patronage politics", with the consequence that rural peoples "can no longer rely on their political representatives to defend their interests". As a matter of fact, "the practice of dealing out logging licences to members of the state legislature to secure their allegiance is so commonplace in Sarawak that it has created a whole class of instant millionaires" [Colchester (1994, pp. 79, 82)]. In such circumstances, it is not surprising that most popular protest movements in Asian forest areas have directed their main criticism at the logging licenses generously distributed by too often corrupt governments.

In lending support to rural elites and powerful urban interests (including its own high-ranking personnel) to help them gain access to valuable natural resources, national states may have recourse to various legal subterfuges. An oft-used method, already applied by the colonial powers to redistribute indigenous lands in favor of white settlers, consists of withdrawing lands from village control by labelling them state property and later awarding them to friends of the regime through de-classifying procedures. Such a method, for example, is currently employed by the Senegalese state in the case of the lands of the Senegal river valley, much to the fury of helpless customary rights-holders.

4. Conclusions

4.1. Incentive systems, decentralization, and co-management

Nowadays, as a response to the numerous excesses of centralization and the ensuing financial crises of specialized administrations, devolution of the management of local resources from state agencies to rural communities is being tried in an increasing number of countries with the active support of bilateral and multilateral donor agencies. It is still too early to have a sound idea about how these programmes of decentralized development can perform and how the above-discussed perverse mechanisms are being surmounted. The step is no doubt in a good direction, yet empowerment of village communities is likely to be a much more difficult task than what many imagine, partly because of the bad habits developed in the past, of the likely resistance of state agents whose responsibilities are going to be encroached upon, and of the inherent difficulties of collective action in resource management matters as underlined in the previous sections.[31] Even in a country like the Philippines where the national irrigation administration has been pioneering efforts of devolution, it appears that there have been more cases of failure than success. The reduction in state agencies' operation and maintenance activities has not been compensated for by the activities of irrigators' associations, with alarming consequences for agricultural production [Fujita, Hayami and Kikuchi (1999, p. 3); see also Lam (1998) for Nepal]. There is therefore an acute need to critically assess ongoing experiences with sound research methodologies.

The challenge ahead lies mainly in finding the right kind of incentives so that both user groups and state agents work effectively and in a coordinated manner to ensure proper conservation of village resource bases. A few examples can illustrate what we have in mind. To begin, if the state does not legally support the actions of user communities by granting them clear and enforceable property rights over the common-property resource, they will have no incentive to guard it against encroachments by external intruders. Marine fishermen communities are thus discouraged when they realize that their efforts to catch industrial vessels found trespassing their fishing territories are in vain because the culprits are released as soon as they are landed or are not required to pay the required fines to the authorities (perhaps because the latter are stakeholders in the business ventures involved). The same problem is often mentioned in connection with protection of village forests or irrigation systems [see, e.g., Sengupta (1991, p. 136)]. The need for user organizations to be free from government and political pressure obeys the same logic of providing them with effective incentives to monitor their resource domain.

The second example is specific to large-scale irrigation systems built by the state. Here, the problem is for the latter to ensure that user communities carry out complementary works and maintenance operations in a reliable manner. This depends on whether the central irrigation administration has devised an appropriate incentive mechanism. Such was not the case, obviously, with the Jamua Irrigation Project in India. Started in 1965 to tap the Jamua River by constructing diversion works and extending the canal network on its left bank, it was able to reach hardly more than 30% of the target area by 1974. The cause of the failure came from the unjustified assumption on the part of the authorities that "once the canal system was constructed, farmers would willingly and jointly contribute their own labor to construct field channels to divert waters from the canal to their field". What they thus overlooked was that "farmers located near the canal would have little incentive to devote their efforts to constructing channels that would deliver water through their own fields into those of others" [Tang (1992, p. 133)]. Here is a delicate problem of imperfect commitment arising from asset specificity that has not been properly perceived by the irrigation department.

---

[31] For a thorough discussion of the limitations of participatory approaches to development arising from community imperfections, see Abraham and Platteau (2001).

In the same vein, sophisticated irrigation technologies involving permanent head-works may run against the interests of downstream farmers because they have the effect of reducing drastically the amount of labor needed for operations and maintenance of the system. As a consequence, the labor contribution of these farmers becomes less critical for the farmers at the head end who can thus afford to ignore the demands for a fair share of the available water by the former [Ostrom and Gardner (1993), Lam (1998, pp. 119–124, 202, 203)]. In other words, by transforming the game from one in which everyone's participation in collective action is required into one in which the participation of only some of the users is sufficient, technological change can lead to a situation where irrigation water becomes appropriated by the owners of strategically located lands.

On the other hand, intervening state agents must also be motivated to perform effec-tively for the benefit of resource users in circumstances where the latter are no more considered as passive subjects. In South Korea, farmers have been integrated at the bot-tom of the formal management hierarchy itself, in the role of patroller. The land of a patroller must lie within the jurisdiction that he irrigates, "so that he experiences irri-gation problems at first hand". Moreover, he is nominated each year by the headmen of the villages within his service area and, if the latter are dissatisfied with his way of handling the task, they nominate someone else. Recruitment and promotion procedures also play an important role inasmuch as they ensure that the senior-level staff are natives of the area in which they work. "Hence the eyes of the irrigation staff are kept firmly on the locality, and identification between their interests and those of farmers is further encouraged" [Wade (1988b, p. 495)]. Elaborating on this theme, Wade (1988b, p. 459) adds:

> Local affiliation of the staff is important because it gives both sides – staff and farmers – a set of shared experiences. This directly assists a sense of mutual oblig-ation between them; and also provides a basis for a shared set of beliefs according to which the existing order is fair and just, and every betrayal is perverse and unjust – including betrayal of the irrigation agency's rules. This is a much more cost-effective method of avoiding free-rider problems than relying on a calculus of punishment.

Much the same picture emerges from the situation in Taiwan where the staff of the Irrigation Associations (IAs) are effectively linked to the local farmers, on the one hand, and to the national agencies, on the other hand. For each rotation area, an irrigation group chief is elected to supervise water distribution and maintenance operations as well as to manage potential conflicts. In particular, these chiefs are in charge of closely monitoring the jointly hired common irrigators who have primary responsibility for the distribution of water and the guarding of the system against frauds and damages. As a matter of principle, they are local farmers and, to avoid undue interference of partisan politics, the process leading to their election is kept separate from elections for other offices. In addition, the irrigation staff themselves are typically recruited from local communities so as to ensure adequate incentives for effective management of the system. In the words of Moore (1989, p. 1742):

> The IAs are overwhelmingly staffed by people who were born in the locality, have lived there all their lives, and in many cases farm there. Further, IA staff are not sharply differentiated from their members in terms of education or income levels. I have a strong overall impression that IA staff are so much part of local society that they can neither easily escape uncomfortable censure if they are conspicuously seen to be performing poorly at their work, nor ignore representations made to them by members in the context of regular and frequent social interactions.

In addition, performance in collecting irrigation fees is an element that enters into the annual evaluation of irrigation officers by their superiors and that indirectly determines salary increases, promotions, and access to additional resources. Each level of the irrigation bureaucracy is thus required at regular intervals to report its collection records to the upper level [Moore (1989, p. 1743)].

Clearly, the above-described mechanisms ensure a good deal of accountability of officials to resource users, which is a critical condition for success in local-level resource management. It is particularly important to control the widespread corrupt practices whereby private contractors are required to pay a certain amount of so-called commissions to officials in order to get a construction or maintenance contract. This forces the contractors to cut expenses by using poor quality, or smaller amounts of, materials in the commissioned works. Since they have taken an illegal commission, the officials are not likely to monitor the quality of these works as they should, and low-quality infrastructure is produced. As a result, resource users are discouraged from supplying effort towards maintaining it properly, both because the incomes derived from its functioning are disappointingly low and because maintaining a poor quality or ill-designed infrastructure is difficult. This sort of problem has been especially emphasized with respect to the management of irrigation systems [see, e.g., Wade (1982), Lam (1998, pp. 179–181)]. Note that resistance against decentralization and democratization efforts in irrigation administrations is partly due to the fear among irrigation staff that they would end the profitable opportunities for illegal profit that the present system allows [Lam (1998, p. 196)].

4.2. Summary of the main points

This chapter calls into question the romantic view according to which small and homogeneous village communities are able on their own to devise rules aimed at the efficient management of their common property resources.

We have first argued that rules devised by communities, wherever they exist, do not necessarily aim to improve the efficiency of use of the resource, but often serve the purpose of regulating (or organizing) access to the resource domain under congested conditions, preventing conflicts, or enhancing the users' market power. In many cases, and contrary to a dominant interpretation in the empirical literature, distributive considerations appear to play a more important role than efficiency considerations in traditional management practices at village level.

Second, it is generally assumed that at local level users have a clear perception of the impact of their behavior on the state of the resource. Careful case studies, however, suggest that this may not be the case, especially with regard to resources that occupy wide territories, are mobile, and are not permanently visible, such as game or fish.

Third, small size and homogeneity do not necessarily facilitate collective management of natural resources. We have thus reviewed a number of analytical arguments and case study materials in which the presence of large, highly motivated agents, or the large size of the user group, promote rather than hinder the efficient management of village-level resources. In particular, we find that inequality is more likely to encourage efficient use of common property resource when it facilitates the establishment of a regulatory authority, and in appropriation problems, when increased inequality reduces the aggregate level of use of the resource, by placing constraints on the individual harvesting efforts of the smaller users. By contrast, when the gamut of available regulatory instruments is limited, inequality between users makes collective agreement and effective enforcement of regulatory schemes more difficult to achieve. In games of voluntary contributions to a common good, the impact of inequality is more ambiguous: while it is generally true that larger users tend to contribute more to the common good, increased inequality also reduces the incentives of small users to contribute.

Lastly, the support of the state is often required to help communities manage their resources. The state can thus play a crucial role in disseminating information about the status of the resources, the relationships between harvesting practices and stocks, and the best management practices available; in imparting skills and administrative capacities needed at the village level; in enforcing community-based property rights; in performing as a mediator of last resort in the event of serious conflicts over resources; etc. Unfortunately, the well-documented experience of recent decades shows that most state interventions have been motivated by the pursuit of private interests that are not compatible with effective support to community-based management. For the success of a co-management approach, it is therefore essential to design and implement appropriate institutional mechanisms that give maximum incentives to both state agents and village communities to act in a way that helps preserve resources in the long term to the greatest benefit of local users. This is a considerable challenge that will necessitate many experiments before it can be effectively met.

Acknowledgements

We are grateful to the Editors as well as David Starrett and Partha Dasgupta for their constructive comments on a earlier draft. Jean-Marie Baland thanks the MacArthur Foundation and the members of the 'Inequality and Economic Performance Group' for support and stimulating conversations.

Appendix. The endogeneity problem in collective action studies

Let us start by describing the model implicitly or explicitly tested by most authors whether they actually use quantitative data or rely on discursive discussions based on qualitative information. Once this is done, we will turn to a critique of the underlying empirical approach. The observation unit is the group of users and the explanandum is the probability that they will collectively organize with a view to regulating access to, and use of, a natural resource or, alternatively, the extent of collective regulation achieved among them. This dependent variable can be hypothesized to depend on the net relative profitability of collective regulation compared to other available institutional arrangements. Any equation attempting to explain success of common property regulation must therefore comprise explanatory variables that bear upon the gross benefits arising from such regulation, the costs involved, and the net benefits achievable under an alternative mode of ownership. The following equation meets such a requirement:

$$Y = Y(\textit{Gains}, \textit{Costs}, \textit{Altern}), \tag{A.1}$$

where Y stands either for the institutional form understood as the extent of collective regulation achieved (or, equivalently, the probability that a set of users collectively regulate the use of a natural resource), or for the degree of effectiveness of collective regulation (i.e., the net gains resulting therefrom) as measured by various performance criteria (incidence of conflicts over use of the resource, incidence of rule conformance, quality of maintenance of the collective infrastructure required for appropriating resource flows, the extent of resource overexploitation, etc.). *Gains* measure the gross benefits achievable with common property regulation, and *Altern* refers to the net gains obtainable under an alternative mode of regulation and ownership, private property or state management and ownership in particular.

The expected gross benefits from collective regulation depend on the attributes of the resource system and the characteristics of the harvesting technology, designated by *Techres*, as well as on other determinants designated by Z^{gains}. For example, it is evident that farmers have not much to gain from coordinating their irrigation efforts if the topography of the service area is unfavorable, or if the basic infrastructure has been poorly devised [see, e.g., Chambers (1988), Sengupta (1991)]. We can therefore write:[32]

$$\textit{Gains} = G\left(\textit{Techres}, Z^{gains}\right). \tag{A.2}$$

On the other hand, the governance costs that common property management entails are influenced by the characteristics of the resource users (in terms of numbers, homogeneity, mobility, previous experiences in community organization, etc.) denoted by *User*,

---

[32] We assume that user characteristics influence net gains through costs rather than through gross benefits.

the aforementioned variable *Techres*, and the official policy and public actions regarding decentralized group initiatives, denoted by *State*. Formally, we have:

$$Costs = C(User, Techres, State). \tag{A.3}$$

Substituting (2) and (3) into (1), we get the following reduced-form equation:

$$Y = Y(Techres, User, State, Altern, Z^{gains}). \tag{A.4}$$

Such is the canonical equation considered in most empirical studies dealing with common property management by user communities. In these studies, the dependent variable is typically defined in terms of performance criteria (see supra).

The model depicted by Equation (A.4) suffers from a major flaw, namely the fact that it overlooks the endogeneity relationships between the organizational form, the expected gains, and the user characteristics. It is indeed difficult to deny that at least some important characteristics of the resource users are not given parameters but rather variables over which the users themselves have some degree of control. More particularly, they may want to modify their own profile and organization so as to make them more conducive to a collective mode of regulation. For example, villagers may control the size of user groups [Wilson and Thompson (1993, pp. 300–312)],[33] or reduce the heterogeneity of group membership along various dimensions such as caste or class composition, wealth, length of residence, location of landholdings, etc. [Sengupta (1991, pp. 114, 119, 128, 167, 189, 192)].

In an analogous manner, it can be argued that users may be able to change some attributes of their resource system or some characteristics of their harvesting technology. This is particularly evident in the case of small-scale irrigation systems, such as storage tanks fed by diversion channels in watershed areas, that can be locally designed according to user-friendly criteria [Sengupta (1991), Chambers (1988)]. In order to avoid confusion and to keep our notation as simple as possible, we assume that the variable *Techres* comprises only attributes and characteristics that are not susceptible of being altered by users. As for those that are susceptible of such alteration, they are subsumed in the *User* variable.

Thus, the variables *Y*, *Gains*, and *User* are endogenous to each other and, as a result, they are simultaneously determined by the user group considered. Analytically, the problem of the user group can be represented as that of maximizing expected net benefits by choosing both the appropriate organizational form and the appropriate user characteristics, to the extent they are manipulable. In formal terms, defining *x* as a particular institutional form belonging to the set *X* (itself a subset of *Y*, since *Y* can also be taken to mean the *performance* of collective regulation), we have:

$$\max_{x \in X, u \in U} Netgains = G(x, Techres, Z^{gains}) - C(x, u, Techres, State) \tag{A.5}$$

---

[33] In their example, liberalization of the *ejidos*'s rules by the Mexican government led members to form grazing coalitions within smaller groups based upon the extended family.

so that

$$x^* = \operatorname*{argmax}_{x \in X} Netgains(u^*, Techres, State, Z^{gains}),$$

$$u^* = \operatorname*{argmax}_{u \in U} Netgains(x^*, Techres, State, Z^{gains}),$$

where the star upperscript indicates that the value of the variable is the equilibrium value. Note that the variable *Altern* does not figure in the above formulation because it is subsumed by the maximization process.

Econometrically, we therefore have a system of three equations to estimate and, owing to the presence of endogeneity, they need to be instrumented for. These three equations are:

$$X = X(User, Gains, Techres, State, Altern, Z^{gains}, R, \varepsilon), \tag{A.6}$$

$$Gains = G(X, User, Techres, State, Altern, Z^{gains}, S, \eta), \tag{A.7}$$

$$User = U(X, Gains, Techres, State, Altern, Z^{gains}, T, \mu), \tag{A.8}$$

where R, S, and T are exogenous variables, and ε, η, and μ random terms, specific to Equations (A.6), (A.7), and (A.8), respectively. Estimating the system (A.6)–(A.8) is obviously more tricky than estimating Equation (A.4), yet it is the only way to ensure that observed facts or relationships are correctly interpreted. In particular, it is essential to measure the effects of user characteristics on collective regulation after having duly controlled for the possible impact of prospective benefits and the organizational form on these characteristics.

Consider the following conclusive statement about the determinants of the relative performances of a large number of irrigation systems located in several (mostly Asian) countries: "Community irrigation systems are likely to be developed and sustained in situations with a reasonable supply of water, no major social cleavages, and low-to-moderate income variance among irrigators. A majority of the bureaucratic cases, on the other hand, are characterized by inadequate supplies of water. Many of them are also characterized by major social cleavages and high income variance" [Tang (1992, p. 124)]. How can we be sure that the state of water supply (a peculiar specification of the *Gain* variable) is not endogenous to the institutional arrangement adopted, rather than being an exogenous condition influencing institutional choice? When the effect of water scarcity is mixed up with that of possible bureaucratic inefficiencies in release of canal water (bureaucratic irrigation systems are always associated with canal irrigation), no definite answer can be provided to that crucial question.[34]

The same problem arises in connection with the user characteristics mentioned, namely social and economic heterogeneity: to what extent is it a given parameter of

---

[34] This difficulty is explicitly mentioned, but not really tackled, in Bardhan (2000).

the institutional choice problem and to what extent an endogenous outcome produced by the operation of the regulatory mode chosen? Thus, for example, collective regulation of irrigation water may involve the scattering of landholdings (particularly those of the economic elite) over the head, tail, and middle of the service area so as to ensure effective maintenance of the entire system and even distribution of water throughout [Coward (1979), Sengupta (1991, pp. 110, 111, 120, 167, 189, 192, 266), Quiggin (1993, p. 1130)]. In so far as this practice forces the big landholders to attend to all parts of the system including the tail-end locations where the poor are likely to have their unique parcel, it may be expected to reduce the effect of inequality in landholdings on income distribution. If the latter is considered as the proper measure of inequality, it is endogenous.

The central difficulty with the estimation of the system (A.6)–(A.8) above is, of course, that instrumentation is bound to be difficult. It is indeed hard to find variables that, for example, influence expected benefits while leaving user characteristics and the organizational form unaffected. This task is even likely to be insurmountable if the sample of resource systems and user groups is large as it is supposed to be in cross-section studies. There remains the solution of estimating a reduced-form equation expressing X as a function of exogenous variables only. Unfortunately, as we know, results thereby obtained do not lend themselves to unambiguous interpretations since they are the outcome of the combined effects of the exogenous variable concerned and all the endogenous variables operating in a hidden way behind the estimated equation.

References

Abraham, A., and J.P. Platteau (2001), "Participatory development in the presence of endogenous community imperfections", Journal of Development Studies (forthcoming).

Abreu, D. (1986), "Extremal equilibria of oligopolistic supergames", Journal of Economic Theory 39(1):191–228.

Abreu, D. (1988), "On the theory of infinitely repeated games with discounting", Econometrica 56(2):383–396.

Agarwal, A., and S. Narain (1989), Towards Green Villages (Centre for Science and the Environment, Delhi).

Aggarwal, R. (2000), "Possibilities and limitations to cooperation in small groups: the case of group owned wells in Southern India", World Development (forthcoming).

Aggarwal, R., and S. Goyal (1999), "Group size and collective action: third-party monitoring in common-pool resources", Comparative Political Studies (forthcoming).

Aggarwal, R., and T. Narayan (1999), "Does inequality lead to greater efficiency in the use of local commons? The role of sunk costs and strategic investments", Mimeo (University of Maryland, College Park, MD).

Alesina, A., and E. La Ferrara (2000), "Participation in heterogeneous communities", Quarterly Journal of Economics, 847–904.

Alexander, P. (1980), "Sea tenure in Southern Sri Lanka", in: A. Spoehr, ed., Maritime Adaptations – Essays on Contemporary Fishing Communities (University of Pittsburgh Press, Pittsburgh) 91–111.

Alexander, P. (1982), Sri Lankan Fishermen – Rural Capitalism and Peasant Society (Australian National University, Canberra).

Allen, R. (1997), "Agriculture and the origins of the state in Ancient Egypt", Explorations in Economic History 34(1):135–154.

Amarasinghe, O. (1989), "Technical change, transformation of risks and patronage relations in a fishing community of South Sri Lanka", Development and Change 20(4):701–733.

Arnold, J.E.M. (2000), "Devolution of control of common-pool resources to local communities: experiences in forestry", in: A. de Janvry, G. Gordillo, J.P. Platteau and E. Sadoulet, eds., Access to Land, Rural Poverty, and Public Action (Clarendon Press, Oxford).

Arnold, J.E.M., and J.G. Campbell (1986), "Collective management of hill forests in Nepal: the Community Forestry Development Project", in: Proceedings of the Conference on Common Property Resource Management, National Research Council (National Academy Press, Washington, DC) 425–454.

Asada, Y., Y. Hirasawa and F. Nagasaki (1983), "Fishery management in Japan", FAO Fisheries Technical Paper No. 238 (FAO, Rome).

Azhar, R.A. (1993), "Commons, regulation, and rent-seeking behavior: the dilemma of Pakistan's *Guzara* forests", Economic Development and Cultural Change 42(1):115–128.

Baland, J.M., and P. Francois (2001), "Commons as insurance and the welfare impact of privatization", Mimeo (University of Namur).

Baland, J.M., and J.P. Platteau (1996), Halting Degradation of Natural Resources – Is there a Role for Rural Communities? (Clarendon Press, Oxford).

Baland, J.M., and J.P. Platteau (1997a), "Coordination problems in local-level resource management", Journal of Development Economics 53:197–210.

Baland, J.M., and J.P. Platteau (1997b), "Wealth inequality and efficiency in the commons. Part I: The unregulated case", Oxford Economic Papers 49(4):451–482.

Baland, J.M., and J.P. Platteau (1998a), "Wealth inequality and efficiency in the commons. Part II: The regulated case", Oxford Economic Papers 50(1):1–22.

Baland, J.M., and J.P. Platteau (1998b), "Dividing the commons – a partial assessment of the new institutional economics of property rights", American Journal of Agricultural Economics 80:644–650.

Baland, J.M., and J.P. Platteau (1999), "The ambiguous impact of inequality on local resource management", World Development 27(5):773–788.

Baland, J.M., and J.A. Robinson (2000), "Is child labor inefficient?", Journal of Political Economy.

Banerjee, A., D. Mookherjee, K. Munshi and D. Ray (2001), "Inequality, control rights and efficiency: a study of sugar cooperatives in Western Maharashtra", Journal of Political Economy (forthcoming).

Bardhan, P. (2000), Irrigation and Cooperation: an Empirical Analysis of 48 Irrigation Communities in South India, Economic Development and Cultural Change (forthcoming).

Bardhan, P., Bowles, S. and H. Gintis (1999), "Wealth inequality, wealth constraints and economic performance", in: F. Bourguignon and A. Atkinson, eds., Handbook on Income Distribution (North-Holland, Amsterdam).

Bardhan, P., M. Ghatak and A. Karaianov (2000), "Inequality and collective action problems", Mimeo (University of California, Berkeley, CA).

Bardhan, P., and C. Udry (1999), Development Microeconomics (Oxford University Press, Oxford).

Barraclough, S.L., and K.B. Ghimire (1995), Forests and Livelihood – The Social Dynamics of Deforestation in Developing Countries (Macmillan Press, London).

Becker, G. (1974), "A theory of social interactions", Journal of Political Economy 82:1063–1093.

Becker, G.S. (1981), A Treatise on the Family (Harvard University Press, Cambridge, MA).

Bergstrom, T., S. Blume and H. Varian (1986), "On the private provision of public goods", Journal of Public Economics 29:25–49.

Berkes, F. (1986), "Local-level management and the commons problem – a comparative study of Turkish coastal fisheries", Marine Policy 10:215–229.

Berkes, F. (1987), "Common-property resource management and cree Indian fisheries in Subarctic Canada", in: B.J. McCay and J.M. Acheson, eds., The Question of the Commons – The Culture and Ecology of Communal Resources (University of Arizona Press, Tucson, AZ) 66–91.

Berry, S. (1984), "The food crisis and agrarian change in Africa: a review essay", African Studies Review 27(2):59–112.

Bhatia, B. (1992), "Lush fields and parched throats: a political economy of groundwater in Gujarat", Economic and Political Weekly 19:A142–170.

Binswanger, H.P. (1991), "Brazilian policies that encourage deforestation in the Amazon", World Development 19(7):821–829.

Binswanger, H., and J. McIntire (1987), "Behavioural and material determinants of production relations in land-abundant tropical agriculture", Economic Development and Cultural Change 36(1):73–99.

Binswanger, H., J. McIntire and C. Udry (1989), "Production relations in semi-arid African agriculture", in: P. Bardhan, ed., The Economic Theory of Agrarian Institutions (Clarendon Press, Oxford) 122–144.

Boserup, E. (1965), The Conditions of Agricultural Growth: The Economics of Agrarian Change under Population Pressure (Allen and Unwin, London).

Bourdieu, P. (1977), Algerie 60: Structures Economiques et Structures Temporelles (Les Editions de Minuit, Paris).

Bourdieu, P. (1980), Le Sens Pratique (Les Editions de Minuit, Paris).

Braverman, A., J.L. Guasch, M. Huppi and L. Pohlmeier (1991), "Promoting rural cooperatives in developing countries: the case of sub-Saharan Africa", World Bank Discussion Papers 121 (World Bank, Washington, DC).

Brightman, R.A. (1987), "Conservation and resource depletion: the case of the boreal forest Algonquians", in: B.J. McCay and J.M. Acheson, eds., The Question of the Commons: The Culture and Ecology of Communal Resources (University of Arizona Press, Tucson, AZ) 121–141.

Bromley, D. (1989), Economic Interests and Institutions – The Conceptual Foundations of Public Policy (Blackwell, Oxford).

Bromley, D.W., and D.P. Chapagain (1984), "The village against the center: resource depletion in South Asia", American Journal of Agricultural Economics 66(5):868–873.

Bruce, J.W. (1986), "Land tenure issues in project design and strategies for agricultural development in Sub-Saharan Africa", Land Tenure Center, LTC Paper No. 128 (University of Wisconsin, Madison, WI).

Carrier, J.G. (1987), "Marine tenure and conservation in Papua New Guinea", in: B.J. McCay and J.M. Acheson, eds., The Question of the Commons – The Culture and Ecology of Communal Resources (University of Arizona Press, Tucson, AZ) 142–167.

Cashdan, E. (1983), "Territoriality among human foragers: ecological models and an application to four Bushman groups", Current Anthropology 24(1):47–66.

Cernea, M. (1985), Putting People First – Sociological Variables in Rural Development (Oxford University Press, Oxford) (published for the World Bank).

Cernea, M. (1989), "User groups as producers in participatory afforestation strategies", World Bank Discussion Papers 70 (World Bank, Washington, DC).

Chamberlin, J. (1974), "Provision of collective goods as a function of group size", American Political Science Review 68:707–716.

Chambers, R. (1988), Managing Canal Irrigation – Practical Analysis from South Asia (Cambridge University Press, Cambridge).

Chan, K.S., S. Mestelman, R. Moir and R.A. Muller (1996), "The voluntary provision of public goods under varying income distributions", Canadian Journal of Economics 19(1):54–69.

Colchester, M. (1994), "Sustaining the forests: the community-based approach in South and South-East Asia", Development and Change 25(1):69–100.

Cordell, J.C., and M.A. McKean (1986), "Sea tenure in Bahia, Brazil", in: National Research Council, Proceedings of the Conference on Common Property Resource Management (National Academy Press, Washington, DC) 85–112.

Cornes, R., and T. Sandler (1985), "The simple analytics of pure public good provision", Economica 52(1):103–116.

Cornes, R., and T. Sandler (1994), "The comparative statics properties of the impure public good model", Journal of Public Economics 54:403–421.

Cornes, R., and T. Sandler (1996), The Theory of Externalities, Public Goods and Club Goods, 2nd edn. (Cambridge University Press, Cambridge).

Cornes, R., S. Mason and T. Sandler (1986), "The commons and the optimal number of firms", Quarterly Journal of Economics 101(3):641–646.

Coward, E.W. (1979), "Principles of social organization in an indigenous irrigation system", Human Organization 38(1):28–36.

Dahlman, C.J. (1980), The Open Field System and Beyond: A Property Rights Analysis of an Economic Institution (Cambridge University Press, Cambridge).

Dasgupta, P. (1993), An Inquiry into Well-Being and Destitution (Clarendon Press, Oxford).

Dasgupta, P., and G. Heal (1979), A Theory of Exhaustible Resources (Nisbet & Co and Cambridge University Press, Cambridge).

Dayton-Johnson, J. (1998), "Rules and cooperation on the local commons: theory with evidence from Mexico", Ph.D. Dissertation (University of California, Berkeley, CA).

Dayton-Johnson, J., and P. Bardhan (1999), "Inequality and conservation on the local commons: a theoretical exercise", Mimeo (University of California, Berkeley, CA).

Di Pasquale, D., and E. Glaeser (1998), "Incentives and social capital: are homeowners better citizens?", NBER Working Paper No. 6363.

Dyson-Hudson, R., and E.A. Smith (1978), "Human territoriality: an ecological reassessment", American Anthropologist 80(1):21–41.

Easter, K., and K. Palanisami (1986), "Tank irrigation in India and Thailand: an example of common property resource management", Mimeo (Department of Agricultural and Applied Economics, University of Minnesota).

Ensminger, J. (1990), "Co-opting the elders: the political economy of state incorporation in Africa", American Anthropologist 92(3):662–675.

Fujita, M., Y. Hayami and M. Kikuchi (1999), "The conditions of collective action for local commons management: the case of irrigation in the Philippines", Mimeo.

Garcia-Barrios, R., and L. Garcia-Barrios (1990), "Environmental and technological degradation in peasant agriculture: a consequence of development in Mexico", World Development 18(11):1569–1585.

Gaspart, F., M. Jabbar, C. Melard and J.P. Platteau (1998), "Participation in the construction of a local public good with indivisibilities: an application to watershed development in Ethiopia", Journal of African Economies 7(2):157–184.

Gaspart, F., and J.P. Platteau (2001), "Collective action for local-level effort regulation: an assessment of recent experiences in Senegalese small-scale fisheries", in: F. Stewart, R. Thorp and J. Heyer, eds., Institutions and Development (Clarendon Press, Oxford).

Gregersen, H., S. Draper and D. Elz (eds.) (1989), "People and trees – the role of social forestry in sustainable development" (World Bank, Washington, DC).

Guha, R. (1985), "Scientific forestry and social change in Uttarakland", Economic and Political Weekly (Special Number) 20(45–47):1939–1952.

Hannesson, R. (1988), "Fishermen's organizations and their role in fisheries management: theoretical considerations and experiences from industrialized countries", FAO Fisheries Technical Paper No. 300 (FAO, Rome).

Hardin, G. (1968), "The tragedy of the commons", Science 162:1243–1248.

Hayami, Y. (1997), Development Economics – From the Poverty to the Wealth of Nations (Clarendon Press, Oxford).

Hayami, Y., and M. Kikuchi (1981), Asian Villages at the Crossroads (University of Tokyo Press, Tokyo, and Johns Hopkins University Press, Baltimore, MD).

Hayami, Y., and V. Ruttan (1985), Agricultural Development – An International Perspective (Johns Hopkins University Press, Baltimore/London).

Heckatorn, D.W. (1993), "Collective action and group heterogeneity: voluntary provision versus selective incentives", American Sociological Review 58:329–350.

Hunt, R.C. (1989), "Appropriate social organization? Water user associations in bureaucratic canal irrigation systems", Human Organization 48:79–90.

Isaac, M., and J.M. Walker (1988), "Group size effects in public goods provision: the voluntary contribution mechanism", Quarterly Journal of Economics 103(1):179–199.

Jessup, T.C., and N.L. Peluso (1986), "Minor forest products as common property resources in East Kalimantan, Indonesia", in: National Research Council, Proceedings of the Conference on Common Property Resource Management (National Academy Press, Washington, DC) 501–531.

Jodha, N.S. (1987), "A case study of the degradation of common property resources in Rajasthan", in: P. Blaikie and P. Brookfield, eds., Land Degradation and Society (Methuen, London and New York).

Jodha, N.S. (1989), "Fuel and fodder management systems in the arid region of Western Rajasthan", International Centre for Integrated Mountain Development, Kathmandu, Nepal.

Johnson, R.N., and G.D. Libecap (1982), "Contracting problems and regulation: the case of the fishery", American Economic Review 72(5):1005–1022.

Kanbur, R. (1992), "Heterogeneity, distribution and cooperation in common property resource management", Policy Research Working Papers (World Bank, Washington, DC).

Kikuchi, M., and Y. Hayami (1980), "Inducements to institutional innovations in an agrarian community", Economic Development and Cultural Change 29(1):21–36.

Kurien, J. (1978), "Entry of big business into fishing – its impact on fish economy", Economic and Political Weekly 13(36):1557–1665.

La Ferrara, E. (2001), "Inequality and participation: theory and evidence from rural Tanzania", Journal of Public Economics (forthcoming).

Laitos, R. (1986), "Rapid appraisal of Nepal irrigation systems", Water Management Synthesis Report No. 43 (Colorado State University, Fort Collins, CO).

Lam, W.F. (1998), Governing Irrigation Systems in Nepal – Institutions, Infrastructure, and Collective Action (ICS Press, Oakland, CA).

Lawry, S.W. (1989a), "Tenure policy and natural resource management in Sahelian West Africa", Land Tenure Center, Research Paper No. 130 (University of Wisconsin, Madison, WI).

Lawry, S.W. (1989b), "Tenure policy toward common property natural resources", Land Tenure Center, Research Paper No. 134 (University of Wisconsin, Madison, WI).

Ledyard, J. (1995), "Public goods: a survey of experimental research", in: J. Kagel and A. Roth, eds., The Handbooks of Experimental Economics (Princeton University Press, Princeton, NJ) 111–194.

Levieil, D.P. (1987), "Territorial use-rights in fishing (TURFS) and the management of small-scale fisheries: the case of lake Titicaca (Peru)", Unpublished Ph.D. Thesis (University of British Columbia, Vancouver).

Levine, H.B. (1984), "Controlling access: forms of 'territoriality' in three New Zealand crayfishing villages", Ethnology 23(2):89–99.

Libecap, G.D. (1989), Contracting for Property Rights (Cambridge University Press, Cambridge).

Mahar, D. (1988), "Government policies and deforestation in Brazil's Amazon region", World Bank Environment Department, Working Paper No. 7.

Mahdi, M. (1986), "Private rights and collective management of water in a High Atlas Berber tribe", in: National Research Council, Proceedings of the Conference on Common Property Resource Management (National Academy Press, Washington, DC) 181–198.

Martin, K.O. (1979), "Play by the rules or don't play at all: space division and resource allocation in a rural Newfoundland fishing community", in: R. Andersen, ed., North Atlantic Maritime Cultures: Anthropological Essays on Changing Adaptations (Mouton, The Hague) 276–298.

Mathew, S. (1990), Fishing Legislation and Gear Conflicts in Asian Countries, Samudra Monograph (Samudra Publications, Brussels).

McCloskey, D.N. (1976), "English open fields as behavior towards risk", in: P. Uselding, ed., Research in Economic History, Vol. 1 (JAI Press, Greenwhich, CT) 124–171.

McKean, M.A. (1986), "Management of traditional common lands (Iriaichi) in Japan", in: National Research Council, Proceedings of the Conference on Common Property Resource Management (National Academy Press, Washington, DC) 533–589.

Menzies, N.K. (1994), Forest and Land Management in Imperial China (Palgrave Macmillan Ltd., New York).

Messerschmidt, D.A. (1986), "People and resources in Nepal: customary resource management systems of the Upper Kali Gandaki", in: National Research Council, Proceedings of the Conference on Common Property Resource Management (National Academy Press, Washington, DC) 455–480.

Moore, M. (1989), "The fruits and fallacies of neoliberalism: the case of irrigation policy", World Development 17(11):1733–1750.

Netting, R.McC. (1972), "Of men and meadows: strategies of alpine land use", Anthropological Quarterly 45:132–144.

Netting, R.McC. (1976), "What alpine peasants have in common: observations on communal tenure in a Swiss village", Human Ecology 4:135–146.

Netting, McC. (1981), Balancing on an Alp (Cambridge University Press, Cambridge).

Netting, R. (1982), "Territory, property and tenure", in: R. Adams, N.J. Smelser and D.J. Treiman, eds., Behavioural and Social Science Research: A National Resource, Part II (National Academy Press, Washington, DC).

Noronha, R. (1997), "Common-property resource management in traditional societies", in: P. Dasgupta and K.-G. Mäler, eds., The Environment and Emerging Development Issues, Vol. 1 (Clarendon Press, Oxford) 48–69.

Nugent, J.B., and N. Sanchez (1993), "Tribes, chiefs, and transhumance: a comparative institutional analysis", Economic Development and Cultural Change 42(1):87–113.

Oliver, P.E., and G. Marwell (1988), "The paradox of group size and collective action: a theory of the critical mass II", American Sociological Review 53:1–8.

Oliver, P.E., G. Marwell and R. Tuxera (1985), "A theory of the critical mass: 1. Interdependence, group heterogeneity and the production of collective action", American Journal of Sociology 91:522–556.

Olson, M. (1965), The Logic of Collective Action (Harvard University Press, Cambridge, MA).

Ostrom, E. (1990), Governing the Commons: The Evolution of Institutions for Collective Action (Cambridge University Press, Cambridge).

Ostrom, E. (1992), Crafting Institutions: Self-governing Irrigation Systems (ICS Press, San Francisco, CA).

Ostrom, E. (2000), "Collective action and the evolution of social norms", Journal of Economic Perspectives 14(3):137–158.

Ostrom, E., and R. Gardner (1993), "Coping with asymmetries in the commons: self-governing irrigation systems can work", Journal of Economic Perspectives 7(4):93–112.

Pagiola, S. (1993), "Soil conservation and the sustainability of agricultural production", Ph.D. Dissertation, Food Research Institute (Stanford University).

Perrings, C. (1996), Sustainable Development and Poverty Alleviation in Sub-Saharan Africa: The Case of Botswana (Macmillan Press, London).

Peters, P.E. (1994), Dividing the Commons – Politics, Policy, and Culture in Botswana (The University Press of Virginia, Charlottesville & London).

Platteau, J.P. (2000), Institutions, Social Norms, and Economic Development (Harwood, London).

Platteau, J.P., and E. Seki (2000), "Coordination and pooling arrangements in Japanese coastal fisheries", in: M. Aoki and Y. Hayami, eds., Community and Market in Economic Development (Clarendon Press, Oxford).

Pomeroy, R.S., B.M. Katon and I. Harkes (1999), "Fisheries co-management: key conditions and principles drawn from Asian experiences", International Center for Living Aquatic Resources Management, Philippines, Mimeo.

Quiggin, J. (1993), "Common property, equality and development", World Development 21(7):1123–1138.

Ribot, J. (1999), "Integral local development: authority, accountability and entrustment in natural resource management", Working Paper, prepared for the Regional Program for the Traditional Energy Sector (RPTES) in the Africa Technical Group of the World Bank, Washington, DC.

Ruddle, K. (1987), "Administration and conflict management in Japanese coastal fisheries", FAO Fisheries Technical Paper No. 273 (FAO, Rome).

Runge, C.F. (1986), "Common property and collective action in economic development", World Development 14(5):623–635.

Schlager, E. (1990), "Model specification and policy analysis: the governance of coastal fisheries", Workshop in Political Theory and Policy Analysis (Indiana University, Bloomington, IN).

Schlager, E. (1994), "Fishers' institutional responses to common-pool resource dilemmas", in: E. Ostrom, R. Gardner and J. Walker, eds., Rules, Games, and Common-Pool Resources (The University of Michigan Press, Ann Arbor, MI) 247–265.

Sengupta, N. (1991), Managing Common Property – Irrigation in India and the Philippines (Sage Publications, New Delhi).

Sethi, R., and E. Somanathan (1996), "The evolution of social norms in common property resource use", American Economic Review 86(4):766–788.

Shanmugaratnam, N. (1996), "Nationalization, privatization and the dilemmas of common property management in Western Rajasthan", Journal of Development Studies 33(2):163–187.

Shanmugaratnam, N., T. Veveld, A. Mossige and M. Bovin (1992), "Resource management and pastoral institution building in the West African Sahel", World Bank Discussion Papers 175 (World Bank, Washington, DC).

Singh, K. (1994), Managing Common Pool Resources (Oxford University Press, Delhi).

Singleton, S. (1999), "Commons problems, collective action and efficiency: past and present institutions of governance in Pacific Northwest Salmon fisheries", Mimeo.

Swallow, B., and D. Bromley (1995), "Institutions, governance and incentives in common property regimes for African rangelands", Environmental and Resource Economics 6(1):99–118.

Tang, S. (1991), "Institutional arrangements and the management of common-pool resources", Public Administration Review 51(1):42–51.

Tang, S. (1992), Institutions and Collective Action – Self-Governance in Irrigation Systems (ICS Press, San Francisco, CA).

Templeton, S., and S. Scherr (1999), "Effects of demographic and related microeconomic change on land quality in hills and mountains of developing countries", World Development 27(6):903–918.

Ternstrom, I. (2001), "Cooperation or conflict in common pools", Stockholm School of Economics, Mimeo.

Thomson, J.T., D.H. Feeny and R.J. Oakerson (1986), "Institutional dynamics: the evolution and dissolution of common property resource management", in: National Research Council, Proceedings of the Conference on Common Property Resource Management (National Academy Press, Washington, DC) 391–424.

Tyler, T.R. (1990), Why People Obey the Law (Yale University Press, New Haven & London).

Wade, R. (1982), "The system of administration and political corruption: canal irrigation in South India", Journal of Development Studies 18(2):287–327.

Wade, R. (1987), "The management of common property resources: finding a cooperative solution", The World Bank Research Observer 2(2):219–234.

Wade, R. (1988a), Village Republics: Economic Conditions for Collective Action in South India (Cambridge University Press, Cambridge).

Wade, R. (1988b), "The management of irrigation systems: how to evoke trust and avoid prisoners' dilemma", World Development 16(4):489–500.

Warr, P.G. (1983), "The private provision of a public good is independent of the distribution of income", Economic Letters 13:207–211.

Weiss L., and J.M. Hobson (1995), States and Economic Development – A Comparative Historical Analysis (Polity Press, Cambridge).

White, T.A., and C.F. Runge (1995), "The emergence and evolution of collective action: lessons from watershed management in Haiti", World Development 23(10):1683–1698.

Wilson, P.N., and G.D. Thompson (1993), "Common property and uncertainty: compensating coalitions by Mexico's pastoral *Ejidatorios*", Economic Development and Cultural Change 41(2):299–318.

Wittfogel, K.A. (1957), Oriental Despotism. A Comparative Study of Total Power (Yale University Press, New Haven, CT).

Chapter 5

POPULATION, POVERTY, AND THE NATURAL ENVIRONMENT

PARTHA DASGUPTA

University of Cambridge, UK
and
Beijer International Institute of Ecological Economics, Stockholm, Sweden

Contents

Handbook of Environmental Economics, Volume 1, Edited by K.-G. Mäler and J.R. Vincent

Abstract

This chapter studies the interface in poor countries of population growth, rural poverty, and deterioration of the local natural-resource base, a subject that has been much neglected by modern demographers and development economists. The motivations for procreation in rural communities of the poorest regions of the world are analyzed, and recent work on the relevance of gender relationships to such motivations is summarized. Four potentially significant social externalities associated with fertility behavior and use of the local natural-resource base are identified. Three are shown to be pronatalist in their effects, while the fourth is shown to be ambiguous, in that it can be either pro- or anti-natalist. It is shown that one of the externalities may even provide an invidious link between fertility decisions and the use of the local natural-resource base. The fourth type of externality is used to develop a theory of fertility transitions in the contemporary world. The theory views such transitions as disequilibrium phenomena.

Keywords

externality, fertility rate, social capital, genuine investment, conformity, gender relations, commons, resilience, bifurcation, social equilibrium, wealth, birth control, education, GNP per head

JEL classification: D62, I3, J13, O12, O13, O47, Q01, Q12, Q2

Prologue

A society's environmental requirements are a function of its demand for goods and services, some of which are produced (e.g., food, housing, transport, education), while others are obtained directly from the natural-resource base (e.g., air for breathing, river water for drinking, microorganisms for treating waste, birds and bees for seed dispersal and pollination). The overall demand made on the environment depends on the average demand per person for goods and services, which in turn depends on existing institutions and the knowledge base, and the technologies that are thereby in use. But population size also contributes to the overall demand. The famous "I = PAT" equation of Ehrlich and Holdren (1971), that Impact on the environment is a function of Population, Affluence and Technology, is a simple metaphor for the complicated set of relationships that exist among the four variables in question.

This chapter is about a more restricted set of relationships than those summarized in the "I = PAT" equation. I do not discuss the ways in which affluence impinges on the natural-resource base,[1] but instead study pathways in which population growth and environmental degradation at the local level today relate to local poverty–pathways that would be expected to perpetuate poverty. The focus here is on the population–poverty–environment nexus in the poorest regions of the contemporary world. I make no attempt to forecast the future, nor do I try to review how societies that are currently affluent grew in population even while accumulating wealth by substituting knowledge, skills, and manufactured capital for natural resources.[2] My sole aim here is to use economic theory and the recent historical experience in poor regions to suggest a way of thinking about the population–poverty–environment nexus in the contemporary world. I do not suggest that the experience I summarize below had anything inevitable about it. There were public choices that could have been made and that would have resulted in superior collective outcomes. I shall argue, however, that such choices were ignored, in part because of inadequate economic analysis.

Table 1 summarizes evidence on poverty, population growth, and one aspect of environmental deterioration in the contemporary world, namely, loss of forest cover. Poverty [sometimes it is called "extreme poverty"; see World Bank (1991)], is taken to be the condition of a person living on less than a dollar a day. As the table shows, there were 1.2 billion poor people in the world at the turn of the century. So, the poor are about a fifth of the world's population. They are concentrated in China, South Asia, and sub-Saharan Africa, numbering in excess of one billion. But there are differences in the incidence of poverty even among those three regions: as proportions of populations, South Asia and sub-Saharan Africa are home to the largest numbers of poor people.

Table 1 also records the high rates of population growth that South Asia and sub-Saharan Africa have experienced in recent decades: population has grown in both regions in excess of 2 percent per year. Although the proportion of poor people declined

---

[1] On which, see McNeill (2000), Arrow et al. (2002), and several of the chapters in this Handbook.

[2] Landes (1969, 1998) are outstanding treatises on that experience.

Table 1
Poverty* and population growth

Region	Number in 1998 (millions)	HI[†] (%)	Annual population growth (%) 1980–1998	Annual deforestation (% of forest area) 1990–2000
East Asia & Pacific (excluding China)	65	11	1.5	
China	213	18	1.3	−1.2
Europe and Central Asia	24	5	0.6	
Latin America and the Caribbean	78	16	1.8	
South Asia	522 (495)[‡]	40		0.1
Bangladesh			2.1	
India			2.0	
Pakistan			2.6	
Sub-Saharan Africa	290 (242)[‡]	46	2.8	0.8
Total	1,192 (1,270)[‡]	24	1.6	

*People living on less than $1 a day in 1998.
[†]Headcount Index (HI): proportion of people that are poor.
[‡]People living on less than $1 a day in 1990.
Source: World Bank (2000a, Table 1.1; 2000b, Table 2.1; 2001, Table 3.4).

almost everywhere during the 1990s, the number of poor people increased in South Asia and sub-Saharan Africa: population growth in the two regions was high relative to the alleviation of poverty. Table 2 offers a picture of the high rates of population growth in terms of crude birth and death rates. The two tables, taken together, show that increases in population size have come about because declines in mortality rates during the second half of the last century (a remarkably good thing) were not matched by reductions in fertility rates. Population increase has brought in its wake additional pressure on local resource bases (a not-so-good thing).

The poor live in unhealthy surroundings, a fact that is both a cause and effect of their poverty.[3] Nearly two million women and children die annually in poor countries from exposure to indoor pollution. (Cooking can be a lethal activity among the poor.) Additionally, over 70 percent of fresh water sources are contaminated or degraded. Moreover, groundwater withdrawal in poor countries exceeds natural recharge rates by a phenomenal 160 billion cubic meters per year. World Bank (2001) suggests that 5–12 million hectares of land are lost annually to severe degradation, and that soil degradation affects 65 percent of African croplands and 40 percent of croplands in Asia. Table 1 also gives

[3] The mutual causation is explored in Dasgupta (1993, 1997).

Table 2
Crude birth and death rates per 1000 people

	B^*		$D^\dagger$		$B-D$	
	1980	1996	1980	1996	1980	1996
China	18	17	6	7	12	10
Bangladesh	44	28	18	10	26	18
India	35	25	13	9	22	16
Pakistan	47	37	15	8	32	29
Sub-Saharan	47	41	18	14	29	27
Africa (Nigeria)	50	41	18	13	32	28
World	27	22	10	9	17	13

*Crude birth rate per 1000 people.
†Crude death rate per 1000 people.
Source: World Bank (1998, Table 2.2).

figures for rates of deforestation relative to forest areas in the poorest regions of the world. Of course, taken alone, the latter figures say nothing about environmental degradation, but subsequently we will see how deforestation can play havoc with the lives of those who inhabit forests.[4]

The emotive term in all this is "population growth"; so emotive, that it elicits widely different responses. Some people believe population growth to be among the causes of world poverty and environmental degradation [e.g., Ehrlich and Ehrlich (1990], while others permute the elements of that causal chain, arguing, for example, that contemporary poverty in poor countries is the cause, rather than the consequence, of rapid population growth and environmental damage. Here is a collection of refrains in the latter category of beliefs: "poverty is the problem, not population"; "poor people cannot afford to conserve their resource base"; "a lack of female autonomy is the problem, not population"; "development is the best form of contraceptive"; "contraceptives are the best form of contraceptive".[5] A recent expression of this line of thinking is World Bank (2000a), which, even though it was devoted to a study of world poverty, took little note of its possible links with population growth and deterioration of the natural environment.

To be sure, there are views about the population–poverty–environment nexus that go well beyond mere agnosticism. Some people have claimed that even in the poorest coun-

---

[4] CES (1982), Agarwal (1986, 1989), Cruz and Repetto (1992), Cleaver and Schreiber (1994), Kalipeni (1994), Seymour and Dubash (2000), and Shapiro (2001) contain accounts of environmental degradation in the poorest regions of the world.

[5] See, for example, Dyson and Moore (1983), World Bank (1984), Birdsall (1988), Robey, Rutstein and Morris (1993), Sen (1994), and Bardhan (1996).

tries today, population growth can be expected to provide a spur to economic progress.[6] Thus, it would seem not only that our attitudes toward population size and its growth differ, but that there is no settled view on how the matter should be studied. As with religion and politics, many people have opinions on population that they cling to with tenacity.

1. Plan of the chapter

I argue below that such divergence of opinion is unwarranted. Differences persist because the interface of population, resources, and welfare at a geographically local level has been a relatively neglected subject. In Section 2 I show that modern demographers and development and environmental economists have neglected crucial aspects of the population–poverty–environment nexus. Neglect by experts is probably the reason why the nexus has attracted popular discourse, which, while often illuminating, is frequently descriptive rather than analytical.

In Section 3 I go into the question of why experts have neglected the nexus. Several reasons are offered, each having to do with a bias in the character of the evidence that has been studied. In Section 4 evidence is produced to show that neglect of the nexus has been a serious mistake. It has not been uncommon, however, among those who *have* written about population, resources, and welfare to adopt a global, future-oriented view: the emphasis frequently has been on the deleterious effects a large and increasingly affluent population would have on Earth in the future. This slant has been instructive, but it has drawn attention away from the economic misery and ecological degradation endemic in large parts of the world today. Disaster is not something for which the poorest have to wait; it is a frequent occurrence. Moreover, among the rural poor in poor countries, decisions on fertility, on allocations concerning education, food, work, health-care, and on the use of the local natural-resource base are in large measure reached and implemented within households that are unencumbered by compulsory schooling and visits from social workers, that do not have access to credit and insurance in formal markets, that cannot invest in well-functioning capital markets, and that do not enjoy the benefits of social security and old-age pension. These features of rural life direct me, in Section 5, to study the interface of population growth, poverty, and environmental stress from a multitude of household, and ultimately individual, viewpoints.

Women's education and reproductive health have come to be seen in recent years as the most effective channels for influencing fertility. Sections 6–8 contain an outline of the theoretical and empirical reasons why they are so seen. An interesting feature of both education and reproductive health is that they can be studied within a framework where households make decisions in isolation of other households. The theory of demand for education and reproductive health can therefore be treated as a branch of the

---

[6]   See, for example, Simon (1981), Bauer (2000), and Johnson (2001). Boserup (1981) is the classic on this, but her work reviewed the past, it was not a commentary on the current state of affairs.

"new household economics", which has been much engaged in the study of the isolated household.[7]

But theoretical considerations imply that interhousehold linkages also influence fertility decisions. Of particular interest are the linkages between market and non-market activities. They give rise to "externalities", by which I mean the side-effects of human activities when the latter are undertaken without mutual agreement. Such activities include those in which women's education and reproductive health play a role. The findings I report are therefore consistent with the contemporary emphasis on women's education and reproductive health.

Four types of externalities are identified. Three are unambiguously pronatalist, while the fourth–conformist behavior–can result in either pro- or anti-natalist behavior. A theory of demographic transitions, based on the fourth mechanism, is then offered. These matters are explored in Section 9 and the Appendix. I conclude that there is substance to what has been called the "population problem". I also argue that in the Indian subcontinent and sub-Saharan Africa, the problem has for a long while been an expression of human suffering.

The population problem facing those two regions was not inevitable: it would have been avoided if social scientists had diagnosed it early enough and if good governance had not been an especially acute commodity there. I argue below that several aspects of the population problem may well persist even when those regions make the transition to low fertility rates. Section 10 summarizes the policy prescriptions that follow from the analysis.

2. Framing links between population, resources, and welfare

It is appropriate first to identify some of the ways social scientists have framed the links between population growth, resources, and human welfare. I review them in this section. The outline will enable us to compare and contrast the way the links have generally been framed with the way I frame them here.

There are three sets of examples to discuss here. They concern (1) the way population growth and economic stress in poor countries are studied by environmental and resource economists, (2) the way modern theories of economic growth view fertility and natural resources, and (3) the way development economists accommodate environmental stress in their analysis of contemporary poverty. The examples are discussed in the next three sub-sections.

2.1. Demography and economic stress in environmental and resource economics

The environmental and resource economics that has been developed in the United States has not shown much interest in economic stress and population growth in poor

---

[7] The modern classic is Becker (1981).

countries. In their survey of the economics of environmental resources, Kneese and Sweeney (1985, 1993), Cropper and Oates (1992), and Oates (1992) altogether by-passed the subject matter of this chapter. They were right to do so, for the prevailing literature regards the environmental-resource base as an "amenity". Indeed, it is today a commonplace that "... (economic) growth is good for the environment because countries need to put poverty behind them in order to care" (*Independent*, 4 December 1999), or that "... trade improves the environment, because it raises incomes, and the richer people are, the more willing they are to devote resources to cleaning up their living space (*The Economist*, 4 December 1999; p. 17).

I quote these views only to show that the natural environment is widely seen as a luxury. This view is hard to justify when one recalls that our natural environment maintains a genetic library, sustains the processes that preserve and regenerate soil, recycles nutrients, controls floods, filters pollutants, assimilates waste, pollinates crops, operates the hydrological cycle, and maintains the gaseous composition of the atmosphere. Producing as it does a multitude of ecosystem services, the natural-resource base is in large part a necessity.[8] A wide gulf separates the perspective of environmental and resource economists in the North (I use the term in its current geopolitical sense) from what would appear to be the direct experience of the poor in the South.[9]

2.2. Population and resources in modern growth theories

In its turn, modern theories of economic growth for the most part assume population change to be a determining factor of human welfare. A central tenet of the dominant theory is that although population growth does not affect the long-run rate of change in living standards, it adversely affects the long-run standard of living [Solow (1956)].

Recent models of economic growth have been more assertive. They lay stress on new ideas as a source of progress, supposing that the growth of ideas is capable of circumventing any constraint the natural environment may impose on the ability of economies to grow indefinitely. Such models note too that certain forms of investment (e.g., research and development) enjoy cumulative returns because the benefits are durable and can be shared collectively. The models also assume that growth in population leads to an increase in the demand for goods and services. An expansion in the demand and supply of ideas implies that in the long run, equilibrium output per head can be expected to grow at a rate that is itself an increasing function of the rate of growth of population (it is only when population growth is nil that the long run rate of growth of output per head is nil). The models regard indefinite growth in population to be beneficial.[10]

---

[8] See also Arrow et al. (1995) and Dasgupta, Levin and Lubchenco (2000), who discuss the implications of the fact that destruction of ecosystems are frequently not reversible.

[9] For first-hand accounts of daily life under the stresses of resource scarcity, see Agarwal (1986, 1989), Narayan (2000), and Jodha (2001). For attempts to develop the economics of such conditions, see Dasgupta (1982, 1993, 1995, 1996, 1997a, 1998a, 2000a).

[10] Jones (1998) contains a review of contemporary growth models.

In their pristine form, contemporary growth models embed an assumed positive link between the creation of ideas (technological progress) and population growth in a world where the natural-resource base comprises a fixed, indestructible factor of production.[11] The problem with the latter assumption is that it is wrong: the natural environment consists of degradable resources (soil, watersheds, fisheries, and sources of fresh water).[12] It may be sensible to make that wrong assumption when studying a period of time when natural-resource constraints did not bite, but it is not sensible for studying development possibilities open to today's poor regions.

Contemporary growth theory does not explicitly model the nature of the new products that embody technological progress. One can only conjecture that it assumes future innovations to be of such a character that indefinite growth in output would make no more than a finite additional demand on the natural-resource base. The assumption is questionable. The point is that if economic growth (i.e., growth in gross national product, GNP) is to be sustainable, capital accumulation and the presumed technological progress must be able to counter a declining resource base. But a vanishing resource base would mean a dwindling supply of the multitude of mostly un-understood ecosystem services upon which life depends. Additionally, property rights to environmental resources are often either vaguely defined or weakly enforced, meaning that environmental services are in all probability underpriced in the market. New technologies would therefore be expected to be rapacious in their use of natural resources: inventors and innovators would have little reason to economize on their use.[13] But this could mean that new ideas do not offer adequate substitutes for local resource bases, in a world where out-migration is frequently not a viable option to what is an intolerable rural situation.[14]

In any event, we should be sceptical of a theory that places such enormous burden on an experience that is not much more than two hundred years old [Maddison (2001)]. Extrapolation into the past is a sobering exercise: over the long haul of history (a 5000 years stretch, say, up to about two hundred years ago), economic growth even in the currently-rich countries was for most of the time not much above zero.[15] The study of

---

[11] Kremer (1993) develops such a model to account for 1 million years of world economic history.

[12] Daily (1997) is a collection of essays on the character of ecosystem services. See also Levin (2001) for an exhaustive collection of summaries of what is currently known about biodiversity and the role it plays as a productive asset.

[13] The literature on environmental externalities is huge. See, for example, Meade (1973), Mäler (1974), and Baumol and Oates (1975). The implications of the underpricing of environmental resources on the direction of technological change are explored in Dasgupta (1996).

[14] See Agarwal (1986), Kalipeni (1994), and Chopra and Gulati (2001) for fine empirical studies on rural poverty and resource depletion in semi-arid regions.

[15] See Fogel (1994, 1999), Johnson (2000), and, especially, Maddison (2001). The claim holds even if the past two hundred years were to be included. The rough calculation is simple enough:
World per capita output today is about 5000 US dollars. The World Bank regards one dollar a day to be about as bad as it can be. People would not be able to survive on anything substantially less than that. It would then be reasonable to suppose that 2000 years ago per capita income was not less than a dollar a day. So, let us assume that it was a dollar a day. This would mean that per capita income 2000 years ago was about

possible feedback loops between poverty, demographic behavior, and the character and performance of both human institutions and the natural-resource base is not yet on the research agenda of modern growth theorists.

2.3. *Population and resource stress in development economics*

Nor is the population-poverty-resource nexus a focus of attention among development economists. Even in studies on the semi-arid regions of sub-Saharan Africa and the Indian sub-continent, the nexus is largely absent. For example, the authoritative surveys by Birdsall (1988), Kelley (1988) and Schultz (1988) on population growth in poor countries fail to touch on environmental matters. Mainstream demography also makes light of environmental stress facing poor communities in sub-Saharan Africa and the Indian sub-continent. Nor does the dominant literature on poverty [e.g., Stern (1989), Dreze and Sen (1990), Bardhan (1996)] take population growth and ecological constraints to be important factors in development possibilities. Textbooks frequently go no further than to point to the Western experience since the Industrial Revolution and conclude that Malthus got it all wrong.

The situation is puzzling. Much of the rationale for development economics is the notion that poor countries suffer particularly from institutional failures. But institutional failures in great measure manifest themselves as externalities. To ignore population growth and ecological constraints in the study of poor countries would be to suppose that demographic decisions and resource use there give rise to no externalities of significance; it would also be to suppose that externalities arising from institutional failure have a negligible effect on resource use and demographic behavior. I know of no body of empirical work that justifies such presumptions.

3. Why the neglect?

How is one to account for these neglects? It seems to me there are four reasons, one internal to the development of the "new household economics", the others arising from limitations in global statistics.

3.1. *Isolated households*

For reasons of tractability, those who developed the new household economics studied choices made by isolated, optimizing households.[16] Such predictions of the theory, as

---

350 dollars a year. Rounding off numbers, this means very roughly speaking, that per capita income has risen about 16 times since then. This in turn means that world income per head has doubled every 500 years, which in its turn means that the average annual rate of growth has been about 0.14 percent per year, a figure not much in excess of zero.

[16] The early works are cited in Becker (1981). Hotz, Klerman and Willis (1997) survey the field by studying fertility decisions in developed countries. Scheffer (1997) makes thorough use of the new household economics for studying the demand for children in poor countries.

that increases in women's labour productivity reduce the household demand for children, are borne out in cross-country evidence [Scheffer (1997)]. Nevertheless, the study of isolated households is not a propitious one by which to explore the possibilities of collective failure among households. For example, there have been few attempts to estimate reproductive externalities. One reason is that the theory of demographic interactions in nonmarket environments is still underdeveloped; and without theory it is hard for the empiricist to know what to look for.[17] I later point to scattered evidence, drawn from anthropology, demography, economics, and sociology, of externalities resulting in pro-natalist attitudes among rural households in poor countries. I also try to develop some of the analytical techniques that would be required for identifying such externalities. The directional predictions of the resulting theory are not at odds with those of the new household economics (e.g., that an increase in women's labour productivity lowers the demand for children); but their predictions differ on the magnitude of household responses.

3.2. Cross-country statistics on the effects of population growth on the standard of living

The second reason for the neglect of the population-poverty-resource nexus is the outcome of an enquiry made some time ago into the economic consequences of population growth [National Research Council (1986)]. Drawing on national time-series and cross-regional data, the investigators observed that population size and its growth can have both positive and negative effects. For the purposes of interpreting the data, population growth was regarded as a causal factor in the study. The investigators concluded that there was no cause for concern over the high rates of growth being experienced in poor countries.[18]

But regression results depend on what is being regressed on what. So, for example, there can be set against National Research Council (1986) and cross-country studies that have found a positive link between population growth and economic development, those by Mauro (1995) and Barro (2001), who have found a negative correlation between population growth and economic growth and a positive correlation between population growth and the magnitude of absolute poverty. In short, cross-country regressions in which population growth is a determining factor have given us mixed messages. Later

---

[17] Surveying the field, Schultz (1988, pp. 417, 418) wrote: "Consequences of individual fertility decisions that bear on persons outside of the family have proved difficult to quantify, as in many cases where social external diseconomies are thought to be important ... The next step is to apply ... microeconomic models (of household behavior) to understand aggregate developments in a general equilibrium framework. But progress in this field has been slow".

[18] Kelley (1988) contains a review of the findings. See also the survey of empirical growth economics by Temple (1999), who adopts a sceptical view regarding the deleterious consequences of population growth in poor countries.

Table 3
Total fertility rates and GNP per head

	TFR		y^*	$g(y)^\dagger$
	(1980)	1998	1998	1965–1998
China	(2.5)	1.9	3,050	6.8
Bangladesh	(6.1)	3.1	1,410	1.4
India	(5.0)	3.2	2,060	2.7
Pakistan	(7.0)	4.9	1,650	2.7
Sub-Saharan Africa	(6.6)	5.4	1,440	−0.3
(Nigeria)	(6.9)	5.3	740	0.0
USA	(1.8)	2.0	29,240	1.6
World	(3.7)	2.7	6,300	1.4

*GNP per head (dollars at purchasing power parity).
†Percentage rate of growth of GNP per head, calculated from constant price GNP in national currency.
Source: World Bank (2000b, Tables 1.1, 1.4, and 2.16).

in this chapter I show that even though we may have learnt something from cross-country regressions, they have frequently misdirected us into asking wrong questions on demographic matters.

3.3. Global statistics on living standards

The third reason stems from a different set of empirical findings. With the exception of sub-Saharan Africa over the past 30 years or so, GNP per head has grown in nearly all poor regions since the end of World War II. In addition, since 1960 growth in world food production has exceeded the world's population growth by an annual rate of approximately 0.6 percent. This has been accompanied by improvements in a number of indicators of human welfare, such as the infant survival rate, life expectancy at birth, and literacy. In poor regions each of the latter improvements has occurred in a regime of population growth rates substantially higher than in the past: excepting for East Asia and parts of South and Southeast Asia, modern-day declines in mortality rates have not been matched by reductions in fertility.

Table 3 presents total fertility rates (*TFR*), GNP per head, and growth in GNP per head in several countries and groups of countries.[19] Between 1980 and 1998 the *TFR* declined everywhere, but very unevenly. Sub-Saharan Africa has displayed the most acute symptoms of poverty: continued high fertility rates allied to declining GNP per head in what is a very poor continent. Nevertheless, as Table 3 confirms, the oft-expressed fear that

---

[19] The total fertility rate (*TFR*) is the number of live births a woman would expect to have if she were to live through her childbearing years and to bear children at each age in accordance with the prevailing age-specific fertility rates. If the *TFR* were 2.1 or thereabouts, population in the long run would stabilize.

rapid population growth will accompany deteriorations in living standards has not been borne out by experience when judged from the vantage of the world as a whole. It is then tempting to infer from this, as does Johnson (2000, 2001) most recently, that in recent decades population growth has not been a hindrance to improvements in the circumstances of living.

3.4. Negative cross-country link between income and fertility

The fourth reason stems from economic theory and cross-country data on the link between household income and fertility. Imagine that parents regard children to be an end in themselves; that is, assume children to be a "consumption good". If, in particular, children are a "normal" consumption good, an increase in unearned income would lead to an increase in the demand for children, other things being equal. This is the "income effect".[20] In his well-known work Becker (1981) argued, however, that if the increase in household income were the result of an increase in wage rates (i.e., an increase in labour productivity), then the cost of children would increase, because time is involved in producing and rearing them. But other things being equal, this would lead to a decrease in the demand for children (this is the "substitution effect"). It follows that a rise in income owing to an increase in labour productivity would lead to a decline in fertility if the substitution effect were to dominate the income effect, a likely possibility.

Figure 1, taken from Birdsall (1988), shows that among countries which in the early 1980s had incomes above 1000 US dollars, those that were richer experienced lower fertility rates. A regional breakdown of even the Chinese experience displays the general pattern: fertility is lower in higher-income regions [Birdsall and Jamison (1983)]. These are only simple correlations and, so, potentially misleading. Moreover, they do not imply causality. But they suggest that growth in income can be relied upon to reduce population growth.

4. Why the neglect is wrong

There are four weaknesses with the reasoning just outlined.

4.1. GNP per head and environmental quality

Earlier, it was noted that the environment is frequently viewed as a luxury. The contemporary source of that perspective is World Bank (1992), which suggested that there is an empirical relationship between GNP per head and concentrations of industrial pollutants. Based on the historical experience of OECD countries, the authors observed that, when GNP per head is low, concentrations of atmospheric pollutants (e.g., sulphur dioxide (SO_2)) increase as GNP per head increases, but when GNP per head is

---

[20] Schultz (1997) confirms this for a pooled set of cross-country data.

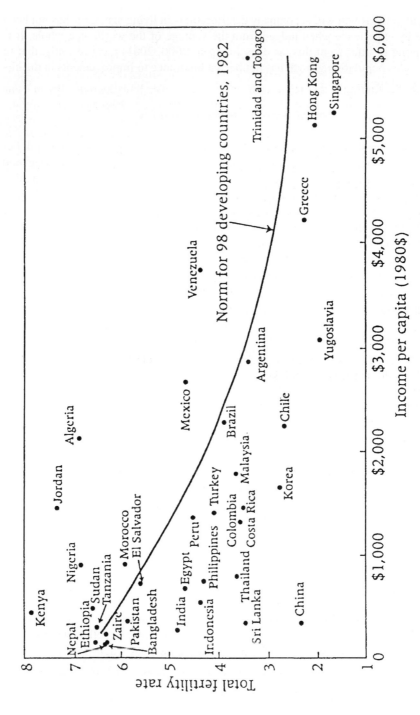

Figure 1. Fertility in relation to income in developing countries, 1982. Source: Birdsall (1988, p. 482).

high, concentrations decrease as GNP per head increases further. In short, it was found that the functional relationship between GNP per head and concentrations of industrial pollutants has an inverted-U shape. Among economists this relationship has been christened the "environmental Kuznets curve".[21] Unfortunately, the moral that has been drawn from the finding is that resource degradation is reversible: degrade all you want now, you can always recover the stock later, because Earth can be relied upon to rejuvenate it.

The science of biodiversity has shown that the presumption is false: the existence of ecological thresholds implies that damage to ecosystems can be irreversible (see below). It should not come as a surprise that the environmental Kuznets curve was detected for mobile pollutants. Mobility means that, so long as emissions decline, the stock at the site of the emissions will decline. As an overarching metaphor for "tradeoffs" between manufactured and natural capital, the relationship embodied in the environmental Kuznets curve has to be rejected.[22]

4.2. GNP growth versus wealth accumulation

The second weakness with the reasoning outlined in the previous section is that conventional indices of the standard of living – GNP and the United Nations' Human Development Index (*HDI*) – reflect commodity production, not the "productive base" upon which life depends. Neither GNP nor *HDI* accounts for the depreciation the productive base may be subjected to. So it can be that an economy's GNP increases for a period even while its productive base shrinks.

An economy's productive base includes not only manufactured capital, human capital, natural capital and knowledge, but also its institutions (public, private, and civic). Together they influence the production, distribution, and use of goods and services. A society's productive base is the source of its well-being. This base is a diverse collection of durable objects, some tangible and alienable (buildings and machinery, land and animals, trees and shrubs), some tangible but non-alienable (human beings, the oceans), some intangible but alienable (codified pieces of knowledge, such as patentable ideas), some intangible and non-alienable (air, skills, the legal framework, and cultural coordinates), and some that are yet to be defined in an acceptable way (social capital).[23]

It is often suggested that poor countries are poor when judged in terms of manufactured and human capital, but are frequently rich in natural and social capital.[24] As

---

[21] It is a misnomer. The original Kuznets curve, which was an inverted U, related income inequality to real national income per head on the basis of historical cross-country evidence.

[22] For more extensive discussions of the environmental Kuznets curve, see Arrow et al. (1995) and the responses it elicited in symposia built round the articles in *Ecological Economics* 15(1) (1995), *Ecological Applications* 6(1) (1996), and *Environment and Development Economics* 1(1) (1996). See also the special issue of *Environment and Development Economics* 2(4) (1997).

[23] On social capital and the difficulties surrounding its definition, see the essays in Dasgupta and Serageldin (2000).

[24] See World Bank (1997) on the share of natural capital in the wealth of poor countries.

illustration, it is frequently observed that a wide variety of coping mechanisms vulnerable people rely upon in poor countries are supplied by what may be called "civic" (or "informal") institutions, not the state, nor markets. This may well be a valid point. On the other hand, informal institutions do not have the reach of either the state or markets to be able to pool risks effectively. Moreover, civic institutions are by their very nature exclusive, not inclusive, and therefore not capable of offering their members the economic opportunities that markets in principle offer. It is possible too that informal institutions are fragile in the face of growing markets, so that they become increasingly undependable during the process of modernization. In any event, what matters is the way an economy's productive base is used for protecting and promoting human welfare. A country could be well endowed in natural assets, but if its institutions are weak (or worse, dysfunctional), not much good would come from them.

It can be shown that the correct measure of a community's welfare over the long run is its wealth. By wealth I mean the social worth of the community's productive base, which includes not only manufactured and human capital, but also natural and institutional, or social, capital. So, wealth is an index of an economy's productive base. It can be shown that a community's long run welfare increases over a period of time if on average the ratio of net investment in its productive base to its wealth exceeds the average population growth rate during the period.[25] This means that if development is to be sustainable, an economy must undertake "genuine" investment in its productive base, relative to population growth. By genuine investment I mean aggregate net investment in its capital assets, meaning thereby net investment (in the aggregate) in its manufactured, human and natural capital, and in changes in the economy's institutions, as measured by the rate at which total factor productivity changes. Since it is possible for both a country's GNP and its agricultural production to increase over a period of time even while its wealth declines, time series of GNP per head and agricultural production can mislead. It is also possible for a country's Human Development Index (*HDI*) to increase even while its wealth declines. This means that time series of *HDI* too can mislead. The point is that, statistics on, say, an increase in GNP per head in a country (or in the country's *HDI*) do not reveal whether or not the increase is being realized by means of a depletion of natural capital, for example, by "mining" soil and water and not replacing natural capital by an adequate accumulation of other forms of capital. In relying on GNP and other current-welfare measures, such as life expectancy at birth and infant survival, we run the danger of ignoring the concerns ecologists have voiced about pathways linking population growth, economic activity, and the state of the natural-resource base.[26]

[25] The exact proposition involves a number of qualifications, which I am ignoring for the moment. See Dasgupta and Mäler (2000), Dasgupta (2001a), and Arrow et al. (2003). For many years economists (including the present author) argued that GNP should be replaced by net national product (NNP) as a measure of social well-being so as to accommodate environmental concerns. But there are problems with NNP [see Dasgupta (2001a)].

[26] For a fuller discussion of this, see Daily et al. (1998) and Dasgupta (2001b).

4.2.1. Public institutions, social capital, and total factor productivity growth in poor countries

How should investment in an economy's productive base be measured? Among the various categories of capital assets, social capital has proved to be conceptually the most problematic. A large contemporary literature interprets social capital to be an economy's informal institutions.[27] However, one of the earliest formulations of the concept [Coleman (1988)] was based on the idea that social capital is an index of interpersonal networks. Notice that, to the extent transactions within such networks do not create externalities, social capital in Coleman's interpretation is a component of human capital. On the other hand, to the extent interpersonal networks are public goods, their efficacy is reflected in macroeconomic accounting by the magnitude of an economy's total factor productivity.

It was once customary to regard total factor productivity as a summary index of public knowledge. In recent years, however, it has become common practice to include in it the character of an economy's institutions. Notice though that to include the latter would be to acknowledge that in imperfect economies total factor productivity can have short bursts in either direction. Imagine, for example, that a government reduces economic inefficiencies by improving the enforcement of property rights, or by reducing centralized regulations (import quotas, price controls, and so forth). We would expect the factors of production to find better uses. As factors realign in a more productive fashion, total factor productivity increases.

In the opposite vein, total factor productivity could decline for a period. Increased government corruption could be a cause, as would civil strife, which destroys capital assets and damages a country's institutions. When institutions deteriorate, assets are used even more inefficiently than previously. This would appear to have happened in sub-Saharan Africa over the past forty years.

Table 4, taken from Collins and Bosworth (1996), gives estimates of the annual rate of growth of GNP per head and its breakdown between two factors of production (manufactured and human capital) in various regions of the world. The estimates are given in the first three columns. The period was 1960–1994. The fourth column represents the percentage rate of change in total factor productivity (the "residual") in each region. This is simply the difference between figures in the first column and the sum of the figures in the second and third columns.[28] Collins and Bosworth did not include natural capital as a factor of production. If the use of environmental services has grown during the period in question (a most likely possibility), we should conclude that the residual is an overestimate. Even so, the residual in Africa was negative (−0.6 percent annually). The true residual was in all probability even lower. The residual in South Asia, the other

---

[27] See, for example, the Special Issue of the *Journal of Interdisciplinary History* 29(3) (1999), on "Patterns of social capital, Part I".

[28] Subject to rounding-off errors.

Table 4
Sources of economic growth, 1960–1994

	$g(Y/L)$	$g(K)$	$g(H)$	$g(A)$
East Asia	4.2	2.5	0.6	1.1
South Asia	2.3	1.1	0.3	0.8
Africa	0.3	0.8	0.2	−0.6

$g(Y/L)$: annual percentage rate of change in GNP per head.
$g(K)$: share of GNP attributable to manufactured capital multiplied by annual percentage rate of change in manufactured capital.
$g(H)$: share of GNP attributable to human capital multiplied by annual percentage rate of change in human capital.
$g(A)$: percentage rate of change in total factor productivity (residual).
Source: Collins and Bosworth (1996).

really poor region of the world, was 0.8 percent annually, but as this is undoubtedly an overestimate, I am unclear about whether or not there has been growth in total factor productivity in that part of the world. One can but conclude from the evidence, such as it is, that two of the poorest regions of the world (the Indian subcontinent and sub-Saharan Africa) have not improved their institutional capabilities over four decades, nor have they been able to improve productivity by making free use of knowledge acquired in advanced industrial nations.

With this evidence as background, I study some crude estimates of genuine investment in the Indian sub-continent and sub-Saharan Africa.

4.2.2. Investment in the productive base

Reporting studies undertaken at the World Bank, Hamilton and Clemens (1999) have provided estimates of genuine investment in a number of countries.[29] The authors included in the list of a country's assets its manufactured, human, and natural capital. There is a certain awkwardness in many of the steps they took to estimate genuine investment. For example, investment in human capital in a given year was taken to be public expenditure on education, which is an overestimate, because each year people die and take their human capital with them. That is depreciation, and should have been

---

[29] That an economy needs to raise the worth of its productive base if social well-being is to increase was the topic of discussion, starting in 1993, among members of an Advisory Council created by Ismail Serageldin, then Vice President for Environmentally Sustainable Development at the World Bank. Serageldin (1995) provided an outline of empirical work on genuine investment that was initiated in his Vice Presidency. For estimates of the depreciation of natural capital on a regional basis, see Pearce, Hamilton and Atkinson (1996) and World Bank (1997).

Table 5
Genuine investment and capital deepening in selected regions: 1970–1993

	I/Y^*	$g(L)^\dagger$	$g(W/L)^\ddagger$	$g(Y/L)^\S$	$\Delta(HDI)^\P$
Bangladesh	−0.003	2.3	−2.40	1.0	+ve
India	0.107	2.1	−0.50	2.3	+ve
Nepal	−0.015	2.4	−2.60	1.0	+ve
Pakistan	0.082	2.9	−1.70	2.7	+ve
Sub-Saharan Africa	0.047	2.7	−2.00	−0.2	+ve
China	0.144	1.7	1.09	6.7	−ve

$^*I/Y$: genuine investment as proportion of GNP. Genuine investment includes total health expenditure (i.e., public plus private), estimated as an average during 1983–1993, World Health Organization. *Source*: Hamilton and Clemens (1999, Tables 3 and 4) and personal communication from Katie Bolt, World Bank.
$^\dagger g(L)$: average annual percentage rate of growth of population, 1965–1996. *Source*: World Bank (1998, Table 1.4).
$^\ddagger g(W/L)$: average annual percentage rate of change in *per capita* wealth. Assumed output-wealth ratio: 0.15.
$^\S g(Y/L)$: average annual percentage rate of change in *per capita* GNP, 1965–1996. *Source*: World Bank (1998, Table 1.4).
$^\P \Delta(HDI)$: sign of change in UNDP's Human Development Index, 1987–1997. *Source*: UNDP (1990, 1999).

deducted. Moreover, they did not include expenditure on health, at least a part of which should be viewed as investment in human capital.

Among the resources making up natural capital, only commercial forests, oil and minerals, and the atmosphere as a sink for carbon dioxide were included. (Not included were water resources, forests as agents of carbon sequestration, fisheries, air and water pollutants, soil, and biodiversity.) So there is an undercount, possibly a serious one. Nevertheless, one has to start somewhere and, despite the limitations in the data, it is instructive to use them to check whether or not in recent decades the representative person in the world's poorest regions has grown wealthier. Table 5 allows us to do that. The account that follows covers sub-Saharan Africa, the Indian sub-continent, and China. Taken together, they contain nearly half the world's population. They also comprise nearly all the world's poorest countries.

The first column of figures in Table 5 contains estimates of genuine investment, as a proportion of GNP, in Bangladesh, India, Nepal, Pakistan, China, and sub-Saharan Africa, averaged over the period 1970–1993.[30] Notice that Bangladesh and Nepal have disinvested: their overall capital stocks shrank during the period in question. In contrast, genuine investment was positive in China, India, Pakistan, and sub-Saharan Africa. This could suggest that the latter countries were wealthier at the end of the period than at the beginning. But when population growth is taken into account, the picture changes.

The second column contains annual growth rates of population in the regions over the period 1965–1996. As Table 5 shows, all but China have experienced growth in

[30] Figures in the first column include total expenditure in health. I am grateful to Katie Bolt of the World Bank for the computations.

excess of 2 percent per year, sub-Saharan Africa and Pakistan having grown in numbers at nearly 3 percent per year. We now want to estimate genuine investment as a ratio of wealth over the period 1970–1993, and compare the cumulative value of that ratio to the change in population size over that same period. To do this, we multiply (genuine) investment as a proportion of GNP by the average output–wealth ratio of an economy to arrive at the (genuine) investment–wealth ratio, and compare that to the rate of growth of population. In adopting this route for estimating changes in the productive base per head, I am regarding the entire period 1970–1993 as a point of time. If the ratio of genuine investment to wealth exceeds the rate of growth of population at a point in time, development is sustainable at that point in time; if, on the other hand, it is less than the rate of growth of population, development is *un*sustainable at that point in time.[31]

Because a wide variety of capital assets (e.g., human and natural capital) are un-accounted for in national accounts, there is an upward bias in published estimates of output–capital ratios, which traditionally have been taken to be something like 0.30. In what follows, I assume the output–wealth ratio to be 0.15, as a check against the upward bias in traditional estimates. This is almost certainly still a conservatively high figure.

The third column in Table 5 contains my estimates of the difference between the (genuine) investment to wealth ratio and the rate of growth of population. The striking message of the column is that all but China have decumulated their capital assets on a per capita basis over the past thirty years or so. The conclusion for sub-Saharan Africa may not cause surprise, since that region is widely known to have regressed in terms of most socio-economic indicators. But the figures in Table 5 for Bangladesh, India, Nepal, and Pakistan should cause surprise. The data, such as they are, say that, relative to its population, the Indian sub-continent's productive base has shrunk. The calculations, taken together, imply that there has been disinvestment on a per capita basis in regions that in total housed over 2 billion people in 1998. Disinvestment has been a common occurrence in the poorest regions of the world.[32]

[31] In the following footnote I offer the complete account of the steps I have taken to construct Table 5.

[32] Formally, the proposition is the following:

Let $K_i(t)$, $k_i(t)$, and $Y(t)$, respectively, be the stock of capital asset i, the per capita stock of capital asset i, and GNP at time t. Let $p_i(t)$ be the accounting price of i. Assume that the period to be studied is $[0, T]$. Let $I(t) = \sum_i \{p_i(t)dK_i(t)/dt\}$ be genuine investment at t. Hamilton and Clemens (1999) provided estimates of $I(t)/Y(t)$. In the third column of Table 5, I have compiled estimates of time averages of $[\sum_i \{p_i(t)dk_i(t)/dt\}]/[\sum_i \{p_i(t)k_i(t)\}]$ for the interval $[0, T]$. The estimates are upper bounds of the true figures. How did I arrive at the estimates?

Consider first a region where $I(t)/Y(t)$ was positive (China, India, and Pakistan). Let $\beta(t) = Y(t)/W(t) > 0$, where $W(t) = \sum_i \{p_i(t)K_i(t)\}$, which is wealth. Define β^* by the equation

$$\int_0^T \left[\frac{I(t)}{W(t)}\right]dt \equiv \int_0^T \left[\frac{\beta(t)I(t)}{Y(t)}\right]dt = \beta^* \int_0^T \left[\frac{I(t)}{Y(t)}\right]dt.$$

Notice that β^* is a time average of $\beta(t)$. Let n be the average growth rate of population. Now suppose that the right-hand side of the above equation is less than nT. I would then be justified in concluding that the

How do changes in the productive base per head compare with changes in conven-tional measures of the quality of life? The fourth column of Table 5 contains estimates of the annual percentage rate of change in per capita GNP during 1965-1996; and the fifth column shows whether UNDP's Human Development Index (*HDI*) has improved or deteriorated over the period 1987–1997.

Notice how misleading our retrospective assessment of long-term economic develop-ment in the Indian subcontinent would be if we were to look at growth rates in GNP per head. Pakistan, for example, would be seen as a country where GNP per head grew at a healthy 2.7 percent per year, implying that the index doubled in value between 1965 and 1993. In fact, the average Pakistani became poorer by a factor of nearly 1.5 during that same period.

Bangladesh too has disinvested in its productive base. The country is recorded as having grown in terms of GNP per head at a rate of 1 percent per year during 1965–1996. In fact, at the end of the period the average Bangladeshi was about half as wealthy as she was at the beginning.

The case of sub-Saharan Africa is, of course, especially sad. At an annual rate of decline of 2.0 percent, the average person in the region becomes poorer by a factor of 2 every thirty-five years. The ills of sub-Saharan Africa are routine reading in today's newspapers and magazines. But they are not depicted in terms of a decline in wealth. Table 5 shows that sub-Saharan Africa has experienced an enormous decline in its pro-ductive base over the past three decades.

India can be said to have avoided a steep decline in its productive base. But it has been at the thin edge of economic development, having managed not quite to maintain its capital assets relative to population size. If the figures in Table 5 are taken literally, the average Indian was slightly poorer in 1993 than in 1970.

Even China, so greatly vaunted for its progressive economic policies, has just man-aged to accumulate wealth in advance of population increase. Moreover, the estimates

productive base per head had declined in the region in question. (More pertinently, we would be justified in concluding that long run welfare had declined.)

The problem is that we do not know $\beta(t)$. Therefore, we cannot estimate β^*. So, I write β_{max} as the maximum value of $\beta(t)$ during $[0, T]$. Obviously, $\beta_{max} > \beta^*$. It would be surprising if β^* were in excess of 0.15 per year: I know of no modern estimate of a country-wide capital–output ratio that is less than 3 years; and as those estimates do not include human capital and natural capital, it would be astonishing if even β_{max} were as high as 0.25 per year. Very conservatively, then, I assumed $\beta^* = 0.15$ per year. Figures for n were given in column 2 of Table 5. It is then simple to confirm that the productive base per capita in India, Pakistan, and sub-Saharan Africa declined, as shown in Table 5. The Hamilton–Clemens data imply that in those three "countries", $\beta(t)$ increased over the period. In India and Pakistan, GNP per head grew, even while the productive base per head declined. Economic growth took place in tandem with a "mining" of the natural-resource base, relative to population growth. In sub-Saharan Africa, even GNP per head declined.

For a region where genuine investment was on average negative (Bangladesh and Nepal), the matter is simpler: given that population had grown in both countries, the productive base per head must have declined there. Now per capita wealth could not possibly have declined at a rate less than n (the rate of decline would equal n only if the output-wealth ratio were zero, a patently absurd figure). So I used 0.15 per year for estimating the average rate of change in per capita wealth even in Bangladesh and Nepal.

of genuine investment do not include soil erosion or urban pollution, both of which are thought to be especially problematic in China.[33]

What of *HDI*? In fact, it misleads even more than GNP per head. As the third and fifth columns show, *HDI* offers precisely the opposite picture of the one we should obtain when judging the performance of countries.

These are all rough and ready figures, but they show how accounting for population growth and natural capital can make for substantial differences in our conception of the development process. The implication should be sobering: over the past three decades the Indian subcontinent and sub-Saharan Africa, two of the poorest regions of the world, comprising something like a third of the world's population, have become poorer. In fact, some of the countries in these regions have become a good deal poorer. The evidence, such as there is, suggests that disinvestment has been a common occurrence in poor countries. We may conclude tentatively that the character of contemporary development in poor countries is unsustainable. Given what we know about the quality of governance in poor countries, we can but conclude that for the most part, people in poor countries both consume and invest too little.

4.3. Weak link between income and fertility within poor countries

The third weakness of the reasoning summarized in the previous section is that among poor countries the relationship between per capita income and fertility is not strong. In Figure 1 countries with GNP per head under $1,000 display nearly the entire range of fertility rates prevailing in the mid-1980s: from 2 to 8 births per woman. Notice that countries lying above the fitted curve are in sub-Saharan Africa, those below are in Asia. I will seek an explanation for this. Admittedly, Figure 1 displays a bivariate distribution, which could be misleading for a problem requiring multi-variate analysis. The figure nonetheless reflects the possibility that among poor households in rural communities the aforementioned substitution effect is not large and cancels the income effect.[34] This could be because responsibility for child-rearing is frequently diffused over the extended family, a matter to which we return below.

4.4. Aggregation can mislead

The fourth weakness of the reasoning summarized in the previous section is that global statistics are overly aggregative. They gloss over spatial variations and disguise the fact that even though the world economy as a whole has enjoyed economic growth over the past fifty years or so, large masses of people in particular regions have remained in

---

[33] Hussain, Stern, and Stiglitz (2000) analyze why China has been the economic success it is widely judged to have been in recent years. However, they do not enquire what has been happening to China's natural-resource base in the process of the country's economic development.

[34] Dreze and Murthi (2001 have found no effect of income on fertility in a pooled set of district level data from India.

poverty (Tables 1 and 2). Economic growth has not "trickled down" consisiently to the poorest, nor have the poorest been inevitably "pulled up" by it.[35]

Landes (1969, 1998) has argued that the discovery of vast numbers of ways of substituting natural resources among one another and of substituting ideas and manufactured capital for natural resources resulted in the Industrial Revolution in the eighteenth century. The extraordinary economic progress experienced in Western Europe and North America since then, and in East Asia more recently, has been another consequence of finding new ways to substitute manufactured goods and services for environmental and natural resources and services, and then of bringing about the substitutions.[36] Spatial dispersion of ecosystems enabled this to happen. The ecological transformation of rural England in the Middle Ages probably reduced the nation's biodiversity, but it increased income without any direct effect on global productivity.

But that was then, and we are in the here and now. The question is whether it is possible for the scale of human activity to increase substantially beyond what it is today without placing undue stress on the major ecosystems that remain. The cost of substituting manufactured capital for natural resources can be high. Low-cost substitutes could turn out to be not so low-cost if accounting prices are used in the costing, rather than market prices. Depleting certain types of natural capital and substituting it with manufactured capital can be socially uneconomic.

5. Population, poverty, and natural resources: Local interactions

In view of the arguments that were offered in the previous section, a few investigators have studied the interface of population, poverty, and the natural-resource base at the local level. The ingredients of their work have been around for some time; what is perhaps new is the way they have been put together. Several models have been constructed to develop the new perspective: we are still far from having an overarching model of the kind economists are used to in the theory of general competitive equilibrium.[37] Some models have as their ingredients large inequalities in land ownership in poor countries and the non-convexities that prevail at the level of the individual person in transforming nutrition intake into nutritional status and, thereby, labour productivity.[38] Others are based on the fragility of interpersonal relationships in the face of an expanding labour market and underdeveloped credit and insurance markets.[39] Yet others are built on possible links between fertility behavior and free-riding on local common-property resources.[40]

---

[35] See the interchange between Johnson (2001) and Dasgupta (2001b) on the usefulness of studying country-level statistics on demographic and environmental matters.

[36] During the past two centuries gross domestic product per head in the currently industrialized countries has increased some nineteen-fold. See Maddison (2001, Table 1-9b).

[37] In this respect, the literature I am alluding to resembles much contemporary economic theory.

[38] Dasgupta and Ray (1986, 1987) and Dasgupta (1993, 1997b).

[39] Dasgupta (1993, 1998a, 1999).

[40] Dasgupta and Mäler (1991, 1995), Nerlove (1991), Cleaver and Schreiber (1994), and Brander and Taylor (1998).

The forces that create poverty traps are accentuated by the fact that ecological processes are usually non-convex.[41] Thus, the biophysical impacts created by the degradation of an ecosystem may be small over a considerable range, but then become immense once a critical threshold is reached. Ecologists have a name for the phenomenon when a system reaches such a threshold: loss of resilience. Resilience means the capacity of the ecosystem to absorb disturbances without undergoing fundamental changes in its functional characteristics. If an ecosystem loses its resilience, it can flip to a wholly new state when subjected to even a small perturbation. Recovery could then be costly; it may even be impossible, which is to say that the flip may be irreversible. Formally, an ecosystem's loss of resilience amounts to moving to a new stability domain—a bifurcation. For example, shallow lakes have been known to flip from clear to turbid water as a consequence of excessive runoff of phosphorus from agriculture [Scheffer (1997), Carpenter, Ludwig and Brock (1999)]. Such flips can occur over as short a period as a month. The transformation of grasslands into shrublands, consequent upon non-adaptive cattle-management practices is another example of a loss of resilience of an ecosystem [Perrings and Walker (1995)]. Human populations have on occasion suffered from unexpected flips in their local ecosystems. Fishermen on Lake Victoria and the nomads in the now-shrublands of southern Africa are examples from recent years. Taken together, the new perspective on population, poverty and the natural-resource base sees the social world as self-organizing itself into an *in*homogeneous whole, so that, even while parts grow, chunks get left behind; some even shrink. To put it colloquially, these models account for locally-confined "vicious circles".[42]

Later in this chapter I present an outline of this work when seen through one particular lens, namely reproductive and environmental externalities, and I report the arguments that have shaped it and on the policy recommendations that have emerged from it. The framework I develop focuses on the vast numbers of small, rural communities in the poorest regions of the world and it identifies circumstances in which population growth, poverty, and resource degradation can be expected to feed on one another, cumulatively, over periods of time. What bears stressing is that my account does not regard any of the three to be the prior cause of the other two: over time each of them influences, and is in turn influenced by, the other two. In short, they are all endogenous variables.

The models under discussion assume that people, when subjected to such "forces" of positive feedback, seek mechanisms to cope with the circumstances they face. The

---

[41] See Chapter 2 (by Levin and Pacala).

[42] Myrdal (1944) called such forms of feedback "cumulative causation". Brock and Durlauf (1999), Levin (1999), and Blume and Durlauf (2001) offer fine accounts of locally interacting structures.

The possible reversal of the Gulf Stream, which now warms northern Europe, represents another example of a potential bifurcation of an ecosystem [Rahmstorf (1995)]. Climate models indicate that such reversals can occur if the rates and magnitude of greenhouse gases increase sufficiently, although the threshold point is not known. It is clear from paleoclimatic history that such events were common. Mastrandrea and Schneider (2001) have employed a linked climate-economy model to investigate the future possibilities of climate thresholds of this type, and assess the implications for climate policy.

models also identify conditions in which this is not enough to lift communities out of the mire. Turner and Ali (1996), for example, have shown that in the face of population pressure in Bangladesh, small land-holders have periodically adopted new ways of doing things so as to intensify agricultural production. However, the authors have shown too that this has resulted in only an imperceptible improvement in the standard of living and a worsening of the ownership of land, the latter probably owing to the prevalence of distress-sales of land. These are the kind of findings that the new perspective anticipated and was designed to meet.

Economic demographers have given scant attention to reproductive externalities. An important exception was an attempt by Lee and Miller (1991) at quantifying the magnitude of reproductive externalities in a few developing countries. The magnitude was found to be small. The authors searched for potential sources of externalities in public expenditures on health, education and pensions, financed by proportional taxation. But such taxes are known to be very limited in scale in poor countries. Moreover, the benefits from public expenditure are frequently captured by a small proportion of the population. So perhaps it should not be surprising that the reproductive externalities consequent upon public finance are small in poor countries. The externalities I study here are of a different sort altogether.

As we would expect from experience with models of complex systems, general results are hard to come by. The models that have been studied analytically are only bits and pieces. But they offer strong intuitions. They suggest also that we are unlikely to avoid having to engage in simulation exercises if we are to study models less specialized than the ones that have been explored so far.[43] This should have been expected. It would seem that for any theoretical inference, no matter how innocuous, there is some set of data from some part of the world over some period that is not consonant with it.[44] Over 40 years of demographic research have uncovered that the factors underlying fertility behavior include not only the techniques that are available to households for controlling their size, but also the household demand for children. The latter in particular is influenced by a number of factors (e.g., child mortality rates, level of education of the parents, rules of inheritance) whose relative strengths would be expected to differ across cultures, and over time within a given culture, responsive as they are to changes in income and wealth and the structure of relative prices. Thus, the factors that would influence the drop in the total fertility rate in a society from, say, 7 to 5 should be expected to be different from those that would influence the drop from 5 to 3 in the same society.

Across societies the matter is still more thorny. The springs of human behavior in an activity at once so personal and social as procreation are complex and interconnected, and empirical testing of ideas is fraught with difficulty. Data often come without appropriate controls. So, what may appear to be a counter-example to a thesis is not necessarily so. Intuition is often not a good guide. For example, one can reasonably imagine

---

[43] Lutz and Scherbov (1999) offer a thoughtful review of why and how.
[44] See Cleland (1996) for a demonstration of this.

that since religion is a strong driving force in cultural values, it must be a factor in fertility behavior. Certainly, in some multivariate analyses [e.g., Dreze and Murthi (2001), in their work on district-level data from India], religion has been found to matter (Muslims are more pronatalist than Hindus and Christians). But in others [e.g., Iyer (2000) in her work on household-level data from a group of villages in the state of Karnataka, India], it has not been found to matter. Of course, the difference in their findings could result from the fact that the unit of analysis in one is the district, while in the other it is the household. But such a possibility is itself a reminder that complicated forms of externalities (e.g., externalities arising from conformist behavior) may be at work in fertility decisions.

6. Education and birth control

Education and reproductive health programmes together are a means for protecting and promoting women's interests. They were the focal points of the 1994 United Nations Conference on Population and Development in Cairo and are today the two pillars upon which public discussion on population is based.[45] Later in this chapter I show that the "population problem" involves a number of additional features. Here I review what is known about the influence of education and reproductive-health programmes on fertility.

6.1. Women's education and fertility behavior

In two classic publications, Cochrane (1979, 1983) studied possible connections between women's education and fertility behavior. She observed that lower levels of education are generally associated with higher fertility. Table 6, based on the Demographic and Health Surveys undertaken in Africa in the 1980s, displays this for Botswana, Ghana, Uganda and Zimbabwe. The finding has proved to be intuitively so reasonable that social scientists have attributed causality: from education to reduced fertility.

What are the likely pathways of the causal chain? Here are some:

Education helps mothers to process information more effectively and so enables them to use the various social and community services that may be on offer more intensively. The acquisition of education delays the age of marriage and so lowers fertility. In populations with generally low levels of education and contraceptive prevalence, literacy and receptiveness to new ideas complement the efforts of reproductive health programmes,

---

[45] To illustrate, with a random, but representative example, I quote from a letter to the *Guardian* newspaper written by Anthony Young of Norwich, UK, on 24 April 2000. Tracing the prevailing famine in Ethiopia to overpopulation relative to Ethiopia's resource base, he writes: "There is an ethically acceptable set of measures for reducing rates of population growth: improvement in the education and status of women, coupled with making family planning services available to all".

Table 6
Women's education and fertility rates, 1

Country	Education level	TFR
Botswana	none	5.8
	1–4 years	5.5
	5–7 years	4.7
	8+ years	3.4
Ghana	none	6.8
	1–4 years	6.6
	5–7 years	6.0
	8+ years	5.5
Uganda	none	7.9
	1–4 years	7.3
	5–7 years	7.0
	8+ years	5.7
Zimbabwe	none	7.2
	1–4 years	6.7
	5–7 years	5.5
	8+ years	3.7

Source: Jolly and Gribble (1993, Table 3.6).

leading to longer birth spacing.[46] This in turn reduces infant mortality, which in its turn leads to a decline in fertility.

Turning to a different set of links, higher education increases women's opportunities for paid employment and raises the opportunity cost of their time (the cost of childrearing is higher for educated mothers). Additionally, educated mothers would be expected to value education for their children more highly. They would be more likely to make a conscious tradeoff between the "quality" of their children and their numbers [Becker (1981)].[47]

Yet Cochrane herself was reluctant to attribute causality to her findings, as have investigators studying more recent data [Cohen (1993), Jolly and Gribble (1993)], for the

---

[46] Above low levels of education and contraceptive use, however, women's education and family planning outreach activities appear to be substitutes.

[47] Subsequent to Cochrane's work, studies have found a positive association between maternal education and the well-being of children, the latter measured in terms of such indicators as household consumption of nutrients, birth spacing, the use of contraceptives, infant- and child-survival rates, and children's height [see Dasgupta (1993, Chapter 12) for references]. As an indication of orders of magnitude, the infant mortality rate in households in Thailand where the mother has had no education (respectively, has had primary and secondary education) was found to be 122 per 1000 (respectively, 39 and 19 per 1000). See World Bank (1991). However, a common weakness of many such empirical studies is their "bivariate" nature.

In a pooled cross-section data-set for poor countries in the 1970s and 80s, Schultz (1997) has found that the total fertility rate is negatively related to women's and men's education (the latter's effect being smaller), as well as to urbanization and agricultural employment; and positively related to unearned income and child mortality. This is what the new household economics would lead one to expect.

reason that it is extremely difficult to establish causality. Women's education may well reduce fertility. On the other hand, the initiation of childbearing may be a factor in the termination of education. Even when education is made available by the state, households frequently choose not to take up the opportunity: the ability (or willingness) of governments in poor countries to enforce school attendance or make available good education facilities is frequently greatly limited. Economic costs and benefits and the mores of the community to which people belong influence their decisions. It could be that the very characteristics of a community (e.g., an absence of associational activities among women, or a lack of communication with the outside world) that are reflected in low education attainment for women are also those giving rise to high fertility. Demographic theories striving for generality would regard both women's education and fertility to be endogenous variables. The negative relationship between education and fertility in such theories would be an association, not a causal relationship. The two variables would be interpreted as "moving together" in samples, nothing more. In a later section I explore a theoretical framework that offers this interpretation.[48]

However, the links between women's education and fertility are not as monotonic as I have reported so far. Set against the positive forces outlined above is a possible effect that runs the other way: taboos against postpartum female sexual activity, where they exist, can be weakened through the spread of education. In sub-Saharan Africa, where polygamy is widely practiced, postpartum female sexual abstinence can last up to three years after childbirth. It is also not uncommon for women to practice total abstinence once they have become grandmothers. The evidence, such as they exist, is consistent with theory: in Latin America and Asia, primary education, when compared to no education, has been found to be associated with lower fertility, but in several parts of sub-Saharan Africa (e.g., Burundi, Kenya and Nigeria) the relationship has been found to be the opposite. Table 7 displays the latter.[49] The conventional wisdom that women's education is a powerful force against pronatalism needs to be qualified: the level of education can matter.

6.2. Family planning

Except under conditions of extreme nutritional stress, nutritional status does not appear to affect fecundity [Bongaarts (1980)]. During the 1974 famine in Bangladesh, deaths in excess of those that would have occurred under previous nutritional conditions numbered round 1.5 million. The stock was replenished within a year [Bongaarts and Cain (1981)]. Of course, undernourishment can still have an effect on sexual reproduction,

---

[48] In their analysis of district-level data in India over the 1981 and 1991 censuses, Dreze and Murthi (2001) have come closer than any other study I know to claiming that a causal link exists between women's education and fertility. But their study was not designed to test the kind of theoretical reasoning I am pursuing here.

[49] Hess (1988) has conducted time-series analysis that attests to a positive association between primary education and fertility in parts of sub-Saharan Africa.

Table 7
Women's education and fertility rates, 2

Country	Education level	*TFR*
Burundi	none	6.9
	1–4 years	7.1
	5–7 years	7.3
	8+ years	5.8
Kenya	none	7.2
	1–4 years	7.7
	5–7 years	7.2
	8+ years	5.0
Nigeria	none	6.5
	1–4 years	7.5
	5–7 years	6.0
	8+ years	4.5

Source: Jolly and Gribble (1993, Table 3.6) and Cohen (1993, Table 2.4).

through its implications for the frequency of stillbirths, maternal and infant mortality, and reductions in ovulation and the frequency of sexual intercourse.

An obvious determinant of fertility is the available technology for birth control. Cross-country regressions [e.g., Pritchett (1994)] confirm that the fraction of women of reproductive age who use modern contraceptives is strongly and negatively correlated with total fertility rates. So it should not be surprising that family planning programmes are often seen as a prerequisite for any population policy. But these regression results mean only that contraception is a proximate determinant of fertility, not a causal determinant. The results could mean, for example, that differences in fertility rates across nations reflect differences in fertility goals, and thus differences in contraceptive use. Of course, the causal route could go the other way. The very existence of family planning programmes might influence the demand for children, as women come to realize that it is reasonable to want a small family [Bongaarts (1997)].

People in all societies practice some form of birth control: fertility everywhere is below the maximum possible. Extended breastfeeding and postpartum female sexual abstinence have been common practices in Africa. Even in poor countries, fertility is not unresponsive to the relative costs of goods and services. In a study on Kung San foragers in the Kalahari region, Lee (1972) observed that the nomadic, bush-dwelling women among them had average birth spacing of nearly four years, while those settled at cattle posts gave birth to children at much shorter intervals. From the viewpoint of the individual nomadic Kung San woman, the social custom is for mothers to nurse their children on demand and to carry them during their day-long trips in search of wild food through the children's fourth year of life. Anything less than a four-year birth interval would increase mothers' carrying loads enormously, threaten their own capacity to survive, and reduce their children's prospects of survival. In contrast to bush dwellers, cattle-post women are sedentary and are able to wean their children earlier.

Traditional methods of birth control include abortion, abstinence or rhythm, coitus interruptus, prolonged breastfeeding, and anal intercourse.[50] These options are often inhumane and unreliable; modern contraceptives are superior. Nevertheless, successful family planning programmes have proved more difficult to institute than could have been thought possible at first [Cochrane and Farid (1989)]. Excepting a few countries, fertility rates in sub-Saharan Africa have not shown significant declines, despite reductions in infant mortality rates over the past decades.

In a notable article, Pritchett (1994) analyzed data from household surveys conducted by the World Fertility Survey and the Demographic and Health Surveys programmes, which included women's responses to questions regarding both their preferences and their behavior related to fertility. Demographers had earlier derived indicators of the demand for children from these data. One such indicator, the "wanted total fertility rate" [Bongaarts (1990)], can be compared to the actual total fertility rate for the purpose of classifying births or current pregnancies in a country or region as "wanted" or "unwanted". Regressing actual fertility on fertility desires in a sample of 43 countries in Asia, Africa, and Latin America, Pritchett found that about 90 percent of cross-country differences in fertility rates are associated with differences in desired fertility. Moreover, excess fertility was found not to be systematically related to the actual fertility rate, nor to be an important determinant of the rate. The figure 90 percent may prove to be an over-estimate, but it is unlikely to prove to be greatly so.[51] Even in poor households the use of modern contraceptives would involve only a small fraction (1 percent or thereabouts) of income.

Pritchett's is a significant finding, if only because it directs us to ask why the household demand for children differs so widely across communities. We turn to this matter next.

7. The household and gender relations

The concept of the household is not without its difficulties. It is often taken to mean a unit of housekeeping or consumption. The household in this sense is the eating of meals together by members, or the sharing of meals derived from a common stock of food [Hajnal (1982)]. This definition has the merit of being in accordance with most modern censuses, but one problem with it is that in rural communities it does not yield exclusive units [Goody (1996)]. A household shares a "table" and may, for example, include live-in servants who do not cook for themselves. In many cases some meals are had in common, while others are not; and often raw and cooked food is passed to parents in adjacent cottages, apartments, or rooms. The boundaries vary with context,

---

[50] Anthropologists have argued, however, that in parts of western sub-Saharan Africa prolonged breastfeeding is not a birth-control measure, but a means of reducing infant mortality: traditionally, animal milk has been scarce in the region.

[51] I am grateful to John Bongaarts for helpful conversations on this matter.

especially where food is not consumed together round a table (as in Europe) but in bowls in distinct groups (as in sub-Saharan Africa). In none of these cases is the housekeeping unit the same as the consumption unit, nor is the consumption unit necessarily clearly defined.

Economists have taken the household to be a well-defined concept, but have debated whether it is best to continue to model it as a unitary entity, in the sense that its choices reflect a unitary view among its members of what constitutes their welfare (the utility maximizing model) or whether instead the household ought to be modelled as a collective entity, where differences in power (e.g., between men and women) manifest themselves in the allocation of food, work, education, and health care.

Of course, one cannot conclude that households are not unitary from the mere observation that intrahousehold allocations are unequal. Poor households would choose to practice some patterns of inequality even if they were unitary. For example, since children differ in their potential, parents in poor households would help develop the most promising of their children even if that meant the remaining ones are neglected. This is confirmed by both theory and evidence [Becker and Tomes (1976), Bledsoe (1994)]. Daughters are a net drain on parental resources in patrilineal and patrilocal communities, such as those in northern India (dowries can be bankrupting). This fact goes some way toward explaining the preference parents show for sons there [Sopher (1980a, b), Dyson and Moore (1983), Cain (1984)] and why girls of higher birth order are treated worse than girls of lower birth order [Das Gupta (1987)]. In northern parts of India the sex ratio is biased in favor of men.[52]

Nevertheless, the magnitude of the inequalities frequently observed is at odds with what would be expected in unitary households. The indirect evidence also suggests that the household is a collective entity, not a unitary one [Alderman et al. (1995)]. For example, if a household were unitary, its choices would be independent of which member actually does the choosing. But recent findings have revealed, for example, that income in the hands of the mother has a bigger effect on her family's health (e.g.,

---

[52] Chen, Huq and D'Souza (1981) is a pioneering quantitative study on the behavioral antecedents of higher female than male mortality from infancy through the childbearing ages in rural Bangladesh. See Dasgupta (1993) for further references. It should be noted that stopping rules governing fertility behavior based on sex preference provide a different type of information regarding sex preference than sex ratios within a population. To see this, suppose that in a society where sons are preferred, parents continue to have children until a son is born, at which point they cease having children. Assume that at each try there is a 50 percent chance of a son being conceived. Now imagine a large population of parents, all starting from scratch. In the first round 50 percent of the parents will have sons and 50 percent will have daughters. The first group will now stop and the second group will try again. Of this second group, 50 percent will have sons and 50 percent will have daughters. The first sub-group will now stop and the second sub-group will have another try. And so on. But at each round the number of boys born equals the number of girls. The sex ratio is 1.

The argument also implies that population remains constant. To confirm this, note that since each couple has exactly one son, couples on average have one son. But as the sex ratio is 1, couples on average have one daughter also. Therefore, the average couple have two children. This means that in equilibrium the size of the population is constant.

Table 8
Fertility rates and indicators of women's status in 79 developing countries

N	TFR	PE	UE	I
9	>7.0	10.6	46.9	65.7
35	6.1–7.0	16.5	31.7	76.9
10	5.1–6.0	24.5	27.1	46.0
25	<5.0	30.3	18.1	22.6

N: number of countries.
TFR: total fertility rate.
PE: women's share of paid employment (%).
UE: percentage of women working as unpaid family workers.
I: women's illiteracy rate (%).
Source: IIED/WRI (1987, Table 2.3).

nutritional status of children) than income under the control of the father [Kennedy and Oniang'o (1990)].

Since gender inequities prevail in work, education, food, and health care, it should not surprise that they prevail in fertility choices as well. Here also, women bear the greater cost. To grasp how great the burden can be, consider that in sub-Saharan Africa the total fertility rate has for long been between 6 and 8 (Figure 1). Successful procreation involves at least a year and a half of pregnancy and breastfeeding. So in societies where female life expectancy at birth is 50 years and the total fertility rate is 7, women at birth can expect to spend about half their adult lives in pregnancy or nursing. And we have not allowed for unsuccessful pregnancies.

In view of this difference in the costs of bearing children, we would expect men to desire more children than women do. On the other hand, if women are economically more vulnerable than men, they could well desire more children than men because children offer an insurance against particularly bad contingencies. Either way, birth rates would be expected to be lower in societies where women are more "empowered". Data on the status of women from 79 so-called Southern countries (Table 8) confirm this and display an unmistakable pattern: high fertility, high rates of female illiteracy, low women's share of paid employment, and a high percentage of women working at home for no pay all go hand in hand. From the data alone it is difficult to discern which measures are causing high fertility and which are merely correlated with it. But the findings are consistent with the possibility that a lack of paid employment and education limits women's ability to make decisions—a condition that promotes high fertility.

Household decisions would assume strong normative significance if the household were unitary, less so if it were not. The evidence is that the unitary household is especially uncommon when the family is impoverished and the stresses and strains of hunger and illness make themselves felt. Despite these caveats, I adopt a unitary view of the household in what follows. Because I am concerned here with reproductive and environmental externalities, assuming a unitary household helps to simplify the exposition without losing anything essential.

8. Motives for procreation

One motive for procreation, common to humankind, relates to children as ends in themselves. We are genetically endowed to want and to value them. It has also been said that children are the clearest avenue open to "self-transcendence" [Heyd (1992)]. Viewing children as ends ranges from the desire to have offspring because they are playful and enjoyable, to a desire to obey the dictates of tradition and religion. One such injunction emanates from the cult of the ancestor, which, taking religion to be the act of reproducing the lineage, requires women to bear many children.[53] The latter motivation has been emphasized by Caldwell and Caldwell (1990) to explain why sub-Saharan Africa has proved so resistant to fertility reduction.

The problem with this explanation is that, although it does well to account for high fertility rates in sub-Saharan Africa, it does not adequately explain why the rates have not responded to declines in infant mortality. The cult of the ancestor may prescribe reproduction of the lineage, but it does not stipulate an invariant fertility rate. Since even in sub-Saharan Africa fertility rates have been below the maximum possible, they should be expected to respond to declines in infant mortality. This is a matter I return to below, where I offer one possible explanation for the resistance that the semi-arid regions of sub-Saharan Africa have shown to fertility reduction.[54]

But for parents, children are not only an end; they can also be a means to economic betterment. In the extreme, they can be a means to survival. Children offer two such means. First, in the absence of capital markets and social security, children can be private security in old age. There is evidence that in poor countries children do offer such security [Cain (1981, 1983), Cox and Jimenez (1992)]. This fact leads to a preference for male offspring if males inherit the bulk of their parents' property and are expected to look after them in their old age.

Secondly, in agriculture-based rural economies children are valuable in household production. Evidence of this is extensive, although such evidence is, of course, no proof that parents have children in order to obtain additional labour. For example, people could have large numbers of offspring by mistake and put them to work only because they cannot afford to do otherwise. Or a large family might be desired as an end in itself,

---

[53] Writing about West Africa, Fortes (1978, pp. 125, 126) " ... a person does not feel he has fulfilled his destiny until he or she not only becomes a parent but has grandchildren ... (Parenthood) is also a fulfillment of fundamental kinship, religious and political obligations, and represents a commitment by parents to transmit the cultural heritage of the community ... Ancestry, as juridically rather than biologically defined, is the primary criterion ... for the allocation of economic, political, and religious status". See also Goody (1976). Cochrane and Farid (1989) remark that both the urban and rural, the educated and uneducated in sub-Saharan Africa have more, and want more, children than their counterparts do in other regions. Thus, even the younger women there expressed a desire for an average of 2.6 more children than women in the Middle East, 2.8 more than women in North Africa, and 3.6 to 3.7 more than women in Latin America and Asia.

[54] Between 1965 and 1987 the infant mortality rate in a number of the poorest countries in sub-Saharan Africa declined from about 200 per 1,000 live births to something like 150 per 1,000 live births [World Bank (1989)].

and putting children to work at an early age might be the only avenue open for financing that end. However, these conjectures are hard to substantiate directly. The former is in any case difficult to believe, since it suggests an inability to learn on the part of parents in a world where they are known to learn in other spheres of activity, such as cultivation. But because the latter is not at variance with any evidence I know, I explore it in a later section.

Caldwell (1981, 1982) put forward the interesting hypothesis that the intergenerational transfer of resources flows from children to their parents in societies experiencing high fertility and high mortality rates, but that it flows from parents to their children when fertility and mortality rates are low. Assuming this to be true, the relationship should be interpreted merely as an association. The direction of intergenerational resource transfers would be endogenous in any general theory of demographic behavior; thus it would not be a causal factor in fertility transitions.

The historical change in the North in parents' attitudes toward their children (from regarding children as a "means" to economic ends, to regarding them simply as an "end") can seem to pose a deep puzzle, as can differences between the attitudes of parents in the North and South today. Some demographers have remarked that a fundamental shift in adults' "world view" must have been involved in such changes in attitudes, a shift that Cleland and Wilson (1987) have called an "ideational change".

These observers may be right. On the other hand, not only is the explanation something of a *deus ex machina*, it is also difficult to test. A different sort of explanation, one that is testable, is that children cease being *regarded* as productive assets when they cease *being* productive assets. When schooling is enforced, children are not available for household and farm chores. If the growth of urban centers makes rural children less reliable as old-age security (children are now able to leave home and not send remittances), children cease being a sound investment for parents' old age.[55] In short, if children were to become relatively unproductive in each of their roles as an economic asset, their only remaining value would be as an end. No change in world view would necessarily be involved in this transformation.

The above argument does not rely on economic growth. It involves a comparison between the productivity of different forms of capital assets. Children could cease being a sound economic investment even if the economy remained poor. One way by which such a change in children's economic worth could come about is through changes in reproductive and environmental externalities through institutional transformations. I turn to this.

9. Reproductive and environmental externalities

What causes private and social costs and benefits of reproduction to differ? One likely source of the distinction has to do with the finiteness of space [World Bank (1984),

---

[55] Sundstrom and David (1988) apply this reasoning to antebellum America.

Harford (1998)]. Increased population size implies greater crowding, and households acting on their own would not be expected to "internalize" crowding externalities. The human epidemiological environment becomes more and more precarious as population densities rise. Crowded centers of population provide a fertile ground for the spread of pathogens; and there are always new strains in the making. Conversely, the spread of infections, such as HIV, would be expected to affect demographic behavior, although in ways that are not yet obvious [Ezzell (2000)].

Large-scale migrations of populations occasioned by crop failure, war, or other disturbances are an obvious form of externality. But by their very nature they are not of the persistent variety. Of those that are persistent, at least four types come to mind. In the remainder of this section I look into them.

9.1. Cost-sharing

Fertility behavior is influenced by the structure of property rights, for instance, rules of inheritance. In his influential analysis of fertility differences between preindustrial seventeenth- and eighteenth-century Northwest Europe, on the one hand, and Asiatic preindustrial societies, on the other, Hajnal (1982) distinguished between "nuclear" and "joint" household systems. He observed that in Northwest Europe marriage normally meant establishing a new household, which implied that the couple had to have, by saving or transfer, sufficient resources to establish and equip the new residence. This requirement in turn led to late ages at marriages. It also meant that parents bore the cost of rearing their children. Indeed, fertility rates in England were a low 4 in 1650–1710, long before modern family planning techniques became available and long before women became widely literate [Coale (1969), Wrigley and Schofield (1981)]. Hajnal contrasted this with the Asiatic pattern of household formation, which he saw as joint units consisting of more than one couple and their children.

Parental costs of procreation are also lower when the cost of rearing the child is shared among the kinship. In sub-Saharan Africa fosterage within the kinship is a commonplace: children are not raised solely by their parents; the responsibility is more diffuse within the kinship group [Goody (1976), Bledsoe (1990), Caldwell and Caldwell (1990)]. Fosterage in the African context is not adoption. It is not intended to, nor does it in fact, break ties between parents and children. The institution affords a form of mutual insurance protection in semi-arid regions. It is possible that, because opportunities for saving are few in the low-productivity agricultural regions of sub-Saharan Africa, fosterage also enables households to smoothen their consumption across time [Serra (1996)].[56] In parts of West Africa up to half the children have been found to be living with kin at any given time. Nephews and nieces have the same rights of accommodation and support as do biological offspring. There is a sense in which children are

---

[56] This hypothesis could be tested by comparing the age structure of households that foster out and those that foster in.

seen as a common responsibility. However, the arrangement creates a free-rider problem if the parents' share of the benefits from having children exceeds their share of the costs. From the point of view of parents, taken as a collective, too many children would be produced in these circumstances.[57]

In sub-Saharan Africa, communal land tenure within the lineage social structure has in the past offered further inducement for men to procreate. Moreover, conjugal bonds are frequently weak, so fathers often do not bear the costs of siring children. Anthropologists have observed that the unit of African society is a woman and her children, rather than parents and their children. Frequently there is no common budget for the man and woman. Descent in sub-Saharan Africa is for the most part patrilineal and residence is patrilocal (an exception are the Akan people of Ghana). Patrilineality, weak conjugal bonds, communal land tenure, and a strong kinship support system of children, taken together, have been a broad characteristic of the region [Caldwell and Caldwell (1990), Caldwell (1991), Bledsoe and Pison (1994)]. They are a source of reproductive externalities that stimulate fertility. Admittedly, patrilineality and patrilocality are features of the northern parts of the Indian sub-continent also,[58] but conjugal bonds are substantially greater there. Moreover, because agricultural land is not communally held in India, large family size leads to fragmentation of landholdings. In contrast, large families in sub-Saharan Africa are (or, at least were, until recently) rewarded by a greater share of land belonging to the lineage or clan.

9.2. Conformity and "contagion"

That children are seen as an end in themselves provides another mechanism by which reasoned fertility decisions at the level of every household can lead to an unsatisfactory outcome from the perspectives of all households. The mechanism arises from the possibility that traditional practice is perpetuated by conformity. Procreation in closely-knit communities is not only a private matter, it is also a social activity, influenced by both

---

[57] To see that there is no distortion if the shares were the same, suppose c is the cost of rearing a child and N the number of couples within a kinship. For simplicity assume that each child makes available y units of output (this is the norm) to the entire kinship, which is then shared equally among all couples, say in their old age. Suppose also that the cost of rearing each child is shared equally by all couples. Let n^* be the number of children each couple other than the one under study chooses to have. (We presently endogenize this.) If n were to be the number of children this couple produces, it would incur the resource cost $C = [nc + (N-1)n^*c]/N$, and eventually the couple would receive an income from the next generation equalling $Y = [ny + (N-1)n * y]/N$. Denote the couple's aggregate utility function by the form $U(Y) - K(C)$, where both $U(\cdot)$ and $K(\cdot)$ are increasing and strictly concave functions. Letting n be a continuous variable for simplicity, it is easy to confirm that the couple in question will choose the value of n at which $yU'(Y) = cK'(C)$. The choice sustains a social equilibrium when $n = n^*$. It is easy to check that this is also the condition that is met in a society where there is no reproductive free-riding. It is a simple matter to confirm that there is free-riding if the parents' share of the benefits from having children exceeds their share of the costs.

[58] Among the prominent Nayyars of the southern state of Kerala, India, descent is matrilineal. Kerala is noteworthy today for being among the poorer of Indian states even while attaining a *TFR* less than 2.

family experiences and the cultural milieu. Formally speaking, behavior is conformist if, other things being equal, every household's most desired family size is the greater, the larger is the average family size in the community [Dasgupta (1993, Chapter 12)]. This is a "reduced form" of the concept, and the source of a desire to conform could lie in reasons other than an intrinsic desire to be like others. For example, similar choices made by households might generate mutual positive externalities, say, because people care about their status; and a household's choice of actions signals its predispositions (e.g., their willingness to belong) and so affects its status [Bernheim (1994), Bongaarts and Watkins (1996)]. In a world where people conform, the desire for children is endogenous.[59]

Whatever the basis of conformism, there would be practices encouraging high fertility rates that no household would unilaterally desire to break. Such practice could well have had a rationale in the past, when mortality rates were high, rural population densities were low, the threat of extermination from outside attack was large, and mobility was restricted. But practices can survive even when their original purposes have disappeared. Thus, as long as all others follow the practice and aim at large family size, no household on its own may wish to deviate from the practice; however, if all other households were to restrict their fertility rates, each would desire to restrict its fertility rate as well. In short, conformism can be a reason for the existence of multiple reproductive equilibria [Dasgupta (1993, Chapter 12)]. The multiple equilibria may even be Pareto rankable, in which case a community could get stuck at an equilibrium mode of behavior even though another equilibrium mode of behavior would be better for all.

These are theoretical possibilities. Testing for multiple equilibria is very difficult. As matters stand, it is only analytical reasoning that tells us that a society could in principle get stuck at a self-sustaining mode of behavior characterized by high fertility (and low educational attainment), even when there is another, potentially self-sustaining, mode of behavior characterized by low fertility (and high educational attainment).

This does not mean that the hypothetical society would be stuck with high fertility rates forever. External events could lead households to "coordinate" at a low fertility equilibrium even if they had earlier "coordinated" at a high fertility equilibrium. The external events could, for example, take the form of public exhortations aimed at altering household expectations about one another's behavior (e.g., family planning campaigns run by women). This is a case where the community "tips" from one mode of behavior to another, even though there has been no underlying change in household attitudes to trigger the change in behavior.

In their aforementioned article Cleland and Wilson (1987, p. 9) argued that the only plausible way to explain the recent onset of fertility transitions among countries at widely different levels of economic development was an ideational change, "... a psychological shift from, *inter alia*, fatalism to a sense of control of destiny, from passivity

---

[59] Household "preferences" embodying such interactions are often called "social preferences". Krishnan (2001) has found evidence of "social preferences" in data from India.

to the pursuit of achievement, from a religious, tradition-bound, and parochial view of the world to a more secular, rational, and cosmopolitan one". The authors may be right that societies have undergone ideational changes, but they are wrong in thinking that ideational change must be invoked to explain recent fertility transitions. The tipping behavior I have just discussed is not a response to ideational changes. This said, I know of no evidence that is able to discriminate between the two types of explanation.

In addition to being a response to external events, the tipping phenomenon can occur because of changes in the peer group on whose behavior households base their own behavior. Inevitably, there are those who experiment, take risks, and refrain from joining the crowd. They subsequently influence others. They are the tradition-breakers, often leading the way. It has been observed that educated women are among the first to make the move toward smaller families [see Farooq, Ekanem and Ojelade, 1987, for a commentary on West Africa]. Members of the middle classes can also be the trigger, becoming role models for others.

A possibly even stronger pathway is the influence that newspapers, radio, television, and now the Internet exert in transmitting information about other lifestyles [Freedman (1995), Bongaarts and Watkins (1996), Iyer (2000)]. The analytical point here is that the media may be a vehicle through which conformism increasingly becomes based on the behavior of a wider population than the local community: the peer group widens.

Such pathways can give rise to demographic transitions, in that fertility rates display little to no trend over extended periods, only to cascade downward over a relatively short interval of time, giving rise to the classic logistic curve of diffusion processes.[60]

In a pioneering article Adelman and Morris (1965) found "openness" of a society to outside ideas to be a powerful stimulus to economic growth. It is possible that the fertility reductions that have been experienced in India and Bangladesh in recent years (Table 3) were the result of the wider influence people have been subjected to via the media or to attitudinal differences arising from improvements in family planning programmes. To be sure, fertility reductions have differed widely across the Indian sub-continent (not much reduction in Pakistan so far, a great deal in southern India), but we should not seek a single explanation for so complex a phenomenon as fertility transition.[61]

Demographers have made few attempts to discover evidence of behavior that is guided in part by an attention to others. Two exceptions are Easterlin, Pollak and Wachter (1980) and Watkins (1990).[62] The former studied intergenerational influence

---

[60] For a more complete account of the theory I am advancing in the text, see Dasgupta (2000a). Formally, the above is a model of demographic transitions viewed as "relaxation phenomena". The mathematical structure I have invoked is similar to one that has recently been used by oceanographers and ecologists in their exploration of tipping phenomena in ocean circulation and lake turbidity, respectively. See Rahmstorf (1995) and Scheffer (1997).

[61] In this connection, the Indian state Andhra Pradesh offers an interesting example. Female illiteracy there is high 55 percent and some 75 percent of the population have access to radio or television. The fertility rate there is now 2.3.

[62] A most recent exception is Krishnan (2001).

in a sample of families in the United States. They reported a positive link between the number of children with whom someone had been raised and the number of children they themselves had.

In her study of demographic change in Western Europe over the period 1870–1960, Watkins (1990) showed that regional differences in fertility and nuptiality within each country declined. In 1870, before the large-scale declines in marital fertility had begun in most areas of Western Europe, demographic behavior differed greatly within countries: provinces (e.g., counties and cantons) differed considerably, even while differences within provinces were low. There were thus spatial clumps within each country, suggesting the importance of the influence of local communities on behavior. By 1960 differences within each country were less than they had been in 1870. Watkins explained this convergence in behavior in terms of increases in the geographical reach national governments enjoyed over the 90 years in question. The growth of national languages could have been the medium through which reproductive behavior spread.

One recent finding could also point to contagious behavior. Starting in 1977 (when *TFR* in Bangladesh exceeded 6), 70 "treatment" villages were served by a massive programme of birth control in Matlab Thana, Bangladesh, while 79 "control" villages were offered no such special service. The prevalence of contraceptive use in the treatment villages increased from 7 to 33 percent within 18 months, and then rose more gradually to a level of 45 percent by 1985. The prevalence also increased in the control villages, but only to 16 percent in 1985. Fertility rates in both sets of villages declined, but at different speeds, with the difference in fertility rates reaching 1.5 births per woman, even though there had been no difference to begin with [Hill (1992)]. If we assume that, although influence travels, geographical proximity matters, we could explain why the control villages followed the example of villages "under treatment", but did not follow them all the way. Contagion did not spread completely.[63]

9.3. Interactions among institutions

Externalities are prevalent when market and nonmarket institutions co-exist. How and why might such externalities affect fertility behavior? A number of pathways suggest themselves [Dasgupta (1993, 1999)].

Long-term relationships in rural communities in poor countries are frequently sustained by social norms–for example, norms of reciprocity. Social norms can be reliably observed only among people who expect to encounter one another in recurring situations.[64] Consider a community of "far-sighted" people who know one another and expect to interact with one another for a long time. By far-sighted, I mean someone who applies a low rate to discount future costs and benefits of alternative courses of

---

[63] I am grateful to Lincoln Chen for a helpful 1996 correspondence on this point. For a formal account of contagion models, see Blume and Durlauf (2001).

[64] This is the setting studied in the theory of repeated games. See Fudenberg and Tirole (1991).

action. Assume that the parties in question are not individually mobile (although they could be collectively mobile, as in the case of nomadic societies); otherwise the chance of future encounters with one another would be low, and people would discount heavily the future benefits of the current costs they incur for the purposes of cooperation.

Simply stated, if people are far-sighted and are not individually mobile, a credible threat by all that they would impose stiff sanctions on anyone who broke the agreement would deter everyone from breaking it. But the threat of sanctions would cease to have bite if opportunistic behavior were to become personally more profitable. The latter would happen if formal markets develop nearby. As opportunities outside the village improve, people with lesser ties (e.g., young men) are more likely to take advantage of them and make a break with those customary obligations that are enshrined in prevailing social norms. People with greater attachments would perceive this and infer that the expected benefits from complying with agreements are now lower. Norms of reciprocity would break down, making certain groups of people (e.g., women, the old, and the very young) worse off. This is a case where improved institutional performance elsewhere (e.g., growth of markets in the economy at large) has an adverse effect on the functioning of a local, nonmarket institution: it is a reflection of an externality.

When established long-term relationships breaks down, people build new ones to further their economic opportunities. Those who face particularly stressful circumstances resort to draconian measures to build new economic channels. Guyer (1994) has observed that in the face of deteriorating economic circumstances, some women in a Yaruba area of Nigeria have borne children by different men so as to create immediate lateral links with them. Polyandrous motherhood enables women to have access to more than one resource network.

In his well-known work Cain (1981, 1983) showed that where capital markets are nonexistent and public or community support for the elderly are weak, children provide security in old age. The converse is that if community-based support systems decline, children become more valuable. But we have just noted that community-based support systems in rural areas may degrade with the growth of markets in cities and towns. So there is a curious causal chain here: growth of markets in towns and cities can lead to an increase in fertility in poor villages, other things being the same. There is evidence of this. In her work on Sarawak, Heyzer (1996) has observed that one half of the total forest area there has now been lost and that this has disrupted the lives of indigenous people in different ways. Communities that lived in the heart of the forest were most severely affected, while others, living near towns, were able to turn from swidden agriculture to wage labour. This transformation, however, involved male migration, leaving women behind to cope with a decreasing resource base. As subsistence alternatives declined, children become one of the few remaining resources that women could control. There was thus a new motivation for having children: to help their mothers with an increased workload. The process involved the creation of new patterns of wealth and poverty, where wealth is based on resource extraction and poverty results from the loss of a community's resource base.

Earlier we noted that growth of markets in towns and cities, by making children less reliable as an investment for old age, can lead to a reduction in fertility. Here we have identified an influence of the growth of markets on fertility that runs in the opposite direction. Only formal modelling of the process would enable us to determine which influence dominates under what conditions.

9.4. Household labour needs and the local commons

The poorest countries are in great part agriculture-based subsistence economies.[65] Much labour is needed even for simple tasks. Moreover, many households lack access to the sources of domestic energy available to households in advanced industrial countries. Nor do they have water on tap. In semi-arid and arid regions water supply is often not even close at hand, nor is fuel-wood nearby when the forests recede. This means that the relative prices of alternative sources of energy and water faced by rural households in poor countries are quite different from those faced by households elsewhere. In addition to cultivating crops, caring for livestock, cooking food and producing simple marketable products, household members may have to spend several hours a day fetching water and collecting fodder and wood. These complementary activities have to be undertaken on a daily basis if households are to survive. Labour productivity is low because both capital and environmental resources are scarce. From an early age, children in poor households in the poorest countries mind their siblings and domestic animals, fetch water, and collect fuelwood, dung (in the Indian sub-continent), and fodder. Mostly, they do not go to school. Not only are educational facilities in the typical rural school woefully inadequate, but parents need their children's labour. Children between 10 and 15 years have been routinely observed to work at least as many hours as adult males [see, for example, Bledsoe (1994), Cleaver and Schreiber (1994), Filmer and Pritchett (2002)].

The need for many hands can in principle lead to a destructive situation when parents do not have to pay the full price of rearing their children, but share such costs with their community. In recent years, social norms that once regulated local resources have changed. Since time immemorial, rural assets such as village ponds and water holes, threshing grounds, grazing fields, swidden fallows, and local forests and woodlands have been owned communally. As a proportion of total assets, the presence of such assets ranges widely across ecological zones. In India the local commons are most prominent in arid regions, mountain regions, and unirrigated areas; they are least prominent in humid regions and river valleys [Agarwal and Narain (1989)]. There is a rationale for this, based on the human desire to reduce risks. Community ownership and control

---

[65] I am thinking of countries in sub-Saharan Africa and the Indian sub-continent. In those countries the agricultural labour force as a proportion of the total labour force is on the order of 60–70 percent, and the share of agricultural-value added in GNP is on the order of 25–30 percent.

enabled households in semi-arid regions to pool their risks.[66] An almost immediate empirical corollary is that income inequalities are less where common-property resources are more prominent. Aggregate income is a different matter though, and the arid and mountain regions and unirrigated areas are the poorest. As would be expected, dependence on common-property resources even within dry regions declines with increasing wealth across households.

Jodha (1986, 1995), studying evidence from over 80 villages in 21 dry districts in India, concluded that, among poor families, the proportion of income based directly on their local commons is for the most part in the range of 15–25 percent. A number of resources (such as fuelwood and water, berries and nuts, medicinal herbs, resin and gum) are the responsibility of women and children. In a study of 29 villages in southeastern Zimbabwe, Cavendish (2000) arrived at even larger estimates: the proportion of income based directly on the local commons is 35 percent, with the figure for the poorest quintile reaching 40 percent. Such evidence does not of course prove that the local commons are well managed, but it suggests that rural households have strong incentives to devise arrangements whereby they would be well managed.

A number of investigators–among them Howe (1986), Wade (1988), Chopra, Kadekodi and Murty (1990), Ostrom (1990, 1992), Baland and Platteau (1996)–have shown that many communities have traditionally protected their local commons from overexploitation by relying on social norms, by imposing fines for deviant behavior, and by other means. I argued earlier that the very process of economic development, as exemplified by urbanization and mobility, can erode traditional methods of control. Social norms are endangered also by civil strife and by the usurpation of resources by landowners or the state. For example, resource-allocation rules practiced at the local level have frequently been overturned by central fiat. A number of states in the Sahel imposed rules that in effect destroyed community management practices in the forests. Villages ceased to have authority to enforce sanctions against those who violated locally instituted rules of use. State authority turned the local commons into free-access resources.[67] As social norms degrade, whatever the cause, parents pass some of the costs of children on to the community by overexploiting the commons. This is another instance of a demographic free-rider problem.

The perception of an increase in the net benefits of having children induces households to have too many. This is predicted by the standard theory of the imperfectly managed commons (see the Appendix). It is also true that when households are further impoverished owing to the erosion of the commons, the net cost of children increases (of course, household size continues to remain above the optimum from the collective point of view). Loughran and Pritchett (1998), for example, have found in Nepal that increasing environmental scarcity lowered the demand for children, implying that the

---

[66] In his work on South Indian villages, Seabright (1997) has shown that producers' cooperatives, unconnected with the management of local commons, are also more prevalent in the drier districts.

[67] See Thomson, Feeny and Oakerson (1986) and Baland and Platteau (1996).

households in question perceived resource scarcity as raising the cost of children. Apparently, increasing firewood and water scarcity in the villages in the sample did not have a strong enough effect on the relative productivity of child labour to induce higher demand for children, given the effects that work in the opposite direction. Environmental scarcity there acted as a check on population growth.

However, theoretical considerations suggest that, in certain circumstances, increased resource scarcity induces further population growth: as the community's natural resources are depleted, households find themselves needing more "hands". No doubt additional hands could be obtained if the adults worked even harder, but in many cultures it would not do for the men to gather fuel-wood and fetch water for household use.[68] No doubt, too, additional hands could be obtained if children at school were withdrawn and put to work. But, as we have seen, mostly the children do not go to school anyway. In short, when all other sources of additional labour become too costly, more children are produced, thus further damaging the local resource base and, in turn, providing the household with an incentive to enlarge yet more. This does not necessarily mean that the fertility rate will increase. If the infant mortality rate were to decline, there would be no need for more births in order for a household to acquire more hands. However, along this pathway poverty, household size, and environmental degradation could reinforce one another in an escalating spiral. By the time some countervailing set of factors diminished the benefits of having further children and, thereby, stopped the spiral, many lives could have suffered by a worsening of poverty. In the Appendix I provide a simple model to illustrate such possibilities.

Cleaver and Schreiber (1994) have provided very rough, aggregative evidence of a positive link between population increase and environmental degradation in the context of rural sub-Saharan Africa; Batliwala and Reddy (1994) for villages in Karnataka, India; and Heyser (1996) for Sarawak, Malaysia. In a statistical analysis of evidence from villages in South Africa, Aggarwal, Netanyahu, and Romano (2001) found a positive link between fertility increase and environmental degradation; while Filmer and Pritchett (2002) have reported a weak positive link in the Sindh region in Pakistan.

None of these investigations quite captures what the theory I am sketching here tells us to study, namely, the link between desired household size and the state of the local natural-resource base. But they come close enough; limitations in existing data prevent investigators from getting closer to the theory.[69] In any event, these studies cannot reveal causal connections, but, excepting the study by Loughran and Pritchett (1998), they are consistent with the idea of a positive-feedback mechanism such as I have described. Over time, the spiral would be expected to have political effects, as manifested by battles for scarce resources, for example, among competing ethnic groups [Durham (1979),

---

[68] Filmer and Pritchett (2002) summarize empirical findings on children's time allocation to household activities in rural areas in poor countries.

[69] However, Deon Filmer has informed me that his colleagues at the World Bank have found in a sample of Nepalese villages a positive relationship between (primary) school attendance and the availability of local natural resources.

Homer-Dixon (1994, 1999)]. The latter connection deserves greater investigation than it has elicited so far.[70]

To be sure, families with greater access to resources would be in a position to limit their size and propel themselves into still higher income levels. Admittedly, too, people from the poorest of backgrounds have been known to improve their circumstances. Nevertheless, there are forces at work that pull households away from one another in terms of their living standards. Such forces enable extreme poverty to persist despite growth in the well-being for the rest of society.

10. Institutional reforms and policies

If in earlier days social scientists looked for policies to shape social outcomes for the better, their focus today is more on the character of institutions within which people make decisions. But if policies that read well often come to naught in dysfunctional institutions, the study of institutions on their own is not sufficient: good policies cannot be plucked from air. There is mutual influence here, and the task of the social scientist is to study it.

Demographers, like economists, seek good news. There is a danger that the recent onset of demographic transitions in parts of the Indian sub-continent and signs of an onset in some of the urban regions of sub-Saharan Africa will make demographers complacent. A distinguished student of demography remarked to me recently that, in view of the signs of demographic transitions everywhere, the "population problem" is now over.

But it is not over. The ultimate size of the world's population, once the transitions have occurred, will matter greatly.[71] There is likely to be a world of a difference between a global population of 11 billion and one of 5 billion, even if we ignored differences in their spatial distributions that would inevitably be implied [Cohen (1995)]. In this connection, it is worth stressing that some of the externalities that I have identified here operate mainly in time, while others operate mainly through time (economists refer to them as "static" and "dynamic" externalities, respectively). So, even if world population were to stabilize, there would remain externalities whose presence calls for public policies.

In this chapter I have identified a number of institutional failures that involve pronatalist reproductive externalities. I have done this by connecting demographic and environmental perspectives. The perspective that emerges tells us that the most potent avenue for reducing the population problem in various parts of the world involves the simultaneous deployment of a number of policies, not a single panacea, and that the relative importance of the several prongs depends on the community in question. Thus,

---

[70] Crook (1996) questions the poverty-population link. But because he treats population density and land productivity as exogenous variables, his is not quite a test of the thesis.

[71] See Bongaarts (2002), who expresses concerns similar to the ones being expressed here.

while family planning services (especially when allied to public health services) and measures that empower women (through both education and improved employment opportunities) are certainly desirable, other policies also commend themselves, such as the provision of infrastructural goods (e.g., cheap sources of household fuel and potable water), changes to property rights (e.g., the rules of inheritance), means of communication with the outside world (e.g., roads, telephones, radios, television, newspapers, and the Internet), and measures that directly increase the economic security of the poor. A number of these policies might well not have come to mind if we studied demographic problems in isolation.

In any event, the aim should not be to force people to change their reproductive behavior. Rather, it should be to identify policies and encourage such institutional changes as those that would "internalize" the externalities I have uncovered here. Recent declines in fertility rates in the Indian sub-continent and in parts of sub-Saharan Africa suggest that outside influence, via the media, may have been powerful. Observing lifestyles elsewhere can no doubt be unsettling to many, but it can give people ideas that are salutary. To the extent that reproductive behavior is based on conformism (I have little notion of what that extent is), modern communication channels, by linking the village to the outside world, have a powerful effect. But the media are likely to be hampered in arbitrary ways except in politically open societies. I have shown elsewhere [Dasgupta (1990), Dasgupta and Weale (1992)] that in poor countries political and civil liberties are congruent with improvements in other aspects of life, such as income per head, life expectancy at birth, and infant survival. Subsequently, Przeworski and Limongi (1995) have shown that these liberties are negatively correlated with fertility rates. We therefore have several reasons for thinking that political and civil liberties have instrumental value, even in poor countries; they are not merely desirable ends. But each of the prescriptions offered by the new perspective presented here is desirable in itself and commends itself even when we do not have fertility rates of poor countries in mind. To me this is a most agreeable fact.

Admittedly, in all this we have looked at matters wholly from the perspective of the parents. This is limiting.[72] But developing the welfare economics of population policies has proved to be extremely difficult.[73] Our ethical intuition at best extends to actual and future people; we do not yet possess a good moral vocabulary for including potential people in the calculus. I have tried to argue in this essay that there is much that we can establish even if we were to leave aside such conceptual difficulties. Population policy involves a good deal more than making family planning centers available to the rural poor. It also involves more than a recognition that poverty is the root cause of high fertility rates. The problem is deeper, but as I have tried to show, it is possible to subject it to analysis.

---

[72] Enke (1966) is a notable exploration of the value of prevented births when the worth of additional lives is judged to be based entirely on their effect on the current generation. As a simplification, Enke took the value of a prevented birth to be the discounted sum of the differences between an additional person's consumption and output over the person's lifetime.

[73] I have gone into some of the difficulties in Dasgupta (1998b).

Acknowledgements

This chapter is a much-revised version of Dasgupta (2000a). I have benefited greatly from discussions with Kenneth Arrow, John Bongaarts, John Caldwell, Jack Goody, Sriya Iyer, and Karl-Göran Mäler.

Appendix. The village commons and household size

The observation that increases in population bring in their wake additional pressures on the local natural-resource base is, no doubt, a banality. So, in what follows I study the reverse influence: the effect of a deterioration of the local natural-resource base on desired household size.

I argued above that villagers' free-riding on the commons can impoverish households in such a way as to create an additional need for household labour. Such a need would translate itself into a demand for more surviving children if having more surviving children were the cheapest means of obtaining that additional labour. Of course, this is only one possibility; another is that the receding commons impoverishes households in such a way that, at the margin, children become too costly, with the result that the number of surviving children declines. In this appendix I offer a formal account of both possibilities. The model enables us to identify parametric conditions under which the various outcomes would be expected to occur. I then compare the non-cooperative village to a cooperative one. The model is timeless. Adjustments over time can then be analyzed in terms of comparative statics.

A.1. The single household

I consider a agriculture-based village economy consisting of N identical households. N is taken to be sufficiently large that the representative household's size does not affect the economy. The model is deterministic. Household size is assumed to be a continuous variable, which is a way of acknowledging that realized household size is not a deterministic function of the size the household sets for itself as a target.

Let n be the size of a household. Members contribute to production, but they also consume from household earnings. I aggregate inputs and outputs and assume that household production possibilities are such that *net income* per household member, $y(n)$, has the quadratic form,

$$y(n) = -\alpha + \beta n - \gamma n^2, \quad \text{where } \alpha, \beta, \gamma > 0 \text{ and } \beta^2 > 4\alpha\gamma. \tag{A.1}$$

The quadratic form enables us to capture certain crucial features of a subsistence economy in a simple way, thereby permitting us to draw conclusions easily. For example, (A.1) presumes that there are fixed costs in running a household, which is altogether realistic: in order to survive, a household must complete so many chores on a daily basis (cleaning, farming, animal care, fetching water and collecting fuel wood, cooking raw ingredients, and so forth), that single-member households are not feasible. Equation (A.1) also presumes that when the household is large, the costs of adding numbers begin to overtake the additional income that is generated. This too is clearly correct.[74]

It follows from (1) that $y(n) = 0$ at

$$\underline{n} = \frac{\beta - \sqrt{\beta^2 - 4\alpha\gamma}}{2\gamma} \quad \text{and} \tag{A.2a}$$

$$\bar{n} = \frac{\beta + \sqrt{\beta^2 - 4\alpha\gamma}}{2\gamma}. \tag{A.2b}$$

$\underline{n}$ is the "fixed cost" of maintaining a household, while $\bar{n}$ could be interpreted to be the environment's "carrying capacity". I assume that the household "chooses" its size so as to maximize net income per head. Let n^* denote the value of n at which $y(n)$ attains its maximum and let y^* denote the maximum. Then

$$n^* = \frac{\beta}{2\gamma} \quad \text{and} \tag{A.3a}$$

$$y^* = -\alpha + \frac{\beta^2}{4\gamma}. \tag{A.3b}$$

$y(n)$ is depicted as the curve ABC in Figure 2, where B is the point $(\beta/2\gamma, -\alpha + \beta^2/4\gamma)$.

Imagine now that the household faces an increase in resource scarcity. We are to interpret this in terms of receding forests and vanishing water-holes. The index of resource scarcity could then be the average distance from the village to the resource base. So, an increase in resource scarcity would mean, among other things, an increase in $\underline{n}$.

But it would typically mean more. For example, equations (A.2a,b) tell us that the household would face an increase in resource scarcity if α, γ, and α/γ were to increase and β were to decline in such a way that $\bar{n}$ declines. Note too that in this case, both n^* and y^* would decline (equations (A.3a,b)). The resulting $y(n)$ is depicted as the curve $A'B'C'$ in Figure 2. In short, the increase in resource scarcity shifts curve ABC to $A'B'C'$.

Consider instead the case where each of α, β, and γ increases, but in such ways that $\underline{n}$ and n^* increase, while $\bar{n}$ and y^* decline. This is the kind of situation in which a household finds that its best strategy against local resource degradation is to increase its size even while finding itself poorer. The resulting $y(n)$ is depicted as the curve

---

[74] The analysis that follows can be developed more generally, without recourse to the quadratic function.

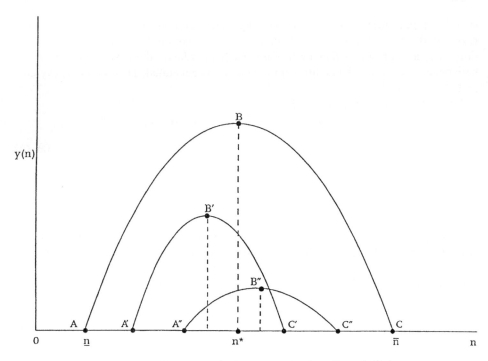

Figure 2. Household income per head, $y(n)$, as a function of household size, n.

$A''B''C''$ in Figure 2. In short, the increase in resource scarcity shifts the curve ABC to $A''B''C''$. This sort of case was noted originally in Dasgupta and Mäler (1991) and Nerlove (1991).

A.2. Social equilibrium

I now construct an equilibrium of the village economy. The state of the local natural-resource base is taken to be a function of the village population, which I write as M. So I assume that α, β, and γ in Equation (A.1) are functions of M. Write $\alpha = \alpha(M)$, $\beta = \beta(M)$, and $\gamma = \gamma(M)$. A symmetrical equilibrium of the village economy is characterized by $M^* = Nn^*$. That is, n^* and y^* are the solutions of

$$n^* = \frac{\beta(Nn^*)}{2\gamma(Nn^*)} \quad \text{and} \tag{A.4a}$$

$$y^* = -\alpha(Nn^*) + \frac{[\beta(Nn^*)]^2}{4\gamma(Nn^*)}. \tag{A.4b}$$

I assume that a solution exists and that $n^* > 1$.

A.3. The optimum village

Consider next an optimizing village community. It would choose n so as to maximize[75]

$$y(n) = -\alpha(Nn) + \beta(Nn)n - \gamma(Nn)n^2. \tag{A.5}$$

Let $\hat{n}$ be the optimum household size. Then $\hat{n}$ is the solution of

$$\left[\beta(Nn) - 2n\gamma(Nn)\right] - N\left[\alpha'(Nn) - n\beta'(Nn) + n^2\gamma'(Nn)\right] = 0. \tag{A.6}$$

A comparison of equations (A.4a) and (A.6) tells us that $\hat{n} < n^*$ if

$$-\alpha'(Nn^*) + n^*\left[\beta'(Nn^*) - n^*\gamma'(Nn^*)\right] < 0. \tag{A.7}$$

That is, if (A.7) holds, the village is overpopulated in social equilibrium. An alternative way of thinking about the matter would be to say that an institutional reform that reduces the "freedom of access" to the commons would lower fertility.

Now (A.7) certainly holds if

$$\alpha', \gamma' > 0 \quad \text{and} \quad \beta' < 0 \quad \text{at } n = n^*. \tag{A.8}$$

But (A.7) holds also if

$$\alpha', \beta', \gamma' > 0 \quad \text{and} \quad \left[-\alpha' + \frac{\beta\beta'}{2\gamma} - \frac{\beta^2\gamma'}{4\gamma^2}\right] < 0 \quad \text{at } n = n^*. \tag{A.9}$$

A.4. The effect of increased resource scarcity

Let us study the implications for equilibrium household size and the standard of living consequent upon small exogenous shifts in the functions $\alpha(M)$, $\beta(M)$ and $\gamma(M)$. We take it that prior to the shifts inequality (A.7) holds. The perturbations will be taken to be sufficiently small so that (A.7) continues to hold in the new equilibrium.

Consider first the case where the perturbation consists of small upward shifts in $\alpha(M)$ and $\gamma(M)$ and a small downward shift in $\beta(M)$. Notice that if (A.8) holds, both n^* and y^* would be marginally smaller in consequence of the perturbation. This is the case we would expect intuitively: a small increase in resource scarcity results in poorer, but smaller, households.

Now consider the case where (A.9) holds. Suppose the perturbation consists of small upward shifts in each of the functions $\alpha(M)$, $\beta(M)$ and $\gamma(M)$. We can so set the relative magnitudes of the shifts that the small increase in resource scarcity results in poorer, but

---

[75] I avoid rigour here and assume (without justification) that the optimum is symmetric in households.

larger, households, that is, y^* declines marginally but n^* increases marginally. This is the timeless counterpart of the positive feedback mechanism between population size, poverty and degradation of the natural-resource base that was discussed in Section 8.4. Such a feedback, while by no means an inevitable fact of rural life, is a possibility. In this chapter I have argued that evidence of the experiences of Sub-Saharan Africa and northern Indian sub-continent in recent decades are not inconsistent with it.

References

Adelman, I., and C.T. Morris (1965), "A factor analysis of the interrelationship between social and political variables and per capita gross national product", Quarterly Journal of Economics 79(2):555–578.

Agarwal, A., and S. Narain (1989), Towards Green Villages: A Strategy for Environmentally Sound and Participatory Rural Development (Centre for Science and Development, New Delhi).

Agarwal, B. (1986), Cold Hearths and Barren Slopes: The Woodfuel Crisis in the Third World (Allied Publishers, New Delhi).

Agarwal, B. (1989), "Rural women, poverty and natural resources: sustenance, sustainability and struggle for change", Economic and Political Weekly 24(43):46–65.

Aggarwal, R., S. Netanyahu and C. Romano (2001), "Access to natural resources and the fertility decision of women: the case of south Africa", Environment and Development Economics 6(2):209–236.

Alderman, H., P.-A. Chiappori, L. Haddad, J. Hoddinot and R. Kanbur (1995), "Unitary versus collective models of the household: is it time to shift the burden of proof?" World Bank Research Observer 10(1):1–20.

Arrow, K.J., B. Bolin, R. Costanza, P. Dasgupta, C. Folke, C.S. Holling, B.-O. Jansson, S. Levin, K.-G. Mäler, C. Perrings and D. Pimentel (1995), "Economic growth, carrying capacity, and the environment", Science 268(5210):520, 521.

Arrow, K.J., G. Daily, P. Dasgupta, P. Ehrlich, L. Goulder, G. Heal, S. Levin, K.-G. Mäler, S. Schneider, D. Starrett and B. Walker (2002), "Are we consuming too much?", Mimeo (Beijer International Institute of Ecological Economics, Stockholm).

Arrow, K.J., P. Dasgupta and K.-G. Mäler (2003), "Evaluating projects and assessing sustainable development in imperfect economies", Environmental and Resource Economics, forthcoming.

Baland, J.-M., and J.-P. Platteau (1996), Halting Degradation of Natural Resources: Is There a Role for Rural Communities? (Oxford University Press, Oxford).

Bardhan, P. (1996), "Research on poverty and development twenty years after redistribution with growth", in: Proceedings of the Annual World Bank Conference on Development Economics, 1995 (Supplement to the World Bank Economic Review and the World Bank Research Observer) 59–72.

Barro, R. (2001), "Human capital and growth", American Economics Review 91:12–17 (Papers & Proceedings).

Batliwala, S., and A.K.N. Reddy (1994), "Energy consumption and population", in: F. Graham-Smith, ed., Population: The Complex Reality (The Royal Society, London).

Bauer, P. (2000), From Subsistence to Exchange and Other Essays (Princeton University Press, Princeton, NJ).

Baumol, W.M., and W. Oates (1975), The Theory of Environmental Policy (Prentice-Hall, Englewood Cliffs, NJ).

Becker, G. (1981), A Treatise on the Family (Harvard University Press, Cambridge, MA).

Becker, G., and N. Tomes (1976), "Child endowments and the quantity and quality of children", Journal of Political Economy, 84(Supplement): 143–162.

Bernheim, B.D. (1994), "A theory of conformity", Journal of Political Economy, 102(4):841–877.

Birdsall, N. (1988), "Economic approaches to population growth", in: H. Chenery and T.N. Srinivasan, eds., Handbook of Development Economics, Vol. 1 (North-Holland, Amsterdam).

Birdsall, N., and D. Jamison (1983), "Income and other factors influencing fertility in China", Population and Development Review 9(4):651–675.

Bledsoe, C. (1990), "The politics of children: fosterage and the social management of fertility among the mende of Sierra Leone", in: W.P. Handwerker, ed., Births and Power: Social Change and the Politics of Reproduction (Westview Press, London).

Bledsoe, C. (1994), "Children are like young bamboo trees': potentiality and reproduction in Sub-Saharan Africa", in: K. Lindahl-Kiessling and H. Landberg, eds., Population, Economic Development and the Environment (Oxford University Press, Oxford).

Bledsoe, C., and G. Pison (eds.) (1994), Nupitality in Sub-Saharan Africa: Contemporary Anthropological and Demographic Perspectives (Clarendon Press, Oxford).

Blume, L., and S.N. Durlauf (2001), "The interactions-based approach to socioeconomic behavior", in: S.N. Durlauf and H. Peyton Young, eds., Social Dynamics (MIT Press, Cambridge, MA).

Bongaarts, J. (1980), "Does malnutrition affect fecundity? A summary of the evidence", Science 208:564–569.

Bongaarts, J. (1990), "The measurement of wanted fertility", Population and Development Review 16:487–506.

Bongaarts, J. (1997), "The role of family planning programmes in contemporary fertility transitions", in: G.W. Jones, R.M. Douglas, J.C. Caldwell and R.M. D'Souza, eds., The Continuing Demographic Transition (Clarendon Press, Oxford).

Bongaarts, J. (2002), "Population: ignoring its impact", Scientific American 286(1):65–67.

Bongaarts, J., and M. Cain (1981), Demographic Responses to Famine (Population Council, New York).

Bongaarts, J., and S.C. Watkins (1996), "Social Interactions and contemporary fertility transitions", Population and Development Review 22(4):639–682.

Boserup, E. (1981), Population Growth and Technological Change (Chicago University Press, Chicago).

Brander, J.A., and M.S. Taylor (1998), "The simple economics of easter Island: A Ricardo–Malthus model of renewable resource use", American Economic Review 88(1):119–138.

Brock, W.A., and S.N. Durlauf (1999), "Interactions-based models" (Social Systems Research Institute, University of Wisconsin, Madison, WI). To appear forthcoming, in: J.J. Heckman and E. Leamer, eds., Handbook of Econometrics, Vol. 5 (North-Holland, Amsterdam).

Cain, M. (1981), "Risk and insurance: perspectives on fertility and agrarian change in India and Bangladesh", Population and Development Review 7(3):435–474.

Cain, M. (1983), "Fertility as an adjustment to risk", Population and Development Review 9(4):688–702.

Cain, M. (1984), "Women's status and fertility in developing countries: son preference and economic security", World Bank Staff Working Paper No. 682 (World Bank, Washington, DC).

Caldwell, J.C. (1981), "The mechanisms of demographic change in historical perspective", Population Studies 35(1):5–27.

Caldwell, J.C. (1982), The Theory of Fertility Decline (Academic Press, New York).

Caldwell, J.C. (1991), "The soft underbelly of development: demographic transition in conditions of limited economic change", in: Proceedings of the Annual Bank Conference on Development Economics 1990 (Supplement to the World Bank Economic Review), 207–254.

Caldwell, J.C., and P. Caldwell (1990), "High fertility in Sub-Saharan Africa", Scientific American 262(5):82–89.

Carpenter, S.R., D. Ludwig and W.A. Brock (1999), "Management of eutrophication for lakes subject to potentially irreversible change", Ecological Applications 9(3):751–771.

Cavendish, W. (2000), "Empirical regularities in the poverty–environment relationships of rural households: Evidence from Zimbabwe", World Development 28(7):1979–2003.

CES (1982), India: State of the Environment Report (Centre for Environment and Science, New Delhi).

Chen, L.C., E. Huq and S. D'Souza (1981), "Sex bias in the family allocation of food and health care in Rural Bangladesh", Population and Development Review 7(1):55–70.

Chopra, K., and S.C. Gulati (2001), Migration, Common Property Resources and Environmental Degradation (Sage, New Delhi).

Chopra, K., G. Kadekodi and M.N. Murty (1990), Participatory Development: People and Common Property Resources (Sage, New Delhi).

Cleaver, K.M., and G.A. Schreiber (1994), Reversing the Spiral: the Population, Agriculture, and Environment Nexus in Sub-Saharan Africa (World Bank, Washington, DC).

Cleland, J. (1996), "Population growth in the 21st century: cause for crisis or celebration?", Tropical Medicine and International Health 1(1):15–26.

Cleland, J., and C. Wilson (1987), "Demand theories of the fertility transition: an iconoclastic view", Population Studies 41(1):5–30.

Coale, A.J. (1969), "The decline of fertility in Europe from the French Revolution to World War II", in: S.J. Behrman, L. Corsa, and R. Freedman, eds., Fertility and Family Planning: A World View (University of Michigan Press, Ann Arbor, MI).

Cochrane, S. (1979), Fertility and Education: What Do We Really Know? (Johns Hopkins University Press, Baltimore, MD).

Cochrane, S. (1983), "Effects of education and urbanization on fertility", in: R. Bulatao and R. Lee, eds., Determinants of Fertility in Developing Countries, Vol. 2 (Academic Press, New York).

Cochrane, S., and S. Farid (1989), "Fertility in Sub-Saharan Africa: analysis and explanation", Discussion Paper No. 43 (World Bank, Washington DC).

Cohen, B. (1993), "Fertility levels, differentials and trends", in: K.A. Foote, K.H. Hill and L.G. Martin, eds., Demographic Change in Sub-Saharan Africa (National Academy Press, Washington, DC).

Cohen, J. (1995), How Many People Can the Earth Support? (Norton, New York).

Coleman, J.S. (1988), "Social capital in the creation of human capital", American Journal of Sociology 94(1):95–120.

Collins, S., and B. Bosworth (1996), "Economic growth in East Asia: accumulation versus assimilation", Brookings Papers on Economic Activity 2:135–191.

Cox, D., and E. Jimenez (1992), "Social security and private transfers in developing countries: the case of Peru", World Bank Economic Review 6(1):155–169.

Crook, N. (1996), "Population and poverty in classical theory: testing a structural model for India", Population Studies 50(2):173–186.

Cropper, M., and W. Oates (1992), "Environmental economics: a survey", Journal of Economic Literature 30(2):675–740.

Cruz, W., and R. Repetto (1992), The Environmental Effects of Stabilization and Structural Adjustment Programmes: The Philippines Case (World Bank, Washington, DC).

Daily, G. (ed.) (1997), Nature's Services: Societal Dependence on Natural Ecosystems (Island Press, Washington, DC).

Daily G., P. Dasgupta, B. Bolin, P. Crosson, J. du Guerny, P. Ehrlich, C. Folke, A.-M. Jansson, B.-O. Jansson, N. Kautsky, A. Kinzig, S. Levin, K.-G. Mäler, P. Pinstrup-Andersen, D. Siniscalco and B. Walker (1998), "Food production, population growth, and the environment", Science 281:1291–1292.

Das Gupta, M. (1987), "Selective discrimination against female children in India", Population and Development Review 13(1):77–100.

Dasgupta, P. (1982), The Control of Resources (Harvard University Press, Cambridge, MA).

Dasgupta, P. (1990), "Well-being and the extent of its realization in poor countries", Economic Journal, 100(Supplement) 1–32.

Dasgupta, P. (1992), "Population, resources and poverty", Ambio 21(1):95–101.

Dasgupta, P. (1993), An Inquiry into Well-Being and Destitution (Clarendon Press, Oxford).

Dasgupta, P. (1995a), "Population, poverty, and the local Environment", Scientific American 272(1):40–45.

Dasgupta, P. (1995b), "The population problem: theory and evidence", Journal of Economic Literature 33(4):1879–1902.

Dasgupta, P. (1996), "The economics of the environment", Environment and Development Economics 1(4):387–428.

Dasgupta, P. (1997a), "Nutritional status, the capacity for work and poverty traps", Journal of Econometrics 77(1):5–38.

Dasgupta, P. (1997b), "Environmental and resource economics in the world of the poor", 15th Anniversary Lecture (Resources for the Future, Washington, DC).

Dasgupta, P. (1998a), "The economics of poverty in poor countries", Scandinavian Journal of Economics 100(1):41–68.

Dasgupta, P. (1998b), "Population, consumption and resources: ethical issues", Ecological Economics 24(2):139–152.

Dasgupta, P. (1999), "Economic progress and the idea of social capital", in: P. Dasgupta and I. Serageldin, eds., Social Capital: A Multifaceted Perspective (World Bank, Washington, DC).

Dasgupta, P. (2000a), "Population, resources, and poverty: an exploration of reproductive and environmental externalities", Population and Development Review 26(4): 643–649.

Dasgupta, P. (2000b), "Valuing biodiversity", in: Encyclopedia of Biodiversity (Academic Press, New York).

Dasgupta, P. (2001a), Human Well-Being and the Natural Environment (Oxford University Press, Oxford).

Dasgupta, P. (2001b), "On population and resources: reply", Population and Development Review 26(4):748–754.

Dasgupta, P. (2002), "Social capital and economic performance", in: E. Ostrom and T.-K. Ahn, eds., Social Capital: A Reader (Edward Elgar, Cheltenham, UK) forthcoming, 2002.

Dasgupta, P., S. Levin and J. Lubchenco (2000), "Economic pathways to ecological sustainability", Bio-Science 50(4):339–345.

Dasgupta, P., and K.-G. Mäler (1991), "The environment and emerging development issues", in: Proceedings of the Annual Bank Conference on Development Economics 1990 (Supplement to the World Bank Economic Review) 101–132.

Dasgupta, P., and K.-G. Mäler (1995), "Poverty, institutions, and the environmental-resource base", in: J. Behrman and T.N. Srinivasan, eds., Handbook of Development Economics, Vol. 3 (North Holland, Amsterdam).

Dasgupta, P., and K.-G. Mäler (2000), "Net national product, wealth, and social well-being", Environment and Development Economics 5(1):69–93.

Dasgupta, P., and D. Ray (1986), "Inequality as a determinant of malnutrition and unemployment, 1: Theory", Economic Journal 96(4):1011–1034.

Dasgupta, P., and D. Ray (1987), "Inequality as a determinant of malnutrition and unemployment, 2: Policy", Economic Journal 97(1):177–188.

Dasgupta, P., and I. Serageldin, eds. (2000), Social Capital: A Multifaceted Perspective (World Bank, Washington, DC).

Dasgupta, P., and M. Weale (1992), "On measuring the quality of life", World Development 20(1):119–131.

Dreze, J., and M. Murthi (2001), "Fertility, education and development: further evidence from India", Population and Development Review 27(1):33–63.

Dreze, J., and A. Sen (1990), Hunger and Public Action (Clarendon Press, Oxford).

Durham, W. (1979), Scarcity and Survival in Central America: Ecological Origins of the Soccer War (Stanford University Press, Stanford, CA).

Dyson, T., and M. Moore (1983), "On kinship structure, female autonomy, and demographic behavior in India", Population and Development Review 9(1):35–60.

Easterlin, R., R. Pollak and M. Wachter (1980), "Toward a more general model of fertility determination: endogenous preferences and natural fertility", in: R. Easterlin, ed., Population and Economic Change in Developing Countries (University of Chicago Press, Chicago, IL).

Ehrlich, P., and A. Ehrlich (1990), The Population Explosion (Simon and Schuster, New York).

Ehrlich, P., and J. Holdren (1971), "Impact of population growth", Science 171:1212–1217.

Enke, S. (1966), "The economic aspects of slowing population growth", Economic Journal 76(1):44–56.

Ezzell, C. (2000), "Care for a dying continent", Scientific American 282(5):72–81.

Farooq, G., I. Ekanem and S. Ojelade (1987), "Family size preferences and fertility in south-western Nigeria", in: C. Oppong, ed., Sex Roles, Population and Development in West Africa (James Currey, London).

Filmer, D., and L. Pritchett (2002), "Environmental degradation and the demand for children: searching for the vicious circle in Pakistan", Environment and Development Economics 7(1):123–146.

Fogel, R.W. (1994), "Economic growth, population theory, and physiology: the bearing of long-term processes on the making of economic policy", American Economic Review 84(3):369–395.

Fogel, R.W. (1999), "Catching up with the economy", American Economic Review 89(1):1–19.

Fortes, M. (1978), "Parenthood, marriage and fertility in West Africa", Journal of Development Studies, 14(4), Special Issue on Population and Development) 121–149.

Freedman, R. (1995), "Asia's recent fertility decline and prospects for future demographic change", Asia–Pacific Population Research Report No. 1 (East–West Center, Honolulu, HI).

Fudenberg, D., and J. Tirole (1991), Game Theory (MIT Press, Cambridge, MA).

Goody, J. (1976), Production and Reproduction (Cambridge University Press, Cambridge).

Goody, J. (1996) "Comparing family systems in Europe and Asia: are there different sets of rules?", Population and Development Review 22(1):1–20.

Guyer, J.L. (1994), "Lineal identities and lateral networks: the logic of polyandrous motherhood", in: C. Bledsoe and G. Pison, eds., Nupitality in Sub-Saharan Africa: Contemporary Anthropological and Demographic Perspectives (Clarendon Press, Oxford).

Hajnal, J. (1982), "Two kinds of preindustrial household formation systems", Population and Development Review 8(3):449–494.

Hamilton, K., and M. Clemens (1999), "Genuine savings rates in developing countries", World Bank Economic Review 13(2):333–356.

Harford, J.D. (1998), "The ultimate externality", American Economic Review 88(1):260–265.

Hess, P.N. (1988), Population Growth and Socioeconomic Progress in Less Developed Countries (Praeger, New York).

Heyd, D. (1992), Genethics: The Morality of Procreation (University of California Press, Los Angeles, CA).

Heyser, N. (1996), Gender, population and environment in the context of deforestation: a Malaysian case study (United Nations Research Institute for Social Development, Geneva).

Hill, K. (1992), "Fertility and mortality trends in the developing world", Ambio 21(1):79–83.

Homer-Dixon, T.E. (1994), "Environmental scarcities and violent conflict: evidence from cases", International Security 19(1):5–40.

Homer-Dixon, T.E. (1999), Environment, Scarcity, and Violence (Princeton University Press, Princeton, NJ).

Hotz, V.J., J.A. Klerman and R.J. Willis (1997), "The economics of fertility in developed countries", in: M.R. Rosenzweig and O. Stark, eds., Handbook of Population and Family Economics (North-Holland, Amsterdam).

Howe, J. (1986), The Kuna Gathering: Contemporary Village Politics in Panama (University of Texas Press, Austin, TX).

Hussain, A., N. Stern and J. Stiglitz (2000), "Chinese reforms from a comparative perspective", in: P.J. Hammond and G.D. Myles, eds., Incentives, Organization, and Public Economics (Oxford University Press, Oxford).

IIED/WRI (International Institute of Environment and Development/World Resources Institute) (1987), World Resources 1987 (Basic Books, New York).

Iyer, S. (2000), Religion and the Economics of Fertility in South India, Ph.D. Dissertation (Faculty of Economics, University of Cambridge).

Jodha, N.S. (1986), "Common property resources and the rural poor", Economic and Political Weekly 21:1169–1181.

Jodha, N.S. (1995), "Common property resources and the environmental context: role of biophysical versus social stress", Economic and Political Weekly 30:3278–3283.

Jodha, N.S. (2001), Living on the Edge: Sustaining Agriculture and Community Resources in Fragile Environments (Oxford University Press, New Delhi).

Johnson, D.G. (2000), "Population, food, and knowledge", American Economic Review 90(1):1–14.

Johnson, D.G. (2001), "On population and resources: a comment", Population and Development Review 27(4):739–747.

Jolly, C.L., and J.N. Gribble (1993), "The proximate determinants of fertility", in: K.A. Foote, K.H. Hill and L.G. Martin, eds., Demographic Change in Sub-Saharan Africa (National Academy Press, Washington, DC).

Jones, C.I. (1998), Introduction to Economic Growth (Norton, New York).

Kalipeni, E., ed. (1994), Population Growth and Environmental Degradation in Southern Africa (Lynne Rienner, Boulder, CO).

Kelley, A.C. (1988), "Economic consequences of population change in the Third World", Journal of Economic Literature 26(4):1685–1728.

Kennedy, E., and R. Oniang'o (1990), "Health and nutrition effects of sugarcane production in south-western Kenya", Food and Nutrition Bulletin 12(4):261–267.

Kneese, A., and J. Sweeney (1985, 1993), Handbook of Natural Resource and Energy Economics, Vols. 1–3 (North-Holland, Amsterdam).

Kremer, M. (1993), "Population growth and technological change: One Million B.C. to 1990", Quarterly Journal of Economics 108(3):681–716.

Krishnan, P. (2001), "Cultural norms, social interactions and the fertility transition in India", Mimeo (Faculty of Economics, University of Cambridge).

Landes, D. (1969), The Unbound Prometheus (Cambridge University Press, Cambridge).

Landes, D. (1998), The Wealth and Poverty of Nations: Why Some Are So Rich and Some So Poor (Norton, New York).

Lee, R. (1972), "Population growth and the beginnings of sedentary life among the Kung Bushmen", in: B. Spooner, ed., Population Growth: Anthropological Implications (MIT Press, Cambridge, MA).

Lee, R.D., and T. Miller (1991), "Population growth, externalities to childbearing, and fertility policy in developing countries", in: Proceedings of the Annual Bank Conference on Development Economics 1990 (Supplement to the World Bank Economic Review) 275–304.

Leisinger, K.M., K. Schmitt and R. Pandya-Lorch (2001), Six Billion and Counting: Population Growth and Food Security in the 21st Century (International Food Policy Research Institute, Washington, DC).

Levin, S.A. (1999), Fragile Dominion: Complexity and the Commons (Addison-Wesley/Longman, Reading, MA).

Levin, S.A. (ed.) (2001), Encyclopedia of Biodiversity (Academic Press, New York).

Loughran, D., and L. Pritchett (1998), "Environmental scarcity, resource collection, and the demand for children in Nepal", Mimeo (World Bank, Washington, DC).

Lutz, W., and S. Scherbov (1999), "Quantifying the vicious circle model: the PEDA model for population, environment, development and agriculture in African countries", Discussion Paper (International Institute for Applied Systems Analysis, Laxenburg).

Maddison, A. (2001), The World Economy: A Millennial Perspective (OECD, Development Research Centre, Paris).

Mäler, K.-G. (1974), Environmental Economics: A Theoretical Enquiry (Johns Hopkins University Press, Baltimore, MD).

Mastrandrea, M., and S.H. Schneider (2001), "Integrated assessment of abrupt climate changes", Climate Policy 1(3):433–449.

Mauro, P. (1995), "Corruption and growth", Quarterly Journal of Economics 110(3):681–712.

McNeill, J.R. (2000), Something New Under the Sun: An Environmental History of the Twentieth-Century World (Norton, New York).

Meade, J.E. (1973), The Theory of Externalities (Institute Universitaire de Hautes Etudes Internationales, Geneva).

Myrdal, G. (1944), An American Dilemma: The Negro Problem and Modern Democracy (Harper & Row, New York).

Narayan, D., R. Patel, K. Schafft, A. Rademacher and S. Koch-Schulte (2000), Voices of the Poor: Can Anyone Hear Us? (World Bank, Washington, DC).

National Research Council (1986), Population Growth and Economic Development: Policy Questions (US National Academy of Sciences Press, Washington, DC).

Nerlove, M. (1991), "Population and the environment: a parable of firewood and other tales", American
    Journal of Agricultural Economics 75(1):59–71.

Oates, W., ed. (1992), The International Library of Critical Writings in Economics: The Economics of the
    Environment (Edward Elgar, Cheltenham).

Ostrom, E. (1990), Governing the Commons: The Evolution of Institutions for Collective Action (Cambridge
    University Press, Cambridge).

Ostrom, E. (1992), Crafting Institutions for Self-Governing Irrigation Systems (ISC Press for Institute for
    Contemporary Studies, San Francisco, CA).

Pearce, D., K. Hamilton and G. Atkinson (1996), "Measuring sustainable development: progress on indica-
    tors", Environment and Development Economics 1(1):85–101.

Perrings, C., and B.W. Walker (1995), "Biodiversity loss and the economics of discontinuous change in semi-
    arid rangelands", in: C. Perrings et al., Biodiversity Loss: Economic and Ecological Issues (Cambridge
    University Press, Cambridge).

Pritchett, L.H. (1994), "Desired fertility and the impact of population policies", Population and Development
    Review 20(1):1–56.

Przeworski, A., and F. Limongi (1995), "Democracy and development", Working Paper 7 (Chicago Center on
    Democracy, University of Chicago).

Rahmstorf, S. (1995), "Bifurcations of the Atlantic thermohaline circulation in response to changes in the
    hydrological cycle", Nature 378:145–149.

Robey, B., S.O. Rutstein and L. Morris (1993), "The fertility decline in developing countries", Scientific
    American 269(6):30–37.

Scheffer, M. (1997), The Ecology of Shallow Lakes (Chapman & Hall, New York).

Schultz, T.P. (1988), "Economic demography and development", in: G. Ranis and T.P. Schultz, eds., The State
    of Development Economics (Blackwell, Oxford).

Scheffer, T.P. (1997), "Demand for children in low income countries", in: M.R. Rosenzweig and O. Stark,
    eds., Handbook of Population and Family Economics (North-Holland, Amsterdam).

Seabright, P. (1997), "Is cooperation habit-forming?", in: P. Dasgupta and K.-G. Mäler, eds., The Environment
    and Emerging Development Issues, Vol. 2 (Clarendon Press, Oxford).

Sen, A. (1994), "Population: delusion and reality", New York Review of Books (September 22) 62–71.

Serageldin, I. (1995), "Are we saving enough for the future?", in: Monitoring Environmental Progress, Report
    on Work in Progress, Environmentally Sustainable Development (World Bank, Washington, DC).

Serra, R. (1996), An Economic Analysis of Child Fostering in West Africa, Ph.D. Dissertation (Faculty of
    Economics, University of Cambridge).

Seymour, F.J., and N.K. Dubash (2000), The Right Conditions: The World Bank, Structural Adjustment, and
    Forest Policy Reform (World Resources Institute, Washington, DC).

Shapiro, J. (2001), Mao's War Against Nature: Politics and the Environment in Revolutionary China (Cam-
    bridge University Press, Cambridge).

Simon, J. (1981), The Ultimate Resource (Princeton University Press, Princeton, NJ).

Solow, R.M. (1956), "A contribution to the theory of economic growth", Quarterly Journal of Economics
    70(1):65–94.

Sopher, D.E. (1980a), "Sex disparity in Indian literacy", in: D.E. Sopher, ed., An Exploration of India: Geo-
    graphical Perspectives on Society and Culture (Cornell University Press, Ithaca, NY, 1980).

Sopher, D.E. (1980b), "The geographical patterning of culture in India", in: D.E. Sopher, ed., An Exploration
    of India: Geographical Perspectives on Society and Culture (Cornell University Press, Ithaca, NY, 1980).

Stern, N. (1989), "The economics of development: a survey", Economic Journal 99(2):597–685.

Sundstrom, W.A., and P.A. David (1988), "Old age security motives, labor markets and farm family fertility
    in antebellum America", Explorations in Economic History 25(2):164–197.

Temple, J. (1999), "The new growth evidence", Journal of Economic Literature 37(1):112–156.

Thomson, J.T., D.H. Feeny and R.J. Oakerson (1986), "Institutional dynamics: the evolution and dissolution of
    common property resource management", in: Proceedings of a Conference on Common Property Resource
    Management, National Research Council (US National Academy of Science Press, Washington, DC).

Turner, B.L., and A.M.S. Ali (1996), "Induced intensification: agricultural change in Bangladesh with implications for Malthus and Boserup", Proceedings of the National Academy of Sciences 93:14984–14991.

Wade, R. (1988), Village Republics: Economic Conditions for Collective Action in South India (Cambridge University Press, Cambridge).

Watkins, S.C. (1990), "From local to national communities: the transformation of demographic regions in Western Europe 1870–1960", Population and Development Review 16(2):241–272.

World Bank (1984), World Development Report (Oxford University Press, New York).

World Bank (1989), Sub-Saharan Africa: From Crisis to Sustainable Development (World Bank, Washington, DC).

World Bank (1991), World Development Report (Oxford University Press, New York).

World Bank (1992), World Development Report (Oxford University Press, New York).

World Bank (1997), Expanding the Measure of Wealth: Indicators of Environmentally Sustainable Development (World Bank, Washington, DC).

World Bank (1998), World Development Indicators (World Bank, Washington, DC).

World Bank (2000a), World Development Report (Oxford University Press, New York).

World Bank (2000b), World Development Indicators (World Bank, Washington, DC).

World Bank (2001), World Development Indicators (World Bank, Washington, DC).

Wrigley, E.A., and R.S. Schofield (1981), The Population History of England 1541–1871: A Reconstruction (Arnold, Cambridge).

Chapter 6

THE THEORY OF POLLUTION POLICY

GLORIA E. HELFAND

School of Natural Resources and Environment, The University of Michigan, Dana Bldg. 430 E. University, Ann Arbor, MI 48109-1115, USA

PETER BERCK

Department of Agricultural and Resource Economics, University of California, 207 Giannini Hall #3310, Berkeley, CA 94720-3310, USA

TIM MAULL

School of Natural Resources and Environment, The University of Michigan, Dana Bldg. 430 E. University, Ann Arbor, MI 48109-1115, USA

Contents

Handbook of Environmental Economics, Volume 1, Edited by K.-G. Mäler and J.R. Vincent

Abstract

Physically, pollution occurs because it is virtually impossible to have a productive process that involves no waste; economically, pollution occurs because polluting is less expensive than operating cleanly. This chapter explores the sources and consequences of, and remedies for, pollution and associated environmental damages. If all goods had well-defined property rights and could be traded in markets, environmental goods would be no different than other goods; however, markets fail for these goods because property rights cannot or do not exist and because of the nonexclusive, nonrival nature of these goods. Thus, environmental goods provide the classic case where government intervention can increase efficiency. Achieving efficient levels of pollution involves charging per unit of pollution based on damages caused by that unit. In practice, this policy can be difficult to achieve, due to difficulties in measuring and differentiating damages by source, difficulties in monitoring and enforcing pollution policies, and the financial and political costs of pollution taxes. Additionally, pre-existing market distortions influence the nature of efficient pollution abatement strategies. Thus, many regulatory approaches that do not achieve first-best outcomes may be used because their technological or political feasibility is superior. Market-based instruments provide flexibility to polluters, while command-and-control (standards-based) approaches limit choice, often through an emissions limit or a technology requirement. Market-based approaches typically achieve a specified level of emissions with lower abatement costs than standards, but their greater efficiency may not hold in the presence of the problems mentioned above. Non-regulatory approaches to pollution control include the use of liability law to define and enforce property rights and some voluntary pollution control initiatives by polluters. While these approaches can play an important role, they are unlikely to achieve adequate provision of environmental goods.

Keywords

pollution, environmental policy, pollution policy, pollution theory, environmental policy instruments, environmental economics

JEL classification: H2, L5, Q2

Introduction

Economic agents that emit effluents harmful to others typically do not bear the full cost of their behavior, because these effluents are seldom traded in markets and are usually unpriced. The final allocation in economies with unpriced effluent is therefore not a Pareto optimum. Pollution control policies seek to increase efficiency by decreasing effluent compared to this suboptimal private outcome. The theory of pollution control would be very short if assigning the correct effluent price to every polluter in every place were feasible and if policy were indifferent to distributional considerations. The complications associated with achieving efficient pollution levels have led to a wide-ranging literature, which this chapter summarizes and reviews.

The chapter begins with a simple model of environmental externalities, to identify the issues whose elaboration will be the subject of the chapter (Section 1). A pollution tax is introduced as the basic form of regulatory intervention. The next two sections discuss effluent generation (Section 2) and fundamental reasons for the lack of markets in effluent (externalities and public goods; Section 3). They are followed by a discussion of complications associated with different formulations of environmental damages (Section 4). The objectives, both in theory and in practice, of environmental policy are then reviewed (Section 5). Although maximizing social welfare is the usual economic objective, other objectives often are more practical or more common in a policy setting; in addition, complications associated with nonconvexities have important implications for the identification of optimal solutions. Different environmental policy instruments are then compared in a variety of settings (Section 6). Various forms of imperfect information that can influence the design of these instruments are discussed in the next section (Section 7). Finally, the chapter examines some non-regulatory approaches to environmental protection (Section 8).

1. A simple model with a Pigouvian tax

When markets are well functioning for all goods and services, the resultant competitive equilibrium is Pareto optimal. When externalities exist, typically due to ill-defined property rights [see Chapter 3 (by David Starrett)], firms commonly emit harmful effluents without making payments for the assimilation services provided by the environment and, implicitly, by those who benefit from a clean environment. The fundamental question for environmental policy is how to get polluters to face the costs of emitting harmful effluents. This question is equivalent to identifying ways to correct for the lack of a working market in assimilation services.

In the simplest model of effluent control, there is a single firm that makes a good, q, and in doing so emits a noxious effluent, a. The effluent causes losses in the amount of $D(a)$ to the single consumer in the model, while consumption of the good benefits the consumer by the amount $U(q)$. For simplicity, U is total willingness to pay and D is measured in commensurate units of currency. The firm's cost function for producing q is $C(q,a)$. Typically, with subscripts referring to partial derivatives, utility from consuming q increases at a diminishing rate ($U_q > 0, U_{qq} \leqslant 0$); damages from a increase at a rising rate ($D_a > 0, D_{aa} \geqslant 0$); marginal production costs are increasing in q ($C_q > 0, C_{qq} \geqslant 0$); at least over a range, production costs increase as effluent decreases ($C_a < 0$); and the marginal costs of production either increase or are unaffected by decreases in effluent ($C_{aq} \leqslant 0$). The change in cost incident upon *decreased* effluent, $-C_a$, is the marginal abatement cost.

In this model, the maximal net surplus – which we hereby define as the maximum of social welfare and refer to as the social optimum – is found by choosing a and q to maximize

$$U(q) - D(a) - C(q,a).$$

Assuming interior solutions, the resulting first-order conditions are

$$U_q = C_q,$$
$$-C_a = D_a.$$

The first condition is that the marginal benefit from consuming one more unit of the good q should equal its marginal cost of production. Because, in decentralized markets, the consumer would consume q until U_q is equal to the price of the good, and the producer would set price equal to marginal cost of production, this condition is equivalent to the one that occurs in the decentralized solution.

The second condition is that the marginal abatement cost should equal the marginal damage. In other words, as long as the cost reduction for the producer from more effluent exceeds the damage to the consumer, then welfare is improved by increasing effluent. Once marginal damage begins to exceed the cost reduction, however, no further effluent should be emitted.

The solutions identified by these conditions, q^* and a^*, maximize welfare. As noted above, the first condition is identical in form to that which would occur in a decentralized system. The second condition, however, will typically not be achieved in a decentralized system. In the classic externality problem, the polluter (here, the producer of the good) does not face the costs its effluent imposes on others. As a result, while the firm sets marginal production cost equal to price (the first condition), instead of the second condition it sets marginal abatement cost equal to zero:

$$-C_a = 0.$$

That is, the firm increases its effluent as long as doing so decreases its production costs, and this results in excess effluent. If the production of q is affected by the amount of effluent (that is, if $C_{qa} \neq 0$), then the market for q will, under decentralization, also be affected by the externality. In particular, if $C_{qa} < 0$ – that is, if marginal costs of producing q decrease as a increases – then excess q will also be produced. As a result, consumers will receive more q at a lower price.

It is often more convenient for theoretical purposes to condense the model so that the only variable is effluent. By solving the first first-order condition for output as a function of effluent, $q^*(a)$, and substituting this result into the cost function, $C(q^*(a), a)$, costs are modeled solely as a function of effluent. The problem is then to

$$\max_a U\big(q(a)\big) - D(a) - C\big(q(a), a\big).$$

By the envelope theorem, this gives the same first-order condition for effluent identified above, $-C_a = D_a$. Substitution of a^* into $q(a)$ yields the resulting level of production of q.

Much of the theory of pollution policy is about feasible ways of achieving the socially optimal level of pollution, or of at least reducing the social costs associated with externalities. Since firms' unrestricted actions are inefficient, pollution policy often focuses on actions and effects that a regulatory or legal system might produce. In this simple model, levying a charge per unit of effluent of $t = D_a(a^*)$ would achieve the social optimum. This effluent charge is commonly called a Pigouvian tax, or simply a tax, which is the term we will use in this chapter. Faced with a tax set at $t = D_a(a^*)$, the firm will now have an incentive to achieve a^* in the pollution market. Achieving a^* in the pollution market will lead to an optimal outcome in the output market as well.

There are many other examples of instruments that a regulator could choose, including standards for effluent, effluent trading schemes, mandates for the use of a particular technology, and the imposition of liability for polluting. In this simple model, mandating that the firm pollute no more than a^* or assigning liability for all damages, $D(a)$, to the firm would also achieve the social optimum. In a model that reflects more of the complexities of reality, however, these policies can have quite different effects. These differences include how cheaply they are able to restrict pollution and who pays the costs of pollution avoidance.

The next two sections will begin the elaboration of this basic model by focusing on the reasons production processes generate effluent (a) and effluent exceeds the socially optimal level $(a > a^*)$. These reasons are related to the firm's cost function $(C(q, a))$. Subsequent sections will elaborate on the damages associated with pollution $(D(a))$ and on the objective of pollution regulation (social welfare or otherwise).

2. The effluent-generating process

This section examines effluent generation from two perspectives, physical science and economics. These two perspectives are intimately related, of course, and that relationship will be identified.

2.1. The physical science of polluting

If people are asked how much pollution should be permitted, the typical impulse is that the amount should be zero. Pollution damages human health and the health of other species, it disrupts the functioning of ecosystems, and it frequently interferes with our use and enjoyment of a number of goods and services. So, why do we pollute? Why is not effluent, a, zero?

One way to answer these questions is that some pollution may be unavoidable. The first and second laws of thermodynamics are relevant to explaining this phenomenon. The first law, conservation of mass and energy, states that mass and energy can be neither created nor destroyed.[1] The second law, entropy, argues that matter and energy tend toward a state in which no useful work can be done, because the energy in the system is too diffuse. Often paraphrased in terms of increasing disorder in a system, the entropy law notes that changes in matter and energy move in only one direction – toward increased entropy – unless a new source of low entropy is used to reverse processes. For instance, solar energy provides new opportunities for order to increase on the earth; otherwise, order would always decrease, and activity on earth would gradually draw to a halt [Ruth (1999)].

Under the first law, if some component of a resource is used, the material that is not used – for example, the sulfur in coal or mine tailings from mineral extraction – must go somewhere; it does not vanish. Under the second law, some of the energy or matter from the production process will be converted to a less ordered form; the final products tend to have higher entropy than the raw materials when all energy and other inputs are considered [Ayres (1999)]. Either of these laws thus suggests the production of pollution, the "ultimate physical output of the economic process" [Batie (1989, p. 1093), discussing Daly (1968)]. Incorporating the physical limits of these laws – for instance, that the ability to dispose of waste products is limited by the assimilative capacity of the environment – into economic analysis has implications for the optimal levels of all goods and services produced [Ayres and Kneese (1969), Mäler (1974)].

In terms of the model, a is positive because physical laws make $a = 0$ virtually impossible. Not all byproducts of production activities have positive market value. Even if they do, the increased entropy associated with collecting those byproducts to bring them to market may make them more costly than the market price will bear. The result is effluent that, if it causes external damages, is considered pollution.

As the above argument suggests, waste (and pollution, if it results) can be reduced in several ways. If disposal of the byproducts becomes more costly, or if the market price for the byproducts increases, firms have more incentive to bring the byproducts to market rather than to dispose of them. Additionally, changes in industrial processes can

---

[1] Relativity argues, via Einstein's famous equation $E = mc^2$, that energy and mass can be converted into each other. Nuclear reactions are the primary example of this. In most other applications, assuming that mass and energy are conserved individually is adequate.

at times lead to less pollution without increasing costs (except, perhaps, for the fixed costs of identifying and putting into place the new processes). The use of just-in-time inventory policies is an example of a management change that dramatically cuts waste in the form of unwanted parts. In recent years, the art of avoiding generation of residuals, known as pollution prevention, has received a great deal of attention as a possible way of achieving environmental gains at no or negative cost (U.S. Environmental Protection Agency). By producing the same output with less input, by substituting less hazardous substances for more damaging ones, or through increased use of recycling methods, waste can be reduced with possible increases of producer profits. Of course, increased profits do not always result from pollution prevention activities, but a large number of firms have discovered that reconfigurations of their processes do bring them both cost and environmental improvements (U.S. Environmental Protection Agency), though typically with up-front engineering costs.

   In sum, pollution can be said to arise from the laws of nature. Byproducts, either materials or wasted energy, are an inevitable part of a production process due to the conservation of mass and energy and the increasing entropy of systems. If these byproducts are undesirable, meaning that they have negative net market value, they become waste (effluent); if they contribute to external damages, they are considered to be pollution.

2.2. The economics of polluting

The above perspective on pollution emphasizes physical relationships. As in duality theory in production, which describes how a physical production process can be described in terms of price and cost information, a primarily economic interpretation can be put on the effluent-generation process. In this interpretation [see, e.g., Baumol and Oates (1988, Chapter 4)], an externality is produced when goods are produced. Production of the externality can be mitigated by expenditures on abatement. Typically, as abatement levels get very high (i.e., as pollution levels get very low), abatement costs increase, possibly exponentially. In terms of the model in Section 1, $C(q, a)$ becomes very large as a becomes small. In other words, pollution can be reduced by abatement expenditures that reduce entropy, but these expenditures increase as entropy is reduced. From an economic perspective, then, there is pollution because it is costly not to pollute.

   The following discussion will link physical aspects of pollution more explicitly to the simple economic model presented earlier. The physical science perspective implies that the activities associated with producing the desired good q also produce effluent a. Formally, let x be a vector of inputs to the production process. These inputs include capital, labor, materials, and energy, as well as inputs specific to pollution abatement, such as scrubbers for smokestacks, filters for wastewater, or equipment to recycle materials. This set of inputs is used to produce the desired good via the production function, $q(x)$. A byproduct of the use of these inputs is the effluent flow, $a(x)$. A change in production technology or in the pollution intensity of production would be reflected as a change in the functions $q(x)$ or $a(x)$.

The classic economic assumption is that firms minimize the costs of producing a specified level of output subject to a restriction on the amount of effluent emitted. The input vector x is purchased at a vector of prices w. Hence, the firm solves the problem

$$\min w'x \quad \text{subject to} \quad q = q(x) \text{ and } a \geqslant a(x).$$

The solution to this problem is the restricted cost function[2] $C(q, a, w)$. For simplicity, we will suppress input costs when they are not at issue and write $C(q, a)$, as in the simple model presented above. If there is no restriction on effluent, the Lagrangian multiplier associated with the second constraint will be zero, and the firm will choose its inputs without regard to their effect on pollution. Suppose, for instance, that x_1 and x_2 are perfect substitutes in the production process, but x_1 costs less and increases pollution more than x_2. Then the solution with unrestricted effluent will involve only x_1. If, on the other hand, pollution is restricted or made costly, such as through the use of a tax, then the firm will readjust its input mix in response to the cost and might either partially reduce its use of x_1 or switch entirely to use of x_2.

The firm is expected to maximize profits. If p is the price of q, and if there is no reason for the firm to pay attention to its effluent, then the firm's profit maximization problem is

$$\max_{q,a} pq - C(q, a),$$

with first-order conditions (assuming an internal solution)

$$p = C_q,$$
$$-C_a = 0.$$

As discussed in Section 1, this solution does not achieve the social optimum in either the q market or the a market as long as $C_{qa} \neq 0$. Because $-C_a$ is positive but decreasing in a, the firm will choose to produce more a than it would if it were forced to pay for the damage it produces: that is, if $-C_a = D_a > 0$ through a tax or other policy method. If $C_{qa} \leqslant 0$ – that is, if production of q is less expensive when a is higher – then q is also higher than the social optimum.[3] Because price is determined by setting $p = U_q$,

---

[2] A more general formulation is the implicit production function $0 = F(q, a, x)$, which implies that it is possible to change q and a while holding x constant. For our exposition, this seeming increase in generality is unlikely to provide additional insight and might in fact produce confusion.

[3] The magnitude of C_{qa} (though not the sign) is restricted by the second-order conditions for profit maximization. If the solution q^*, a^* is a unique maximum to the profit maximization problem, then $C_{qq}C_{aa} - (C_{qa})^2 \geqslant 0$. If this condition does not hold, there may not be a unique solution to this maximization problem.

a higher q and the assumption of decreasing marginal utility combine to lead to a lower price for q than the price associated with the social optimum.

As mentioned above, technological change appears in this model through changes in the production function $q(x)$ or the effluent function $a(x)$. These functions become embedded in the cost function and can be recovered through duality methods. Typically, technological change leads to lower costs of production of q. With no incentive to reduce a, the effects of technical change on a can be either positive or negative, but in recent years more research and development effort has been aimed at reducing effluent. As a result, pollution abatement has often turned out to be less expensive than predicted. For instance, abatement of sulfur dioxide emissions under the Clean Air Act Amendments of 1990 were originally predicted to cost \$250–\$350 per ton. In fact, abatement costs in the late 1990s were closer to \$100 per ton, although much of this reduction came from lower costs of low-sulfur coal than expected [Schmalensee et al. (1998)]. We refer the reader to Chapter 11 (by Adam Jaffe, Richard Newell, and Robert Stavins) for a review of theoretical and empirical studies on technological change and the environment.

3. Economic reasons for excess effluent

Unregulated firms set $-C_a = 0$ rather than $-C_a = D_a$ because their effluent is an externality: it creates an effect external to the firm, and there is no market transaction associated with it. Looked at this way, the root cause of pollution is the lack of markets in effluent. There are two good and related reasons for this lack of markets. The first is the lack of property rights for a clean environment. The second is the public good nature of effluents. We sketch the main arguments here; Chapter 3 (by David Starrett) provides a more detailed exposition.

3.1. Property rights and Coase

Coase (1960) analyzed the cases where assigning property rights was and was not a solution to an effluent problem. His analysis is notable for calling attention to transaction costs as a reason why property rights might not be sufficient to solve externality problems.

Consider two agents, one who emits effluent and the other who is damaged by it. The payoff to the first agent, as a function of its own effluent a, is $\pi^1(a)$, while the payoff to the second agent, who is harmed by the effluent, is $\pi^2(-a)$. The maximum amount of effluent discharged by the first agent is T. With the assumptions that the marginal payoff to the first agent is decreasing in a and the marginal damage to the second agent is increasing in a, the unique (interior) point at which the sum of the two agents' payoffs is maximized, a^*, is given by $\partial \pi^1 / \partial a = \partial \pi^2 / \partial a$, where the marginal payoff to the first

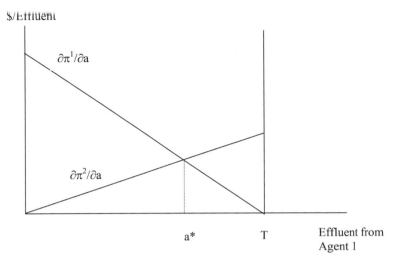

Figure 1. Gains from Coasean negotiations.

agent no longer exceeds the marginal damages imposed on the second agent. These marginal payoff curves are shown in Figure 1.[4]

Coase's theorem, which echoes Edgeworth and Pareto, states that if transaction costs are small enough, then the agents will trade to the efficient solution, a^*, regardless what value of a is initially assigned to them. For concreteness, consider an initial allocation to the left of a^* in Figure 1. A small increase in effluent increases the first agent's payoff at a rate faster than it decreases the second agent's payoff. If the first agent pays the second agent an amount per unit of additional effluent that is between $\partial \pi^1/\partial a$ and $\partial \pi^2/\partial a$, then both agents will experience increases in their payoffs. Further exchanges of effluent for money will continue until the agents reach a^*. A similar argument works for an initial effluent allocation in excess of a^*. If there are income or wealth effects, however, then the solution a^* will indeed be affected by the initial allocation of rights.

An obvious condition that must hold for a Coasean solution to be efficient is that there must be no effects on third parties, i.e., any parties that do not negotiate. That is, there can be no effects external to the negotiators. Yet, making agreements more inclusive is likely to increase transaction costs. Large transaction costs, relative to the gains of transacting, are likely when there is a great number of agents, each of whom receives little benefit from the transaction. For instance, if the rights to clean water for a particular river were equally distributed to all citizens, then a potentially effluent-emitting firm would need to buy miniscule amounts of effluent rights from thousands of

---

[4] The same diagram and conclusions apply to the case of two polluters. In that case, T is the total allowable effluent from the two producers, a is the first polluter's initial share, and $T - a$ is the second polluter's share. The benefits to trading are the integral of the difference between the two marginal payoff curves, taken between the initial allocation of rights and a^*.

people. The costs of finding the people, contacting them, and getting them to respond would likely dwarf the value to the firm (or to the citizens) of the clean water in question.

Transactions costs are thus a reason for the lack of effluent markets between consumers and producers. Similar problems can arise if the number of polluters is large, such as cars producing air pollution or farms contributing runoff that influences a river or estuary. On the other hand, it is becoming clear that transaction costs do not always preclude the creation of effluent markets between producers [see Chapter 9 (by Robert Stavins) on experience with market-based instruments].

If the transaction costs are larger than the benefits from trading, then the agents will not trade. Indeed, they should not trade. Thus, the efficiency of the system may be sensitive to the initial allocation of rights, as Coase discussed. If, for instance, initial rights are allocated so that the first player gets a^*, then the efficient solution will result even if no trades are made. If, however, the initial allocation is anywhere other than a^*, and if transaction costs prevent trading, then the result could be improved by a different initial allocation of rights.

Under many circumstances the initial allocation is an all-or-nothing grant to one agent or the other. Then, the better allocation is the one that causes the lower deadweight loss, given by the triangles to the right and left of a^*. In Figure 1, deadweight loss is lower if the first agent is given all the rights: although pollution is higher than optimal, the net gains from the pollution exceed the damages caused.

The initial allocation may come from legal precedent or legislation. For instance, common law typically permits people a right to be free from a nuisance. If one party harms another party's health, then the affected party can seek compensation for the harm through legal action. Knowing that suit can be brought should therefore induce a polluter either to avoid damage or to negotiate an agreement in advance. Rights are often not clearly defined, however, and expensive litigation might be required to determine them.

Issues of rights frequently arise in the context of land-use decisions. Because a landowner does not have complete rights to use private property for any conceivable purpose, the degree of rights can often lead to controversy. Can a government require access to a public beach through private property? Can a government limit activities on private land that might affect habitat for an endangered species? If the rights are not defined, markets cannot be used to achieve a Pareto-improving allocation. Section 8.2 contains further discussion on rights-based approaches for managing environmental externalities.

3.2. Public goods

In many cases, defining the rights for environmental goods is insufficient for efficient markets because disposal of the effluents in public media leads to a public good (or, more to the point, a public bad). Effluents cause ambient pollution, and this ambient pollution might damage all who come in contact with it. If a single individual were to purchase from a polluter the rights to emit effluent and retire those rights, then that individual would benefit not only herself but also all other people who would otherwise have come in contact with the resultant pollutant. The individual would be unable to prevent others from costlessly benefiting from the reduction (from "free riding"). Since the individual's willingness to pay would be determined only by the benefits she receives, not the benefits others receive, the amount she would pay for the rights would be lower than the social benefit (i.e., the willingness to pay summed across all individuals).

In the limit, with a great many people and each unit of effluent causing only a little bit of damage, each individual would buy no rights to emit effluent. In other words, even if property rights are well-defined and transaction costs are low, markets will provide insufficient levels of environmental quality, because any individual who buys an effluent right receives only a small fraction of the benefits associated with the transaction but bears all the costs.

Conversely, if individual consumers were initially allocated the rights to emit effluent – that is, if they owned the rights to pollute and could either retire those rights or sell them to polluters – then each consumer would sell all of her rights to potentially emitting firms. Each individual would reason that the benefits to selling the rights are entirely private and received by her, while the damage of a little more pollution is spread across many parties, with the damage to her being small. If all consumers behave this way, then the result is an inefficient outcome in which firms end up owning all the rights. Defining property rights in effluent does not in this case lead to an efficient market in pollution. Instead, the nonexclusive and nonrival nature of the pollutant leads to excess provision of the public bad.

If the total level of effluent is capped, and if effluent from one polluter has the same effect on damages as effluent from another polluter, then trading between polluters does not have a public goods problem. Damage done by effluent remains constant, because the effluent quantity remains constant. When one firm sells a unit of effluent and another firm purchases it, each firm fully bears the consequences of its actions. There is no reason, therefore, to expect that trading of effluent rights between firms will lead to an inefficient outcome if the effluent cap is set optimally. The setting of the effluent cap requires a regulatory authority, however, and implicitly it defines a rights allocation between polluters and those who suffer the damage.

In sum, markets are unlikely to achieve efficient outcomes for environmental goods. Several factors contribute to explaining this result. First, rights for these goods are typically not defined or allocated adequately. Additionally, transaction costs inhibit the efficient functioning of markets. Finally, the nonrival and nonexclusive nature of many environmental goods creates a divergence between the private and the social effects of a transaction.

While these characteristics describe why markets do not work efficiently, this discussion of market shortcomings has not identified the efficient outcome. The next section provides further elaboration of the nature of environmental damages, before an examination of how the efficient level of pollution might be identified.

4. The damage function

The key to the existence of market failure for pollution is that pollution imposes damages, or costs, that are not incorporated into decentralized market decisions. The simple model in Section 1 assumed that damages were directly and only caused by the polluter's effluent. There are several important generalizations to that basic damage function, including multiple individuals damaged by pollution, heterogeneity in effluents and their spatial dispersion, pollution affecting multiple environmental media (air, water, land), and the ability of individuals to undertake defensive activities to avoid damage.

4.1. Multiple individuals damaged by pollution

Environmental damages are best measured at the level of the affected individual. Individuals have different preferences and susceptibilities toward pollution: for instance, dislike of limited visibility due to smog varies because of people's attitudes toward scenic views, and effects on health will vary across individuals due to genetics, lifestyle, and other factors. In a general formulation, each person i gets utility from a vector of market goods, q^i, as well as disutility from the pollutant. The resulting utility function $U^i(q^i, a)$ reflects the individual's preferences over goods and pollution. The lack of superscript on a reflects the public good nature of pollution, discussed above. A separate individual damage function $D^i(a)$ can be written if a is separable from other goods in an individual's utility function. Although this is not necessarily a good assumption – for instance, see the discussion of weak complementarity in the Handbook chapter on nonmarket valuation (by Nancy Bockstael and A. Myrick Freeman III) – empirical analyses of pollution frequently focus on the effects of pollution without considering spillover effects into other markets.

An aggregate damage function $D(a)$, representing damages to all affected individuals, is typically constructed by summing the effects of pollution across individuals. This aggregation process suffers from the same difficulties that any aggregation of individual preferences faces, such as whether to weight the preferences of all individuals the same, or whether a gain to one individual offsets a loss to another. Because all individuals are expected to be harmed, or at least not benefited, by pollution, and because of the public good nature of pollution, aggregating individual damages into one function $D(a)$ is theoretically less controversial than aggregating individuals' utilities into a social welfare function. Although the aggregate measure will differ if individuals' damage functions do not have the same weights, all individuals will benefit from a decrease in a.

4.2. Multiple effluents and ambient quality

A proper model of pollution should have damage dependent upon ambient quality and ambient quality a function of effluent. The link between damage and ambient quality

is discussed below, under the heading "Damage Avoidance". The link between effluent and ambient quality is less direct than typically modeled in the literature. Damage,
as written above, is directly caused by a single effluent. In practice, effluents of many
types and from many sources combine to lower the ambient quality of air and water.
For instance, in the presence of sunlight, a series of chemical reactions involving oxides
of nitrogen from combustion activities and reactive organic gases (ROG) from a variety of sources leads to ozone formation. The ambient air quality indicator would be the
concentration of ozone, while the emissions data would be tons of nitrogen oxides and
tons of ROG. Additionally, ozone's effects on human health or the natural environment
might be affected by interactions with other pollutants. Although the damage function
should reflect the interactive effects of all pollutants, pollutants are almost always modeled individually, rather than interactively.[5]

The effects of a unit of pollution also vary with geography, weather, and other factors.
For instance, different soil types influence nonpoint source pollution runoff [Helfand
and House (1995)]. Air pollution is highly sensitive to weather conditions and other
factors. As noted above, ozone formation is strongly influenced by the presence of
sunlight, and acid deposition is enhanced by the use of tall smokestacks that put the
pollution higher into the atmosphere. Thus, the effect of one more unit of pollution on
ambient quality will vary based on the spatial arrangement of pollution sources as well
as other characteristics associated with that spatial arrangement. This spatial aspect of
pollution has long been noted in the environmental literature [e.g., Montgomery (1972)]
and has been incorporated in many empirical studies [e.g., O'Neil et al. (1983), Oates,
Portney and McGartland (1989)], but much theory and some environmental policies assume that one more unit of effluent will contribute the same marginal damage regardless
of source or ambient conditions.

Formally, let $f(A)$ give ambient quality as a function of the matrix of effluent flows.
The rows of A are the flows of J different pollutants, indexed by j. The columns of
A, a^i, are the pollutants produced by firm i; there are I different firms, indexed by i.
Ambient quality weakly decreases with an increase in any one effluent ($f_{a_{ij}} \leqslant 0$ for
all i, j), and damage decreases with an increase in ambient quality ($D_f < 0$). The
damage function as a function of effluents is $D(A) = D^*(f(A))$, where D^* is damage as a function of ambient quality. It follows immediately that the solution to the
welfare maximization problem is a multidimensional generalization of the first-order
conditions from the simple model. For each effluent the marginal rule remains, set marginal abatement cost equal to marginal damage. To write this problem compactly, define
$C^*(q, A) = \sum_i C^i(q^i, a^i)$, where the summation is over the I firms. Let $Q = \sum_i q^i$.
Now the optimization problem is again

$$\max_{q^i, a_{ij}} U(Q) - D^*(f(A)) - C^*(q, A),$$

---

[5] Ozone might be the exception that proves the rule, in that it cannot be modeled other than as the result of
interactions of nitrogen oxides and ROG.

with one set of first-order conditions being that price equals marginal production cost for each firm's output q^i, and the second set being

$$-C_{a_{ij}} = D^{*'} f_{a_{ij}}.$$

This again has the interpretation that marginal abatement cost equals the marginal value of damage. The marginal value of damage is now composed of two pieces: the contribution of effluent to ambient quality ($f_{a_{ij}}$), and the marginal contribution of ambient quality to damage ($D^{*'}$).

Setting a vector of taxes equal to $D^{*'} f_{a_{ij}}$, evaluated at the optimal quantity, for each a^i will again lead to an optimal solution in a decentralized economy. Now, however, achieving the optimum requires a separate tax for each i, j combination, reflecting the marginal damage caused by an additional unit of a particular effluent from a particular polluter. We discuss the feasibility of taxing firms based upon ambient quality rather than effluent in the next section.

4.3. Effluent transport and spatial heterogeneity in ambient quality

In the above formulation, multiple firms contribute to one index of ambient environmental quality. A further generalization of the damage function is implied by the recognition that ambient environmental quality is also subject to spatial heterogeneity. Consider a firm i emitting a single effluent a^i. Now, use j to denote the place where the effluent is deposited and causes damage. The amount of effluent arriving in place j is given by the transport function $T_{ij}(a^i)$. Assuming that damage in one place is due to the aggregate effluent arriving at that place, the damage caused in place j is $D_j(\sum_i T_{ij}(a^i))$. If total damages can be considered the sum of damages at individual receptor sites, then the simple model modified for spatial heterogeneity is

$$\max_{q,a} U(Q) - \sum_j D_j \left(\sum_i T_{ij}(a^i) \right) - C^*(q,a),$$

where a is a vector of effluent, indexed by firm, and $C^*(q,a) = \sum_i C^i(q^i, a^i)$.

The first-order conditions for a maximum, assuming an interior solution, are again "price equals marginal production cost" and an expanded version of "marginal abatement cost equals marginal damage":

$$\frac{\partial C_i}{\partial a_i} = \sum_j D_j' \left(\sum_i T_{ij}(a^i) \right) T_{ji}'(a^i).$$

The marginal damage now affects many locations, but only in the amount of the marginal effluent transported to them. Again a tax based on marginal damage will produce an optimum, but now marginal damage is calculated as the sum of effects of one more unit of a_i at each of the receptor points. One tax will be needed for each of the firms, and in general the tax will be at a different rate for each firm, because each firm's effluents reach the locations in different amounts. The calculation of a different tax or standard for each source can be administratively challenging. Below we will examine the case where a single standard or tax is used even though many taxes are necessary for the first-best solution.

4.4. Effluent disposal in multiple media

Further complications can arise when a firm has a choice of emitting its pollution into different media. The "multimedia problem" refers to the possibility that "abating" a pollutant merely involves transferring it to another medium rather than eliminating it. A method to address one environmental problem, such as reducing air pollution through the gasoline additive MBTE, can lead to other environmental problems, such as MBTE getting into water supplies. Another example is the disposal of sludge from sewage treatment plants in landfills. These multimedia problems have typically not been addressed. Instead, pollutants are usually analyzed and regulated individually.

In terms of the simple model, a production process with residuals that can be disposed into different media can be represented by a cost function that includes two types of effluent, a^1 and a^2, that cause damage in two different media, D^1 and D^2. If one regulator is responsible for both media, then the regulation problem is just a multidimensional version of the basic problem discussed earlier. The first-order conditions for an optimum imply that marginal cost of abatement should equal marginal damage for each of the effluents. When each of the two types of pollution has a separate regulator, however, the problem is one of common agency. For instance, a regulator responsible for air quality might require the addition of MBTE to gasoline to reduce air pollution. If MBTE ends up in groundwater, it is very hard to remove. A water regulator faced with the same choice would therefore presumably choose a different gasoline additive, ethanol, which pollutes air and not water.

Dumas (1997, p. 160) considers these issues in a formal model as follows. The regulator of the first medium solves the problem

$$\max_{a^1} U\big(q\big(a^1,a^2\big)\big) - D^1\big(a^1\big) - C\big(q\big(a^1,a^2\big),a^1,a^2\big).$$

Note that the first regulators' objective function omits the damages of pollution in the second medium. The regulator of the second medium solves an analogous problem. If each regulator makes the Nash assumption that the other regulator's decision will not change as a function of his own choice, then the first-order conditions are the same as those for a full first-best optimum. If, instead, the first regulator acts as a Stackelberg leader and has a first-mover advantage, then the first regulator is in a position to set a tougher standard – a lower a^1 – than he would in the first-best outcome and to force the second regulator to accept a greater a^2. Two conditions are necessary to have a suboptimal outcome in this game: one regulator must be the leader, and the two regulators cannot have identical objective functions. If both regulators use the full social objective function, then the first-best optimum again obtains even with common agency in a Stackelberg game.

Real outcomes may well be better approximated by the Stackelberg model. Regulation is expensive and time-consuming to make and revise. The regulatory process is characterized by concern for different media in different periods of time, leading to sequential decisions. As long as regulatory action by the first agent is slow to change, the resulting sequence of equilibria will be suboptimal.

4.5. Damage avoidance

The link between ambient environmental quality and damage depends on an individual's exposure to the pollutant. Exposure is influenced by many human choices, such as whether to exercise on a day with high ozone levels or to install a filter on a water tap. The effluent-producing firm is not necessarily the party that should take action to reduce damage. Sometimes consumers can avoid damage at a lower cost than firms can abate the flow of effluent. Shibata and Winrich (1983) argue that the ability of individuals to undertake defensive activities has the potential to complicate greatly any policy activities involving pollution abatement. The ability of individuals to undertake defensive activities is likely to vary a great deal by pollutant, by medium, and by personal preference, as suggested by such examples as people who buy filters for water taps or who choose to live in areas with lower pollution levels. Although individual opportunities for defensive behavior may be more available than many individuals think, the public good nature of pollution probably influences the likelihood that individuals actually undertake these activities. Courant and Porter (1981) show that individuals' expenditures on averting activities are in general not a good measure of willingness to pay for improved environmental quality.

To model this behavior formally, though simplistically, assume that the consumer can make a damage-reducing effort that reduces his consumption of the consumer good. Let θ be the quantity of consumption given up to reduce damages, so a social optimum results from solving the problem

$$\max_{q,a,\theta} U(q - \theta) - D(a, \theta) - C(q, a).$$

Here D is decreasing in effort θ. The new first-order condition, with respect to θ, is $U_q = -D_\theta$. In the special case $D(a, \theta) = D(a - \theta)$, $U_q = -D_\theta = D_a = -C_a$: the marginal cost of damage prevention equals the marginal cost of abatement.[6] If the regulator chooses the optimal standard or price for a, then the firm will also choose the proper q, and the consumer will choose the proper θ. It can be shown that, when $D_{a\theta} < 0$ and

---

[6] Shibata and Winrich (1983) noted that the equality was a special case and used a very different model from the one presented here.

$D_{\theta\theta} \geqslant 0$ and, for instance, when D is of the form $D(a - \theta) - d\theta/da$ is positive;[7] a greater ability to undertake defensive activity implies a lower tax or a higher (i.e., more lax) effluent standard, due to the stronger consumer incentive to undertake defensive activities. This formulation is, however, only a special case.

The optimal solution can be a corner solution instead of involving both defensive activities and abatement – either the consumer defends against pollution that is freely emitted, or the firm abates pollution and the consumer undertakes no defensive activity. Which of these two local optima is the global optimum is an empirical matter. At the extreme, consumers might decide to leave the area in which the effluent is emitted if the polluter undertakes no abatement. The effect of such an outcome is that marginal damages suddenly go from highly positive to zero. This issue is discussed again in Section 5.6 below.

5. The objective function

As discussed above, getting to zero pollution is typically not feasible, and it might be so costly that it is socially undesirable. At the same time, as also discussed, unregulated pollution imposes damages that should not be ignored by those who generate the pollution. The optimization approach that has been presented so far in this chapter is one of economic efficiency: effluent is abated until the marginal costs of abatement equal the marginal benefits of abatement (marginal damages avoided). The socially optimal level of pollution is thus likely to be somewhere between zero pollution and unregulated pollution.

This section examines various ways of specifying the objective function for the pollution control problem, starting with some considerations related to implementation of the economic efficiency approach. While economists focus on this approach, it is often deliberately not chosen for public policy-making in practice.

5.1. Monetary measures of damage

In the preceding sections, both damages and costs were implicitly defined in utility terms, since the cost function was subtracted from the utility function. Actual determination of the damages associated with pollution typically involves at least two steps. In the first step, the physical effects of pollution are identified, with reliance on the appropriate sciences (for instance, medicine and epidemiology for human health effects).[8] The second step is to assign a monetary value to these damages. Putting damages into

---

[7] Let Δ be the second-order condition for the maximization of welfare over θ and q with a fixed. This second-order condition is positive and the same as the determinant of the Jacobian matrix of partials from the first-order conditions $p = C_q$ and $D_\theta - p = 0$. The expression $d\theta/da = [(p' - C_{qq})D_{a\theta} + p'C_{qa}]/\Delta > 0$.

[8] The Handbook chapter on valuation of health risks by W. Kip Viscusi and Ted Gayer reviews methods for estimating the impacts of pollution on physical measures of health.

monetary units has the significant advantage that damages can then be compared directly and commensurately to the costs of pollution control. As the theory discussed above indicates, direct comparison of marginal damages and marginal costs is necessary for identifying the optimal level of pollution control.

At the same time, monetizing the damages associated with pollution is subject to a great deal of controversy associated with the technical, political, and moral problems of this approach. Valuation of nonmarket goods, such as protection of ecological functions and reductions in harm to human health, has received a great deal of attention from environmental economists (see the overview chapter by Bockstael and Freeman in this Handbook, and the subsequent chapters on specific methods by other authors), and some general principles have evolved for how to conduct these studies [e.g., National Oceanic and Atmospheric Administration (1993) on one particular method, contingent valuation)]. Yet, a number of concerns remain about how well specific valuation methods work [e.g., Diamond and Hausman (1994), again on the contingent valuation method]. Some question whether assigning price tags to nonmarket goods is an appropriate basis for public policy [Batie (1989)]. One concern is that the very act of assigning a dollar value to these goods cheapens them by making them substitutable with other goods.

Economists' response is that society often makes tradeoffs involving protection of environmental goods, and that identifying monetary values for these goods makes the tradeoffs more systematic. Some alternatives to this approach are discussed in the following sections.

5.2. Health-based standards

While, in principle, social welfare would be maximized if a benefit–cost approach were used in setting environmental standards, determining environmental quality through the use of benefit–cost analysis is, at best, controversial. Legislation in the U.S. for pollution control often does not develop targets for ambient quality based on this approach. Indeed, in setting the National Ambient Air Quality Standards, the U.S. Clean Air Act does not specify consideration of economic tradeoffs. Instead, the U.S. Environmental Protection Agency (EPA) bases the standards on protection of public health. Other environmental legislation as well bases targets on achievement of health-based standards. The use of a health-based standard is currently being argued before the U.S. Supreme Court in the case of proposed new air quality standards for ozone and particulate matter [American Trucking Associations, Inc., et al. v. United States Environmental Protection Agency]. If there is no threshold level of pollution below which there are no effects – and there does not appear to be a threshold for ozone – then the health-based standard might require an ambient standard of zero ozone.[9]

A difficulty with the use of a health-based standard, then, is that it does not consider the feasibility of attaining that standard. It also leads to a different damage function.

---

[9] The standard proposed by the U.S. EPA was actually not zero. EPA argued that the health effects of pollution below its standard were not very high.

Typically, damages from pollution are due not just to health effects on people, but also to effects on other species, ecosystems, and commercial activities. For instance, ozone inhibits crop growth and damages structures, in addition to affecting human health [Kim, Helfand and Howitt (1998)]. Let damages D be a function of human health $h(a)$ and other effects $e(a)$, where both health and other effects are determined by effluent levels: that is, $D = D(h(a), e(a))$. Then, marginal damages, D_a, equal the sum, $D_h h_a + D_e e_a$. Under the economic efficiency approach, these marginal damages are then set equal to marginal costs of abatement to find the optimal level of pollution a^*, where a^* solves $D_h h_a + D_e e_a = -C_a$.

A health-based standard, in contrast, sets $S = a^h$, where a^h is the solution to $D_h h_a = 0$. With this formulation, it is easy to see that, unless $D_e e_a = -C_a$ at $a = S$, these approaches will lead to different optimal levels of pollution. If the other effects of pollution are large relative to the costs of abatement, then a health-based standard can lead to underregulation of pollution. Alternatively, if the costs of abatement exceed the other effects of pollution, then a health-based standard might permit too little pollution [e.g., Krupnick and Portney (1991)].

The inflexibility of health-based standards with respect to local conditions is one frequent criticism of the approach. Consider regions in a country, each of which is individually subject to the same health-based standard, S. Ambient quality in each region f is determined by the interaction of its effluent (a) and a region-specific variable (ε) that reflects heterogeneous local conditions, with higher levels of ε leading to lower ambient quality. The optimization problem for each region is

$$\max_{q,a} U(q) - C(q,a) \quad \text{subject to} \quad f(a, \varepsilon) \leqslant S.$$

For regions where $f < S$, the constraint will not bind, and polluters will base their activities only on the condition that product price equals marginal production cost. For many of these regions, the unconstrained level of pollution is unlikely to be optimal; some level of pollution control may be efficient, although perhaps not as much as S. The uniform standard does not provide polluters with an incentive to reduce pollution below S.

For regions where this constraint binds, the a that solves $S = f(a, \varepsilon)$ will be the chosen level of pollution, and q will be affected through the function $q(a)$, discussed in Section 1. The solution to this problem for regions where effluent avoidance is difficult is to choose the least cost way to produce ambient quality S. This formulation leads to one set of regions just attaining the standard and another set, the regions with lower abatement costs, not being constrained and overachieving the standard. In the law and economics literature this is the problem of a "due care standard". The phenomenon of many of the agents choosing to just meet the standard was described by Diamond (1974). For many of these regions, marginal damages might be lower than the marginal costs associated with the standard; as a result, the standard may be too stringent for them.

This problem is subject to further generalization by adding a random element, most easily to the level of effluent, a. In that case the objective function takes the form of achieving the standard with given probability at minimum cost. For example, the weather is a random variable that determines how much of the effluent in agricultural runoff, such as fertilizer or pesticides, ends up in streams [Shortle and Dunn (1986), Segerson (1988)].

5.3. Cost-effectiveness: Least cost achievement of a policy target

A health-based standard, or any standard for ambient quality, merely sets the target to be achieved without necessarily suggesting how to achieve the target. Baumol and Oates (1988, Chapter 11) suggest the approach of "efficiency without optimality", more commonly known as cost-effectiveness [e.g., Kneese (1971)]. Here, the role of economic analysis begins after the target of a policy has been set. Regardless of whether the target is socially optimal, economic analysis can identify tools that can achieve it in the least costly way.

This approach recognizes the political reality that factors other than maximizing net benefits to society contribute to environmental policy, as suggested by Arrow et al. (1996). Indeed, achieving any specified target at minimum cost is a prerequisite to achieving the social optimum; regulatory approaches that achieve cost-effectiveness also achieve efficiency if the target is optimal. This issue is discussed further in the context of different regulatory instruments for pollution control, in Section 6.

5.4. Political goals

All of the objective functions discussed so far have been based on a scientific method for determining the target to be attained: specifically, either benefit–cost analysis or health-based analysis. Environmental laws are not written in a vacuum, however, and the targets required by these laws are not implemented without external input and review. Those affected by a proposed environmental law are typically less concerned with its overall target than with its effects on themselves. This and the following section discuss two approaches that explicitly consider the distributional effects of environmental policy. While in principle an efficient policy would be capable of achieving any feasible distributional target through reallocation of net gains, in practice such reallocation rarely occurs. For that reason, distributional effects frequently have a significant impact on the shape of environmental policies.

If one group suffers disproportionately from a socially optimal policy, then that group has a strong incentive to work against it. Political processes are often affected by organized protest [Peltzman (1976)]. As a result, policies based on benefit cost analysis or any other "objective" process can be changed by the political process to reflect the influence of groups affected by the policies. Political economy models explicitly recognize the influence of interest groups and model their efforts to gain advantages for them-

selves [see Chapter 8 (by Wallace Oates and Paul Portney) for a review of these models
as applied to environmental policy-making].

One specific way to characterize political goals is to assume that politicians chose
a position based both upon its popularity and upon the amount of campaign contributions that the position will generate. Campaign contributions are used to increase the
likelihood of maintaining office. In these circumstances, which closely approximate political reality, politicians will often adopt positions that are not favored by the majority.
It is not surprising that this process does not maximize net social benefits. On the other
hand, if contributions to politicians are viewed as reflecting the intensity of preferences,
then the political equilibrium can be viewed as a kind of market equilibrium, where the
equilibrium balances the interests of political constituencies. The likelihood that this
equilibrium matches the one that maximizes social welfare is small, however.

Partly in response to these concerns, Arrow et al. (1996) advocate benefit–cost analysis for all major regulatory decisions. They argue that such an analysis can make better
policies, by helping decision-makers understand better the consequences of their actions
and by making explicit the gains and losses, as well as the gainers and losers, associated
with a policy. At the same time, because of the uncertainties involved with quantifying
many factors in the analysis and because other factors, such as distributional effects,
can be important in the policy process, Arrow et al. acknowledge that decision-makers
should consider additional factors when choosing a policy.

5.5. Distribution and environmental justice

Distributional effects of environmental policy can show up in other ways as well. In recent years, the accusation has been made that poor and minority groups disproportionately face environmental damages [e.g., Bryant and Mohai (1992)]. Public and political
interest in this issue is strong, at least in the U.S. Indeed, it has been strong enough to
lead the U.S. government to pay attention to the issue through the issuance of Executive
Order 12898. While there is not one definition of "environmental justice" [Helfand and
Peyton (1999)], the existence of disparities in exposure to pollution is frequently cited
as evidence of injustice.

The distributional consequences of environmental policy come most strongly to the
fore when a facility with a noxious effluent is being sited. An aggregate measure, such as
net social benefits, does not take into account the distributional effects of this decision.
Benefit–cost analysts typically believe that all affected parties should be weighted according to their willingness to pay or willingness to accept, with distributional impacts
handled in a separate analysis. Yet, it is easy enough to produce a formal model that
includes distributional elements. Expand the number of effluents to two, a_1 and a_2, and
the number of consumers to two, each of whom is affected by only one of the two effluents. Similarly, let q^1 and q^2 be the quantities consumed by each of the two consumers.
The problem of finding a social optimum now is

$$\max_{a,q} U^1(q^1) + U^2(q^2) - D^1(a_1) - D^2(a_2) - C(q^1 + q^2, a_1, a_2).$$

The first-order conditions are again that price, which is now common to both consumers, equals marginal cost, and that marginal cost of abatement equals marginal damage averted for both types of consumers.

Though simple, this model suggests that environmental disparities can result either from normal market forces or from social injustice. If the utility functions or the damage functions differ between the individuals, or if the cost of abating differs across the effluent streams, then the first-order conditions imply that the optimal levels of effluent exposure will be different between the two individuals. For example, there is indeed reason to expect the function D to differ for rich and poor individuals: if environmental quality is a normal good, then those with higher income will have a higher willingness to pay for pollution reduction. Scale economies in pollution control might also lead to disparities. For example, the economics of waste disposal might be such that only one of a_1 and a_2 will be nonzero in the efficient solution, because one large dump is less expensive than two small dumps. Now the rules for marginal equality of damage between the two consumers will not hold.

On the other hand, observed disparities might occur because damages to minority groups receive less weight in the policymaking process due to discrimination. The outcome could be very different if the social objective function weighted the poor more strongly than the rich. In the polar case in which the social objection function is the minimum of $(U^1 - D^1, U^2 - D^2)$ and goods are distributed by the market, the allocation of effluent would be the only way to correct for differences in income. The rich would be allocated effluent until their utility net of damage was the same as that of the poor.

5.6. Nonconvexities

Most of the above discussion explicitly assumed a unique interior solution, reflecting a convex production set. In many cases, these assumptions are very reasonable. Kim, Helfand, and Howitt (1998) provide one empirical example. In other cases, however, the social production set might not, in fact, be convex. Indeed, Baumol and Bradford (1972) argue that nonconvexities in the production set are inevitable for a sufficiently strong externality: as mentioned in the discussion of averting behavior (Section 4.5), at some point the party suffering the harm from pollution will act to avoid the pollution altogether, either by moving away (people), by shutting down (firms), or by dying off (ecosystems). In all these cases, marginal costs of pollution go from a high positive number to zero. Nonconvexities can arise from other causes as well [Helfand and Rubin (1994)], including increasing returns to scale in production (as in the waste-disposal example two paragraphs above), convex utility functions, and decreasing marginal pollution damages.

If the production set becomes nonconvex, the social optimum is more difficult to locate, because the net social benefit curve of an activity might have multiple local maxima and minima. In these cases, each possible local optimum, as well as each boundary point, needs to have its net benefits calculated so that the highest of these local optima can be identified. Obviously, this complicates pollution regulation. Is it less expensive to control pollution from a source than it is to relocate all those suffering harm? Should pollution damages be concentrated in one area so that other areas can be left undamaged? No general rule for the social optimum can be stated in these cases.

While these cases might appear to be exceptions, there are likely to be a large number of situations where nonconvexities enter into an analysis. For instance, Repetto (1987) argues that nonconvexities influence the choice of least-cost control strategy for ozone, a common urban pollutant. Ozone formation requires (relatively) fixed proportions of hydrocarbons and nitrogen oxides; increasing one precursor while holding the other constant results in diminishing increases in ozone, generating diminishing marginal damages. Whether nuclear waste should be concentrated at one place or handled in a more decentralized fashion depends on whether the social choice set is convex or not. Finally, as noted above, damages from a polluting facility could become large enough to lead to victims either dying or moving away; in either case, marginal damages suddenly go from a very high value to zero. In these cases of nonconvex social choice sets, the optimal level of pollution could be the unregulated level, zero, or some other amount.

Even if a nonconvexity is present, however, the social production set could still be convex. For instance, in Repetto's case, although diminishing marginal damages imply that regulating either nitrogen oxides or hydrocarbons could be more efficient than regulating both, the optimal solution can involve reduction of both precursors when costs are taken into consideration. In many cases, Baumol and Bradford's shutdown point will not be achieved. Thus, while nonconvexities can affect the nature of the calculation of the optimal level of pollution, they do not inevitably change the standard result that marginal benefits should equal marginal costs at an interior solution.

5.7. Dynamic considerations in pollution control

Many pollutants, particularly those responsible for global warming, persist long after they have been emitted. These pollutants are called stock pollutants, and the theory of their control is necessarily cast in a dynamic setting, because present activity has both present and future consequences. Conrad examines optimal pollution strategies for a stock pollutant, using a simple linear and quadratic model. If the level of the stock pollutant is increased by a constant fraction of the economy's output and naturally decaying at a given rate, the optimal policy is to have the maximum output and run up the amount of the stock pollutant until it reaches an optimal level, at which point output is curtailed and the pollutant stock is held constant. In a dynamic and stochastic version of the model, Conrad (1992) adds a result that an increase in the instantaneous variance of the uncertain environmental cost leads to a lower optimal steady state level of the pollutant. Falk and Mendelsohn (1993) specify a stock pollutant model closer to the static models of pollution discussed in the rest of this chapter: Damage is a function of the stock of pollution, and buildup of stock can be avoided by a costly abatement process. The objective is to minimize the discounted value of the damage and abatement costs. Again, an optimal policy is to let the quantity of the stock pollutant increase in the beginning. The authors apply their model to global warming and come up with a price for carbon emissions of between \$1 and \$6/ton in the immediate future, and \$4 and \$167/ton one hundred years hence.

5.8. Summary

This section has reviewed the source of the objective function for identifying the socially optimal level of pollution. Economists typically prefer to start with individual utility functions and aggregate them, via a social welfare function, to calculate net social benefits. The damage function – the effects of pollution on welfare – is frequently modeled as being additively separable from the remainder of the utility function. Economic efficiency is the natural rule that economists advocate for identifying the optimal level of pollution, because it leads to the greatest net benefits for society, which can be reallocated, through distributional policies, as society sees fit. In actual policies, however, the legal target level of pollution is often determined by other objectives, such as health. The distributional effects of the policies – either through adverse effects on relatively powerless groups or favors granted to relatively powerful groups – influence what might ideally be considered a scientific process. Understanding the political-economic forces that are rooted in distributional considerations provides insight into why inefficient policies are often enacted. For more on this matter, see Chapter 8 (by Oates and Portney).

Nonconvexities add complications even if environmental policy is set by trading off the benefits and costs of environmental protection. In such cases, the benefits and costs at multiple local equilibria and boundary points must be compared in order to find the global optimum of the objective function. Finally, if pollutants persist over time, their management has a dynamic element that must be included in the decision over pollution levels.

6. Alternative regulatory instruments

So far this chapter has highlighted the use of a Pigouvian tax to internalize pollution externalities. The important conclusion, which is a direct application of the First Welfare Theorem, is that a tax, or more generally a set of taxes, on effluent is sufficient for an efficient outcome. An economy with a single externality is an economy with one missing market. Because a Pigouvian tax equals the price that effluent would have if an effluent market existed, it causes the economy to behave exactly as if all markets were present. Therefore, with the optimal tax on pollution, the equilibrium is a Pareto optimum. If there are many effluents, each with a different contribution to damage, then, as we have seen, a tax for each effluent is needed to make the economy act as if it had complete markets. Similarly, if there arc many time periods, then a tax is needed for each time period.

This section describes other instruments to reduce effluent, and it compares their performance to that of the tax. The ability of these other instruments to achieve the Pareto optimum does not follow as directly as for the tax, especially when they are applied to goods other than effluent. The instruments also have differing effects on firms' costs and profits and on the entry and exit of firms.

The models in this section are based on an assumption of perfect information about the benefits and costs of abating pollution. Imperfect information, which has important impacts on instrument choice, is taken up in Section 7.

6.1. Non-tax instruments applied to effluent

As mentioned above, effluent taxes lead to Pareto-optimal outcomes. At the same time, they create substantial additional costs for polluters: not only must polluters pay for abatement, but they must also pay the tax on any units of effluent they continue to emit. Other instruments have been used to reduce pollution. Some achieve the same allocational effects as effluent taxes but with different distributional consequences, and thus different long-run and general-equilibrium effects. Others lead to different allocations. The most common of these instruments are effluent and other standards ("command-and-control" instruments), abatement subsidies, and marketable effluent permits.

6.1.1. Uniform effluent standards

One possible way of regulating firms is to restrict each firm's effluent to a specified level A. This approach, typically referred to as a uniform effluent or emissions standard, provides a firm with no choice in its maximum level of effluent, although it does allow the firm to emit less than the standard. The problem for each firm is to maximize profits subject to the effluent constraint. With λ denoting the Lagrangian multiplier on the pollution constraint, the maximization problem is

$$\max_{q^j, a^j} \min_{\lambda} U(q^j) - C(q^j, a^j) + \lambda^j [A - a^j].$$

Assuming an interior solution, the first-order conditions are: (i) marginal utility equals marginal production cost, and (ii) for each firm ($j = 1, \ldots, N$ firms),

$$-C_{a^j} = \lambda^j,$$

along with the constraint, $a^j \leqslant A$. The Lagrangian multiplier, λ^j, can be interpreted as the shadow value to the firm of being able to emit one more unit of pollution. Phrased another way, it is the reduction in the cost of abatement associated with a one-unit increase in allowable effluent.

If $-C_{a^j} = \lambda^j = \partial D/\partial a^j = \partial D/\partial a^k = \lambda^k = -C_{a^k}$ for all firms, then the conditions for social optimality are achieved. This condition requires, however, firstly that a marginal unit of effluent has the same effect on damages regardless of which polluter emitted it, and secondly that all firms have the same marginal cost of abatement (the shadow cost of the constraint at A is the same for all firms). The first condition has been discussed in the context of spatial effects in Section 4.3; it holds for some global pollutants and for other pollutants at a local level. The second condition is even more difficult to achieve. In most cases, firms will differ in their abatement costs, due to differences in technologies, goods produced, and other factors. If marginal costs differ, then society can gain by reducing effluent from firms with low costs of abatement and reallocating that effluent to firms with higher marginal costs. A uniform effluent standard does not permit this reallocation. For this reason, uniform standards are unlikely to be cost-effective if firms are heterogeneous.

The marginal cost of production can be shown to be different for an effluent standard compared to a tax. When a firm pays an effluent charge of t, its costs are

$$C(q,t) = \min_a C(q,a) + ta.$$

The quantity of effluent chosen is the conditional factor demand for effluent, $a(q,t)$. It is reasonable to assume that a is increasing in q and decreasing in t. Thus, it is also possible to write $C(q,t) = C(q,a(q,t)) + ta(q,t)$. Taking the derivative with respect to q yields

$$\frac{dC}{da} = C_q + (C_a + t)a_q.$$

If this expression is evaluated at $C(q,a(q,t))$ using the first-order condition $-C_a = t$, the result is $C_q(q,a(q,t)) = C_q(q,t)$: marginal costs under the tax and the effluent standard regimes are the same when the effluent standard a is allowed to vary as a function of q. Of course, the effluent standard is typically assumed to be fixed; changes in q cannot lead to changes in a. If q changes while a is held constant, marginal costs will differ for a standard and a tax. For a given tax, t^*, there is an optimal output q^*, with a corresponding standard at $a^* = a(q^*, t^*)$. At this single point, the marginal cost curves with the standard, a^*, and with the tax, t^*, are identical. Therefore the quantities chosen by the firm are the same, and the firm's costs and profits differ only by the cost associated with the effluent tax, a^*t^*.

The output supply curves coincide only at this one point. We now show that for $q < q^*$ (respectively $> q^*$), the polluter's supply curve is more steeply sloped with a standard than with a tax: $C_q^{\text{standard}} < C_q^{\text{tax}}$ (respectively $> C_q^{\text{tax}}$). To begin, consider marginal cost at two different prices, t^* and $t_0 < t^*$, and any specific quantity, $q'' < q^*$. C_t is the factor demand for effluent, and $C_{tq} = C_{qt}$ is positive by the assumption that factor demand increases in output. This establishes that $C_q(q'', t^*) > C_q(q'', t_0)$. Since $a(q,t)$ is assumed to be increasing in q, $a(q'', t^*) < a^*$. Let t_0 solve $a(q'', t_0) = a^*$. ($t_0 < t^*$ because a is decreasing in t.) Again using the equivalence of taxes and subsidies, this time at q'', t_0 yields $C_q^{\text{tax}}(q'', t_0) = C_q^{\text{standard}}(q'', a^*)$. This result, combined with the above, gives $C_q^{\text{tax}}(q'', t^*) > C_q^{\text{standard}}(q'', a^*)$. The marginal cost with a standard is less than the marginal cost with a tax, when effluent is a normal factor of production and the quantity produced is less than the quantity where the tax and standard are equivalent. For $q'' > q^*$, the proof is similar, and the result is the opposite. The marginal cost with a standard is greater than with the tax.

Since marginal cost is supply for the single price-taking firm, the standard and the tax result in the same outcome for an industry made up of a fixed number of price-taking firms. However, that result is not true either with entry of new firms or with a shift in the demand curve for the final good with a fixed number of firms. In either of these cases, q will change for each firm, with different cost structures leading to different levels of q under a standard or a tax.

6.1.2. Other standards

It should be noted that the highly stylized version of effluent standards presented in the previous section is not a good representation of reality. First, as Helfand (1991) notes, pollution standards take many forms, including restrictions on effluent (as above), restrictions on effluent per unit of output (or per unit of an input), restrictions on polluting inputs, or requirements for specific abatement technology ("technology standards"). For a given level of total effluent from a single firm, an effluent standard is the most cost-effective among these, because it provides the most flexibility to the firm.

Secondly, firms in different industries, and sometimes different firms within the same industry, are often subjected to different levels of standards. In other words, A is not the same for all polluters. Kling (1994a), for instance, found very low potential gains (between 1 and 20 percent) from a marketable permit system for automobile pollution, because the existing command-and-control regulatory system was not very cost-ineffective. In theory, if standards are individuated – that is, if polluters with different abatement costs have standards based on their characteristics – then they can achieve the socially optimal allocation of effluent across polluter. This requires a great deal of information on the part of the regulator, however [Griffin and Bromley (1982)].

6.1.3. Subsidies

A subsidy is in many ways the mirror image of a tax. Now, polluters receive a payment of s for each unit of pollution they abate below a specified level S. S could be polluter-specific or general; we assume the latter here. The objective function for the firm becomes

$$\max_{q^j, a^j} pq^j - C^j(q^j, a^j) + s[S - a^j],$$

with the first-order conditions being "price equals marginal production cost" and

$$-C_a^j = s.$$

Again, if the per-unit subsidy is set equal to marginal damages ($s = D_{a^j}$), the marginal conditions for social optimality are achieved. The subsidy creates an opportunity cost, rather than an explicit cost, at the margin for the firm: the firm must decide whether to abate another unit of pollution and receive a subsidy payment, or to emit the unit and forgo the payment. At the margin, therefore, the subsidy produces the same pollution per firm as a tax. Of course, as will be discussed below, the total effects are quite different from those of a tax.

6.1.4. Marketable permits

A marketable permit scheme combines features of taxes, standards, and subsidies. In this scheme, which was originally discussed by Dales (1968) and Crocker (1966) and whose properties were analyzed by Montgomery (1972), the regulator makes a specified number of effluent permits A^* available to polluters. The initial allocation can be similar to the standard, by allowing polluters a specified amount of pollution, or the permits can be auctioned to polluters or allocated in other ways. Marketable permits differ from standards, however, in that polluters can buy and sell the permits. In some cases, consumers can also purchase permits and can choose to retire them if they wish.

When a fixed number of permits is either given or auctioned to firms, the program is termed a "cap-and-trade" program, because the total effluent is "capped" by the number of permits made available. In contrast, an "effluent reduction credit" program does not explicitly cap the number of permits. Instead, if a firm abates more than it is required, then it may sell the excess "credits" to another polluter. While the two approaches are very similar in that a firm faces either an explicit cost (if it is considering buying a permit or credit) or an opportunity cost (if it is considering selling a permit or credit) of polluting, credit programs have raised concerns about whether actual or only "paper" reductions will be achieved, because they lack an explicit cap on total effluent.

With a cap-and-trade program, the social maximization problem involves two preliminary steps. The first, the identification of the optimal level of pollution, has been discussed above. It fixes the aggregate level of permits, A^*. The following problem does not require that A^* be optimal, but if A^* is not chosen optimally, then the solution is only cost-effective, not efficient. The second step is the initial allocation of permits to firms. The optimization problem is then

$$\max_{q^j, a^j} \min_{\lambda} \sum_j \left[pq^j - C^j(q^j, a^j) \right] + \lambda \left[A^* - \sum_j a^j \right].$$

The first-order conditions are "price equals marginal production cost" and

$$-C_a^j = \lambda.$$

The Lagrangian multiplier λ is now the market-clearing permit price. Polluters with high marginal costs of abatement will purchase permits from those with low marginal costs of abatement (who get paid per unit of abatement, as they would under a subsidy), just as they would pay the pollution tax. The price of a permit therefore serves the same function at the margin as the tax. For this reason, pollution taxes and marketable permits are often termed "market-based instruments".[10]

If the number of permits A^* is set optimally, then the permit price λ equals marginal damages, and the conditions for optimality are achieved. In a cap-and-trade program with a suboptimal A^*, the polluters still efficiently allocate effluent permits among themselves; thus, a cap-and-trade program will always produce the least cost way of meeting a given effluent cap A^*. Because they are cost-minimizing without the need for extensive regulatory knowledge of polluters' cost structures, cap-and-trade programs are attractive for a regulator.

Because a permit system is based on quantity, while a tax system is based on price, these instruments respond differently when market conditions change – due, for instance, to changing demand conditions, changing preferences, or inflation. A cap-and-trade permit system will not allow the level of pollution to change, though the market price for the permit will respond to changes in these conditions. In contrast, a tax will allow the level of pollution to change but will hold constant the price. Thus, if the demand for polluting goods increases over time, a permit system, keeping pollution levels fixed, might result in inefficiently low pollution levels; in contrast, a tax system might allow inefficiently high levels of pollution. The increased certainty of achieving a specified emissions level under a permit system, considered by regulators one of its more desirable characteristics, may or may not be more efficient than the increased price certainty (often appreciated by polluters) of a tax. The optimal approach for a regulator is to re-evaluate the levels of either of these instruments when conditions change. Because conditions constantly change, and because regulatory systems typically cannot respond quickly, a set level of either a tax or total effluent is likely to become inefficient over time. Understanding that these instruments respond differently to changing conditions might influence a regulator's choice of policy instrument.

Marketable permits have also been discussed in a dynamic setting, where polluters can either save permits for the future (i.e., bank them) or borrow and pollute more now. Rubin (1996) finds that banking and borrowing lead to a least-cost solution for polluters. Kling and Rubin (1997), though, show that the private incentives for banking and borrowing will not necessarily lead to the socially optimal flows of pollutants, and they propose a modified trading scheme to correct this temporal misallocation. Leiby and Rubin (2001) extend the stock pollutant model to the case of bankable permits, which have been suggested as a tool to control buildup of greenhouse gases. They find the rate at which the banking mechanism should allow permits to be withdrawn at a later date for each permit deposited today. This rate of exchange and a total sum of permits are what is required for an efficient system.

---

[10] The term "market-based instruments" usually encompasses a range of instruments in addition to pollution taxes and marketable permits (e.g., taxes on polluting inputs). See Chapter 9 (by Robert Stavins).

6.1.5. Hybrid instruments

To provide some additional flexibility to respond to uncertainty, some authors have suggested combinations of instruments. Roberts and Spence (1976) introduce a hybrid policy of permits, subsidies and taxes when the regulator does not know the abatement costs of firms. If a firm abates more than the level of its permit, the firm gets a subsidy, and if a firm pollutes more than its permit, the firm pays a tax. They argue that permits can guard against very high levels of emissions, while subsidies can promote more pollution reduction if abatement costs are low. McKibbin and Wilcoxen (1997) argue that a permit and tax hybrid instrument for carbon dioxide emissions would be more flexible, encourage enforcement and monitoring, and be less stressful on world trade than just a emissions permit system. Pizer (1997) uses a global integrated climate economy model and concludes that a hybrid tax and permit device will perform marginally better than the optimal tax and far better than optimal permit system. Additionally, this scheme would allow policy makers to balance competing interests of revenue, equity, and political feasibility. Multiple instruments thus provide, at least in theory, for additional opportunities to improve efficiency.

In sum, in principle both permits and subsidies have the same ability to achieve an efficient allocation of pollution among a fixed set of polluters as does a tax. Clearly their distributional effects are different, however. Even when these regulatory approaches have the same marginal effects, they lead to very different flows of money among firms. The next section examines the implications of these differences.

6.2. Distributional effects of instruments applied to effluent

6.2.1. Effects on a firm's costs and profits

These different instruments can be viewed as implicitly or explicitly assigning different initial allocations of rights to pollute. The effluent tax and the subsidy provide the extremes, with standards and freely allocated marketable permits as intermediate cases. Under the tax, a firm will abate its effluent as long as its marginal costs of abatement are less than the tax; once the marginal costs of abatement exceed the tax, it is cheaper for the firm to pay the tax than to abate. Thus, its total costs under this regime are its abatement costs plus the tax payment. In contrast, under the subsidy, the firm is paid not to pollute. It receives a payment for each unit of abatement; as long as that payment exceeds the marginal costs of abatement, not only will the firm abate, but it will also earn positive amounts of money. The tax and the subsidy therefore differ quite substantially in terms of lump-sum payments.

A standard, if set at the level of pollution that a firm will attain under a tax or subsidy, requires a firm to pay for abatement, but it requires no additional payments (unlike a tax) and provides no additional payments (unlike a subsidy). Unlike a tax, discharge of effluent up to A is free. Even though a standard has greater aggregate costs of abatement than a tax across a set of firms, polluters might actually earn greater profits with a stan dard. Helfand and House (1995) provide an example of this situation. Indeed, Buchanan and Tullock (1975) argue that a restriction on pollution per firm results in profits being higher than under a tax that would achieve the same level of pollution. By restricting the level of output, the standard acts like a government-imposed cartel. For this rea son, firms might be expected not only to prefer a standard to a tax, but also to prefer a standard to a situation of no regulation. Maloney and McCormick (1982) provide some empirical evidence of this phenomenon.

A permit system can be considered to begin with a standard – the initial allocation of permits among firms – before buying and selling begin. Because firms will only buy or sell when it is advantageous for them to do so, marketable permits will reduce the total costs of abatement relative to a uniform standard. If the permits are granted to polluters at no cost, even firms that buy permits will still have costs no higher than they would have had under the tax, because they receive some effluent rights free of charge. Firms that sell permits will end up earning more money than they would under the standard but less than they would under the subsidy, unless they receive as many permits as their unrestricted level of effluent. In contrast, a system of auctioned permits behaves like a tax, since none of the permits are free of charge.

These significant distributional consequences may account, in the U.S., for the dom inance of standards and marketable permits in much of the country's pollution policy, and for subsidies in the realm of agriculturally-related environmental policy. Using these policies, instead of taxes, has lowered the costs of abatement to polluters while provid ing abatement to consumers. Providing gains to both parties makes these policies more politically feasible than using taxes that impose significant costs on polluters. The net gains are not as large as might be achieved via taxes with redistribution, but redistri bution could be very difficult to achieve. Howe (1994) notes that pollution taxes are more commonly used in Europe,[11] but less to act as a disincentive to pollute than as a source of funding to subsidize pollution abatement equipment. Chapter 8 (by Oates and Portney) discusses further these and other issues related to the political economy of environmental policy.

6.2.2. Effects on entry and exit of firms

Because the different instruments imply different levels of rents flowing to firms, they can be expected to affect the total number of firms and thus the total amount of pollution produced. The impact on entry of new firms into an industry was first examined by Kneese (1971), who asserted that a subsidy for abatement and a tax on effluent could lead to the same results. Subsequently, Kneese and Mäler (1973) argued that a subsidy and a tax give the same result as long as *potential* entrants are subsidized in the same

---

[11] Chapter 9 (by Robert Stavins) reviews global experience with pollution taxes and other market-based instruments.

way as actual entrants, which removes the incentive to enter. In practice, of course, subsidizing potential entrants is infeasible, since anyone can claim to be a potential entrant.

Mäler (1974, p. 218) demonstrated that the average production cost for a typical polluter is lower under a standard than under a tax. The difference in average production costs under a tax and a standard is

$$\frac{C(q,t^*)}{q} - \frac{C(q,a(q,t^*))}{q} = \frac{a(q,t^*)t^*}{q}.$$

Since lower costs imply entry, an industry regulated with a standard will have more firms and thus more effluent than one regulated with a tax. A standards-based approach requires two instruments – one to limit entry, and one to set the standard per firm – to achieve the optimum. A tax will achieve both conditions without the need for additional instruments.

Spulber (1985) analyzed the effects of taxes, standards, marketable permits, and subsidies on entry and exit. Consistent with Mäler (1974), he found that taxes provide the conditions associated with the optimal number of firms and the optimal aggregate level of effluent. Auctioned permits provide the same effects. The other instruments lead to a greater number of firms, because they assign more rights and thus more rents to the firms. Even if individual firms pollute the same amount under these instruments as under taxes or auctioned permits, the greater number of firms results in an overall level of pollution that is greater than the social optimum. Achieving the same aggregate pollution level would require more restrictive policies under these other instruments.

Actual standards often differentiate between firms already in the industry and new entrants, with new entrants being held to higher standards. The New Source Performance Standards for air pollution in the U.S. are a good example. This differentiation is typically intended to reflect the fact that new sources of pollution are likely to be designed with environmental protection in mind and are therefore likely to be able to achieve better environmental performance than existing sources, but it can also act as a barrier to entry. While these differentiated standards are unlikely to get entry/exit conditions right, they may partially address the issue of standards encouraging entry relative to taxes.

On the other hand, differentiated standards can create perverse incentives for retention of older facilities, which remain under the older, less stringent standards, when replacement of these older facilities could have improved production methods and decreased pollution. Ackerman and Hassler (1981) examined the requirement that new coal fired power plants install scrubbers, while older plants are not required to install this technology. The regulation makes new plants more expensive relative to older plants; as a result, they believe that the induced retention of older, much more polluting plants could easily outweigh the pollution control benefits of making new plants cleaner. Gruenspecht (1982) notes the existence of a similar problem with automobile emissions standards. The possibility of reclassification to a new facility when a plant is upgraded provides a powerful incentive for firms to fail to upgrade their plants as new technology becomes available. For example, in a recent lawsuit against large coal-fired electric plants, the EPA alleges that the plants engaged in so much technological change that they now qualify as new sources, subject to much more stringent pollution control requirements ("U.S. Sues 7 Utilities Over Air Pollution"). Hahn and Hester (1989) discuss a controversy over "shutdown credits" – allowing firms that go out of business to sell their emissions rights. While environmentalists criticized allowing shutdown credits for discouraging older plants from cleaning up, not allowing them could lead to older plants staying in business instead of being replaced by newer, less pollution-intensive facilities.

6.3. Regulatory instruments applied to goods other than effluent

The discussion of the damage function in Section 4 suggests that taxing effluent can be a complex matter. If sources differ in their marginal effects on damages, or if different effluents are emitted, then socially optimal taxes need to be adjusted to reflect these differential impacts. Similar principles apply to subsidies and permits when damages vary by source or by effluent: different polluters will need to face differentiated instruments. Additional complexities can arise if the good being taxed is not effluent. The following discussion will examine the implications of imposing a tax on something other than effluent.

6.3.1. Instruments applied to ambient quality

Ambient quality, such as the concentration of pollution in the air or water at a particular monitoring point, is often easier to observe than effluent. It is also much more directly related to damages than measures of effluent, since ambient quality incorporates the effects of transport of effluent as well as meteorological, geographical, and other conditions that complicate the relationship between effluent and damages.

Given that the intent of pollution policy is to lower damage, taxing firms based on ambient quality might therefore appear to be preferable to taxing effluent [Montgomery (1972)]. Difficulties arise with this approach, however, because ambient quality is produced collectively by polluters. If one polluter reduced effluent, then all firms would receive a tax break. In this situation, abatement is analogous to a public good, and the tax scheme that solves the problem looks a great deal like mechanisms that solve the public goods problem.

If each firm i $(i = 1, \ldots, N)$ is taxed t on some share s^i of total damages, with other firms' damages assumed to be constant, then each firm will maximize its profits, given by:

$$\max_{q^i, a^i} pq^i - C^i(q^i, a^i) - ts^i D(a^1, \ldots, a^N).$$

The first-order conditions are: (i) price equals marginal cost of production, and (ii)

$$-C_{a^i} = ts^i \frac{\partial D}{\partial a^i} \quad (i = 1, \ldots, N \text{ firms}).$$

The social optimum will be achieved as long as $ts^i = 1$ – that is, as long as each firm is taxed as though it were fully responsible for damages, with the actions of other firms taken as fixed. While this is technically a solution to the problem of controlling pollution, it requires each firm to make a potentially very large payment, and it is therefore unlikely to be politically acceptable. If the polluter is assigned only a fraction of responsibility for damages – that is, if $ts^i < 1$ – then too little abatement will take place to achieve the optimal level of ambient quality.

It is possible to add lump-sum side payments to this mechanism to reduce the size of the firms' tax payouts. For instance, Segerson (1988) proposes that firms be taxed if ambient water quality is worse than a specified target, but that they receive a subsidy for every unit that ambient water quality is better than that target. Her solution is reminiscent of the pivotal mechanism for public goods. On average, this mechanism should result in no net payments if the tax is set appropriately.

6.3.2. Instruments applied to inputs

Just as ambient quality is more easily observed than effluent in some cases, application of inputs might also be more observable. Holterman (1976) and Griffin and Bromley (1982) have suggested that inputs could be the subject of regulatory action instead of effluent when inputs are more easily monitored. This approach is more easily understood by modeling a firm's input choices directly, through the primal formulation, than by the cost-function approach. Here, as in Section 2.2, let $q^i(x^i)$ be firm i's output as a function of its inputs, and let $a^i(x^i)$ be its effluent. The firm's profit-maximization problem when faced with a vector of input taxes, t^i, on inputs $j = 1, \ldots, J$, is

$$\max_{x^i} pq^i(x^i) - (w + t^i)'x^i.$$

The first-order conditions are

$$p\left(\frac{\partial q^i}{\partial x_j^i}\right) = w_j + t_i^j \quad (i = 1, \ldots, N \text{ polluters}; \ j = 1, \ldots, J \text{ inputs}).$$

The social optimum will be achieved only if $t_i^j = (\partial D / \partial a^i)(\partial a^i / \partial x_j^i)$. Hence, an input should remain untaxed only if it has no effect on effluent – that is, only if $\partial a^i / \partial x_j^i = 0$. If an element of x^i abates pollution, rather than increases it, then the tax on that input should be negative (a subsidy) rather than positive. If all inputs from all polluters influence effluent in positive or negative ways, then the optimum requires $J \cdot N$ different taxes or subsidies.

Input taxes should be uniform across firms only if use of a given input by one firm has the same effect on marginal damages as use of that input by another firm. Except by coincidence, this condition will arise if effluent from one polluter is a perfect substitute for effluent from other polluters from the standpoint of pollution damages, *and* if use of an input by one polluter leads to the same level of effluent as use of that input by other polluters. The first condition has previously been discussed in Section 4, in reference to the different spatial effects of pollution. The second condition refers to the pollution intensity of different sources' use of inputs. Because one polluter is likely to use an input in a different way than another, the marginal effects of input use on effluent levels are likely to vary by firm, and the second requirement for optimality is likely to be at least as difficult to achieve as the first. (Examples where the second condition holds include chlorofluorocarbons in refrigerants, which deplete stratospheric ozone, and the carbon content of fossil fuels, which contribute to the buildup of greenhouse gases [Helfand (1999)].) Thus, taxing inputs to achieve the social optimum will typically require a separate tax for each input that affects pollution from each firm.

Fullerton and West (1999) examine the case where the effects of an emissions tax on cars can be achieved instead by taxing gasoline and engine size while subsidizing pollution control equipment. The efficient solution can only be achieved if a gasoline tax can be differentiated by the characteristics of the car. Fullerton and Wolverton (1999) show that a presumptive tax (on output, not effluent) combined with an environmental subsidy (for clean production) can also achieve the efficient solution and may actually be preferable because of greater ability to measure and enforce these policies, lower administrative costs, and possibly greater political appeal.

6.3.3. Deposit-refund schemes for waste

Municipal waste might appear to be equivalent to effluent in the models above, with the results for the various regulatory instruments carrying through. In fact, some incentive schemes for solid-waste management can lead to perverse incentives.

Incentive schemes have been used most commonly to reduce the flow of solid waste by encouraging recycling. The use of deposits, payable upon the return of an item, is a common way to induce the public to choose an environmentally less destructive manner of disposal. Prime examples include deposits on bottles and automobile batteries, which regulatory authorities wish to see recycled rather than disposed in landfills or dumped randomly. Bohm (1981) provides the pure theory of deposit-refund systems and compares them to alternatives such as a tax on all beverage containers produced. Fullerton and Kinnaman (1995) explore the optimal policy and point out that a higher deposit results in both a higher percentage return and also a higher percentage of theft of recyclable material. Fullerton and Kinnaman (1996) examine charging for garbage pickup. A higher price for garbage reduces landfilling, but it also increases illegal dumping. Thus, as will be discussed further below, the enforceability of a policy influences the optimal design of the policy.

6.4. General-equilibrium effects of pollution policies

All the analysis presented above has been in a partial-equilibrium context. Implicit in the calculation of the optimal Pigouvian taxes has been an assumption that there are no market distortions other than the environmental distortions these taxes address. Of course, that setting describes no known world inhabited by humans. As Lipsey and Lancaster (1956–1957) have shown, information on other distortions in an economy should influence the way in which any one distortion is addressed. Otherwise, it is possible for the isolated correction of one market failure to *decrease* social welfare.

General-equilibrium analysis has relatively recently been applied to examine whether environmental policies that appear to be optimal in a partial-equilibrium context might in fact need to be adjusted in response to the existence of other distortions. The chapter by Lars Bergman in this Handbook describes methods for conducting applied general-equilibrium analysis of environmental policies, and the chapter by Raymond Kopp and William Pizer compares partial- and general-equilibrium estimates of the impacts of environmental regulations on production and abatement costs. Here, we highlight theoretical results related to the interactions between pollution control instruments, especially market-based instruments, and taxes on goods other than pollution or polluting inputs. The chapter by Markandya provides a more comprehensive review of the literature on interactions between environmental and nonenvironmental policies.

In a general-equilibrium setting, the level of the optimal pollution tax depends on the levels of other taxes. As a result, the optimal general-equilibrium pollution tax is likely to differ from a partial-equilibrium Pigouvian tax that is used only to correct the pollution externality.[12] This is because a tax that is levied to reduce pollution has the added effect of raising revenue. If that revenue is used to reduce or replace other taxes, then there is the potential for a "double dividend": less pollution and less deadweight loss from other taxes.

The taxing of pollution has three effects on welfare. First, it decreases the production of the dirty good that creates the externality. Although an effluent tax can be viewed as just one of many commodity taxes, unlike most taxes it increases efficiency through its allocational effects. Second, in what is called the "recycling effect", it raises revenue that can be used to lower other, more distortionary taxes. Third, it is itself a distorting tax, through its impacts on other markets.

Bovenberg and de Mooij (1994) carry out an optimal tax exercise in a simple analytical model that includes a clean good and a dirty good, a tax on income (labor), and a tax on the dirty good. They conclude that the optimal tax on the dirty good is less than the rate that would be used to correct the externality. Taxing income causes agents to work less and consume more leisure than they would in a first-best world. As a result, there is a marginal cost to raising public funds, which is likely to be greater than one. The optimal tax on the dirty good is shown to be the Pigouvian tax divided by the marginal cost of raising public funds, so it is less than the Pigouvian tax.

---

[12] A formal analysis of this issue is provided by Diamond and Mirrlees's model of optimal taxation with a public good. Bovenberg and Goulder (1996) carry out the algebra for the specific case of a pollution tax, and they also provide some numerical estimates from a computable general-equilibrium model. Their estimates show that, in the presence of an income tax, the optimal tax rate is far less than the Pigouvian tax rate.

Fullerton (1997) reinterprets this conclusion. Because income in Bovenberg and de Mooij's model is all from labor and is all spent on the clean and dirty good, the same budget constraint results whether labor is taxed at rate t or both goods are taxed at the rate $(1 - t)^{-1}$. That is, the tax on income is equivalent to a uniform tax on both goods. Viewed with that normalization, the total tax on the dirty good – the sum of the income tax expressed as the equivalent uniform tax on both goods and the additional tax on just the dirty good – is greater than the Pigouvian tax.[13]

These double dividend models depend in part upon an assumption about preferences, specifically that the marginal rate of substitution between labor (leisure) and consumption goods is not changed by the amount of pollution, as in the separable utility function in the models presented in this chapter. If the marginal utility of leisure is related to environmental quality, then the optimal general-equilibrium tax needs to reflect this relationship. If pollution and leisure are complements, then taxing pollution reduces leisure (increases labor) and results in less distortion in the labor market. In this case, the optimal tax on the dirty good would be above the Pigouvian tax. Since the effects of pollution taxes on the labor–leisure choice have not been empirically investigated, it is not yet possible to conclude that the optimal pollution taxes in an economy with prior distortions should be lower than the Pigouvian level.

Nevertheless, it is clear in this literature that, in a comparison between pollution taxes and equivalent pollution standards, taxes have a significant advantage. Both taxes and standards raise the price of the dirty good by the same amount and so have the same pollution abatement and labor–leisure distorting effects. Only the tax has the revenue-recycling effect in its favor [Parry, Williams III and Goulder (1999)]. The tax generates government revenue, while a standard creates rents that the firms capture.

7. Imperfect information

In the simple model in Section 1, the optimal quantity of pollution was determined by a straightforward maximization problem. That model implicitly assumed that the regulator knows the costs of abating pollution and the damages associated with pollution. In fact, regulators rarely know any of this information with certainty. Uncertainty associated with the damages of pollution contributes to skepticism toward a benefit–cost rule for determining optimal abatement levels, and it also contributes to the development of the other possible objective functions discussed in Section 4. Uncertainty associated with the costs of abatement has contributed to economists' preference for price-based instruments like effluent taxes over quantity-based instruments like effluent standards: polluters will reveal their marginal costs associated with a particular level of abatement in response to the tax but not the standard.

---

[13] Because income is all from labor and is spent on the clean and dirty good, the same budget constraint results whether labor is taxed at rate t or both other goods are taxed at the rate $(1 - t)^{-1}$.

This section discusses the implications of imperfect information for environmental regulation. In particular, uncertainty about benefits and costs can affect the choice of price instruments vs. quantity instruments; the inability to link nonpoint pollution to specific sources compels the use of regulatory targets other than effluent; and polluters' awareness that regulators' monitoring and enforcement activities are imperfect affects compliance rates and has implications for the efficient design of those activities.

7.1. Uncertainty about the benefits and costs of pollution control

Weitzman (1974) and Adar and Griffin (1976) analyzed the effects of uncertainty about the aggregate marginal benefits and costs of pollution abatement on the choice of a price-based or a quantity-based regulatory instrument.[14] These analyses found that uncertainty in the marginal benefits of abatement (damages avoided) does not affect the choice of regulatory instrument, whereas uncertainty in marginal abatement costs can lead to either a price or a quantity instrument being more desirable, depending on the relative slopes of the marginal benefit and marginal cost curves.

Weitzman's framework considers a regulator who does not entirely know the costs of the regulated firm. Suppressing for now issues associated with the goods (q) market, let $C(a, \varepsilon)$ be the firm's cost of abatement, where ε is a parameter known to the firm but not to the regulator.[15] The regulator is assumed to be able to measure the effluent stream and to enforce pollution standards or taxes. Weitzman's model also includes uncertainty in the damage function, $D(a, \delta)$, where δ is a random variable whose distribution is known to the regulator. The problem is to determine the advantage of taxes over standards[16] in maximizing the expected value of $-D - C$, which is equivalent to minimizing social costs.[17]

Weitzman proceeded by first finding the standard a^* that maximized $E[-D - C]$. He then expanded D and C in a quadratic Taylor series about a^*, so that D_a and $-C_a$ were linear in a and the uncertain elements served to shift the linear functions D_a and $-C_a$ without changing their slopes. This approximation is shown in Figure 2. The horizontal axis of the figure has abatement increasing (effluent decreasing) in the x direction. D_a is drawn as decreasing, because the marginal damage from pollution decreases with more

---

[14] In this context, a cap-and-trade marketable permit program is a quantity-based instrument, because it fixes the total quantity.

[15] The asymmetry in information can come about because the regulator cannot learn everything there is to know about the cost function or because the regulator does not think this knowledge is worth the cost of acquiring. A solution identical to the asymmetric information solution can also come about when the regulator is compelled to treat a class of firms alike, even if it knows they are not alike.

[16] While the argument is phrased as referring to a standard, it is perhaps more accurate to consider it a marketable permit scheme with a fixed total level of effluent (abatement). Because most standards systems result in higher aggregate costs of abatement than incentive approaches, the abatement cost curve is likely to be the same under a tax and a permit scheme but different under a standard.

[17] Weitzman's paper is written in terms of a general planning problem. His notion of benefit is $-D$ here. The diagrammatic exposition that follows is used in Adar and Griffin (1976) and in Stavins (1996).

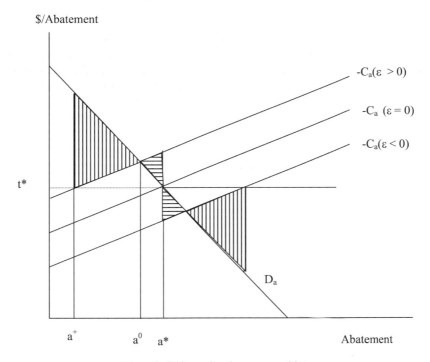

$/Abatement

$-C_a(\varepsilon > 0)$

$-C_a \ (\varepsilon = 0)$

$-C_a(\varepsilon < 0)$

t^*

D_a

a^+ a^0 a^* Abatement

Figure 2. Weitzman's prices vs. quantities.

abatement (less pollution), while $-C_a$ increases with the level of abatement. To keep the graph simple, we have assumed that $\delta = 0$ and that ε takes on only two possible values, one positive and one negative. The marginal abatement cost curves have been drawn for each of the two possible values of ε, along with the curve for its expected value of zero.

The intersection of the curves at a^* gives the expected optimal level of abatement. Because the unknown elements enter in an additive fashion and are assumed to have means of zero, the expected optimal tax t^* also occurs at the intersection of the marginal curves. If the actual value of the firm's unknown parameter is positive, then, in response to the tax, the firm will set the marginal cost of abatement equal to t^* and abate at a^+. This is too little, because marginal abatement cost with a positive ε intersects marginal damage at a^0. The deadweight loss associated with abating at a^+ instead of a^0 is given by the vertically striped area in the diagram. In contrast, in response to a standard of a^* the firm will abate too much ($a^* > a^0$), leading to the horizontally striped deadweight loss. The advantage of a tax over a standard for an ε-positive firm is the amount by which the vertically striped area exceeds the horizontal striped area.

The same calculation can be made for an ε-negative firm. The probability-weighted average of the two calculations gives the total advantage of a tax over a standard. As the D_a curve approaches the vertical, the difference between the optimal abatement level and a^* becomes small, implying that a standard is better than a tax when the marginal damages curve is steep. This comports with common sense: if one cares very much about getting the damages right, then getting the quantity of abatement right is more socially valuable than giving the polluter flexibility in how it responds to the instrument. Indeed, this view suggests why the health-based approach focuses on standards rather than effluent charges. In contrast, if the marginal damage curve is shallowly sloped but the marginal cost curve for abatement is steep, then permitting price flexibility via a tax improves efficiency over a standard.

Weitzman, along with the others who have used this framework, carried out this calculation algebraically. His most general expression for the benefit of taxes over standards is

$$\frac{\text{var}(\varepsilon)}{C_{aa}} \left[\frac{1}{2} - \frac{D''}{2C_{aa}} - \frac{\text{cor}(\varepsilon, \delta)\text{sd}(\delta)}{\text{sd}(\varepsilon)} \right],$$

where var denotes variance, cor denotes correlation, and sd denotes standard deviation. This formula indicates that variation in the marginal damage curve $-\delta \neq 0$ instead of $\delta = 0$, as assumed for simplicity above – does not affect the choice of a tax versus a standard as long as ε and δ are uncorrelated. When this correlation exists, however, the uncertainty in marginal benefits ($\text{sd}(\delta)$) does affect the choice. In particular, a positive correlation increases the likelihood that a quantity-based instrument is more efficient than a tax. Stavins (1996) showed that the correlation between damages and costs is in fact positive in many examples.

Mendelsohn (1986) redefined Weitzman's analysis to consider heterogeneity of benefits and costs instead of uncertainty. In his model, the variation in marginal benefits does not directly influence the choice between a price instrument and a quantity instrument, but the variation in marginal costs and the covariance of marginal benefits and marginal costs do. This result illustrates that the Weitzman approach, although framed in terms of uncertainty, can be applied to other contexts.

In theory, at least, there is an alternative to setting a single regulation, be it price or quantity, and allowing polluters to respond. Ellis (1992) noticed the possibility of using mechanism design for pollution control.[18] Instead of a tax or standard, a regulator could allow the polluters to chose from a menu of abatement standards and lump-sum subsidies. Under some circumstances, this mechanism will cause the truthful revelation of the firm's private information about costs and also lead to the first-best level of effluent.

The preceding articles are premised on the assumptions that the objective of pollution control is to maximize the expected surplus net of expected costs and that all decisions are made before the state of nature is known. In that case, expected surplus less costs is the correct measure. When financing for projects is state-contingent or some decisions can be made after the state of nature is revealed, however, the objective function is

---

[18] Chapter 7 (by Sandeep Baliga and Eric Maskin) reviews the mechanism design literature from the standpoint of environmental regulation.

quite different. With state contingent funding – for instance, a consumer pays different amounts for a flood control project depending on the weather – the choice of funding mechanism contributes to the value of the project. Graham (1981) advocated measuring benefits under the optimal choice of funding mechanism, rather than the actual choice [see also Graham (1984), Mendelsohn and Strang (1984), and Smith (1987)].

With the ability to act after the state of nature is known, the optimal time to act is delayed [MacDonald and Segal (1986)], while the value of the action is increased over what it would be if the decision were made before the information were revealed. The value specifically attributable to the ability to act after information becomes known is termed quasi-option value [Arrow and Fisher (1974), Hanemann (1986)]. The Handbook chapter by Anthony Fisher and Michael Hanemann reviews this and related concepts of risk and uncertainty used in environmental economics. The ability to change policy in response to new information can influence both the level of policy and the timing of its implementation. For instance, if the expected costs and benefits, based upon current information, were slightly positive for hard-to-reverse actions to prevent global warming, then waiting for further information on the benefits and costs of control might be preferable to regulating immediately.[19]

Papers by Newell and Pizer (1998), Hoel and Karp (1998), and Karp and Zhang (1999) all contain generalizations of the Weitzman "prices versus quantities" framework to a dynamic setting. They maintain the quadratic objective and linear state equation setting of the previous models. The conclusions are, as in the prices versus quantities literature, that steeper marginal damage or flatter marginal abatement costs favor quotas. New to the dynamic setting is that a higher discount factor or a lower decay rate for the stock pollutant favors the use of quotas. Hoel and Karp, in reviewing these other papers, note that the conclusions are ordained by the linear quadratic framework with an additive uncertainty in the marginal abatement cost. Therefore, they investigate a model in which the slope of the marginal abatement cost is the random element. They then simulate their model with data for global warming and come to the conclusion that, just as in the additive model, taxes dominate quotas.

In sum, if a regulator has incomplete information about the benefits and costs of pollution control, then the resulting uncertainty can influence the choice of optimal regulatory instrument. While a standard provides more certainty over the level of damages, a tax provides more flexibility in abatement and thus lower costs. The relative slopes of the marginal damage and marginal cost curves, along with the correlation between the uncertainty in their positions, influence the choice of instrument. As noted above, this problem generalizes to situations where a uniform tax or standard must be applied in heterogeneous conditions [as in Mendelsohn (1986)] or to dynamic situations. Finally, being able to change policy over time in response to new information increases the options available for policy and can thus increase welfare.

---

[19] See the Handbook chapter on the economics of climate change by Charles Kolstad and Michael Toman.

7.2. Control of nonpoint pollution sources

Pollution control is commonly divided into the control of point and of nonpoint sources of pollution. This distinction is made on the basis of the possibility of monitoring effluent. For point sources, effluent can be observed and linked directly to the polluter responsible for it. Nonpoint source pollution is characterized by an inability either to observe the effluent – for instance, when it enters a river through subsurface flows – or to link the effluent directly to a source – for instance, when subsurface drains collect runoff from farms in addition to the one on which the drains are located. The point/nonpoint distinction fundamentally depends on the cost of obtaining information.

The ability to regulate in the manner assumed so far in the chapter is greatly limited when information is available only on, e.g., average effluent by source. Some of the proposals for nonpoint source pollution focus on regulating observable goods, including inputs and ambient quality. As discussed above, optimal regulation based on inputs requires tremendous information about individual polluters, in particular the effect of using one more unit of each input by each polluter. With N polluters each using some subset of M inputs, optimal regulation might entail $N \cdot M$ taxes or standards [Griffin and Bromley (1982)]. In a case study of two inputs on two soil types, however, Helfand and House (1995) found that a uniform input regulation (either a tax or a standard) was not very inefficient if it was chosen carefully. Segerson (1988) identified a tax/subsidy scheme targeted toward ambient water quality that could achieve efficient abatement. It required the same penalty for all polluters, regardless of the marginal damages each caused, because of the public bad characteristics of pollution (being a relatively clean polluter would provide no advantage unless ambient quality improved sufficiently). The simplicity of the measure, while desirable from an administrative standpoint, could also make the measure difficult to implement politically, given the equal penalization of big and small polluters.

A regulator is obviously at an informational disadvantage relative to the nonpoint polluter, who knows more about its behavior. Incentive-compatible mechanisms can be designed to encourage polluters to reveal polluting behavior. Wu and Babcock (1996) suggest a "green payments" scheme that provides polluters (farmers, in their model) with that incentive. Subsidies are necessary in their model because participation is voluntary, and the payments must increase with the level of restriction imposed. Farmers who declare that they have more productive land must receive lower payments but must be allowed to use more of the polluting inputs than those who claim less productive land, to avoid moral hazard. While the scheme is not as efficient as a pollution tax, the inability to implement a pollution tax in this context necessarily forces consideration of second-best approaches.

7.3. Imperfect monitoring and enforcement

Monitoring is rarely perfect even for pollution from point sources,[20] because it is costly to regulators. As a result, firms do not always comply with environmental requirements. Given that it is reasonable to assume that polluters know more about their behavior than regulators, it is also reasonable to expect that polluters might at times choose to violate an environmental policy in order to reduce costs, under the assumption either that they are unlikely to be caught or that the penalty will not be severe if they are caught.

Becker (1968) provides an early, static enforcement model. The model examines the implications of penalties and probabilities of detection on violations. It assumes that higher penalties and higher probabilities of detection both decrease the likelihood of violations. This model does not capture the realities of environmental enforcement very well, however. Harrington (1988) points out three common observations in the empirical literature. First, regulators consider pollution sources to be in compliance most of the time. Second, they monitor firms infrequently. Finally, when regulators find violations, they rarely impose monetary fines or penalties. A goal of the environmental enforcement literature has been to explain these phenomena.

Harrington (1988) presents a game-theoretic model to explain regulators' behavior. His enforcement model is a repeated game with restricted penalties and binary compliance. He shows that a regulator can maximize steady-state compliance by using a state-dependent enforcement plan. The regulator puts firms into two groups: firms that complied in the previous period (group one), and firms that were not in compliance in the previous period (group two). Group one firms are not penalized if they are found in violation, while firms in group two receive the maximum penalty. This state-dependent enforcement system provides the regulator with "penalty leverage" over firms in group two. The benefit of compliance for a group two firm is two-fold: not being penalized in the current period, and the leniency that results from being moved into group one in the next period. In state-independent enforcement models, this penalty leverage does not exist.

Several studies have expanded Harrington's model and offered different explanations of the enforcement situation. Raymond (1999) adds asymmetric information about compliance costs to Harrington's model. He shows that additional assumptions must be made for Harrington's model to hold when compliance costs differ among firms and are unknown to the regulator. He shows that, if the number of firms with high compliance costs is large, then the zero penalty for group one firms will induce some high-cost violators in both groups to switch to being compliers in the second group. If instead the number of firms with low compliance costs is large, then, unless the group one penalty is set at the maximum, some firms will shift from complying in both groups to cheating in group one and complying in group two. Raymond's model is thus consistent with the empirical literature if many firms have high compliance costs.

---

[20] The ability to monitor sulfur dioxide emissions continuously has contributed to the success of the marketable permit program for this pollutant in the U.S. [Schmalensee et al. (1998)].

Heyes and Rickman (1999) explain the empirical evidence by using a slightly different model than Harrington's. In their model, the regulator enforces multiple rules on the same firm. They show that the regulator can increase compliance by some firms by allowing violations without penalties for some regulations in exchange for compliance with other regulations. They call this "regulatory dealing". They find that citizen suits brought against noncompliant firms have an ambiguous impact on regulatory dealing: the suits reduce the potential "bribe" that the regulator must offer to induce compliance, but they also reduce the number of firms that will increase violation.

As an alternative or a supplement to monitoring by the regulator, many environmental policies require self-reporting, including the U.S. Clean Air and Clean Water Acts. Malik (1993) explores these self-reporting requirements. He uses a principle-agent framework and compares the optimal regulatory policies with and without self-reporting. He finds that, with self-reporting, firms can be monitored less often, but they must be punished more frequently when they are caught violating. He finds that self-reporting ambiguously affects the social costs of enforcement, depending on the cost and accuracy of monitoring and the cost and magnitude of penalties. Self-reporting is more likely to reduce the social costs of enforcement when monitoring accuracy is low or the maximum penalty is low.

Extending both Harrington's and Malik's models, Livernois and McKenna (1999) show that self-reporting requirements can enable a regulator to obtain higher compliance rates with lower penalties. Lowering the penalty reduces the number of firms that comply, but it increases the number of firms that report truthfully. The firms who truthfully report violations can then be ordered to comply. Depending on the number of high- and low-cost compliance firms, setting the penalty to zero can minimize the cost of enforcing a given level of pollution when the cost of switching between compliance and violation is high.

Enforcement issues are further reviewed in Heyes (2000). The general finding, consistent with the results reported above, is that the careful design of monitoring procedures and penalties can reduce the costs associated with those activities while still leading to high levels of compliance.

8. Non-regulatory strategies

The correction of externalities as discussed so far in this chapter has focused on the government intervening in private markets through such regulatory approaches as taxes, permits, and standards. If the government operates with objectives other than maximizing social welfare [Peltzman (1976)], however, then there is no guarantee that government intervention will achieve the social optimum. Moreover, the second-best arguments of Lipsey and Lancaster (1956–1957) and the "double dividend" literature suggest that government interventions might decrease social welfare even if the government's objective is to improve welfare. It is therefore useful to consider other possible avenues for achieving environmental protection. Two possibilities include voluntary environmental compliance programs and the use of courts.

8.1. Voluntary programs

The premise underlying environmental economic analysis is that pollution occurs because polluting is less costly than not polluting. If that premise is true, then it is hard to imagine that polluters will change their behavior in the absence of legal requirements to do so. In recent years, however, evidence has emerged that businesses sometimes exceed legal requirements in their environmental performance. Some empirical studies have investigated the prevalence of such behavior, and a variety of hypotheses have been developed to explain it.

Empirical studies of voluntary overcompliance are relatively recent and gradually growing in number, and they are driving the theoretical discussions. Some papers, such as Arora and Cason (1995), examine the choice of firms to participate in government-sponsored voluntary overcompliance programs, such as the 33/50 Program sponsored by the U.S. Environmental Protection Agency. They find that firms with large levels of toxic emissions are more likely to participate in this program. They consider this result a hopeful sign, since those firms have the greatest potential for reductions in emissions.

Other papers, such as Hamilton (1995), Konar and Cohen (1997), and Khanna, Quimio and Bojilova (1998), look at the effect of publicizing information on firms' emissions. The Toxics Release Inventory (TRI) in the United States provides information on the emissions of a large list of toxic substances by each source over a specified size. These papers all examine the effects of the release of this information on firms' stock market performance. Hamilton found that release of TRI information reduced high-polluting firms' stock market values. Konar and Cohen found that firms that experienced a strong stock market decline after the release of TRI data subsequently reduced their emissions more than other firms. Khanna, Quimio and Bojilova (1998) found that, in response to stock market losses due to TRI information, firms reduced their on-site emissions by transferring the waste to other facilities (off-site transfers). The net effect was little reduction in total waste, although off-site transfers were likely to be for safer recycling or treatment.

Lyon and Maxwell (1999) offer several reasons for firms voluntarily undertaking environmental protection beyond their requirements. First, firms might be able to cut costs by improving their environmental performance. Porter and van der Linde (1995) argue that pollution is a signal that firms are inefficient, because it indicates that polluters are not getting the greatest output from their production practices. They suggest that firms have ample opportunities to improve their environmental performance and to improve their profitability at the same time, and they argue that firms who do not take advantage of these opportunities will be driven out of business by firms that do. Palmer, Oates and Portney (1995) dispute this theory, arguing that businesses are in fact very smart about their resource allocations.

Second, firms might benefit from the favorable public image that being "greener" provides. In some cases businesses might find a marketing advantage: if consumers are willing to pay a premium for goods produced in more environmentally sensitive ways, then there is likely to be a profitable niche for greener businesses.

Third, firms might be playing a strategic game with regulators. In Segerson and Miceli's (1998) view, firms undertake voluntary effluent reductions to avoid the imposition of mandatory controls, which are presumed to be more costly for a given level of abatement. Wu and Babcock (1999) add the possibility that the regulator might provide a positive incentive to cooperate, such as the provision of technical expertise in pollution reduction. When the regulator provides such a service at a lower cost than the firm can, then the firm has further reason to join a voluntary agreement.

Fourth and finally, firms might be playing a strategic game with their competitors, using overcompliance as a strategic tool. For instance, suppose a firm develops a new technology to achieve higher environmental standards. Even if that technology is more expensive than existing ones, it might confer on the early innovator an advantage over other firms if it becomes the basis of a regulatory requirement. This hypothesis is an example of Salop and Scheffman's (1983) argument that an action, even if costly, can benefit a firm if it makes the firm's competitors even worse off.

Voluntary environmental performance is a tantalizing notion. For a variety of reasons, some firms appear to be doing on their own what, in the past, they would have done only under threat of law. Indeed, as noted, one reason for voluntary action might be the fear of more stringent regulation. Firms' actions can perhaps be considered experiments. After many years of actively opposing many environmental requirements with limited success, some businesses appear to be wondering if there are entrepreneurial opportunities to be found in being green. While some anecdotal evidence indicates that firms have increased profits through improvements in environmental performance, other anecdotes suggest that these "win-win" opportunities are limited [Lyon and Maxwell (1999)]. On balance, the evidence at this point does not support the notion that polluters will consistently reduce their effluent without government regulations and programs to encourage this behavior.

8.2. Using courts to enforce rights

As described in Section 3.2, Coase argued that defining rights for environmental goods might be sufficient for achieving pollution abatement without environmental regulation. As long as all parties recognize and honor the initial allocation of rights and can negotiate, and as long as the costs of negotiation are smaller than the gains from negotiation, then negotiations should yield the social optimum. No government role should be necessary other than defining and enforcing rights; in particular, courts would be needed to enforce property rights if disputes arose.

The list of qualifications above suggests the problems with this approach. As described in Section 3, the receivers of pollution are usually consumers of a public bad, which is nonrival in consumption; what one receives, all receive. In this case, negotiations among effluent receivers will not achieve an efficient allocation among receivers: some will want lower pollution levels than others and will be willing to pay, but the fact that pollution is nonrival will prevent that outcome from occurring. Similarly, if a receiver pays a polluter for additional abatement, all receivers will benefit equally. This nonrivalry in consumption explains why free trade in a public bad will not achieve the social optimum.

Additionally, it becomes very difficult for negotiations to occur without significant costs, either among recipients of pollution or between recipients and polluters, because of the large numbers of people involved. When it is costly to conduct the negotiations or to enforce a right, Coase himself points out that the initial allocation of rights can influence the overall efficiency of the system. Farrell (1987) notes that even costless negotiation is unlikely to achieve the social optimum because of gaming by participants. The easiest case to analyze is one in which: (i) the costs of changing the allocation of pollution are greater than the maximum area between the two marginal payoff curves, (ii) property rights can be assigned with only very crude instruments (e.g., the property rights all go either to agent 1 or agent 2), and (iii) reassignment (but not negotiation) is costly. Under these assumptions, the initial allocation influences the welfare loss. If the deadweight loss from assignment to agent 1 is much less than that of assignment to agent 2, then the socially efficient outcome is to permit pollution rather than to ban pollution.

A less crude instrument is the assignment of liability. If a polluting firm were liable for the damages from pollution, it would pollute until the marginal costs of abatement exceed the marginal damages it imposes on others, and it would pay compensation for the damage resulting from that amount of effluent. This solution requires well-defined rights and low-cost enforcement of those rights. For most environmental goods, rights are not well defined and cannot be well defined, due to the problem of nonrival bads. Enforcement, typically via the courts, is rarely costless and is not even certain, due to imperfections in the legal system [Shavell (1984)].

Additional problems can arise due to the combination of nonconvexities and bankruptcy. The Coasean solution relies on the desirability, to both parties, of negotiating to the optimum. In the case of a nonconvexity (see Section 5.6), the optimal solution might be at a corner – zero pollution or the unregulated, maximum amount – instead of an interior solution. Assume that zero pollution is optimal and that the victim of the pollution has an unambiguous right to a clean environment. Suppose also that the polluter can declare bankruptcy if marginal costs become too high. This last assumption makes the firm "judgment proof", because its assets are worth less than the damages for which it is liable. In that case, rather than negotiate, the firm might choose to pollute as much as it likes; when the victim of the pollution seeks compensation for damages, it could declare bankruptcy, leaving a legacy of pollution without recompense. Bankruptcy allows the firm to avoid paying the full consequences of its actions, and so pollution exceeds the social optimum. Increasing the level of liability can, in some situations, decrease the effort of a judgment-proof firm to reduce damages [Pitchford (1995)], for example, when the firm is financed by debt and liability is increased for the lender.

9. Conclusion

Environmental economics theory grows out of the analysis of the failure of private markets to provide efficient amounts of environmental goods. As a result, it pays much attention to ways that governments can improve welfare by intervening in markets. Many of the regulatory approaches proposed in the environmental economics literature seek to correct market failures by creating markets where they did not previously exist (e.g., by using taxes to stand in for missing prices), thus simulating the effects of markets. These market-based approaches are often discussed without differentiation as clearly superior to standards. Their advantages are well-known and strongly advocated. They all create incentives for firms to abate without either mandating a specific abatement method or allocating firm-specific amounts of abatement. By leaving this flexibility, they all permit achievement of an aggregate pollution target at lower aggregate cost than a uniform standard (i.e., they are more cost-effective). Nevertheless, different market-based instruments do differ from each other in ways that influence both partial-equilibrium and general-equilibrium outcomes. Moreover, standards are frequently not as poorly designed as the economics literature suggests.

One of the prerequisites for well-functioning markets is the explicit creation and allocation of property rights. Markets have failed for environmental goods because of the inability, in most cases, to define rights, due to the nonrival or nonexclusive nature of the goods. Pollution regulations, whether market-based or command-and-control, implicitly define rights and thus have significant distributional consequences. They affect polluters' profits and thus entry/exit in a regulated industry. While distribution can therefore affect efficiency, its political effects are probably more important, by influencing the acceptability and feasibility of regulatory programs. Distribution influences the politics of instrument choice and might even influence the target level of pollution. Models that focus only on the short-term allocational effects of regulatory instruments might not explain either longer run efficiency impacts or the political process.

Relying on firms to reduce pollution voluntarily or, following the Coase Theorem, through negotiations to allocate pollution rights are two approaches that do not require explicit government regulatory programs. The former requires the existence of ways for businesses to improve both environmental performance and profits, while the latter requires explicit definitions of rights and a legal system to enforce those rights. While both are applicable in some circumstances, neither can likely serve as an adequate substitute for government programs, because the assumptions required to make them work do not hold very broadly.

For the most part, this chapter has reviewed the effects of different pollution policies in a context where markets are otherwise functioning well. The requirements for inferring conclusions about global efficiency from this scenario of limited market failure are, in fact, very steep: as Lipsey and Lancaster point out, imperfections in one market can cause correction of a market failure in another sector to make society worse off. Indeed, the double dividend literature has emphasized that, through their impacts on flows of government funds and relative prices, pollution control instruments influence and are influenced by the existence of other taxes.

Because of the number, variety, and interconnectedness of environmental problems, second-best problems and other policy spillovers must be considered a likely possibility. For instance, regulation of one pollution medium, such as water, might lead to increased pollution in another medium, such as land or air; or, regulating one toxic substance might lead to substitution toward an unregulated but possibly worse toxic substance. These concerns suggest that regulating at a very aggregate level, such as pollution damages, might be preferable to regulating at a lower level, such as effluent or polluting inputs. On the other hand, the feasibility and enforceability of regulating at these lower levels is often much higher. Design of environmental policy must consider tradeoffs among the various monitoring and enforcement difficulties involved in implementation.

Acknowledgements

Senior authorship is shared by Berck and Helfand. We are grateful for the very helpful suggestions of Wallace E. Oates and Robert N. Stavins, and the comments, encouragement, and editorial advice and assistance of Jeffrey Vincent.

References

Ackerman, B., and W.T. Hassler (1981), Clean Coal/Dirty Air (Yale University Press, New Haven, CT).

Adar, Z., and J.M. Griffin (1976), "Uncertainty and the choice of pollution control instruments", Journal of Environmental Economics and Management 3:178–188.

American Trucking Associations, Inc., et al. v. United States Environmental Protection Agency, United States Court of Appeals for the District of Columbia Circuit, No. 97-1440, Decided May 14, 1999. http://search.cadc.uscourts.gov/P:/opinions/199905/97-1440a.txt.

Arora, S., and T.N. Cason (1995), "An experiment in voluntary environmental regulation: Participation in EPA's 33/50 Program", Journal of Environmental Economics and Management 28:271–286.

Arrow, K.J., M.L. Cropper, G.C. Eads, R.W. Hahn, L.B. Lave, R.G. Noll, P.R. Portney, M. Russell, R. Schmalensee, V.K. Smith and R.N. Stavins (1996), "Is there a role for benefit–cost analysis in environmental, health, and occupational regulation?", Science 272:221–222.

Arrow, K.J., and A.C. Fisher (1974), "Environmental preservation, uncertainty, and irreversibility", Quarterly Journal of Economics 88:312–319.

Ayres, R.U. (1999), "Materials, economics and the environment", in: J. van den Bergh, ed., Handbook of Environmental and Resource Economics (Edward Elgar, Cheltenham) 867–894.

Ayres, R.U., and A.V. Kneese (1969), "Production, consumption, and externalities", American Economic Review 59:282–297.

Batie, S.S. (1989), "Sustainable development: Challenges to the profession of agricultural economics", American Journal of Agricultural Economics 71:1083–1101.

Baumol, W.J., and D.F. Bradford (May, 1972), "Detrimental externalities and non-convexity of the production set", Economica 39:160–176.

Baumol, W.J., and W.E. Oates (1988), The Theory of Environmental Policy, 2nd edn. (Cambridge University Press, Cambridge).

Becker, G.S. (1968), "Crime and punishment: An economic approach", Journal of Political Economy 76:169–217.

Bohm, P. (1981), Deposit Refund Systems (The Johns Hopkins University Press for Resources for the Future, Baltimore, MD).

Bovenberg, A.L., and L.H. Goulder (1996), "Optimal environmental taxation in the presence of other taxes: General-equilibrium analyses", American Economic Review 86:985–1000.

Bovenberg, L., and R.A. de Mooij (1994), "Environmental levies and distortionary taxation", American Economic Review 84:1085–1089.

Bryant, B., and P. Mohai (1992), Race and the Incidence of Environmental Hazards: a Time for Discourse (Westview, Boulder, CO).

Buchanan, J.M., and G. Tullock (1975), "Polluters' profits and political response: Direct controls versus taxes", American Economic Review 65:139–147.

Coase, R. (1960), "The problem of social cost", Journal of Law and Economics 3:1–44.

Conrad, J. (1992), "Stopping rules and the control of stock pollutants", Natural Resource Modeling 6:315–328.

Courant, P.N., and R.C. Porter (1981), "Averting expenditure and the cost of pollution", Journal of Environmental Economics and Management 8:321–329.

Crocker, T. (1966), "The structuring of atmospheric pollution control systems", in: H. Wolozin, ed., The Economics of Air Pollution (Norton, New York) 61–86.

Dales, J.H. (1968), Pollution, Property, and Prices (University of Toronto Press, Toronto).

Daly, H.E. (1968), "On economics as a life science", Journal of Political Economy 76:392–406.

Diamond, P.A. (1974), "Accident law and resource allocation", Bell Journal of Economics 5:366–405.

Diamond, P.A., and J.A. Hausman (1994), "Contingent valuation: Is some number better than no number?", Journal of Economic Perspectives 8:45–64.

Diamond, P.A., and J.A. Mirrlees (1971), "Optimal taxation and public production II: Tax rules", American Economic Review 61:261–278.

Dumas, C. (1997), Cross-Media Pollution and Common Agency, Ph.D. Dissertation (Department of Agricultural and Resource Economics, University of California at Berkeley).

Ellis, G.M. (1992), "Incentive compatible environmental regulations", Natural Resource Modeling 6:225–256.

Falk, I., and R. Mendelsohn (1993), "The economics of controlling stock pollutants: an efficient strategy for greenhouse gases", Journal of Environmental Economics and Management 25:76–88.

Farrell, J. (1987), "Information and the Coase theorem", Journal of Economic Perspectives 1:113–129.

Freeman, A.M. (1993), The Measurement of Environmental and Resource Values: Theory and Methods (Resources for the Future, Washington, DC).

Fullerton, D. (1997), "Environmental levies and distortionary taxation: Comment", American Economic Review 87:245–251.

Fullerton, D., and T.C. Kinnaman (1995), "Garbage, recycling, and illicit burning or dumping", Journal of Environmental Economics and Management 29:78–91.

Fullerton, D., and T.C. Kinnaman (1996), "Household response to pricing garbage by the bag", American Economic Review 86:971–984.

Fullerton, D., and S. West (1999), "Can taxes on cars and on gasoline mimic an unavailable tax on emissions?", Working Paper No. 7059 (National Bureau of Economic Research).

Fullerton, D., and A. Wolverton (1999), "The case for a two-part instrument: presumptive tax and environmental subsidy", in: A. Panagariya, P.R. Portney and R.M. Schwab, eds., Environmental and Public Economics: Essays in Honor of Wallace E. Oates (Edward Elgar, Cheltenham) 32–57.

Graham, D.A. (1981), "Cost-benefit analysis under uncertainty", American Economic Review 71:715–725.

Graham, D.A. (1984), "Cost-benefit analysis under uncertainty: Reply", American Economic Review 74:1100–1102.

Griffin, R.C., and D.W. Bromley (1982), "Agricultural runoff as a nonpoint externality: A theoretical development", American Journal of Agricultural Economics 64:547–552.

Gruenspecht, H.K. (1982), "Differentiated regulation. The case of auto emissions standards", American Economic Review 72:328–331.

Hahn, R.W., and G.L. Hester (1989), "Where did all the markets go? An analysis of EPA's emissions trading program", Yale Journal on Regulation 6:109–153.

Hamilton, J.T. (1995), "Pollution as news: Media and stock market reactions to the toxics release inventory data", Journal of Environmental Economics and Management 28:98–113.

Hanemann, W.M. (1986), "On reconciling different concepts of option value", Working Paper No. 295 (Department of Agricultural and Resource Economics, University of California at Berkeley).

Harrington, W. (1988), "Enforcement leverage when penalties are restricted", Journal of Public Economics 37:29–53.

Heyes, A. (2000), "Implementing environmental regulation: enforcement and compliance", Journal of Regulatory Economics 17:107–129.

Heyes, A., and N. Rickman (1999), "Regulatory dealing – revisiting the Harrington paradox", Journal of Public Economics 72:361–378.

Helfand, G.E. (1991), "Standards versus standards: the effects of different pollution restrictions", American Economic Review 81:622–634.

Helfand, G.E. (1999), "Controlling inputs to control pollution: when will it work?", AERE Newsletter (Newsletter of the Association of Environmental and Resource Economists) 19:13–17.

Helfand, G.E., and B.W. House (1995), "Regulating nonpoint source pollution under heterogeneous conditions", American Journal of Agricultural Economics 77:1024–1032.

Helfand, G.E., and L.J. Peyton (1999), "A conceptual model of environmental justice", Social Science Quarterly 80:68–83.

Helfand, G.E., and J. Rubin (1994), "Spreading versus concentrating damages: Environmental policy in the presence of nonconvexities", Journal of Environmental Economics and Management 27:84–91.

Hoel, M., and L. Karp (1998), "Taxes versus quotas for a stock pollutant", Working Paper No. 855 (Department of Agricultural and Resource Economics, University of California at Berkeley).

Hoel, M., and L. Karp (2000), "Taxes and quotas for a stock pollutant with multiplicative uncertainy", Working Paper No. 870 (Department of Agricultural and Resource Economics, University of California at Berkeley), forthcoming in Journal of Public Economics.

Holtermann, S. (1976), "Alternative tax systems to correct for externalities, and the efficiency of paying compensation", Economica 43:1–16.

Howe, C.W. (1994), "Taxes versus tradable discharge permits: A review in the light of the US and European experience", Environmental and Resource Economics 4:151–169.

Karp, L., and J. Zhang (1999), "Regulation of stock externalities with learning", Working Paper No. 892 (Department of Agricultural and Resource Economics, University of California at Berkeley).

Khanna, M., W.R.H. Quimio and D. Bojilova (1998), "Toxics release information: a policy tool for environmental protection", Journal of Environmental Economics and Management 36:243–266.

Kim, H.J., G.E. Helfand and R.E. Howitt (1998), "An economic analysis of ozone control in California's San Joaquin Valley", Journal of Agricultural and Resource Economics 23:55–70.

Kling, C. (1994), "Emission trading vs. rigid regulations in the control of vehicle emissions", Land Economics 70:174–188.

Kling, C. (1994), "Environmental benefits from marketable discharge permits, or an ecological vs. an economical perspective on marketable permits", Ecological Economics 11:57–64.

Kling, C., and J. Rubin (1997), "Bankable permits for the control of environmental pollution", Journal of Public Economics 64:101–115.

Kneese, A.V. (1971), "Environmental pollution: Economics and policy", American Economic Review 61:153–166.

Kneese, A., and K.G. Mäler (1973), "Bribes and charges in pollution control: An aspect of the Coase controversy", Natural Resources Journal 3:705–716.

Konar, S., and M.A. Cohen (1997), "Information as regulation: the effect of community right to know laws on toxic emissions", Journal of Environmental Economics and Management 32:109–124.

Krupnick, A.J., and P.R. Portney (1991), "Controlling urban air pollution: a benefit–cost assessment", Science 252:522–528.

Larson, D.M., G.E. Helfand and B.W. House (1996), "Second-best tax policies for controlling nonpoint source pollution", American Journal of Agricultural Economics 78:1108–1117.

Leiby, P., and J. Rubin (2001), "Bankable permits for the control of stock and flow pollutants: Optimal intertemporal greenhouse gas emission trading", Environmental and Resource Economics 19:229–256.

Lipsey, R.G., and K. Lancaster (1956–1957), "The general theory of the second best", Review of Economic Studies 24:11–32.

Livernois, J., and C.J. McKenna (1999), "Truth of consequences: Enforcing pollution standards with self-reporting", Journal of Public Economics 71:415–440.

Lyon, T.P., and J.W. Maxwell (1999), "'Voluntary' approaches to environmental regulation: A survey", Forthcoming in: M. Franzini and Nicita, eds., Environmental Economics: Past, Present, and Future (Ashgate, Aldershot).

Mäler, K.G. (1974), Environmental Economics: A Theoretical Inquiry (Johns Hopkins University Press Resources for the Future, Baltimore, MD).

Malik, A.S. (1993), "Self-reporting and the design of policies for regulating stochastic pollution", Journal of Environmental Economics and Management 24:241–257.

Maloney, M.T., and R.E. McCormick (1982), "A positive theory of environmental quality regulation", Journal of Law and Economics 25:99–123.

McKibbin, W.J., and P.J. Wilcoxen (1997), "A better way to slow global climate change", Brookings Policy Briefs No. 17.

Mendelsohn, R. (1986), "Regulating heterogeneous emissions", Journal of Environmental Economics and Management 13:301–312.

Mendelsohn, R., and W.J. Strang (1984), "Cost-benefit analysis under uncertainty: Comment", American Economic Review 74:1096–1099.

Millock, K., and F. Salanie (2000), "Are collective environmental agreements ever effective?", Paper presented to the Association of Environmental and Resource Economists Summer Meeting, La Jolla, CA.

Montgomery, W.D. (1972), "Markets in licenses and efficient pollution control programs", Journal of Economic Theory 5:395–418.

National Oceanic and Atmospheric Administration (1993), "Report of the NOAA panel on contingent valuation", Federal Register 58:4602–4614.

Newell, R., and W. Pizer (1998), "Stock externality regulation under uncertainty", Working Paper 99-10-REV (Resources for the Future, Washington, DC).

Oates, W.E., P.R. Portney and A.M. McGartland (1989), "The net benefits of incentive-based regulation: A case study of environmental standard setting", American Economic Review 79:1233–1242.

O'Neil, W. et al. (1983), "Transferable discharge permits and economic efficiency: The Fox River", Journal of Environmental Economics and Management 10:346–355.

Palmer, K., W.E. Oates and P.R. Portney (1995), "Tightening environmental standards: The benefit–cost or the no-cost paradigm?", Journal of Economic Perspectives 9:119–132.

Parry, I.W.H., R.C. Williams III and L.H. Goulder (1999), "When can carbon abatement policies increase welfare? The fundamental role of distorted factor markets", Journal of Environmental Economics and Management 37:52–84.

Peltzman, S. (1976), "Toward a more general theory of regulation", Journal of Law and Economics 19:211–240.

Pitchford, R. (1995), "How liable should a lender be? The case of judgment-proof firms and environmental risk", American Economic Review 85:1171–1186.

Pizer, W. (1997), "Prices vs. quantities revisited: The case of climate change", Discussion Paper 98-02 (Resources for the Future, Washington, DC).

Porter, M.E., and C. van der Linde (1995), "Toward a new conception of the environment–competitiveness relationship", Journal of Economic Perspectives: 97–118.

Raymond, M. (1999), "Enforcement leverage when penalties are restricted: a reconsideration under asymmetric information", Journal of Public Economics 73:289–295.

Repetto, R. (1987), "The policy implications of non-convex environmental damages: A smog control case study", Journal of Environmental Economics and Management 14:13–29.

Roberts, M.J., and M. Spence (1976), "Effluent charges and licenses under uncertainty", Journal of Public Economics 5:193–208.

Rubin, J.D. (1996), "A model of intertemporal emission trading, banking, and borrowing", Journal of Environmental Economics and Management 31:269–286.

Ruth, M. (1999), "Physical principles and environmental economic analysis", in: J. van den Bergh, ed., Handbook of Environmental and Resource Economics (Edward Elgar, Cheltenham) 855–866.

Salop, S.C., and D.T. Scheffman (1983), "Raising rivals' costs", American Economic Review 73:267–271.

Schmalensee, R., P.L. Joskow, A.D. Ellerman, J.P. Montero and E.M. Bailey (1998), "An interim evaluation of sulfur dioxide emissions trading", Journal of Economic Perspectives 12:53–68.

Segerson, K. (1988), "Uncertainty and incentives for nonpoint pollution control", Journal of Environmental Economics and Management 15:87–98.

Segerson, K., and T.J. Miceli (1998), "Voluntary environmental agreements: Good or bad news for environmental protection?", Journal of Environmental Economics and Management 36:109–130.

Shavell, S. (1984), "A model of the optimal use of liability and safety regulation", Rand Journal of Economics 15:271–280.

Shibata, H., and J.S. Winrich (1983), "Control of pollution when the offended defend themselves", Economica 50:425–437.

Shortle, J.S., and J.W. Dunn (1986), "The relative efficiency of agricultural source water pollution control policies", American Journal of Agricultural Economic 68:668–677.

Smith, V.K. (1987), "Uncertainty, benefit–cost analysis, and the treatment of option value", Journal of Environmental Economics and Management 14:283–292.

Spulber, D.F. (1985), "Effluent regulation and long-run optimality", Journal of Environmental Economics and Management 12:103–116.

Stavins, R.N. (1996), "Correlated uncertainty and policy instrument choice", Journal of Environmental Economics and Management 30:218–232.

"U.S. sues 7 utilities over air pollution; Reno, EPA blame companies for smog, acid rain in east", The Baltimore Sun, November 4, 1999, Section Telegraph, 1A.

Weitzman, M.L. (1974), "Prices vs. quantities", Review of Economic Studies 41:477–491.

Wu, J., and B.A. Babcock (1996), "Contract design for the purchase of environmental goods from agriculture", American Journal of Agricultural Economics 78:935–945.

Wu, J., and B. Babcock (1999), "The relative efficiency of voluntary vs. mandatory environmental controls", Journal of Environmental Economics and Management 38:58–76.

Chapter 7

MECHANISM DESIGN FOR THE ENVIRONMENT

SANDEEP BALIGA

Kellogg Graduate School of Management (M.E.D.S.), Northwestern University, Evanston, IL, USA

ERIC MASKIN

Institute for Advanced Study and Princeton University, Princeton, NJ, USA

Contents

Handbook of Environmental Economics, Volume 1, Edited by K.-G. Mäler and J.R. Vincent

Abstract

We argue that when externalities such as pollution are nonexcludable, agents must be compelled to participate in a "mechanism" to ensure a Pareto-efficient outcome. We survey some of the main findings of the mechanism-design (implementation-theory) literature – such as the Nash implementation theorem, the Gibbard–Satterthwaite theorem, the Vickrey–Clarke–Groves mechanism, and the Arrow/d'Aspremont–Gerard-Varet mechanism – and consider their implications for the environment, in particular the reduction of aggregate emissions of pollution. We consider the cases of both complete and incomplete information.

Keywords

implementation, mechanism design, public goods, asymmetric information

JEL classification: D6, D7, H4, Q2, Q3

1. Introduction

Economists are accustomed to letting the "market" solve resource-allocation problems. The primary theoretical justification for this laissez-faire position is the "first fundamental theorem of welfare economics" [see Debreu (1957)], which establishes that, provided all goods are priced, a competitive equilibrium is Pareto efficient. Implicit in the "all-goods-priced" hypothesis, however, is the assumption that there are no significant externalities; an externality, after all, can be thought of as an unpriced commodity.

Once externalities are admitted, the first welfare theorem no longer applies. Thus, a school of thought dating back to Pigou (1932), if not earlier, calls for government-imposed "mechanisms" (e.g., taxes on pollution) as a way of redressing the market failure.[1]

In opposition to the Pigouvian school, however, proponents of the Coase Theorem [Coase (1960)] argue that, even in the presence of externalities, economic agents should still be able to ensure a Pareto-efficient outcome without government intervention provided that there are no constraints on their ability to bargain and contract. The argument is straightforward: if a prospective allocation is inefficient, agents will have the incentive to bargain their way to a Pareto improvement. Thus, even if markets themselves fail, Coasians hold that there is still a case for laissez-faire.

The Coasian position depends, however, on the requirement that any externality present be *excludable* in the sense that the agent giving rise to it has control over who is and who is not affected by it. A pure public good, which, once created, will be enjoyed by everybody, constitutes the classic example of a nonexcludable externality.[2]

To see what goes wrong with nonexcludable externalities, consider pollution. For many sorts of pollution, particularly that of the atmosphere or sea, it is fairly accurate to say that a polluter cannot choose to pollute one group of agents rather then another, that is, pollution can be thought of as a pure public bad and hence pollution reduction as a public good.

Now imagine that there is a set of communities that all emit pollution and are adversely affected by these emissions. Suppose, however, that reducing pollution emission is costly to a community (say, because it entails curtailing or modifying the community's normal activities). It is clear that if communities act entirely on their own, there will be too little pollution reduction, since a community shares the benefit of its reduction with the other communities but must bear the full cost alone. A Coasian might hope, however, that if communities came together to negotiate a pollution-reduction agreement – in which each community agrees to undertake some reduction in exchange for other communities' promises to do the same – a Pareto-efficient reduction might be attainable. The problem is, however, that any given community (let us call it "C") will calculate

---

[1] For more on Pigouvian taxes and other regulatory responses to pollution externalities, see Chapter 6 (by Gloria Helfand, Peter Berck, and Tim Maull).

[2] For more on the theory of externalities and public goods, see Chapter 3 (by David Starrett).

that if all the other communities negotiate an agreement, it is better off not participating. By staying out, C can enjoy the full benefits of the negotiated reduction (this is where the nonexcludibility assumption is crucial) without incurring any of the cost. Presumably, the agreed reduction will be somewhat smaller than had C participated (since the benefits are being shared among only $N - 1$ rather then N participants). However, this difference is likely to be small relative to the considerable saving to C from not bearing any reduction costs (we formalize this argument in Section 2 below).[3]

Hence, it will pay community C to *free-ride* on the others' agreement. But since this is true for *every* community, there will end up being no pollution-reduction agreement at all, i.e., the only reduction undertaken will be on an individual basis. We conclude that, in the case of nonexcludable public goods, even a diehard Coasian should agree that outside intervention is needed to achieve optimality. The government – or some other coercive authority – must be called on to *impose* a method for determining pollution reduction. We call such a method a *mechanism* (or game form). Devising a suitable mechanism may, however, be complicated by the fact that the authority might not know critical parameters of the problem (e.g., the potential benefits that different communities enjoy from pollution reduction).

Because environmental issues often entail nonexcludable externalities, the theory of mechanism design (sometimes called "implementation theory") is particularly pertinent to the economics of the environment. In this short survey, we review some of the major concepts, ideas, and findings of the mechanism-design literature and their relevance for the environment.

We necessarily focus on only a few topics from a vast field. Those interested in going further into the literature are referred to the following other surveys and textbooks: Corchon (1996), Fudenberg and Tirole (1991, Chapter 7), Groves and Ledyard (1987), Jackson (2001a, 2001b), Laffont and Martimort (2002), Maskin (1985), Maskin and Sjöström (2001), Moore (1992), Myerson (1991, Chapters 6 and 10), Palfrey (1992, 2001).

2. The model

There are N players or *agents*, indexed by $j \in \{1, 2, \ldots, N\}$, and a set of *social choices* (or *social decisions*) Y with generic element y. Agents have preferences over the social

---

[3] Implicit in this argument is the assumption that the other communities cannot, in effect, *coerce* community C's participation by threatening, say, to refrain from negotiating any agreement at all if C fails to participate. What we have in mind is the idea that any such threat would not be credible, i.e., it would not actually be carried out if push came to shove. Also implicit is the presumption that community C will not be offered especially favorable terms in order to persuade it to join. But notice that if communities anticipated getting especially attractive offers by staying out of agreements, then they would *all* have the incentive to drag their heels about negotiating such agreements and so the same conclusion about the inadequacy of relying on negotiated settlements would obtain. For further discussion of these points see Maskin (1994) and Baliga and Maskin (2002).

choices, and these depend on their preference parameters or types. Agent i of type $\theta_j \in \Theta_j$ has a utility function $U_j(y, \theta_j)$ (the interpretation of agent j as a firm is one possibility, in which case U_j is firm j's profit function). Let $\theta \equiv (\theta_1, \ldots, \theta_N) \in \Theta \equiv \prod_{i=1}^{N} \Theta_i$ be the *preference profile* or state. A choice y is *(ex-post) Pareto-efficient* for preference profile θ if there exists no other decision y' such that, for all $i = 1, \ldots, N$,

$$U_i(y', \theta_i) \geqslant U_i(y, \theta_i)$$

with strict inequality for some i. A *social choice function* (or *decision rule*) f is a rule that prescribes an appropriate social choice for each state, i.e., a mapping $f : \Theta \to Y$. We say that f is *efficient* if $f(\theta)$ is Pareto efficient in each state θ.

We illustrate this set-up with an example based on the discussion of pollution in the Introduction. Suppose that N communities (labelled $i = 1, \ldots, N$) would like to reduce their aggregate emission of pollution. Suppose that the gross benefit to community j of a pollution reduction r is $\theta_j \sqrt{r}$ where $\theta_j \in [a, b]$, and that the cost per unit of reduction is 1. If r_j is the reduction of pollution by community j, $r = \sum_{i=1}^{N} r_i$, and t_j is a monetary transfer to community j, then an social choice y takes the form $y = (r_1, \ldots, r_N, t_1, \ldots, t_N)$, and

$$U_j(y, \theta_j) = \theta_j \sqrt{r} - r_j + t_j.$$

We will assume that there is no net source of funds for the N agents, and so for *feasibility* it must be the case that

$$\sum_{i=1}^{N} t_i \leqslant 0.$$

The stronger requirement of *balance* entails that

$$\sum_{i=1}^{N} t_i = 0.$$

To see why Coasian bargaining will *not* lead to Pareto-efficient pollution reduction, observe first that because preferences are quasi-linear, any efficient social choice function that does not entail infinite transfers (either positive or negative) to some communities must implicitly place equal weight on all communities. Hence, the Pareto-efficient reduction $r^*(\theta_1, \ldots, \theta_N)$ will maximize

$$\left(\sum_{i=1}^{N} \theta_i \right) \sqrt{r} - r,$$

and so

$$r^*(\theta_i, \ldots, \theta_n) = \frac{(\sum \theta_i)^2}{4}. \tag{1}$$

However, if there is no reduction agreement, community j will choose $r_j = r_j^{**}(\theta_j)$ to maximize $\theta_j \sqrt{r_j + \sum_{i \neq j} r_i(\theta_i)} - r_j$. Thus, if none of the θ_i's are equal, we have

$$r_j^{**}(\theta_j) = \begin{cases} \dfrac{\theta_j^2}{4}, & \text{if } \theta_j \text{ is maximal in } \{\theta_1, \ldots, \theta_n\}, \\ 0, & \text{otherwise,} \end{cases}$$

and so the total reduction is

$$r^{**}(\theta_1, \ldots, \theta_n) = \sum_{i=1}^{N} r_i^{**}(\theta_i) = \max_j \frac{\theta_j^2}{4}. \tag{2}$$

Note the sharp contrast between (1) and (2). In particular, if all the θ_i's are in a small neighborhood of z, then (1) reduces approximately to $n^2 z^2/4$, whereas (2) becomes $z^2/4$. In other words, the optimum reduction differs from the reduction that will actually occur by a factor n^2.

Now, suppose that the communities attempt to negotiate the Pareto-efficient reduction (1) by, say, agreeing to share the costs in proportion to their benefits. That is, community j will pay a cost equal to $\theta_j \sum_{i=1}^{N} \theta_i/4$, so that its net payoff is

$$\theta_j \sqrt{\frac{(\sum \theta_i)^2}{4}} - \frac{\theta_j(\sum_{i=1}^{N} \theta_i)}{4} = \frac{\theta_j(\sum_{i=1}^{N} \theta_i)}{4}. \tag{3}$$

If instead, however, community j stands back and lets the others undertake the negotiation and costs, it will enjoy a pollution reduction of

$$r^*(\theta_{-j}) = \frac{(\sum_{i \neq j} \theta_i)^2}{4}$$

and, hence, realize a net payoff of

$$\frac{\theta_j(\sum_{i \neq j} \theta_i)}{2}. \tag{4}$$

But provided that

$$\sum_{i \neq j} \theta_i > \theta_j, \tag{5}$$

(4) exceeds (3), and so community j does better to free-ride on the others' agreement. Furthermore, as we have assumed that all the θ_i's are distinct, notice that (5) must hold for *some* j, and so a Pareto-efficient agreement is not possible. Indeed, the same argument shows that any agreement involving two or more communities is vulnerable to free-riding. Thus, despite the possibility of negotiation, pollution reduction turns out to be no greater than in the case where negotiation is ruled out.

We conclude that some sort of government intervention is called for. Probably the simplest intervention is for the government to impose a vector of quotas $(q_1, \ldots, q_N)$, where for each j, community j is required to reduce pollution by at least the amount q_j. If $q_j = \theta_j (\sum_{i=1}^{N} \theta_i)/4$, then the resulting outcome will be Pareto efficient.

Another familiar kind of intervention is for the government to set a vector of subsidies $(s_1, \ldots, s_N)$, where, for each j, community j is paid s_j for each unit by which it reduces pollution (actually this is not quite complete: to finance the subsidies – and thereby ensure feasibility – each community must also be taxed some fixed amount). If $s_j = 1 - \theta_j / \sum_{i=1}^{N} \theta_i$, then the outcome induced by the subsidies will be Pareto efficient.

Notice that both these solutions rely on the assumption that the state is verifiable to the government.[4] But the more interesting – and typically harder – case is the one in which the preference profile is not verifiable. In that case, there are two particular information environments that have been most intensely studied: first, the preference profile could, although unobservable to the government, be observable to all the agents (*complete information*); or, second, each agent j could observe only his *own* preference parameter θ_j (*incomplete information*). In either case, the government typically "elicits" the true state by having the agents play a game or mechanism.

Formally, a *mechanism* is a pair (M, g) where M_i is agent i's *message space*, $M = \prod_{i=1}^{N} M_i$ is the product of the individual message spaces with generic element m, $g: M \rightarrow Y$ is an *outcome function*, and $g(m) \in Y$ is the social choice.

Returning to our pollution example, we note that if each community j observes only its own type θ_j, the government might have the community "announce" its type so that $M_j = \Theta_j$. As a function of the profile of their announcements $\hat{\theta}$,[5] the government chooses the reduction levels and transfers:

$$g(\hat{\theta}) = \left(r_1(\hat{\theta}), \ldots, r_N(\hat{\theta}), t_1(\hat{\theta}), \ldots, t_N(\hat{\theta})\right).$$

To predict the outcome of the mechanism, we must invoke an equilibrium concept. Because which equilibrium concept is appropriate depends on the information environment, we study the complete and incomplete information settings separately.

---

[4] They also depend on the assumption that each community's reduction is verifiable. If only a noisy signal of a reduction is verifiable, then there is said to be moral hazard. However, we will assume throughout that the social choice is indeed verifiable so that the issue of moral hazard does not arise.

[5] We write the profile of *announced* parameters as $\hat{\theta}$, to distinguish it from the *actual* parameters θ.

3. Complete information

We begin with complete information. This is the case in which all agents observe the preference profile (the state) θ but it is unverifiable to the mechanism-imposing authority. It is most likely to be a good approximation when the agents all know one another well, but the authority is a comparative outsider.

Let S be a equilibrium concept such as Nash equilibrium, subgame perfect equilibrium, etc. Let $O_S(M, g, \theta)$ be the set of equilibrium outcomes of mechanism (M, g) in state θ.

A social choice function f is *implemented by the mechanism* (M, g) in the solution concept S if $O_S(M, g, \theta) = f(\theta)$ for all $\theta \in \Theta$. In that case, we say f is *implementable* in S. Notice that, in every state, we require that *all* the equilibrium outcomes be optimal (we will say more about this below).

3.1. Nash implementation

Suppose first that S is Nash equilibrium. A message profile m is a *Nash equilibrium* in state θ if

$$U_i\big(y(m), \theta_i\big) \geqslant U_i\big(y(m_i', m_{-i}), \theta_i\big)$$

for all $i = 1, \ldots, N$, and all $m_i' \in M_i$ where m_{-i} is the profile of messages $(m_1, \ldots, m_{i-1}, m_{i+1}, \ldots, m_N)$ that excludes m_i.

We note that it is easy to ensure that at least one equilibrium outcome coincides with what the social choice function prescribes if there are three or more agents ($N \geqslant 3$): let all agents announce a state simultaneously. If $N - 1$ or more agree and announce the same state $\hat{\theta}$, then let $g(\hat{\theta}) = f(\hat{\theta})$; define the outcome arbitrarily if fewer than $N - 1$ agents agree. Notice that, if θ is the true state, it is an equilibrium for every agent to announce $\hat{\theta} = \theta$, leading to the outcome $f(\theta)$, since a unilateral deviation by any single agent will not change the outcome. However, it is equally well an equilibrium for agents to unanimously announce any other state (and there are many nonunanimous equilibria as well). Hence, uniqueness of the equilibrium outcome is a valuable property of an implementing mechanism.

To ensure that it is possible to construct such a mechanism, we require the social choice function to satisfy monotonicity. A social choice function f is *monotonic* if for any $\theta, \phi \in \Theta$ and $y = f(\theta)$ such that $y \neq f(\phi)$, there exists an agent i and outcome y' such that $U_i(y, \theta_i) \geqslant U_i(y', \theta_i)$ but $U_i(y', \phi_i) > U_i(y, \phi_i)$. That is, a social choice function is monotonic if whenever there is an outcome y that is optimal in one state θ but not in another ϕ, there exists an agent i and an outcome y' such that agent i strictly prefers y' to y in state ϕ but weakly prefers y to y' in state θ. This is a form of "preference reversal".

The other condition on social choice functions we will impose to guarantee implementability is no veto power. A social choice function f satisfies *no veto power* if whenever agent i, state θ and outcome y are such that $U_j(y, \theta_j) \geqslant U_j(y', \theta_j)$ for all agents $j \neq i$ and all $y' \in y$, then $y = f(\theta)$. That is, if in state θ, $N - 1$ or more agents agree that the best possible outcome is y, then y is prescribed by f in state θ. Notice that in our pollution example, there is *no* alternative that any agent thinks is best: an agent would always prefer a bigger monetary transfer. Hence, no veto power is automatically satisfied.

THEOREM 1 [Maskin (1999)]. *If a social choice function is implementable in Nash equilibrium, then it is monotonic. If $N \geqslant 3$, a social choice function that satisfies monotonicity and no veto power is Nash implementable.*

PROOF. *Necessity*: Suppose f is Nash implementable using the mechanism (M, g). Suppose m is a Nash equilibrium of (M, g) in state θ, where $f(\theta) = y$. Then, $g(m) = y$. But, if $f(\theta) \neq f(\phi)$, m cannot be a Nash equilibrium in state ϕ. Therefore, there must exist an agent i with a message m'_i and an outcome $y' = g(m'_i, m_{-i})$ such that

$$U_i(y', \phi_i) = U_i\big(g(m'_i, m_{-i}), \phi_i\big) > U_i\big(g(m), \phi_i\big) = U_i(y, \phi_i).$$

But because m is a Nash equilibrium in state θ, agent i must be willing to send the message m_i rather than m'_i in state θ. Hence,

$$U_i(y, \theta_i) \geqslant U_i(y', \theta_i),$$

implying that f is monotonic.
    Sufficiency: See Maskin (1999). □

It is not hard to verify that in out pollution example, the efficient social choice function $f(\theta) = (r_1(\theta), \ldots, r_N(\theta), t_1(\theta), \ldots, t_N(\theta))$, where, for all j,

$$r_j(\theta) = \frac{\theta_j \sum_{i=1}^{N} \theta_i}{4} \tag{6}$$

and

$$t_j(\theta) = 0, \tag{7}$$

is monotonic and hence Nash implementable. To see this, choose θ and θ', and let $y = (r_1, \ldots, r_N, t_1, \ldots, t_N) = f(\theta)$. Then, from (6) and (7), $r_j = \theta_j \sum_{i=1}^{N} \theta_i / 4$ and $t_j = 0$ for all j. For concreteness, suppose that, for some j, $\theta_j < \theta'_j$. Note that

$$U_j(y, \theta_j) = \frac{\theta_j \sum_{i=1}^{N} \theta_i}{2} - \frac{\theta_j \sum_{i=1}^{N} \theta_i}{4}. \tag{8}$$

Choose $y' = (r'_1, \ldots, r'_N, t'_1, \ldots, t'_N)$ such that,

$$\sum_{i=1}^{N} r'_i = \left(\sum_{i=1}^{N} \theta_i \right)^2, \tag{9}$$

$$r'_j = r_j = \frac{\theta_j \sum_{i=1}^{N} \theta_i}{4}, \tag{10}$$

and

$$t'_j = \frac{-\theta_j \sum_{i=1}^{N} \theta_i}{2}. \tag{11}$$

From (6)–(11), we have

$$U_j(y', \theta_j) = U_j(y, \theta_j).$$

But because $\theta'_j > \theta_j$ and $\sum_{i=1}^{N} r'_i > \sum_{i=1}^{N} r_i$ we have

$$U_j(y', \theta'_j) > U_j(y, \theta'_j),$$

as monotonicity requires.

Here is an alternative but equivalent definition of monotonicity: a social choice function is *monotonic* if, for any θ, ϕ, and $y = f(\theta)$ such that

$$U_i(y, \theta_i) \geqslant U_i(y, \theta_i) \implies U_i(y, \phi_i) \geqslant U_i(y, \phi_i) \quad \text{for all } i,$$

we have $y = f(\phi)$. This rendition of monotonicity says that when the outcome that was optimal in state θ goes up in everyone's preference ordering when the state becomes ϕ, then it must remain socially optimal. Although this may seem like a reasonable property, monotonicity can be quite a restrictive condition:

THEOREM 2 [Muller and Satterthwaite (1977)]. *Suppose that Θ consists of all strict preference orderings on the social choice space Y. Then, any social choice function that is monotonic and has a range including at least three choices is dictatorial (i.e., there exists an agent i^* such that in all states agent i^*'s favorite outcome is chosen).*[6]

---

[6] Monotonicity is a good deal less restrictive if one considers implementation of social choice *correspondences* rather than functions [see Maskin (1999)].

3.2. Other notions of implementation

One way to relax monotonicity is to invoke *refinements* of Nash equilibrium, which make it easier to knock out unwanted equilibria while retaining optimal ones. Let us, in particular, explore the concept of *subgame perfect equilibrium* and the use of *sequential mechanisms*, i.e., mechanisms in which agents send messages one at a time. We maintain the assumption that the preference profile is common knowledge among the agents but is unverifiable by an outside party. Therefore, we consider mechanisms of *perfect information* and (this is the subgame perfection requirement) strategies that constitute a Nash equilibrium at *any* point in the game.

Rather than stating general theorems, we focus immediately on our pollution example. For simplicity, restrict attention to the case of two communities ($N = 2$). We shall argue that *any* social choice function in this setting is implementable in subgame perfect equilibrium using a sequential mechanism.

We note first that, for $i = 1, 2$ and any $\theta_i, \theta_i' \in (a, b)$ there exist $(r_1^o(\theta_i, \theta_i'), r_2^o(\theta_i, \theta_i'), t_i^o(\theta_i, \theta_i'))$ and $(r_1^{oo}(\theta_i, \theta_i'), r_2^{oo}(\theta_i, \theta_i'), t_i^{oo}(\theta_i, \theta_i'))$ such that

$$\theta_i \sqrt{r_1^o(\theta_i, \theta_i') + r_2^o(\theta_i, \theta_i')} - r_i^o(\theta_i, \theta_i') + t_i^o(\theta_i, \theta_i')$$
$$> \theta_i \sqrt{r_1^{oo}(\theta_i, \theta_i') + r_2^{oo}(\theta_i, \theta_i')} - r_i^{oo}(\theta_i, \theta_i') + t_i^{oo}(\theta_i, \theta_i') \tag{12}$$

and

$$\theta_i' \sqrt{r_1^{oo}(\theta_i, \theta_i') + r_2^{oo}(\theta_i, \theta_i')} - r_i^{oo}(\theta_i, \theta_i') + t_i^{oo}(\theta_i, \theta_i')$$
$$> \theta_i' \sqrt{r_1^o(\theta_i, \theta_i') + r_2^o(\theta_i, \theta_i')} - r_i^o(\theta_i, \theta_i') + t_i^o(\theta_i, \theta_i'). \tag{13}$$

Formulas (12) and (13) constitute a *preference reversal condition*. The condition says that for any two types θ_i and θ_i' we can find choices (r_1^o, r_2^o, t_i^o) and $(r_1^{oo}, r_2^{oo}, t_i^{oo})$ such that the former is preferred to the latter under θ_i and the latter is preferred to the former under θ_i'.

In view of preference reversal, we can use the following mechanism to implement a given social choice function f:

Stage 1.

Stage 1.1. Agent 1 announces a type $\hat{\theta}_1$.

Stage 1.2. Agent 2 can *agree*, in which case we go to Stage 2, or disagree by announcing some $\hat{\theta}_1' \neq \hat{\theta}_1$, in which case we go to Stage 1.3.

Stage 1.3. Agent 1 is fined some large amount p^* and then chooses between $(r_1^o(\hat{\theta}_1, \hat{\theta}_1'), r_2^o(\hat{\theta}_1, \hat{\theta}_1'), t_i^o(\hat{\theta}_1, \hat{\theta}_i'))$ and $(r_1^{oo}(\hat{\theta}_1, \hat{\theta}_1'), r_2^{oo}(\hat{\theta}_1, \hat{\theta}_1'), t_i^{oo}(\hat{\theta}_1, \hat{\theta}_i'))$. If he chooses the former, agent 2 is also fined p^*; if he chooses the latter, agent 2 receives p^*. The mechanism stops here.

Stage 2. This is the same as Stage 1.2 except the roles are reversed: agent 2 announces $\hat{\theta}_2$, and agent 1 can either agree or disagree. If he agrees, we go to Stage 3. If he disagrees, then agent 2 is fined p^* and must choose between $(r_1^o(\hat{\theta}_2, \hat{\theta}_2'), r_2^o(\hat{\theta}_2, \hat{\theta}_2'), t_2^o(\hat{\theta}_2, \hat{\theta}_2'))$ and $(r_1^{oo}(\hat{\theta}_2, \hat{\theta}_2'), r_2^{oo}(\hat{\theta}_2, \hat{\theta}_2'), t_2^{oo}(\hat{\theta}_2, \hat{\theta}_2'))$. If he chooses the former, agent 1 is also fined p^*; if he chooses the latter, agent 1 receives p^*.

Stage 3. If $\hat{\theta}_1$ and $\hat{\theta}_2$ have been announced, the outcome $f(\hat{\theta}_1, \hat{\theta}_2)$ is implemented.

We claim that, in state (θ_1, θ_2), there is a unique subgame perfect equilibrium of this mechanism, in which agent 1 truthfully announces $\hat{\theta}_1 = \theta_1$ and agent 2 truthfully announces $\hat{\theta}_2 = \theta_2$, so that the equilibrium outcome is $f(\hat{\theta}_1, \hat{\theta}_2)$. To see this, note that in Stage 2, agent 1 has the incentive to disagree with any untruthful announcement $\hat{\theta}_2 \neq \theta_2$ by setting $\hat{\theta}_2' = \theta_2$. This is because agent 1 forecasts that, by definition of $(r_1^o(\hat{\theta}_2, \theta_2), r_2^o(\hat{\theta}_2, \theta_2), t_2^o(\hat{\theta}_2, \theta_2))$ and $(r_1^{oo}(\hat{\theta}_2, \theta_2), r_2^{oo}(\hat{\theta}_2, \theta_2), t_2^{oo}(\hat{\theta}_2, \theta_2))$ and from (13), agent 2 will choose the latter, and so 1 will collect the large sum p^*. By contrast, agent 1 will *not* disagree if $\hat{\theta}_2$ is truthful – i.e., $\hat{\theta}_2 = \theta_2$ – because otherwise (regardless of what $\hat{\theta}_2'$ he announces) (12) implies that agent 2 will choose $(r_1^o(\theta_2, \hat{\theta}_2'), r_2^o(\theta_2, \hat{\theta}_2'), t_2^o(\theta_2, \hat{\theta}_2'))$, thereby requiring 1 to pay a large fine himself. But this, in turn, means that agent 2 will announce truthfully because by doing so he can avoid the large fine that would be entailed by 1's disagreeing. Similarly, agent 1 will be truthful in Stage 1, and agent 2 will disagree if and only if 1 is untruthful. Because both agents are truthful in equilibrium, the desired outcome $f(\theta_1, \theta_2)$ results in Stage 3.

Herein we have examined only one simple example of implementation in a refinement of Nash equilibrium. For more thorough treatments, see the surveys by Moore (1992), Palfrey (2001), or Maskin and Sjöström (2001).

4. Incomplete information

We next turn to incomplete information. This is the case in which agent i observes only his own type θ_i.

4.1. Dominant strategies

A mechanism (M, g) that has the property that each agent has a dominant strategy – a strategy that is optimal regardless of the other agents' behavior – is clearly attractive since it means that an agent can determine his optimal message without having to calculate those of other agents, a calculation may be particularly complex under incomplete information.

Formally, a strategy μ_i for agent i is mapping from his type space Θ_i to his message space M_i. A strategy, $\mu_i : \Theta_i \to M_i$, is *dominant* for type θ_i if:

$$U_i\big(g\big(\mu_i(\theta_i), m_{-i}\big), \theta_i\big) \geq U_i\big(g\big(m_i', m_{-i}\big), \theta_i\big)$$

for all $m'_i \in M_i$, $m_{-i} \in M_{-i}$. A strategy profile $\mu = (\mu_1, \ldots, \mu_N)$ is a *dominant strategy equilibrium* if, for all i and θ_i, $\mu_i(\theta_i)$ is dominant for θ_i.

A social choice function f is *implemented in dominant strategy equilibrium* by the mechanism (M, g) if there exists a dominant strategy equilibrium μ for which $g(\mu(\theta)) = f(\theta)$ for all $\theta \in \Theta$.[7]

Of course, implementation in dominant strategy equilibrium is a demanding requirement, and so perhaps not surprisingly it is difficult to attain in general:

THEOREM 3 [Gibbard (1973) and Satterthwaite (1975)]. *Suppose that Θ consists of all strict preference orderings. Then, any social choice function that is implementable in dominant-strategy equilibrium and whose range includes at least three choices is dictatorial.*

PROOF. Suppose that f is implementable in dominant-strategy equilibrium and that the hypotheses of the theorem hold. Consider $\theta, \theta' \in \Theta$ such that $f(\theta) = y$ and, for all i,

$$U_i(y, \theta_i) \geqslant U_i(y', \theta_i) \quad \text{implies} \quad U_i(y, \theta'_i) \geqslant U_i(y', \theta'_i) \tag{14}$$

for all y'. By assumption, there exists a mechanism (M, g) with a dominant-strategy equilibrium μ such that $g(\mu(\theta)) = y$. We claim that

$$g(\mu(\theta')) = y. \tag{15}$$

To see why (15) holds, suppose that

$$g(\mu_1(\theta'_1), \mu_2(\theta_2), \ldots, \mu_N(\theta_N)) \neq g(\mu(\theta)) = y.$$

Then

$$U_1(g(\mu_1(\theta'_1), \mu_2(\theta_2), \ldots, \mu_N(\theta_N)), \theta'_1) > U_1(y, \theta'_1), \tag{16}$$

a contradiction of the assumption that $\mu_1(\theta_1)$ is dominant for θ_1. Hence,

$$g(\mu_1(\theta'_1), \mu_2(\theta_2), \ldots, \mu_N(\theta_N)) = y$$

after all. Continuing iteratively, we obtain

$$g(\mu_1(\theta'_1), \mu_2(\theta'_2), \mu_3(\theta_3), \ldots, \mu_N(\theta_N)) = y,$$

---

[7] Notice that, unlike with implementation in Nash equilibrium, we require only that *some* dominant strategy equilibrium outcome coincide with $f(\theta)$, rather then that there be a unique equilibrium outcome. However, multiple equilibria are not typically a serious problem with dominant strategies. In particular, when preferences are *strict* (i.e., indifference is ruled out), the dominant-strategy equilibrium outcome is, indeed, unique.

and

$$g\left(\mu\left(\theta'\right)\right) = y. \tag{17}$$

But (17) implies that $f(\theta') = y$. We conclude that f is monotonic, and so Theorem 2 implies that it is dictatorial. $\qquad\square$

In contrast to the pessimism of Theorem 3, Vickrey (1961) and, more generally, Clarke (1971) and Groves (1973) have shown that much more positive results are obtainable when agents' preferences are quasi-linear. Specifically, suppose that we wish to implement a social choice function $f(\theta) = (r_1(\theta), \ldots, r_N(\theta), t_1(\theta), \ldots, t_N(\theta))$ entailing Pareto-efficient pollution reduction, i.e., such that

$$\sum_{i=1}^{N} r_i(\theta) = r^*(\theta), \tag{18}$$

where $r^*(\theta)$ solves

$$r^*(\theta) = \arg\max \sum_{i=1}^{N} \theta_i \sqrt{r} - r. \tag{19}$$

If community j is not allocated any transfer by the mechanism, then j solves

$$\max \theta_j \sqrt{\sum_{i \neq j} r_i + r_j} - r, \tag{20}$$

which clearly does not result in the total reduction being $r^*(\theta)$. To bring the maximands of individual communities and overall society into line, we shall give community j a transfer equal to the sum of the other communities' payoffs (net of transfers):

$$t_j\left(\hat{\theta}\right) = \sum_{i \neq j} \left(\hat{\theta}_i \sqrt{r^*\left(\hat{\theta}\right)} - r_i\left(\hat{\theta}\right)\right) + \tau_j\left(\hat{\theta}_{-j}\right), \tag{21}$$

where $\tau_j(\cdot)$ is an arbitrary function of θ_{-j}. A mechanism in which each agent j announces $\hat{\theta}_j$ and the outcome is $(r_1(\hat{\theta}), \ldots, r_N(\hat{\theta}), t_1(\hat{\theta}), \ldots, t_N(\hat{\theta}))$ where $(r_1(\cdot), \ldots, r_N(\cdot))$ satisfies (18) and (19), and $(t_1(\cdot), \ldots, t_N(\cdot))$ satisfies (21), is called a *Groves scheme* [see Groves (1973)].

We claim that, in a Groves scheme, community j's telling the truth (announcing $\hat{\theta}_j = \theta_j$) is dominant for θ_j for all j and all θ_j. Observe that in such a mechanism, community j's overall payoff if it tells the truth and the other communities announce $\hat{\theta}_{-j}$

is

$$\theta_j \sqrt{r^*\left(\theta_j, \hat{\theta}_{-j}\right)} - r\left(\theta_j, \hat{\theta}_{-j}\right) + \sum_{i \neq j}\left(\hat{\theta}_i \sqrt{r^*\left(\theta_j, \hat{\theta}_{-j}\right)} - r_i\left(\theta_j, \hat{\theta}_{-j}\right)\right) + \tau_j\left(\hat{\theta}_{-j}\right)$$

$$= \left(\theta_j + \sum_{i \neq j}\hat{\theta}_i\right)\sqrt{r^*\left(\theta_j, \hat{\theta}_{-j}\right)} - r^*\left(\theta_j, \hat{\theta}_{-j}\right) + \tau_j\left(\hat{\theta}_{-j}\right).$$

But from (19),

$$\left(\theta_j + \sum_{i \neq j}\hat{\theta}_i\right)\sqrt{r^*\left(\theta_j, \hat{\theta}_{-j}\right)} - r^*\left(\theta_j, \hat{\theta}_{-j}\right) + \tau_j\left(\hat{\theta}_{-j}\right)$$

$$\geqslant \left(\theta_j + \sum_{i \neq j}\hat{\theta}_i\right)\sqrt{r'} - r' + \tau_j\left(\hat{\theta}_{-j}\right) \tag{22}$$

for all r'. In particular, (22) holds when $r' = r^*(\hat{\theta}_j, \hat{\theta}_{-j})$, which then implies that taking $\hat{\theta}_j = \theta_j$ is dominant as claimed.

Thus, with one proviso, a Groves scheme succeeds in implementing the Pareto-efficient pollution reduction. The proviso is that we have not yet ensured that the transfer functions (21) are feasible. One way of ensuring feasibility is to take

$$\tau_j\left(\hat{\theta}_{-j}\right) = -\max_r \sum_{i \neq j}\left(\theta_i \sqrt{r} - r\right)$$

for all j.

Then, community j's transfer becomes

$$t_j\left(\hat{\theta}\right) = \sum_{i \neq j}\left(\hat{\theta}_i \sqrt{r^*\left(\hat{\theta}\right)} - r_i\left(\hat{\theta}_i\right)\right) - \max_r\left(\sum_{i \neq j}\hat{\theta}_i \sqrt{r} - r\right). \tag{23}$$

When transfers take the form (23), a Groves scheme is called a *pivotal mechanism* or a *Vickrey–Clarke–Groves mechanism*. Notice that the transfer (23) is always (weakly) negative, ensuring feasibility.

The logic underlying (23) is straightforward. If community j's announcement has no effect on the social choice, the community pays nothing. However, if it *does* change this choice (i.e., it is "pivotal"), j pays the corresponding loss imposed on the rest of society. Although the pivotal mechanism is feasible, it is not balanced, i.e., the transfers do not sum to zero. Indeed, as shown by Green and Laffont (1979), *no* Groves scheme is balanced. Furthermore, arguments due to Green and Laffont (1977) imply that in a slightly more general version of our pollution example, Groves schemes are essentially

the *only* mechanisms that implement social choice functions with Pareto-efficient pollution reductions. This motivates the search for balanced mechanisms that invoke a less
demanding notion of implementation than in dominant-strategy equilibrium, a question
we turn to in the next subsection.

We have been assuming that each community j's payoff depends directly only on its
own preference parameter θ_j. Radner and Williams (1988) extend the analysis to the
case when j's payoff may depend on the entire profile θ. We have also been concentrating on the case of *Pareto-efficient* social choice functions (or at least social choice
functions for which the pollution reduction is Pareto-efficient); Dasgupta, Hammond
and Maskin (1980) examine dominant-strategy implementation of more general social
choice functions.

4.2. Bayesian equilibrium

Dominant-strategy equilibrium requires that each agent be willing to use his equilibrium strategy whatever the behavior of the other agents. Bayesian equilibrium requires
only that each agent be willing to use his equilibrium strategy when he expects other
agents to do the same. A couple of points are worth noting here. First, because agents'
equilibrium strategies depend on their types but, given the incomplete information, an
agent does not know others' types, we must specify his *beliefs* about these types to complete the description of the model. Second, if a social choice function is implementable
in dominant-strategy equilibrium, then it is certainly implementable in Bayesian equilibrium, so by moving to the latter concept, we are weakening the notion of implementation.

We assume that agents' types are independently distributed; the density and distri-
bution functions for agent i of type $\theta_i \in [a, b]$ are $p_i(\theta_i)$ and $P_i(\theta_i)$, respectively. We
suppose that these distributions are common knowledge amongst the agents. Hence,
the c.d.f. for agent i's beliefs over the types of the other agents is given by $F_i(\theta_{-i}) \equiv
\prod_{j \neq i} P_j(\theta_j)$.

There are two critical conditions that a social choice function must satisfy to ensure that it is implementable in Bayesian equilibrium [see Postlewaite and Schmeidler (1986), Palfrey and Srivastava (1987) and Jackson (1991)]. The first is Bayesian
incentive-compatibility. A social choice function f is *Bayesian incentive compatible*
(BIC) if

$$E_{\theta_{-i}}\big[U_i\big(f(\theta_i, \theta_{-i}), \theta_i\big)\big] \geq E_{\theta_{-i}}\big[U_i\big(f(\theta_i', \theta_{-i}), \theta_i\big)\big]$$

for all i, and $\theta_i, \theta_i' \in \Theta_i$, where

$$E_{\theta_{-i}}\big[U_i\big(f(\theta_i, \theta_{-i}), \theta_i\big)\big] = \int_{\Theta_i} U_i\big(f(\theta_i, \theta_{-i}), \theta_i\big)\mathrm{d}F_i(\theta_{-i}).$$

The second condition is the incomplete-information counterpart to monotonicity. For
this purpose, we define a *deception for agent j* to be a function $\alpha_j : \Theta_j \to \Theta_j$. A *decep-*

tion α is a profile $u = (u_1, \ldots, u_N)$. A social choice function f is *Bayesian monotonic* if for all deceptions α such that $f \circ \alpha \neq f$ there exist j and a function $\gamma : \Theta_{-j} \to \Upsilon$ such that

$$EU_j\big(f(\theta_j, \theta_{-j}), \theta_j\big) \geq EU_j\big(\gamma(\theta_{-j}), \theta_j\big)$$

for all $\theta_j \in \Theta_j$, and

$$EU_j\big(f\big(\alpha(\theta'_j, \theta_{-j})\big), \theta'_j\big) < EU_j\big(\gamma\big(\alpha_{-j}(\theta_{-j})\big), \theta'_j\big)$$

for some $\theta'_j \in \Theta_j$.

Jackson (1991) shows that in quasi-linear settings, such as our pollution example, BIC and Bayesian monotonicity are not only necessary but sufficient for a social choice function to be implementable in Bayesian equilibrium.

Let us return to our pollution example. We noted in the previous subsection that a social choice function entailing Pareto-efficient pollution reduction (i.e., reduction satisfying (18) and (19)) cannot be implemented in dominant-strategy equilibrium if it is balanced. However, this negative conclusion no longer holds with Bayesian implementation.

To see this, consider a pollution reduction profile $(r_1^o(\theta), \ldots, r_N^o(\theta))$ that is Pareto-efficient (i.e., $\sum_{i=1}^N r_i^o(\theta) = r^*(\theta)$, where $r^*(\cdot)$ satisfies (19)). Consider the mechanism in which each agent j announces $\hat{\theta}_j$ and the outcome is $(r_1^o(\hat{\theta}), \ldots, r_N^o(\hat{\theta}), t_1^o(\hat{\theta}), \ldots, t_N^o(\hat{\theta}))$, where $t_j^o(\hat{\theta})$ satisfies

$$t_j^o(\hat{\theta}) = \int_{\Theta_{-j}} \sum_{i \neq j} \left(\hat{x}_i \sqrt{r^*(\hat{\theta}_j, x_{-j})} - r_i(\hat{\theta}_j, x_{-j}) \right) dF_j(x_{-j})$$

$$- \frac{1}{N-1} \sum_{i \neq j} \int_{\Theta_{-i}} \sum_{k \neq i} \left(x_k \sqrt{r^*(\hat{\theta}_i, x_{-i})} - r_k(\hat{\theta}_i, x_{-i}) \right) dF_i(x_{-i}). \qquad (24)$$

Notice that the first term (integral) on the right-hand side of (24) is just the expectation of the sum in (21). Furthermore the other terms in (24) do not depend on $\hat{\theta}_j$. Hence, this mechanism can be thought of as an "expected Groves scheme". It was first proposed by Arrow (1979) and d'Aspremont and Gérard-Varet (1979).

The terms after the first integral in (24) are present to ensure balance. If all communities tell the truth (we verify below that the social choice function satisfies BIC $f(\theta) = (r_1^o(\hat{\theta}), \ldots, r_N^o(\hat{\theta}), t_1^o(\hat{\theta}), \ldots, t_N^o(\hat{\theta})))$, then observe that

$$\sum_{j=1}^N t_j^o(\theta) = \sum_{j=1}^N \int_{\Theta_{-j}} \sum_{i \neq j} \left(\theta_i \sqrt{r^*(\theta_j, \theta_{-j})} - r_i^o(\theta_j, \theta_{-j}) \right) dF_j(\theta_{-j})$$

$$- \frac{1}{N-1} \sum_{j=1}^{N} \sum_{i \neq j} \int_{\Theta_{-i}} \sum_{k \neq i} \left(\theta_k \sqrt{r^*(\theta_i, \theta_{-i})} - r_k^o(\theta_i, \theta_{-i}) \right) dF_i(\theta_{-i})$$

$$= \sum_{j=1}^{N} \int_{\Theta_{-j}} \sum_{i \neq j} \left(\theta_i \sqrt{r^*(\theta_j, \theta_{-j})} - r_i^o(\theta_j, \theta_{-j}) \right) dF_j(\theta_{-j})$$

$$- \sum_{j=1}^{N} \int_{\Theta_{-i}} \sum_{i \neq j} \left(\theta_i \sqrt{r^*(\theta_j, \theta_{-j})} - r_i^o(\theta_j, \theta_{-j}) \right) dF_j(\theta_{-j})$$

$$= 0,$$

as desired.

To see that BIC holds (so that truth-telling is an equilibrium) note that if $f(\theta) = (r_1^o(\theta), \ldots, r_N^o(\theta), t_1^o(\theta), \ldots, t_N^o(\theta))$, then, for all j, θ_j, θ_j', and θ_{-j},

$$E_{\theta_{-j}} \left[U_j \left(f(\theta_j', \theta_{-j}), \theta_j \right) \right]$$

$$= E_{\theta_{-j}} \left[\theta_j \sqrt{r^*(\theta_j', \theta_{-j})} - r_j^o(\theta_j', \theta_{-j}) + t_j^o(\theta_j', \theta_{-j}) \right]$$

$$= E_{\theta_{-j}} \left[\theta_j \sqrt{r^*(\theta_j', \theta_{-j})} - r_j^o(\theta_j', \theta_{-j}) \right.$$

$$\left. + E_{\theta_{-j}} \sum_{i \neq j} \left(\theta_i \sqrt{r^*(\theta_j', \theta_{-j})} - r_i^o(\theta_j', \theta_{-j}) \right) \right], \tag{25}$$

where the last line of the right-hand side of (25) corresponds to the first term of $t_j^o(\theta_j', \theta_{-j})$ as given by the right-hand side of (24), but with all but the first term omitted (since the other terms on the right-hand side of (24) do not depend on θ_j' and hence do not affect incentive compatibility for community j). But the last line of the right-hand side of (25) can be rewritten as

$$E_{\theta_{-j}} \left[\left(\sum_{j=1}^{N} \theta_i \right) \sqrt{r^*(\theta_j', \theta_{-j})} - r^*(\theta_j', \theta_{-j}) \right]. \tag{26}$$

By definition of $r^*(\theta)$, the square-bracketed expression in (26) is maximized when $\theta_j' = \theta_j$. Hence from (25) and (26), we have

$$E_{\theta_{-j}} \left[U_j \left(f(\theta_j', \theta_{-j}), \theta_j \right) \right] \leqslant E_{\theta_{-j}} \left[U_j \left(f(\theta_j, \theta_{-j}), \theta_j \right) \right],$$

as required for BIC.

One can readily show that f also satisfies Bayesian monotonicity (but we will refrain from doing so here). Hence, we conclude that it is implemented by the Groves mechanism (actually, it turns out that the equilibrium outcome of the expected Groves mechanism is not unique, so, without modification, *that* mechanism does not actually implement f). Thus relaxing the notion of implementability from dominant-strategy to Bayesian equilibrium permits the implementation of balanced social choice functions. On the downside, however, note that the very construction of the expected Groves mechanism requires common knowledge of the distribution of θ.

Acknowledgements

We would like to thank Jeffrey Vincent, Karl-Göran Mäler, David Starrett and Theodore Groves for their comments on an earlier version of this chapter.

References

Arrow, K. (1979), "The property rights doctrine and demand revelation under incomplete information", in: M. Boskin, ed., Economies and Human Welfare (Academic Press, New York).
Baliga, S., and E. Maskin (2002), "The Free-Rider problem and the Coase theorem", work in progress.
Clarke, E. (1971), "Multi-part pricing of public goods", Public Choice 11:17–33.
Coase, R. (1960), "The problem of social cost", Journal of Law and Economics 3:1–44.
Corchon, L. (1996), The Theory of Implementation of Socially Optimal Decisions in Economics (St. Martin's Press, New York).
Dasgupta, P., P. Hammond and E. Maskin (1980), "On imperfect information and optimal pollution control", Review of Economic Studies 47:857–860.
d'Aspremont, C., and L.A. Gérard-Varet (1979), "Incentives and incomplete information", Journal of Public Economics 11:25–45.
Debreu, G. (1957), Theory of Value (Wiley, New York).
Fudenberg, D., and J. Tirole (1991), Game Theory (MIT Press, Cambridge, MA).
Green, J., and J.-J. Laffont (1977), "Characterization of satisfactory mechanism for the revelation of preferences for Public goods", Econometrica 45:727–738.
Green, J., and J.-J. Laffont (1979), Incentives in Public Decision Making (North-Holland, Amsterdam).
Groves, T. (1973), "Incentives in teams", Econometrica 41:617–663.
Groves, T., and J. Ledyard (1987), "Incentive compatibility since 1972", in: T. Groves, R. Radner and S. Reiter, eds., Information, Incentives and Economic Mechanisms (University of Minnesota Press, Minneapolis, MN).
Jackson, M. (1991), "Bayesian implementation", Econometrica 59:461–477.
Jackson, M. (2001a), "A crash course in implementation theory", Social Choice and Welfare 18:655–708.
Jackson, M. (2001b), "Mechanism theory", forthcoming in: Encyclopedia of Life Support Systems.
Laffont, J.-J., and D. Martimort (2002), The Theory of Incentives: The Principal-Agent Model (Princeton University Press, Princeton, NJ).
Laffont, J.-J., and E. Maskin (1979), "A differentiable approach to expected utility maximizing mechanisms", in: J.-J. Laffont, ed., Aggregation and Revelation of Preferences (North-Holland, Amsterdam) 289–308.
Mas-Colell, A., M. Whinston and J. Green (1995), Microeconomic Theory (Oxford University Press, Oxford).
Maskin, E. (1985), "The theory of implementation in Nash equilibrium: a survey", in: L. Hurwicz, D. Schmeidler and H. Sonnenschein, eds., Social Goals and Social Organization (Cambridge University Press, Cambridge).
Maskin, E. (1994), "The Invisible Hand and externalities", American Economic Review 84(2):333–337.

Maskin, E. (1999), "Nash equilibrium and welfare optimality", Review of Economic Studies 66:23–38.

Maskin, E., and T. Sjöström (2001), "Implementation theory", forthcoming in: K. Arrow, A. Sen and K. Suzumura, eds., Handbook of Social Choice and Welfare (North-Holland, Amsterdam).

Moore, J. (1992), "Implementation in environments with complete information", in: J.-J. Laffont, ed., Advances in Economic Theory: Proceedings of the Sixth World Congress of the Econometric Society (Cambridge University Press, Cambridge) 182–282.

Myerson, R. (1991), Game Theory (Harvard University Press, Cambridge).

Myerson, R., and M. Satterthwaite (1983), "Efficient mechanisms for bilateral trade", Journal of Economic Theory 29:265–281.

Palfrey, T. (1992), "Implementation in Bayesian equilibrium: the multiple equilibrium problem in mechanism design", in: J.J. Laffont, ed., Advances in Economic Theory (Cambridge University Press, Cambridge).

Palfrey, T. (2001), "Implementation theory", forthcoming in: R. Aumann and S. Hart, eds., Handbook of Game Theory, Vol. 3, Mastermind (North-Holland, Amsterdam).

Palfrey, T., and S. Srivastava (1987), "On Bayesian implementation allocations", Review of Economic Studies 54:193–208.

Pigou, A.C. (1932), The Economics of Welfare (Macmillan & Co., London).

Postlewaite, A., and D. Schmeidler (1986), "Implementation in differential information economics", Journal of Economic Theory 39:14–33.

Radner, R., and S. Williams (1988), "Informational externalities and the scope of efficient dominant strategy mechanisms", Mimeo.

Vickrey, W. (1961), "Counterspeculation, auctions, and competitive sealed-tenders", Journal of Finance 16:8–37.

Chapter 8

THE POLITICAL ECONOMY OF ENVIRONMENTAL POLICY

WALLACE E. OATES

Department of Economics, University of Maryland, College Park, MD, and
Resources for the Future, 1616 P Street N.W., Washington, DC, USA

PAUL R. PORTNEY

Resources for the Future, 1616 P Street N.W., Washington, DC, USA

Contents

Handbook of Environmental Economics, Volume 1, Edited by K.-G. Mäler and J.R. Vincent

Abstract

This chapter provides a review and assessment of the extensive literature on the political determination of environmental regulation. A promising theoretical literature has emerged relatively recently that provides models of the political interaction of government with various interest groups in the setting of environmental standards and the choice of regulatory instruments. A large empirical literature supports such models, finding evidence of the influence of interest groups but also evidence that net social benefits are often an important determinant of environmental policy choices. A later section of the paper takes up the issue of environmental federalism and the large and growing theoretical literature that addresses the so-called competitive "race to the bottom" as various jurisdictions attempt to use environmental policy as an instrument of economic competition. The evidence on all this is sparse, although some recent work in the U.S. is unable to find any support for the race-to-the-bottom hypothesis. The paper concludes with a brief look at the evolution of environmental policy and finds that economics has come to play a growing role both in the setting of standards for environmental quality and in the design of regulatory measures. There seems to be a discernible trend toward more efficient decision-making for environmental protection.

Keywords

political economy, interest groups, environmental taxes, tradeable permits, command-and-control policies, environmental federalism

JEL classification: Q2, H4, H7

The term "political economy" has a long and rich history. In its earliest manifestations, it meant essentially economics; indeed the two terms were basically synonyms in
the nineteenth and early twentieth centuries [Groenewegen (1987)]. However, as "economics" came to denote the discipline, the term "political economy" has come to take
on a variety of shades of meaning. It is now, in fact, a rather elusive term that typically
refers to the study of the collective or political processes through which public economic
decisions are made.

For our purposes in this essay, we shall settle for this admittedly broad and somewhat
vague definition. Our concern here is with the determinants of actual decisions on environmental programs. Environmental economics (as the other chapters in this Handbook
reveal) has much to say about the design of efficient and effective policy measures for
protection of the environment. But when we turn to actual policy, we find, often to our
dismay, that existing measures or institutions do not stack up at all well in terms of these
guidelines. How are we to understand such "political failure"?

In this chapter, we shall explore the various political, or collective choice, facets of
environmental policy-making. Our treatment will focus on the political economy of domestic environmental policy, for there is another chapter by Scott Barrett in this Handbook that takes up the issue of international environmental agreements. In addition, our
study is more or less limited to the experience of countries with elected governments –
in particular, the United States and the nations of Western Europe. Both existing theory
and more rigorous empirical work tend to relate to these countries.[1]

We begin the chapter with a few preliminary and general observations on the various
theoretical and empirical approaches that have been employed to study regulatory behavior. Our sense is that certain of them are more directly relevant to understanding the
determination of environmental policy than others. More specifically, a framework in
which various interest groups vie with one another in a political setting seems to us the
most promising approach to a positive theory of environmental regulation. In Section 3
we review the fairly extensive theoretical literature that sets forth this approach. This
body of work draws heavily on the recent work on the positive theory of international
trade, which has developed models of competing interest groups that seek to restrict
trade in ways that promote their own interests. The parallels to the determination of
environmental policy are straightforward.

The chapter then turns to a survey of the wide-ranging empirical work on the political
economy of environmental regulation. This body of research encompasses a vast array
of studies that range from largely qualitative, case studies to more formal and rigorous
econometric investigations. These studies provide support for the view that not only
specific interest groups influence environmental measures, but that (at least in some
cases) the social benefits and costs also play a role in determining outcomes.

---

[1]   For a useful study of the political economy of environmental policy in the developing world, see World
Bank (2000). David Wheeler and his colleagues at the World Bank provide numerous insights into the complex interaction between economic development and environmental protection with important implications
for the design of environmental policies.

Section 5 moves on to a description and assessment of the now vast and rich literature on environmental federalism. The issue of the respective roles of various levels of
government is a contentious issue on both sides of the Atlantic. Our sense is that the
"race-to-the bottom" arguments that are the basis for current pressures for the harmonization (or centralization) of environmental policies across jurisdictional boundaries
are not so compelling as some argue. There are important responsibilities for governments at all levels in the design and implementation of environmental programs. And
it is important to get these "functions" aligned properly in the vertical structure of public decision-making. We conclude the chapter with some reflections on trends in the
political economy of environmental decision-making – most notably on the encouraging tendency in recent years to give more weight to economic analysis in the design of
environmental policy.

1. On theories of regulation: Some preliminaries

There are a number of distinct approaches to understanding regulatory activity. The
traditional neoclassical and normative approach sees regulatory measures as one means
for correcting allocative distortions in a market system. In the case of environmental
policy, the standard theory of externalities provides a basic explanation for tendencies
in a market economy toward excessive levels of pollution. From this approach follows a
clearcut prescription for correction of this distortion: the internalization of the external
costs through either a system of taxes on polluting activities equal to marginal social
damage or a system of tradable emission permits that restricts aggregate pollution to
the efficient level and, at the same time, guides abatement activities into a least-cost
pattern.[2]

But this is the normative theory of environmental regulation. It emerges from an
analytical exercise involving the maximization of social welfare. As such, it presumes
implicitly an enlightened public sector that designs and implements social programs
for environmental protection with the sole objective of promoting the well-being of the
polity as a whole (i.e., some weighted average of individual utilities). This, we know, is
not how social policies typically come into being, which leads to the search for positive
models that can describe the actual determination of social policy.

One such conceptual construct is the widely employed median-voter model. In this
framework, social choices that are made directly by voters or through their elected representatives reflect the median of the most preferred outcomes of the individuals in the

---

[2] Alternatively, we may see such public intervention in terms of its capacity to reduce the "transactions
costs" associated with achieving an efficient outcome. See, for example, Zerbe, Jr., and McCurdy (1999).
As Coase (1960) has shown us, there may well be cases where voluntary negotiations among a small group
of affected parties can effectively resolve an externality with no need for public regulatory measures. Or, if
the transactions costs are sufficiently low and the link between cause and effect sufficiently clear, a well-
functioning tort system that makes polluters liable for the costs they impose on society can constitute an
efficient system of pollution control.

relevant social group.[3] One familiar and popular adaption [Downs (1957)] extends the model to a setting of two political parties in which competition for votes among these parties leads to outcomes that converge on the preferred outcome of the median-voter. Moreover, as Bergstrom (1979) has shown, under certain conditions, the median-voter outcome will satisfy the first-order conditions for Pareto-efficiency. This is of particular interest, since it provides us with a case where the actual outcome of a plausible process of political decision-making satisfies the conditions from our normative model for efficient social choice. One cannot push this too far, for the Bergstrom conditions for such a coincidence of outcomes are admittedly restrictive; moreover, there are many realistic complications (such as a multi-dimensional policy space) that create serious problems for the model. But in its defense, it has had some success in explaining a substantial range of social choice outcomes.[4]

Let us note one obvious prediction of our normative and median-voter models. Since environmental externalities can involve significant social damages and economic distortions, one would expect policies to emerge to restrict polluting activities – as has been the case in most places. In other words, and at this most simple level, the predictions of models that lead to efficient (or "quasi-efficient") outcomes find some support in real-world outcomes. But again, we must not push too hard on this. When we ask the harder questions concerning the stringency of these programs and the choice of policy instruments, things quickly become more complex and less clear. But it is worthwhile to note at the outset that our basic normative framework (linked to a positive model through something like the median-voter model) does have some, if limited, explanatory power.

There are other theories of social choice and, in particular, of regulation that are potentially important here. A major (and radical) attempt to provide a positive theory of regulation originated with Stigler (1971) with subsequent development by Peltzman (1976) and others. This approach sees regulation not as a means to promote the general welfare by mitigating efficiency losses from market failure, but rather as a form of wealth transfers. In Stigler's view, "regulation is acquired by the industry and is designed and operated primarily for its benefit" [Stigler (1971, p. 3)]. This so-called "capture theory" of regulation sees regulated industries, not as the victims of regulatory measures, but rather as their beneficiary. Measures enacted by "captured" agencies may take the form, for example, of direct monetary subsidies or, alternatively, of less direct assistance in the form of barriers to entry into the regulated industry.

---

[3] Black (1948) is the source for the first modern treatment of the median-voter theorem. See Mueller (1989) for an excellent survey of the more recent literature on the median-voter model.

[4] One such application involves the use of the model to provide a framework for the estimation of demand functions for local public goods [Borcherding and Deacon (1972), Bergstrom and Goodman (1973)]. Assuming that the outcome in each community represents a point on the demand curve of the median voter, this literature has generated plausible econometric estimates of the demand functions for a number of different public services provided by local governments. For surveys that explain this application and describe the findings, see Rubinfeld (1987) and Oates (1996).

While the capture theory may describe some classes of regulatory activity reasonably well, it does not seem to us that it is very successful as a positive theory of environmental policy. As we have suggested, environmental measures come about largely as a result of the real or perceived social damages that are borne quite widely across the social spectrum from polluting activities. A theory that relates regulation directly to social welfare maximization thus does not seem too far off the mark. Environmental measures typically impose costs – sometimes quite significant costs-on the sources of polluting activities. For this reason, it seems misleading at best to describe environmental measures as instigated by regulatees – that is by polluting industries. We surely find certain cases (which we will examine later) where such measures have been manipulated into a form that provides specific benefits to at least some regulatees, but to argue that environmental policy has its basic impetus in the designs of polluting industries seems misplaced.[5]

A more attractive model of environmental regulatory choice is one in which various interest groups vie with one another through a political process to determine the extent and form of environmental policies. The problem with this approach, as Stigler observed, is that such a view can lead to the position that a regulatory outcome "defies rational explanation", being the result of "an imponderable, a constantly and unpredictably shifting mixture of forces of the most diverse nature" [Stigler (1971, p. 3)].

Fortunately, as Becker (1983) showed in his seminal paper, things are not this intractable. In fact, it is not even the case that such political processes lead invariably to distorted, inefficient outcomes. Becker's analysis finds that competition among interest groups for political influence can have some important efficiency-enhancing properties. Moreover, some recent work on interest group politics pushes this farther; in the Becker spirit, this work finds that such processes lead to political equilibria that can be economically efficient. Aidt (1998), for example, lays out an interesting model in which competition among interest groups leads to an efficient internalization of detrimental externalities. In Aidt's model (which we shall examine in more detail later), government pursues its own goals, seeking a mixture of political contributions and social welfare. So long as the interest groups represent the interests of their constituencies faithfully, their contributions induce public decision-makers both to select efficient levels of externality-generating activities and to employ efficient regulatory instruments. Such models at least remind us that the outcomes from the political interplay of diverse interest groups need not be inherently distortionary, although fully efficient outcomes are admittedly special cases.

It seems to us that approaches that explicitly recognize this interaction of different interest groups are the most promising for an understanding of environmental policy. The stage upon which the environmental policy process plays itself out is typically one

---

[5] Interestingly, however, it has been argued that the Clean Air Amendments in 1970 in the United States resulted from pressures from industry for federal standards as a means for inhibiting states from setting yet more stringent (and nonuniform) standards! See Elliott, Ackerman and Millian (1985).

in which environmental advocacy groups and potentially regulated parties (which can include corporations, other levels of government, and even individuals) push their cases, and where regulators may even bring to bear basic measures of social costs and benefits.

In this process, it seems clear that institutions matter. Environmentalists, business trade organizations, and other interest groups interact first in the determination of environmental legislation. But this is not the end of the story. The implementation of such legislation by environmental agencies provides another arena in which divergent interests must be reconciled in the actual design, administration, and enforcement of policy. Through the selective enforcement of specific environmental measures, regulatory agencies may either weaken the measures or expand their scope and effectiveness. Some statutes actually allow the regulator to negotiate with the source to determine the form and extent of compliance. Finally, key decisions often end up in the hands of the judiciary, as the courts interpret the intent of legislation and the faithfulness of its implementation by administrative agencies. The analysis of the political economy of environmental policy must thus encompass the institutional setting in which the interplay of interest groups takes place.

2. On the empirical study of the political economy of environmental regulation: A few more preliminaries

To assess the role of different interests in the determination of policy measures, one can look in a relatively informal, historical way at various policy decisions through qualitative case studies. Such studies abound in the literature – and frequently provide valuable insights into the political economy of particular environmental programs. Ackerman and Hassler (1981), for example, provide a penetrating account of the evolution of the Clean Air Act in the United States with particular attention to the crucial role played by coal interests. More recently, Leveque (1996) has assembled a series of case studies in Europe that describe different dimensions of the environmental policy process in the emerging European Union. At the same time, there are available somewhat more formal approaches that allow us to make statistical inferences about the groups or issues that figured significantly in the policy-making process. And these have been widely used in the environmental literature to shed light on how environmental decisions have, in fact, been made.

One such approach examines the "revealed preferences" of the regulatory agency. McFadden (1975, 1976) set forth such a method in which the actual decisions of a public agency can be used to infer the criteria that gave rise to these choices. Making use of a multinomial logit model, the approach essentially selects statistically a set of decision rules that can explain the observed choices. As an application, McFadden used the model to explore the determinants of freeway planning decisions by the California Division of Highways.

More generally, one can posit a relationship between a class of decisions regarding, say, the regulation of a set of industries or pollutants, as the dependent variable and as explanatory variables the behavior (e.g., contributions, lobbying activities, testifying before regulatory bodies) of various interest groups, along with other factors such as the toxicity of the pollutants and the cost of controlling them. With the needed data, we can then estimate econometrically the impact of each of these determinants on a series of environmental decisions. As we shall see, this general approach has been used quite effectively for the study of a variety of environmental programs – it has allowed us to explore ex post who has "called the tune" for various kinds of environmental measures.

An alternative approach involves the examination of voting behavior. Some studies, for example, look at the voting records of individual legislators and then relate these records back to the characteristics of the politicians themselves or their constituencies. In other cases, environmental measures have been determined by direct vote in a referendum; under this form of decision-making, we can look directly at the pattern of votes across precincts and relate these patterns to the characteristics of the voters to find out what effectively determined the outcomes. Thus, we find in the sections that follow a number of different approaches to uncovering the role of various interest groups and decision-making procedures in the setting of enviromental policy.

3. Toward a positive theory of environmental regulation[6]

3.1. The early literature

As we suggested earlier, the most promising approach to understanding the actual form and stringency of environmental measures appears to us to be one which tries to understand how various interest groups interact in a specified political setting with environmental policies as the outcome. This general approach has its roots in some early pieces that sought to explain why existing environmental policies had taken an inefficient and inferior form rather than the kinds of measures suggested by economic analysis.

In one of these early papers, Buchanan and Tullock (1975), drawing on basic models of the firm, showed that emissions standards (or, more precisely, quotas on polluting outputs) would generally be preferred to effluent taxes by firms themselves, where these measures take a form that effectively limits entry. In such a setting, environmental regulations can produce a cartel-type outcome with increased profits for existing firms. For example, it is easy to see that environmental measures that prescribe more stringent standards for new, than for existing, plants (as is often the case since retrofitting can be quite expensive) may be welcomed by industrial interests as a newly created barrier-to-entry into the polluting industry.

---

[6] There is now a substantial literature addressing the political economy of environmental regulation, some, but not all, of which we shall draw on explicitly in our treatment. For five useful books on this issue, see Magat, Krupnick and Harrington (1986), Congleton (1996), Dijkstra (1999), Svendsen (1998), and Wallart (1999).

This general line of analysis was pursued in some subsequent papers. Dewees (1983), for example, laid out very nicely a systematic treatment of how various policy instruments affected the well-being both of industrial interests (including shareholders) and workers in the affected industries. Dewees confirmed and extended the findings of Buchanan and Tullock; he showed, for example, that industrial interest groups could well prefer systems of marketable emissions permits to effluent standards *if* the permits were distributed free of charge to existing sources.[7]

This general line of analysis thus examined the implications of different policy instruments for the welfare of various interest groups, and in this effort generated a number of insights into just why we would expect to find opposition in certain quarters to efficient and effective policy measures. It is straightforward, for example, to show that a system of pollution taxes (or tradeable permits distributed by an initial auction) is likely to prove more costly to polluting industries than a less efficient assignment of emissions quotas – or even the required adoption of a specified control technology. The point here is that under a system of taxes (or auctioned tradeable permits) polluting firms must bear not only the costs of their pollution control activities, but, in addition, must pay taxes on (or buy permits for) their remaining discharges. And some empirical studies have suggested that even where a command-and-control (CAC) program produces a quite inefficient pattern of abatement efforts, the extra control costs may be dwarfed, in comparison, by the taxes that must be paid under a regime of pollution levies [e.g., Seskin, Anderson and Reid (1983)]. In consequence, it should not be surprising to find that industrial interests have often shown little enthusiasm for the systems of environmental taxes championed by economists.[8]

These first-generation studies of positive theories of environmental policy thus sought to explain how various policy measures affected the different interest groups. But they did not take the next step of actually predicting outcomes. It is one thing to show that policy measure A will be favored by interest group B, but it is much more complicated to show how this measure will be received by the various interest groups (some of which may support it and others not) and then how this will play out in a process of interaction among these groups to produce a policy outcome Hahn (1990).

3.2. *The theory of interest groups and environmental outcomes*

The second generation of work on the positive theory of environmental regulation has taken up this challenging issue. The basic approach involves setting out a public-choice

---

[7] For another extension of the Buchanan–Tullock analysis, see Leidy and Hoekman (1996) who treat the issue in the context of an open economy and find further reasons for various interest groups to prefer direct regulation to emissions taxes.

[8] Opposition from industrial (and other) interests in the early days of environmental legislation also had its source in a failure to understand and appreciate the virtues of a market-based system. In the case of the U.S., for example, Kelman (1981) found that in the late 1970s hardly anyone in the policy-making community could even explain clearly the rationale for incentive-based environmental measures!

or political setting in which competing interest groups, taking the form of lobby groups, provide support in one form or another (often monetary support for their preferred candidate), and then, making use of game-theoretic analytical techniques, characterizing outcomes under differing conditions. Such models can, for example, provide an explicit rationale for the choice of command-and-control instruments over more efficient incentive-based measures under certain specified circumstances.

This body of work draws heavily on recent research into the positive theory of international trade – research that seeks to explain the introduction of tariffs and other impediments to free trade through the political interplay of various interest groups [Hillman (1989), Grossman and Helpman (1994)].[9] It is useful, following Grossman and Helpman (1994), to distinguish between two strands in this literature. The first envisions the political setting as one of political competition between opposing candidates (or parties). The competing candidates announce the policy measures that they will introduce if elected, and then organized interest groups make their decisions concerning which candidate to support [e.g., Hillman and Upsprung (1992)].

The second approach to the study of endogenous policy determination involves a setting in which an incumbent government seeks to maximize its political support through the choice of policy measures. Under this "political support" type of model, the various interest or lobby groups offer contributions, and the government determines policy so as to maximize the likelihood of being re-elected. This typically involves the maximization of an objective function that includes as arguments both the general welfare of the electorate and the contributions from the various interest groups [e.g., Aidt (1998)]. Under this latter approach involving the so-called "common agency model of politics", one of the intriguing findings (mentioned earlier) is that if all agents have their interests represented accurately by an interest group, then the political equilibrium is socially efficient. All external effects become effectively internalized through the political process with the result that the policy-maker chooses both the efficient policy instrument and the efficient level of regulation; in the case of environmental policy, this is a Pigouvian tax.

To get a better sense of these quite striking results, it may prove helpful to treat all this a bit more formally. We shall follow Aidt (1998) here; his formulation builds on Grossman and Helpman (1994). In Aidt's model, the government's objective function encompasses both social welfare goals and political contributions:

$$G(p, q, t^e) = \Theta W(p, q, t^e) + \sum C^i(p, q, t^e), \tag{1}$$

---

[9] The seminal paper by Grossman and Helpman (1994) on tariff policies provides the foundation for much of this work on environmental regulation. The fundamental contribution of this paper is to show that the political support for various policy measures has its source in well-defined preferences of individuals that manifest themselves in a political process that can be described in an explicit, precise manner with a resulting equilibrium policy outcome.

where W is a Benthamite social welfare function, Θ is a weighting parameter, and C^i is the contribution from interest group i.[10] The variables p, q, and t^e represent producer prices, consumer prices, and emissions taxes, respectively. Each citizen is a generalist consumer, a shareholder in one industry (product), and is adversely affected by polluting emissions associated with production.

Aidt limits the government to only two policy instruments: taxes on emissions and product taxes/subsidies. The government can use these instruments both to control emissions and to redistribute income. Each citizen receives an equal lump-sum share of the tax revenue collected.

In the Aidt model, each interest group represents all those citizens that hold shares of a particular industry. However, rather than simply focusing on increasing the profit earned by the industry, the interest group faithfully represents all of its members' interests. Thus, the interest group is concerned with each of the elements affecting its members' welfare. The objective function for each interest group, $W^i(p, q, t^e)$, is thus the sum of its members' utility. Following Grossman and Helpman (1994), it can be shown that the optimal contribution for each interest group is equal to its objective function minus a constant, K^i:

$$C^i(p, q, t^e) = W^i(p, q, t^e) - K^i.$$ (2)

With this in place, the insight of concern here, namely the existence of an efficient lobbying outcome, follows in a straightforward manner.[11] If all N industries are represented by an interest group (and because each citizen holds shares of only one industry), then all citizens are represented by an interest group. In this case, the government's objective function collapses to:

$$G(p, q, t^e) = (\Theta + 1)W(p, q, t^e) - \sum K^i$$ (3)

and the optimal tax levels for the government are the same as those for the social welfare function: product taxes that equal zero and emissions taxes that equal marginal social damage.[12]

---

[10] We note here that Aidt's model (like many others in this literature) treats government as a single unit by characterizing it in terms of a single, well-defined objective function. This effectively abstracts from some of the richness of a more realistic setting involving both legislative and bureaucratic activities in the public sector and its multilevel structure.

[11] Note that Equation (2) implies that in the vicinity of the equilibrium, the lobby group is willing to contribute its full value of the incremental change in the activity. Lobbies thus reveal their true preferences in the neighborhood of the equilibrium. The constant term reflects the division of the rent between the lobby and the government. See Grossman and Helpman (1994) for a careful explication of this point.

[12] Distortions in both the emissions taxes and product taxes/subsidies typically arise when not all industries are represented by a lobby group. In this more realistic setting, Aidt finds another interesting result. If only some interest groups contribute to campaigns, marginal damages continue to enter only the argument for the optimal (from the government's perspective) emissions taxes. For an interesting application of this framework to pollution taxes in an open-economy setting, see Fredriksson (1997).

Deviations from efficient outcomes in this framework result from the failure of lobby groups to emerge to represent certain interests. Aidt does not examine the formation of interest groups. He simply takes their existence as given and assumes that they have overcome the free-rider and associated challenges that confront the organization of these groups. However, the basic theory of public goods leads us, in fact, to expect such failures in organization. In his classic work, Olson (1965) laid out a theory of special interest groups in which he explored the conditions under which effective lobby groups were likely to emerge. As Olson taught us, the basic free-rider problem limits the capacity for individuals with common interests to organize to obtain a collective benefit. Powerful lobbies are typically those that perform some function in addition to providing purely collective goods: they provide direct services to their members or have various tools of "coercion" at their disposal to enforce membership on those who benefit from their activities. So it comes as no surprise to find that certain interest groups – business trade associations, for example, that encompass relatively small and fairly homogeneous groups – are able to organize and represent their collective interests effectively, as compared to larger and more diffuse groups like consumers. Thus, it is easy to see how inefficient policy outcomes can emerge as a result of incomplete representation through interest groups.[13]

In fact, from this perspective what does seem surprising is the extent to which environmental advocacy groups have mobilized their constituencies so effectively. The benefits from programs to improve air quality on a national scale, for example, would appear to represent an Olsonian "large-group" case, where it would be extremely difficult to organize environmental interests. But in seeming contradiction to the prediction of the theory, environmental groups have proved to be a very powerful force in the policy arena. In the case of air quality management in the United States, for example, the efforts of these groups were clearly very important in obtaining at least some standards that appear to be more stringent than the economically efficient ones. Likewise in Europe, a variety of environmental groups have had great influence on measures for environmental protection. In several northern European countries, green interest groups have formed their own political parties and have become part of a governing majority.[14]

3.3. The range and interaction of environmental interest groups

The issue of interest groups is a complicated one in the context of environmental policy. At the level of pure theory, one can finesse this issue with a very general framework

---

[13] Boyer and Laffont (1999) take a somewhat different theoretical approach in which inefficient environmental choices are not the result of incomplete representation. In their framework, inefficient constitutional "constraints" on policy instruments arise from the limitations that these constraints impose on the capacity of politicians to distribute rents.

[14] More generally, Ostrom (1990, 2000) and others have enriched our understanding of organizing behavior, most notably with a wide range of empirical studies that find numerous instances where individuals, in fact, eschew free-rider opportunities and voluntarily band together for purposes of mutual advantage.

that includes n interest (or lobby) groups, each of which contributes money or other efforts to influence policy decisions. In more concrete applications, we find in some simple models cases of two opposing groups: environmental advocacy organizations in opposition to trade associations representing business interests. But the interplay of interests is often much more complex than this; in some instances, the same individuals may find that they wear different hats in that they may be part of several different interest groups. Moreover, there may exist a substantial number of groups with a stake in the choice of policy instruments and their level of stringency: environmental organizations, business interests, labor unions, administrative and trial lawyers, government agencies themselves, as well as the general public. For example, in one applied study of forest-service decision-making, Martin et al. (1996) identified seven separate interest groups who had a stake in oil and gas leasing on federal lands: the oil company seeking the leases, local environmental organizations, the local tourist industry, the local timber industry, local retail/wholesale merchants, local government units, and the federal government (Forest Service). In any particular application (and we shall examine several in the empirical section of our survey), the identification and characterization of the relevant interest groups is an essential and challenging part of the analysis.

The actual choice of regulatory instruments is thus an outcome of a process of interaction between policy-makers and the various interest groups that bring pressure to bear on these decisions. In a recent and intriguing approach to characterizing this process, Keohane, Revesz and Stavins (1998) have suggested that we envision a "political market" in which various interest groups provide a demand for environmental measures and where legislators, offering levels of support for various competing policy instruments, constitute the supply side of the market. In this framework, the legislative outcome (i.e., the choice of policy instrument) is determined by an equilibrium between the aggregated demands of the interest groups and the aggregate political-support supply function of the legislators. The "political currency" in this market encompasses not only monetary contributions but other forms of support for the legislator's re-election.

While the recent and more formal theoretical work on the political economy of environmental policy is impressive and promising, it would seem to be subject to certain limitations. The formal models typically treat government as a monolithic entity in the sense that they characterize "the" public decision-maker in terms of a single objective function. One might interpret this objective function as somehow representing the collection of public-sector "interests", but this is not fully satisfactory. As we have noted, the process that generates environmental outcomes is typically a complex and rich one that involves not only legislation, but administrative implementation at the bureaucratic level (often a complicated process in itself), and sometimes judicial review. It is hard to see how strictly formal modelling can ever capture the full range of this complexity. This most assuredly does not imply that these theoretical exercises are without value; we have already noted some of the important insights that they have provided. Rather it should remind us that case studies of particular environmental programs must go beyond the basic theory to consider the course of the regulatory program through the maze of the institutional structure that produces the ultimate outcome.

As a transition to the next section on empirical studies, it will be useful here to summarize in more concrete form some of the important insights that the existing positive theory of environmental regulation has provided into policy choices. As we noted earlier, a primary motivation for this literature has been the observed divergence of actual environmental policy from the efficient measures suggested by economic theory. And this has been done largely by showing that less efficient (so-called "command-and-control") instruments can, in quite realistic circumstances, be more beneficial to certain important interest groups than more efficient, incentive-based policy measures.[15]

1. As we have discussed, certain kinds of command-and-control policies can provide effective barriers-to-entry. It is, in fact, quite possible for such measures to raise the profits of existing firms above those that would exist in the absence of *any* environmental measures.[16] But even where this is not the case, it is likely that relatively inefficient control measures will be less costly to the polluting firm than a system of effluent taxes involving both control costs *and* tax payments (or a system of tradeable permits where firms must purchase permits to validate their residual waste discharges).

2. Environmental organizations may also look unfavorably on certain incentive-based instruments.[17] Many environmentalists object to such instruments on philosophical grounds, espousing the view that pollution taxes or systems of tradeable emission permits (TEP) involve "putting the environment up for sale" and are, for this reason, immoral and unacceptable. Environmental organizations must thus be careful about alienating their members by supporting such policy measures.[18] Moreover, environmental groups may have serious reservations about such policy instruments in practice. If, for example, the environmental authority sets too low a tax rate, then the environmental objective will not be realized. And it may not be an easy matter to raise tax rates where needed.

3. The literature has also clarified some important differences among various incentive-based instruments. In the United States, some environmentalists, for example, have shown much more interest in quantity instruments (systems of tradeable permits) than in price instruments (pollution taxes). They have found that a policy instrument which explicitly limits levels of polluting activities can more reliably achieve environmental goals than a price instrument the response to which is uncertain. And taking this a step further, polluting industries have been more receptive to such quantity instruments *if* the permits are allocated initially free of charge (through some kind of grandfathering

---

[15] See Keohane, Revesz and Stavins (1998) for an extended treatment of the various ways in which different policy instruments are likely to affect the welfare of the basic interest groups involved in environmental policy-making. Schneider and Volkert (1999) describe why various interest groups are likely to oppose incentive-based policy measures (and environmental policies in general!).

[16] Maloney and McCormick (1982) show the precise conditions under which regulation will increase profits.

[17] See Keohane, Revesz and Stavins (1998) for a good treatment of this.

[18] In the United States, Environmental Defense (formerly the Environmental Defense Fund or EDF) provides an interesting counter-example in which an environmental organization has actively supported a trading system for airborne sulfur emissions and other pollutants as well. Other advocacy groups have begun to follow the lead of Environmental Defense. See [Keohane, Revesz and Stavins (1998, p. 354)].

scheme) rather than auctioned off. In this case, rather than having to purchase permits to validate emissions, the firms receive without cost a valuable asset that can be used either to validate their own emissions or can be sold for a profit. Systems of tradeable permits with a free initial distribution can thus achieve support from various interest groups that may not be forthcoming for other forms of incentive-based instruments.[19]

4. Empirical studies of the political economy of environmental protection

There is a wide array of empirical work that explores the actual determination of environmental standards in different places and at different times. It is impossible to characterize this work in any very simple way, but one theme does emerge in nearly all these studies – namely, that actual environmental measures bear the imprint in various ways of the interest groups that have taken part in the debate and design of these measures. Even where, for example, very stringent policies have been adopted in response to environmental concerns, there are typically provisions in the legislation (or subsequent implementing regulations) to accommodate the particular interests of those who must bear the costs. To take one broad case, Ekins and Speck (1999), in their comprehensive survey of the use of environmental taxes in Europe, find that in nearly all the European nations, the implementation of such taxes includes a wide array of significant exemptions and tax relief for particular sectors – often to allay concerns about the adverse effects of the taxes on competitiveness.

The first environmental application of the formal revealed preference approach was that of Magat, Krupnick, and Harrington (1986). They analyzed the technologies that a very diverse set of industries were required to install by the U.S. Environmental Protection Agency to control emissions of two water pollutants: "total suspended solids" (TSS) and "bio-chemical oxygen demand" (BOD). This work was prompted by the observation that in some industries (or sub-categories within an industry), some firms were required to spend a great deal per unit of BOD or TSS removed, while others were asked to spend much less. Among many other things in this comprehensive study, the analyses suggested that the EPA – at least in its standard setting for TSS and BOD – gave very little weight to economic efficiency and appeared not to be influenced by either industry participation in the rule-making process or by the number of plants that might be shut down as a result of the control requirements. What did appear to matter was the strength of the trade association that represented the affected industries and also the profitability of the industries.

As we saw in the preceding section on positive theories of environmental regulation, a number of recent contributions envision an objective function for the policy-maker

---

[19] However, as recent theoretical work has made clear, the failure of such systems to raise revenues that can be used to reduce other taxes can seriously undermine their efficiency properties. This issue has arisen in the so-called "double-dividend" debate. See, for example, Parry and Oates (2000).

(or legislator) that consists of a weighted average of two kinds of arguments: the contributions of the various interest groups *and* a term reflecting the general social welfare. In this regard, it is interesting to find in several recent empirical studies that measures of social welfare have a significant impact on regulatory outcomes as well as variables indicating the influence of specific interest groups. For example, in an econometric study of pesticide regulation in the United States, Cropper et al. (1992) found that the probability that the U.S. EPA disallowed the continued use of a particular ingredient used in pesticides depended significantly on the estimated benefits and costs of such a restriction; more specifically, both the economic benefits of the ingredient to producers and the degree of health risk it posed were significant determinants of EPA decisions. The explanatory power of the model increased markedly with the addition of some interest group variables (representing business and environmental groups), suggesting that both net social value and interest group pressures mattered in decisions concerning discontinuing the use of particular ingredients.[20]

On a quite different issue, Hoagland and Farrow (1996) likewise found that the planning decisions made by the U.S. Secretary of the Interior concerning the sale of leases for offshore gas and oil drilling depended not only on political variables, but also on the estimated net social value attached to the various sites. And, in a third case, Hird (1990), in a study of Superfund expenditures to clean up hazardous waste sites in the United States, found that the chief determinant of the pace and funding of cleanup at particular sites was the site's hazard ranking – a measure of its public health risk, with only more modest influence from interested legislators. Such studies thus provide some support for the view that many environmental policy decisions represent a kind of amalgam of group interests *and* general social welfare maximization.

In certain of the theoretical political-support models that we have examined, the formulation is one in which policy decisions depend significantly on financial contributions from interest groups. There is some evidence to support this view in studies of environmental policies. Coates (1996), for example, examined the impact of campaign contributions on the voting behavior of members of the U.S. House of Representatives on a set of amendments to wilderness-designation legislation for federal lands in California and Oregon. Such legislation effectively protects these lands from commercial development so that the issue is essentially one of jobs versus wilderness protection. Coates' estimation of a series of probit equations suggests that campaign contributions had some effect on legislators' positions and voting patterns on the issue. There are some tricky issues of interpretation here, however. As Stratmann (1991) has pointed out, contributions may have the purpose of helping the re-election of a candidate whose position coincides with that of the contributor or they may have the intent of changing the position of a legislator with a view opposing that of the contributor. At any rate,

---

[20] Nadai (1996) provides a more descriptive history of pesticide regulation in the European Union. He finds that various interest groups have been deeply engaged in the evolution of EU policy measures. This is a "case where interest groups have clearly influenced the final content of a regulation" [Nadai (1996, p. 71)]. There is no explicit investigation here of the importance of net social value in the decision process.

Coates finds that contributions had their intended effect although these effects were not sufficiently large to alter the overall outcome.

Interest groups can also form along regional lines. And there is some evidence to suggest that support for certain environmental measures has, to some extent, reflected the economic self-interest of specific regions. Two studies in the U.S., for example, have found that provisions in the Clean Air Act that were especially costly in rapidly growing areas received disproportionate support from areas that stood to lose economic activity to these areas. Crandall (1983) found evidence in Congressional voting patterns that reflected much stronger support in northern jurisdictions for measures that placed more stringent control requirements on new sources and that limited the growth in pollution in relatively clean areas; such measures resulted in relatively higher control costs in the more rapidly growing southern and western areas in the U.S. Pashigian (1985) likewise found some support for the "locational competition hypothesis". His analysis suggests (like that of Crandall) that the policy of limiting incremental pollution in clean areas (known as the "prevention of significant deterioration") derived significant support from those regions that would gain a competitive advantage from the measure.

Finally, we note that our treatment has been wholly in the context of democratic systems where interest groups can express their preferences through various political processes. The setting is obviously quite different in more autocratic systems. Congleton (1992) has, in fact, found that this is an important distinction for purposes of environmental management. He suggests that we should expect more stringent environmental regulations in democratic than in authoritarian regimes, and his findings support this proposition. More specifically, his estimates indicate that democratic countries were much more likely to support stringent limitations on CFC emissions under the Montreal Protocol and actually to reduce emissions of CFC gases. Likewise, Murdoch and Sandler (1997) find that the extent of political and civil freedoms had a positive impact on reductions in CFC emissions in the late 1980s. Political systems clearly influence the extent of environmental protection.

5. Environmental federalism

5.1. The assignment of environmental management to different levels of government

As we have discussed, institutional structure is of central importance in the process of environmental decision-making. One key dimension of this structure is the vertical division of policy-making responsibilities among the different levels of government. This brings us to the issue of the respective roles in theory and in practice of central and decentralized public agencies, both in the design and the implementation of environmental measures.[21]

---

[21] For three volumes containing collections of useful papers on environmental policy-making in a federal system, see Braden, Folmer and Ulen (1996), Braden and Proost (1997), and Proost and Braden (1998). They

Here again, there is a body of normative theory from which to derive some basic precepts. Within the field of public economics, the subfield of "fiscal federalism" addresses this set of issues, in particular the distribution of functions among levels of government [for example, see Oates (1999)]. The term "fiscal" here is unduly restrictive; in fact, the so-called principles of fiscal federalism extend to regulatory matters as well. From a normative perspective, the issue here is one of aligning specific responsibilities and regulatory instruments with the different levels of government so as best to achieve our social objectives.

The basic idea that runs through this literature is that the responsibility for providing a particular service should be placed with the smallest jurisdiction whose boundaries encompass the various benefits and costs associated with the provision of the service.[22] By structuring decision-making in this way, the levels of public services can be tailored to the specific circumstances – the tastes of residents, the costs of production, and other peculiar local conditions – of each jurisdiction. It is straightforward to show that the pattern of outputs that emerges from allowing efficient decentralized choice in this way increases social benefits relative to a centralized solution that imposes more uniform levels of outputs across all jurisdictions.[23]

This "principle" thus establishes a general presumption in favor of decentralized decisions where the benefits and costs are limited primarily to a particular jurisdiction or locality. Moreover, this general prescription has received widespread acceptance. In Europe, the case for decentralization is known as the "principle of subsidiarity"; as such, it is explicitly integrated into the Maastrict Treaty for European Union.[24] In the U.S., it is recognized more colloquially as an aversion to the "one size fits all" approach.

From this perspective, we can envision a system of environmental policy-making in which the central government sets standards and oversees measures to address explicitly national pollution problems and intervenes in cases where (like acid rain) polluting activities in one jurisdiction impose substantial damages elsewhere. In addition, the

---

include both theoretical and empirical studies, several of which compare experiences in the U.S. and Western Europe with multi-level environmental management.

[22] In a more realistic sense, it would probably be better to say that responsibility should be placed with the smallest jurisdiction that spatially encompasses the lion's share of the benefits and costs. There are nearly always some small benefits and costs that will escape over jurisdictional boundaries because of people passing through or perhaps even some existence values that accrue elsewhere. If these are large, of course, then a more encompassing presence in environmental decision-making is called for.

[23] For a formal treatment of this proposition (known as the "Decentralization Theorem") see Oates (1972, Chapter 2, and 1997).

[24] The principle of subsidiarity is broadly based and basically states that the responsibility for addressing a particular public issue should rest with the lowest level of government capable of handling the problem. More explicitly, the Maastricht Treaty in 1992 allows action at the Union level "only and insofar as the objectives of the proposed action cannot be sufficiently achieved by the Member States and can, therefore, by reason of the scale or effects of the proposed action, be better achieved by the Community" (E.C. Treaty, Art. 3B). It is interesting that the intellectual source of the principle of subsidiarity is Papal social teaching; Pius XI held it to be morally wrong "... to assign to a larger and higher society what can be performed successfully by smaller and lower communities" [(quote taken from Inman and Rubinfeld (1998, p. 545)].

central government would provide basic support for research and the dissemination of information on environmental problems, since these are activities that benefit everyone. At the same time, decentralized levels of government would set their own standards and establish their own programs for managing those dimensions of environmental quality that are primarily contained within their own boundaries (for instance, the standards that a local landfill might have to meet).

5.2. The issue of a "race to the bottom"

This basic view of environmental federalism has, however, been the subject of a fundamental challenge, both at the theoretical and policy levels. The source of this challenge is the claim that "local" officials, in their eagerness to encourage new business investment and economic growth, will set excessively lax environmental standards to hold down the costs of pollution control for existing and prospective firms. The result will be a "race to the bottom" with inefficiently high levels of polluting activities.

This turns out to be a quite complicated, as well as contentious, issue. There has emerged a large theoretical literature that explores interjurisdictional competition and its welfare implications.[25] It has two sides. It is not difficult, on the one hand, to describe a world in which competition among governments for new business investment is welfare-enhancing, where it leads to Pareto-efficient choices involving, among other things, levels of local environmental quality. Oates and Schwab (1988, 1991, 1996) have constructed a series of such models in which local jurisdictions compete for mobile firms both to increase levels of wage income and to enlarge the local tax base. These models generate a set of "invisible-hand" outcomes in which such competition induces local decision-makers to select efficient levels of local outputs (including environmental quality).

These models are, however, quite demanding in terms of some essential conditions: governments are small in the sense of being price takers in a large capital market and not engaging in strategic behavior in response to the policy choices of other competing governments; they have access to the full range of policy instruments they need for efficient fiscal and regulatory decisions; and public outputs are wholly self-contained – they have no external effects on other jurisdictions. Within such a setting, it is not difficult to construct a quite rich model in which governments compete with one another for mobile firms making use of expenditure, tax, and environmental policy instruments, and where the outcome for all these policy measures is Pareto-efficient. Competition in such a framework is efficiency-enhancing; in a kind of analogue to the case of perfect competition in the private sector, it guides public decisions into efficient outcomes.

On the other hand, if any of these conditions are relaxed (often in quite realistic ways), the efficiency properties of these models of interjurisdictional competition can be

---

[25] Wilson (1996, 1999) provides two excellent surveys and assessments of this literature. The former paper (1996) focuses explicitly on the race-to-the-bottom issue in environmental management. In a recent book, Wellisch (2000) presents a thorough and rigorous review of the fiscal competition literature.

compromised. Governments may not, for example, have access to the tax and regulatory instruments they need for efficient public management. An important line of work in the fiscal competition literature examines the case where "local" governments can tax only mobile capital so that all public services must be financed by a tax on local firms [Zodrow and Mieszkowski (1986), Wilson (1986), Wildasin (1989)]. In this setting, a kind of fiscal externality arises in that local officials do not take into account the impact that their policy decisions have on tax bases in other jurisdictions. The typical outcome in such models is one in which public services are underprovided. Distortions can also occur where governments are large in the sense of having an impact on the price of capital – or where they behave in strategic ways in the setting of policy parameters.

Of direct relevance here is a public-choice setting in which public agencies have their own set of objectives, including such things as budget maximization. If local officials seek to enlarge the size of the public sector, it is straightforward to show that they will tend to set overly lax environmental measures in order to attract more capital and enlarge the local tax base [Oates and Schwab (1988)]. More generally, there exists a large and rich literature, much of it drawing on game-theoretic models, that explores these issues and describes the sorts of allocative distortions that competition can generate in a variety of settings [Wellisch (2000)].

The theory of environmental federalism thus leaves us in an uneasy position. While there are clearly ways in which economic competition among governments can encourage good fiscal and environmental decisions, there are also circumstances where things can go awry. And the important issue here is really one of magnitude: how large are the kinds of distortions that this literature describes? If they represent only small deviations from efficient outcomes, they may not be of much consequence. The problem is that we have little evidence on this. There is plenty of evidence that governments actively engage in various forms of economic competition. But this really does not address the issue. Such competition *may*, as we have discussed, be healthy in the sense of encouraging good public decisions. Thus, the discovery that governments introduce policy measures to influence industrial location really does not tell us much about any distortions that may be present or their magnitude [Courant (1994)].

5.3. Environmental federalism in practice: Some evidence

When we look at the actual practice of environmental decision-making, we find that it has often tended to be quite centralized. In the United States, for example, the Clean Air Act Amendments of 1970, one of the cornerstones of federal legislation emerging from the environmental movement of the 1960s, directs the central government [more precisely, the Environmental Protection Agency (EPA)] to set uniform standards for air quality – standards that must be met (or exceeded) in every part of the country. Moreover, these standards (in the form of maximum allowable pollutant concentrations) are to be set so as to protect the health of the most sensitive residents with little regard to their cost or other mitigating circumstances. In addition, Congress itself established tailpipe emissions standards that were to apply uniformly to all new vehicles sold in the U.S. (with the exception of California, which was granted the right to set more stringent standards). In contrast, it is interesting that two years later, the U.S. Congress introduced sweeping measures for water quality management that call for the states to set their own standards for water quality. But, at the same time, Congress directed the EPA to issue technology-based discharge standards for all publicly-owned sewage-treatment plants and for virtually all industrial sources of water pollution. There is thus a real ambivalence that runs through U.S. environmental federalism.

Likewise, the emerging European Union is struggling with the extent to which environmental measures should be harmonized across Europe and the extent to which such decisions should remain with the member states [Leveque (1996), Pfander (1996)]. As we mentioned, there is a general recognition of the principle of subsidiarity, but various types of arguments have convinced many that Union-wide standards are needed to address a range of environmental problems. In particular, there has been support in Europe for harmonization for purposes of encouraging the development of a common market. This makes some sense for the case of product standards. Without such standards, one member state may exclude products of others if they do not meet its own health, safety, and environmental standards. From this perspective, a harmonized set of standards can facilitate the free movement of goods and services within Europe. This case is, however, much less compelling for so-called process standards that relate to the conditions under which products are manufactured. The need for setting uniform standards for ambient environmental quality or uniform emissions standards at the European Union level is much less clear. In fact, proponents of such measures in Europe, like their counterparts on the other side of the Atlantic, have relied heavily on "race-to-the-bottom" arguments.

The Council of the European Union has extensive powers for environmental governance. It is authorized to issue "directives" to the member states on environmental matters, and, according to Article 189 of the Maastricht Treaty, such directives "shall be binding, as to the result to be achieved, upon each Member State to which it is addressed, but shall leave to the national authorities the choice of form and methods". The Treaty thus gives the Union the power to command the member states to meet centrally defined standards for environmental quality.[26] However, the Council's decisions still require de facto unanimity so that its powers are, in fact, quite circumscribed.[27] Moreover, there is evidence that the member states have not always complied with these directives; the problem of enforcement remains a basic concern.

In terms of existing policy, there seems to be a good deal of confusion on both sides of the Atlantic. The degree of decentralization of environmental management differs significantly for various pollutants, often with little justification in terms of any apparent principles. Moreover, in Europe, attempts to centralize standard-setting have been frequently undercut by the reluctance of member states to comply with the directives.

---

[26] This structure bears some similarity to the Clean Air Act in the United States, where the EPA sets the air-quality standards to be attained, and the states are then directed to develop plans to reach these standards. For an excellent comparison of environmental federalism in Europe and the United States, see Pfander (1996).

[27] Braden and Proost (1996) describe and assess this issue in the context of a comparison of policies in Europe and in the U.S. for controlling tropospheric ozone.

Environmental federalism thus remains a highly contentious issue, both in terms of theory and practice. The case for centralization relies heavily on the adverse effects of competition with a resulting race to the bottom. But does such a "race" really exist? The support for an affirmative answer to this question is largely anecdotal; to our knowledge, there is little systematic evidence in its support.

On the other side, there is some, if admittedly limited, evidence suggesting that there is no widespread race to the bottom. If there were fierce and distorting economic competition, we might expect to find few instances in which decentralized jurisdictions introduce environmental measures that are more stringent than the centrally determined standards. Yet in the United States at least, we find plenty of instances where states have introduced regulations for the control, for example, of pesticides and hazardous wastes that go well beyond federal requirements. The one instance where the states have not gone beyond federal standards relates to the ambient air quality standards under the Clean Air Act. But our sense of this case is that the legislation calls for such stringent measures (standards so tough that there are no adverse health effects from air pollution *irrespective* of the costs of control) that few would want anything tougher.[28]

Using another approach, three recent studies in the U.S. have examined the impact on environmental outcomes of the devolution of responsibilities during the Reagan years for certain aspects of environmental management. Although this covers an admittedly short time span, it is interesting that none of the three studies finds any evidence of a race to the bottom. List and Gerking (2000), using state-level data, have estimated a fixed-effects model that looks at both levels of environmental quality and abatement expenditures. They find no evidence of any deterioration in environmental quality or decline in abatement efforts; on the contrary, they find some instances of improvements leading them to conclude that "... in this instance, the race to the bottom did not appear to materialize". In another assessment of the experience in the Reagan era, Millimet (2000) has studied airborne emissions of sulfur dioxide and industry spending on pollution abatement. He finds that actual emissions were lower and abatement spending higher than forecast by his model, suggesting a race to the top rather than the reverse. Finally, Fredriksson and Millimet (2002) likewise find little impact of Reagan devolution on environmental policy; their results, in fact, provide evidence for a strategic race to the top among U.S. states.[29]

---

[28] In this context, Goklany (1999) provides a provocative account of the history of air quality management in the U.S. in which he documents the sometimes quite extensive and effective measures introduced by state and local governments prior to federal intervention with the Clean Air Act Amendments of 1970.

[29] In another piece that extends the Grossman and Helpman (1994) model of competing interest groups, Fredriksson and Gaston (2000) find that, in a model where capital is perfectly mobile across domestic jurisdictions but immobile internationally, centralized and decentralized environmental management yield identical environmental regulations. In their model, the stringency of environmental measures is independent of institutional structure. This interesting result emerges from an internalization of the social costs of emissions through differing efforts (i.e., campaign contributions) of capital owners, environmentalists, and labor under the two regimes. The authors suggest that EU recycling policies support this result in that the centralization of these measures neither increased nor decreased their stringency, but represented a rough average of the policies in member countries [Paul (1994/95)].

The efficiency gains from environmental measures that are tailored to local circumstances may be quite substantial. In one study, Dinan, Cropper and Portney (1999) have examined the case of drinking water standards in the United States. This is an interesting case for two reasons. First, the purity of local drinking water (with the exception of a couple of contaminants) is mainly of interest to local users; any adverse effects manifest themselves only after prolonged exposure. We can thus reasonably characterize this as a local public good. Second, there exist large economies of scale in the treatment of drinking water such that the costs of additional purification per household can be much higher in smaller, than in larger, jurisdictions. The Safe Water Drinking Act of 1974 instructed the U.S. EPA to set national standards for drinking water. But as the authors find, a set of uniform national standards can be quite inefficient. They examine the case of a particular class of contaminants, a class of radionuclides known as "adjusted gross alpha emitters". And they find that the quite stringent EPA standard can be justified on benefit–cost grounds for only the very largest districts where the costs can be shared among several hundred thousand households. For smaller districts, the standard has (often quite large) negative benefits.

There remains, in our view, a strong case for "localized" environmental management where the benefits and costs of such measures are themselves localized. The potential efficiency gains may be large; moreover, there is little evidence of a destructive "race-to-the-bottom" in environmental regulation.[30] At the same time, we stress that there remains a crucial role for central government. In addition to addressing "national" pollution problems (where emissions spillover across jurisdictional boundaries), a central environmental agency can provide essential information and research support. We can envision a system of environmental management in which a very active central agency not only supports research into environmental issues, but offers guidance to "local" authorities in the form of recommended standards and levels of treatment that effectively lay out the menu of choices available to local decision-makers. In such a setting, local agencies could then select the parameters for environmental programs that best suit their local constituencies.

6. Concluding observations on some recent trends

In concluding our survey of the political economy of environmental policy, we want to call attention to the quite striking and fascinating evolution of environmental management over the past few decades – and, in particular, to the role that economics plays in the design of new policy measures. In the early days of the environmental movement in the 1960s and early 1970s, there existed a strong disposition toward command-and-control approaches to regulation. Under these approaches, environmental agencies set standards with little regard to their economic implications and then issued directives

---

[30] For an excellent review and critical assessment of this issue, see Revesz (2001).

to polluters limiting their levels of waste emissions and often specifying the control technology. The economic prescriptions for the setting of standards by balancing benefits and costs at the margin and for the use of incentive-based policy instruments to achieve these standards were largely ignored on both sides of the Atlantic. But things have changed in some quite dramatic ways.

First, new procedures have been introduced requiring the systematic measurement and sometimes consideration of the benefits and costs associated with policy measures. Pearce (1998) provides a careful description of the evolution of environmental appraisal procedures in the European Union. The Fifth Environmental Action Plan, promulgated in 1992 by the European Community calls explicitly for ". . . the development of meaningful cost/benefit analysis methodologies and guidelines in respect of policy measures and actions which impinge on the environment and natural resource stock" [Pearce (1998, p. 490)]. Pearce finds that "Since the early 1990s formal appraisal procedures have improved and are applied more widely" [Pearce (1998, p. 498)].

Likewise, in the United States, various Presidential Executive Orders have called for benefit–cost analyses of all major environmental regulations.[31] The most widely publicized of these was Reagan's Executive Order 12291 in 1981 which required not only that benefit–cost studies be carried out for all major regulatory programs but that, to the extent permitted by law, regulations be undertaken only if the benefits exceed the costs. The U.S. EPA has been making benefit–cost studies of environmental regulations since the mid-1970s, but has been limited in applying them to actual decisions by various statutes. We can do no more than note that in key parts of various environmental laws in the U.S., regulators are prohibited from even considering costs in setting ambient standards, while under other statutes, they are almost required to strike a balance between benefits and costs in standard setting There is little consistency here. As Morgenstern (1997) puts it, "Various statutes forbid, inhibit, tolerate, allow, invite, or require the use of economic analysis in environmental decisionmaking" [Morgenstern (1997, p. 20)].

Second, there has been increased interest in, and some use of, incentive-based policy instruments for the attainment of environmental standards.[32] The economic prescriptions for policy measures that were essentially ignored in the early days of environmental legislation are getting a much wider hearing in the current policy arena and are actually appearing in practice. The use of taxes to discourage polluting activities and the introduction of systems of tradeable emissions allowances are now more than just ideas appearing in textbooks on the subject.

What accounts for these modifications in the direction of environmental policy? This is not an easy question to answer.[33] To some extent, the deficiencies associated with

---

[31] Hahn (1998) provides a concise and insightful history and assessment of the U.S. experience with benefit–cost analyses of environmental and other forms of regulation in the U.S.

[32] For an excellent treatment of this issue, see Chapter 9 (by Stavins). Hahn (1989) provides an insightful history of the early efforts in the U.S. and Europe to introduce incentive-based policy instruments for environmental protection.

[33] Oates (2000) tries to answer this question in the context of the United States. It is clear in retrospect that some serendipitous events helped set the process in motion.

command-and-control techniques have become more apparent over time, and the resulting dissatisfaction has stimulated the search for alternatives. In this context, the growing number of economists who have turned their attention to environmental issues have played an important role both in educating policy-makers and in taking a more active role in the design and implementation of feasible policy measures. As we move farther down the path of environmental protection, we are finding that yet tighter controls on polluting activities are becoming increasingly expensive. And this puts a higher premium on finding efficient means for regulating them. But more generally, the last two decades have been a period of renewed "faith in market forces" [Kay (1988)] in which perceived "government failures" have ushered in a setting in the Western world that is much more receptive to market-based forms of regulation.

An intriguing anomaly in this process of evolution is the rather different paths taken on the two sides of the Atlantic. In Europe, the tendency has been to turn to environmental taxes to provide incentives for reducing pollution.[34] In contrast, regulators in the United States have adopted systems of tradeable emissions allowances to control airborne emissions of sulfur dioxide, as well as some other air pollutants and other forms of damaging emissions.[35] The reasons for this divergence in the choice of policy instruments is not altogether clear. But it does seem to represent to some extent the rather extreme aversion to new forms of taxation in the United States and perhaps some historical accidents as well [Hahn (1989), Oates (2000)].[36]

At any rate, the atmosphere for environmental regulation has changed quite dramatically. There is now attention given to incentive-based policy measures in many countries around the world. Indeed, there is even serious consideration being given to systems of tradeable allowances on a global scale to address the problem of climate change [Hahn and Stavins (1995)]. The various interest groups have found that there are ways in which incentive-based policy instruments can be constructed and implemented that make them suitable for their support (especially when their cost-minimizing properties can be used to fight for a more stringent standard than would be politically possible

---

[34] For useful treatments of the European experience with environmental taxes, see Brannlund and Gren (1999) on the Scandinavian countries, Smith (1995) on Britain and Germany, and Dijkstra (1999) on the Netherlands. The papers in Bluffstone and Larson (1997) provide an extensive description and analysis of the use of taxes and charges for pollution in the transition economies of Eastern Europe.

[35] Tietenberg (1985) provides a careful description and assessment of the U.S. Emissions Trading Program, a program that evolved in interesting ways to facilitate the trading of emissions allowances within air quality control regions. More recently, Ellerman et al. (2000) have described and analyzed the U.S. experience with sulfur allowance trading on a national scale to address the acid-rain problem. A smoothly functioning and efficient market has developed for the trading of sulfur allowances.

[36] There are some exceptions. The U.S. did introduce at the federal level a tax on CFC's; in addition, there is some use at state and local levels of unit charges on municipal solid waste and peak-period tolls on some freeways. Wallart (1999, p. 104) suggests interestingly that this difference in approaches may be in part a cultural phenomenon. He notes the impact in America of the Coasian perspective which suggests that distortions from externalities result from the non-existence of markets. In contrast, in Europe, the Pigouvian tradition is the predominant one with its focus of the malfunctioning of markets in the presence of externalities and the need for corrective taxes.

under a command-and-control approach). In addition, as we saw in several empirical studies, the relevant benefits and costs of environmental measures are by no means without their influence on policy decisions; with the accumulating experience with environmental programs, there seems to be a discernible trend toward more efficient decision-making for environmental protection.

Acknowledgements

The authors thank Emily Arnow for her valuable research assistance. For many helpful comments on earlier drafts of the chapter, they are grateful to Amy Ando, David Evans, Per Fredricksson, Lawrence Goulder, Karl-Göran Mäler, Bernardo Mueller, Ian Parry, Stef Proost, and Jeffrey Vincent.

References

Ackerman, B.A., and W.T. Hassler (1981), Clean Coal/Dirty Air (Yale University Press, New Haven, CT).

Aidt, T.S. (1998), "Political internalization of economic externalities and environmental policy", Journal of Public Economics 69:1–16.

Becker, G. (1983), "A theory of competition among pressure groups for political influence", Quarterly Journal of Economics 98:371–400.

Bergstrom, T.C. (1979), "When does majority rule supply public goods efficiently?", Scandinavian Journal of Economics 8:216–227.

Bergstrom, T.C., and R.P. Goodman (1973), "Private demands for public goods", American Economic Review 63:280–296.

Black, D. (1948), "On the rationale of group decision making", Journal of Political Economy 56:23–34.

Bluffstone, R., and B. Larson (1997), Controlling Pollution in Transition Economies: Theories and Methods (Elgar, Cheltenham).

Borcherding, T.E., and R.T. Deacon (1972), "The demand for the services of non-federal governments", American Economic Review 62:891–901.

Boyer, M., and J. Laffont (1999), "Toward a political theory of the emergence of environmental incentive regulation", Rand Journal of Economics 30:137–157.

Braden, J., H. Folmer and T. Ulen (eds.) (1996), Environmental Policy with Political and Economic Integration: The European Union and the United States (Elgar, Cheltenham).

Braden, J., and S. Proost (1996), "Economic assessment of policies for combating tropospheric ozone in Europe and the United States", in: J. Braden, H. Folmer and T. Ulen, eds., Environmental Policy with Political and Economic Integration: The European Union and the United States (Elgar, Cheltenham) 365–413.

Braden, J., and S. Proost (1997), The Economic Theory of Environmental Policy in a Federal System (Elgar, Cheltenham).

Brannlund, R., and I.M. Gren (1999), Green Taxes: Economic Theory and Empirical Evidence from Scandinavia (Elgar, Cheltenham).

Buchanan, J.M., and G. Tullock (1975), "Polluters' profits and political response: direct controls versus taxes", American Economic Review 65:139–147.

Coase, R.H. (1960), "The problem of social cost", Journal of Law and Economics 3:1–44.

Coates, D. (1996), "Jobs versus wilderness areas: the role of campaign contributions", in: R. Congleton, ed., The Political Economy of Environmental Protection (University of Michigan Press, Ann Arbor, MI) 69–96.

Congleton, R. (1992), "Political Institutions and pollution control" Review of Economics and Statistics 74:412–421.

Congleton, R. (1996), The Political Economy of Environmental Protection (University of Michigan Press, Ann Arbor, MI).

Courant, P. (1994), "How would you know a good economic policy if you tripped over one?", National Tax Journal 47:863–881.

Crandall, R.W. (1983), Controlling Industrial Pollution: The Economics and Politics of Clean Air (Brookings Institution, Washington, DC).

Cropper, M.L., W.N. Evans, S.J. Berardi, M.M. Ducla-Soares and P.R. Portney (1992), "The determinants of pesticide regulation: a statistical analysis of EPA decision making", Journal of Political Economy 100:175–197.

Dewees, D. (1983), "Instrument choice in environmental policy", Economic Inquiry 21:53–71.

Dijkstra, B. (1999), The Political Economy of Environmental Policy: A Public Choice Approach to Market Instruments (Elgar, Cheltenham).

Dinan, T., M. Cropper and P. Portney (1999), "Environmental federalism: welfare losses from uniform national drinking water standards", in: A. Panagariya, P. Portney and R. Schwab, eds., Environmental and Public Economics: Essays in Honor of Wallace E. Oates (Elgar, Cheltenham) 13–31.

Downs, A. (1957), An Economic Theory of Democracy (Harper and Row, New York).

Ekins, P., and S. Speck (1999), "Competitiveness and exemptions from environmental taxes in Europe", Environmental and Resource Economics 13:369–396.

Ellerman, A.D., P. Joskow, R. Schmalensee, J. Montero and E. Bailey (2000), Markets for Clean Air: The U.S. Acid Rain Program (Cambridge University Press, Cambridge).

Elliott, E.D., B.A. Ackerman and J.C. Millian (1985), "Toward a theory of statutory evolution: the federalization of environmental law", Journal of Law, Economics, and Organization 1:313.

Fredriksson, P. (1997), "The political economy of pollution taxes in a small open economy", Journal of Environmental Economics and Management 33:44–58.

Fredriksson, P., and N. Gaston (2000), "Environmental governance in federal systems: the effects of capital competition and lobby groups", Economic Inquiry 38:501–514.

Fredriksson, P., and D. Millimet (2002), "Strategic interaction and the determination of environmental policy across U.S. states", Journal of Urban Economics 51:101–122.

Goklany, I. (1999), Clearing the Air: The Real Story of the War on Air Pollution (Cato Institute, Washington, DC).

Goulder, L., I. Parry and D. Burtraw (1997), "Revenue-raising versus other approaches to environmental protection: the critical significance of preexisting tax distortions", Rand Journal of Economics 28:708–731.

Groenewegen, P. (1987), "Political economy and economics", in: J. Eatwell et al., eds., The New Palgrave: A Dictionary of Economics, Vol. 3 (Macmillan & Co., London) 904–907.

Grossman, G.M., and E. Helpman (1994), "Protection for sale", American Economic Review 84:833–850.

Hahn, R. (1989), "Economic prescriptions for environmental problems: how the patient followed the doctor's orders", Journal of Economic Perspectives 3:95–114.

Hahn, R. (1990), "The political economy of environmental regulation: towards a unifying framework", Public Choice 65:21–47.

Hahn, R. (1998), "Government analysis of the benefits and costs of regulation", Journal of Economic Perspectives 12:201–210.

Hahn, R., and R. Stavins (1995), "Trading in greenhouse permits: a critical examination of design and implementation issues", in: H. Lee, ed., Shaping National Responses to Climate Change (Island Press, Washington, DC) 177–217.

Hillman, A. (1989), The Political Economy of Protection (Harwood Academic, Zürich, Switzerland).

Hillman, A., and H. Upsprung (1992), "The influence of environmental concerns on the political determination of international trade policy", in: R. Blackhurst and K. Anderson, eds., The Greening of World Trade Issues (Harvester Wheatsheaf, New York) 195–220.

Hird, J. (1990), "Superfund expenditures and cleanup priorities: distributive politics or the public interest?", Journal of Policy Analysis and Management 9:455–483.

Hoagland, P., and S. Farrow (1996), "Planning versus reality: political and scientific determinants of outer continental shelf lease sales", in: R. Congleton, ed., The Political Economy of Environmental Protection (University of Michigan Press, Ann Arbor, MI) 145–166.

Inman, R.P., and D.L. Rubinfeld (1998), "Subsidiarity and the European Union", in: P. Newman, ed., The New Palgrave Dictionary of Economics and the Law, Vol. 3 (St. Martins Press, New York) 545–551.

Kay, J.A. (1988), "Faith in market forces", in: H. Stein, ed., Tax Policy in the Twenty-First Century (Wiley, New York) 203–220.

Kelman, S. (1981), What Price Incentives? Economists and the Environment (Auburn House, Boston, MA).

Keohane, N.O., R. Revesz and R.N. Stavins (1998), "The choice of regulatory instruments in environmental policy", Harvard Environmental Law Review 22:313–367.

Leidy, M.P., and B.M. Hoekman (1996), "Pollution abatement, interest groups, and contingent trade policies", in: R. Congleton, ed., The Political Economy of Environmental Protection (University of Michigan Press, Ann Arbor, MI) 43–68.

Leveque, F. (1996), Environmental Policy in Europe: Industry Competition and the Policy Process (Elgar, Cheltenham).

List, J., and S. Gerking (2000), "Regulatory federalism and U.S. environmental policies", Journal of Regional Science 40:453–471.

Magat, W.A., A.J. Krupnick and W. Harrington (1986), Rules in the Making: A Statistical Analysis of Regulatory Agency Behavior (Resources for the Future, Wasington, DC).

Maloney, M.T., and R.E. McCormick (1982), "A positive theory of environmental quality regulation", Journal of Law and Economics 25:99–123.

Martin W.E., D.J. Shields, B. Tolwinski and B. Kent (1996), "An application of social choice theory to U.S.D.A. forest service decision making", Journal of Policy Modeling 18:603–621.

McFadden, D. (1975), "The revealed preferences of a government bureaucracy: theory", Bell Journal of Economics and Management Science 6:401–416.

McFadden, D. (1976), "The revealed preferences of a government bureaucracy: empirical evidence", Bell Journal of Economics and Management Science 7:55–72.

Millimet, D. (2000), "Assessing the empirical impact of environmental federalism", unpublished paper.

Morgenstern, R. (1997), Economic Analyses at EPA: Assessing Regulatory Impact (Resources for the Future, Washington, DC).

Mueller, D.C. (1989), Public Choice II (Cambridge University Press, London).

Murdoch, J., and T. Sandler (1997), "The voluntary provision of a pure public good: the case of reduced cfc emissions and the Montreal Protocol", Journal of Public Economics 63:331–350.

Nadai, A. (1996), "From environment to competition – the EU regulatory process in pesticide regulation", in: F. Leveque, ed., Environmental Policy in Europe (Elgar, Cheltenham) 53–73.

Oates, W.E. (1972), Fiscal Federalism (Harcourt, Brace, Jovanovich, New York).

Oates, W.E. (1996), "Estimating the demand for public goods: the collective choice and contingent valuation approaches", in: D. Bjornstad and J. Kahn, eds., The Contingent Valuation of Environmental Resources (Elgar, Aldershot) 211–230.

Oates, W.E. (1997), "On the welfare gains from fiscal decentralization", Journal of Public Finance and Public Choice 2–3:83–92.

Oates, W.E. (1999), "An essay on fiscal federalism", Journal of Economic Literature 37:1120–1149.

Oates, W.E. (2000), "From research to policy: the case of environmental economics", University of Illinois Law Review 2000:135–153.

Oates, W.E., and R.M. Schwab (1988), "Economic competition among jurisdictions: efficiency enhancing or distortion inducing?", Journal of Public Economics 35:333–354.

Oates, W.E., and R.M. Schwab (1991), "The allocative and distributive implications of local fiscal competition", in: D. Kenyon and J. Kincaid, eds., Competition Among States and Local Governments (Urban Institute, Washington, DC) 127–145.

Oates, W.E. and R.M. Schwab (1996), "The theory of regulatory federalism: the case of environmental management", in: W. Oates, ed., The Economics of Environmental Regulation (Elgar, Cheltenham) 319–331.

Olson, M. (1965), The Logic of Collective Action (Harvard University Press, Cambridge, MA).

Ostrom, E. (1990), Governing the Commons: The Evolution of Institutions for Collective Action (Cambridge University Press, Cambridge).

Ostrom, E. (2000), "Collective action and the evolution of social norms", Journal of Economic Perspectives 14:137–158.

Parry, I., and W. Oates (2000), "Policy analysis in the presence of distorting taxes", Journal of Policy Analysis and Management 19:606–613.

Pashigian, B.P. (1985), "Environmental regulations: whose self interest are being protected?", Economic Inquiry 23:551–584.

Paul, J. (1994/95), "Free trade, regulatory competition and the autonomous market fallacy", Columbia Journal of European Law 1:29–62.

Pearce, D.W. (1998), "Environmental appraisal and environmental policy in the European Union", Environmental and Resource Economics 11:489–501.

Peltzman, S. (1976), "Toward a more general theory of regulation", Journal of Law and Economics 19:211–240.

Pfander, J.E. (1996), "Environmental federalism in Europe and the United States: a comparative assessment of regulation through the agency of member states", in: J.B. Braden, H. Folmer and T. Ulen, eds., Environmental Policy with Political and Economic Integration: The European Union and the United States (Elgar, Cheltenham) 59–131.

Proost, S., and J. Braden (1998), Climate Change, Transport and Environmental Policy: Empirical Applications in a Federal System (Elgar, Cheltenham).

Revesz, R.L. (2001), "Federalism and regulation: some generalizations", in: D. Esty and D. Geradin, eds., Regulatory Competition and Economic Integration: Comparative Perspectives (Oxford University Press, Oxford) 3–29.

Rubinfeld, D.L. (1987), "The economics of the local public sector", in: A. Auerbach and M. Feldstein, eds., Handbook of Public Economics, Vol. 2 (North-Holland, Amsterdam) 571–645.

Schneider, F., and J. Volkert (1999), "No chance for incentive-oriented environmental policies in representative democracies? A public choice analysis", Ecological Economics 31:123–138.

Seskin, E.P., R.J. Anderson, Jr., and R.O. Reid (1983), "An empirical analysis of economic strategies for controlling air pollution", Journal of Environmental Economics and Management 10:112–124.

Smith, S. (1995), "Green" Taxes and Charges: Policy and Practice in Britain and Germany (The Institute for Fiscal Studies, London).

Stigler, G. (1971), "The theory of economic regulation", Bell Journal of Economics and Management Sciences 2:3–21.

Stratmann, T. (1991), "What do campaign contributions buy? Deciphering causal effects of money and votes", Southern Economic Journal 57:606–620.

Svendsen, G.T. (1998), Public Choice and Environmental Regulation (Elgar, Cheltenham).

Tietenberg, T. (1985), Emissions Trading: An Exercise in Reforming Pollution Policy (Resources for the Future, Washington, DC).

Wallart, N. (1999), The Political Economy of Environmental Taxes (Elgar, Cheltenham).

Wellisch, D. (2000), Theory of Public Finance in a Federal State (Cambridge University Press, Cambridge).

Wildasin, D. (1989), "Interjurisdictional capital mobility: fiscal externality and a corrective subsidy", Journal of Urban Economics 25:192–212.

Wilson, J. (1986), "A theory of interregional tax competition", Journal of Urban Economics 19:296–315.

Wilson, J. (1996), "Capital mobility and environmental standards: is there a theoretical basis for a race to the bottom?", in: J. Bhagwati and R. Hudec, eds., Fair Trade and Harmonization: Prerequisites for Free Trade? (MIT Press, Cambridge, MA) 393–427.

Wilson, J. (1999), "Theories of tax competition", National Tax Journal 52:269–304.

World Bank (2000), Greening Industry: New Roles for Communities, Markets, and Governments (Oxford University Press, Oxford).

Zerbe, Jr., R.O., and H.E. McCurdy (1999), "The failure of market failure", Journal of Policy Analysis and Management 18:558–578.

Zodrow, G., and P. Mieszkowski (1986), "Pigou, Tiebout, property taxation, and the underprovision of public goods", Journal of Urban Economics 19:356–370.

Chapter 9

EXPERIENCE WITH MARKET-BASED ENVIRONMENTAL POLICY INSTRUMENTS

ROBERT N. STAVINS

John F. Kennedy School of Government, Harvard University, Cambridge, MA and Resources for the Future, Washington, DC, USA

Contents

Handbook of Environmental Economics, Volume 1, Edited by K.-G. Mäler and J.R. Vincent

Abstract

Environmental policies typically combine the identification of a goal with some means to achieve that goal. This chapter focuses exclusively on the second component, the means – the "instruments" – of environmental policy, and considers, in particular, experience around the world with the relatively new breed of economic-incentive or market-based policy instruments. I define these instruments broadly, and consider them within four categories: charge systems; tradable permits; market friction reductions; and government subsidy reductions. Within charge systems, I consider effluent charges, deposit-refund systems, user charges, insurance premium taxes, sales taxes, administrative charges, and tax differentiation. Within tradeable permit systems, I consider both credit programs and cap-and-trade systems. Under the heading of reducing market frictions, I examine market creation, liability rules, and information programs. Finally, under reducing government subsidies, I review a number of specific examples from around the world. By defining market-based instruments broadly, I cast a large net for this review of applications. As a consequence, the review is extensive. But this should not leave the impression that market-based instruments have replaced, or have come anywhere close to replacing, the conventional, command-and-control approach to environmental protection. Further, even where these approaches have been used in their purest form and with some success, such as in the case of tradeable-permit systems in the United States, they have not always performed as anticipated. In the final part of the chapter, I ask what lessons can be learned from our experiences. In particular, I consider normative lessons for design and implementation, analysis of prospective and adopted systems, and identification of new applications.

Keywords

marked-based policies, economic-incentive instruments, pollution taxes, tradeable permits, deposit-refund systems

JEL classification: H23, H41, K32, Q25, Q28

1. What are market-based policy instruments?

Environmental policies typically combine the identification of a goal (either general or specific) with some means to achieve that goal. In practice, these two components are often linked within the political process. This chapter focuses exclusively on the second component, the means – the "instruments" – of environmental policy, and considers, in particular, experience around the world with the relatively new breed of economic-incentive or market-based policy instruments.[1]

1.1. Definition

Market-based instruments are regulations that encourage behavior through market signals rather than through explicit directives regarding pollution control levels or methods.[2] These policy instruments, such as tradable permits or pollution charges, are often described as "harnessing market forces"[3] because if they are well designed and implemented, they encourage firms (and/or individuals) to undertake pollution control efforts that are in their own interests and that collectively meet policy goals.

By way of contrast, conventional approaches to regulating the environment are often referred to as "command-and-control" regulations, since they allow relatively little flexibility in the means of achieving goals. Such regulations tend to force firms to take on similar shares of the pollution-control burden, regardless of the cost.[4] Command-and-control regulations do this by setting uniform standards for firms, the most prevalent of which are technology- and performance-based standards. Technology-based standards specify the method, and sometimes the actual equipment, that firms must use to comply with a particular regulation. A performance standard sets a uniform control target for firms, while allowing some latitude in how this target is met.

Holding all firms to the same target can be expensive and, in some circumstances, counterproductive. While standards may effectively limit emissions of pollutants, they typically exact relatively high costs in the process, by forcing some firms to resort to unduly expensive means of controlling pollution. Because the costs of controlling emissions may vary greatly among firms, and even among sources within the same firm,

---

[1] There is considerable overlap between environmental and natural resource policies. This chapter focuses on market-based policy instruments in the environmental realm, chiefly those that reduce concentrations of pollution, as opposed to those that achieve various goals of natural resource management. This means, for example, that tradeable development rights [Field and Conrad (1975), Bellandi and Hennigan (1980), Mills (1980)] are not reviewed, nor are tradeable permit systems used to govern the allocation of fishing rights [Batstone and Sharp (1999)].

[2] This section of the chapter draws, in part, on: Hockenstein, Stavins and Whitehead (1997); and Stavins (2000).

[3] See: OECD (1989, 1991, 1998a), Stavins (1988, 1991), and U.S. Environmental Protection Agency (1991). Another strain of literature – a known as "free market environmentalism" – focuses on the role of private property rights in achieving environmental protection [Anderson and Leal (1991)].

[4] But various command-and-control standards do this in different ways [Helfand (1991)].

the appropriate technology in one situation may not be appropriate (cost-effective) in another. Thus, control costs can vary enormously due to a firm's production design, physical configuration, age of its assets, or other factors. One survey of eight empirical studies of air pollution control found that the ratio of actual, aggregate costs of the conventional, command-and-control approach to the aggregate costs of least-cost benchmarks ranged from 1.07 for sulfate emissions in the Los Angeles area to 22.0 for hydrocarbon emissions at all domestic DuPont plants [Tietenberg (1985)].[5]

Furthermore, command-and-control regulations tend to freeze the development of technologies that might otherwise result in greater levels of control.[6] Little or no financial incentive exists for businesses to exceed their control targets, and both technology-based and performance-based standards discourage adoption of new technologies. A business that adopts a new technology may be "rewarded" by being held to a higher standard of performance and not given the opportunity to benefit financially from its investment, except to the extent that its competitors have even more difficulty reaching the new standard.

1.2. Characteristics of market-based policy instruments

In theory, if properly designed and implemented, market-based instruments allow any desired level of pollution cleanup to be realized at the lowest overall cost to society, by providing incentives for the greatest reductions in pollution by those firms that can achieve these reductions most cheaply.[7] Rather than equalizing pollution levels among firms (as with uniform emission standards), market-based instruments equalize the incremental amount that firms spend to reduce pollution – their marginal cost [Montgomery (1972), Baumol and Oates (1988), Tietenberg (1995)]. Command-and-control approaches could – in theory – achieve this cost-effective solution, but this would require that different standards be set for each pollution source, and, consequently, that policy makers obtain detailed information about the compliance costs each firm faces. Such information is simply not available to government. By contrast, market-based instruments provide for a cost-effective allocation of the pollution control burden among sources without requiring the government to have this information.

In contrast to command-and-control regulations, market-based instruments have the potential to provide powerful incentives for companies to adopt cheaper and better

---

[5] One should not make too much of these numbers, since actual, command-and-control instruments are being compared with theoretical benchmarks of cost-effectiveness, i.e., what a perfectly functioning market-based instrument would achieve in theory. A fair comparison among policy instruments would involve either idealized versions of both market-based systems and likely alternatives; or realistic versions of both [Hahn and Stavins (1992)].

[6] For more on technological change and the environment, see Chapter 11 (by Adam Jaffe, Richard Newell, and Robert Stavins).

[7] Under certain circumstances, substituting a market-based instrument for a command-and-control instrument can lower environmental quality, because command-and-control standards tend to lead to over-control [Oates, Portney and McGartland (1989)].

pollution-control technologies. This is because with market-based instruments, particularly emission taxes, it always pays firms to clean up a bit more if a sufficiently low-cost method (technology or process) of doing so can be identified and adopted [Downing and White (1986), Malueg (1989), Milliman and Prince (1989), Jaffe and Stavins (1995), and Jung, Krutilla and Boyd (1996)].

Most environmental policy instruments, whether conventional or market-based, can be directed to one of a range of "levels" of regulatory intervention: inputs (for example, a tax on the leaded content of gasoline); emissions (following the same example, a tax on emissions); ambient concentrations; exposure (whether human or ecological); and risk or damages. In general, administrative costs increase as one moves further along this set of points of regulatory intervention, but it is also the case that the instrument is more clearly addressing what is presumably the real problem.

One important characteristic of individual pollution problems that will affect the identification of the optimal point of regulatory intervention is the degree of mixing of the pollutant in the receiving body (airshed, watershed, or ground). At one extreme, uniformly mixed pollution problems (in their purest form, global commons problems such as stratospheric ozone depletion and global climate change) can be efficiently addressed through input or emissions interventions. At the other extreme, it would be problematic to address a highly non-uniformly mixed pollution problem through such an approach; instead, an intervention that focused on ambient concentrations, at a minimum, would be preferable.

Most applications of market-based instruments have been at the input or emission point of regulatory intervention, although a few have focused on ambient concentrations. Much the same can be said of nearly all conventional, command-and-control policy instruments in the environmental realm.

1.3. Categories of market-based instruments

I consider market-based instruments within four major categories: pollution charges; tradable permits; market friction reductions; and government subsidy reductions [OECD (1994a, 1994b, 1994c, 1994d)].[8]

Pollution charge systems assess a fee or tax on the amount of pollution that a firm or source generates [Pigou (1920)]. Consequently, it is worthwhile for the firm to reduce emissions to the point where its marginal abatement cost is equal to the tax rate. A challenge with charge systems is identifying the appropriate tax rate. Ideally, it should be

---

[8] A significant recent trend in environmental policy has been the increased use of voluntary programs for the purpose of achieving various environmental objectives. Because voluntary actions can offer firms rewards such as public recognition, some observers have characterized these voluntary programs as incentive-based instruments for environmental protection. Having already cast an exceptionally large net for this review of experience, I do not include this approach to environmental management in my review of market-based instruments. For a review of the use of voluntary initiatives in the United States, see: U.S. Environmental Protection Agency (2001).

set equal to the marginal benefits of cleanup at the efficient level of cleanup, but policy makers are more likely to think in terms of a desired level of cleanup, and they do not know beforehand how firms will respond to a given level of taxation. A special case of pollution charges is a *deposit-refund system*, where consumers pay a surcharge when purchasing potentially polluting products, and receive a refund when returning the product to an approved center, whether for recycling or for disposal [Bohm (1981), Menell (1990)].[9]

Tradable permits can achieve the same cost-minimizing allocation of the control burden as a charge system, while avoiding the problem of uncertain responses by firms.[10] Under a tradable permit system, an allowable overall level of pollution is established and allocated among firms in the form of permits.[11] Firms that keep their emission levels below their allotted level may sell their surplus permits to other firms or use them to offset excess emissions in other parts of their facilities.

Market friction reductions can also serve as market-based policy instruments. In such cases, substantial gains can be made in environmental protection simply by reducing existing frictions in market activity. Three types of market friction reductions stand out: (1) *market creation* for inputs/outputs associated with environmental quality, as with measures that facilitate the voluntary exchange of water rights and thus promote more efficient allocation and use of scarce water supplies; (2) *liability rules* that encourage firms to consider the potential environmental damages of their decisions; and (3) *information programs*, such as energy-efficiency product labeling requirements.

Government subsidy reductions are the fourth category of market-based instruments. Subsidies, of course, are the mirror image of taxes and, in theory, can provide incentives to address environmental problems.[12] In practice, however, many subsidies promote economically inefficient and environmentally unsound practices.

---

[9] A deposit-refund system can also be viewed as a special case of a "performance bond".

[10] Thirty years ago, Crocker (1966) and Dales (1968) independently developed the idea of using transferable discharge permits to allocate the pollution-control burden among sources. Montgomery (1972) provided the first rigorous proof that such a system could provide a cost-effective policy instrument. A sizeable literature has followed, much of it stemming from Hahn and Noll (1982). Early surveys were provided by Tietenberg (1980, 1985). Much of the literature may be traced to Coase's (1960) treatment of negotiated solutions to externality problems.

[11] Allocation can be through free distribution (often characterized as "grandfathering") or through sale, including by auction. The program described above is a "cap-and-trade" program, but some programs operate as "credit programs", where permits or credits are assigned only when a source reduces emissions below what is required by existing, source-specific limits.

[12] In many countries, subsidies have been advocated (and sometimes implemented) as means of *improving* environmental quality. Although such subsidies could, in theory, advance environmental quality [see, for example, Jaffe and Stavins (1995)], it is also true that subsidies, in general, have important and well-known disadvantages relatives to taxes [Baumol and Oates (1988)]. They are not considered as a distinct category of market-based instruments in this chapter. Although the prevalence of subsidies intended to improve environmental quality is not very great in developed market economies, they are more common in transition and, to a lesser extent, developing economies [Żylicz (2000)]. Most environmental funds in transition economies, however, fail to select efficient projects or calculate efficient subsidies [Anderson and Żylicz (1999), Peszko and Żylicz (1998)].

1.4. Scope of the Chapter

This chapter focuses on market-based policy instruments in the environmental realm, chiefly those that reduce concentrations of pollution, as opposed to those that operate in the natural resources realm and achieve various goals of resource management. This means, for example, that tradeable development rights, wetlands mitigation banking, and tradeable permit systems used to govern the allocation of fishing rights are not reviewed in this chapter.[13]

Sections 2–5 review experiences around the world with the four major categories of market-based instruments for environmental protection: charge systems; tradeable permit systems; market-friction reductions; and government subsidy reductions. Section 6 examines lessons that can be learned from these experiences.

Although much of the chapter is descriptive in nature, normative analysis of the implementation of market-based instruments is surveyed in those cases in which evidence is available. That normative analysis focuses on the criteria of static and dynamic cost-effectiveness; little or no attention is given to efficiency *per se*. In other words, in this chapter, the targets of respective environmental policies are taken as given, and are not subjected to economic analyses.

Despite the chapter's expressed purpose of reviewing and providing some understanding about experiences with market-based instruments, virtually no attention is given to the important set of positive political economy questions that are raised by the increasing use of these instruments, such as the following. Why was there so little use of market-based instruments, relative to command-and-control instruments, over the 30-year period of major environmental regulation that began in 1970, despite the apparent advantages in many situations of the former? Why has the political attention given to market-based environmental policy instruments increased dramatically in recent years? Such questions of the positive political economy of instrument choice are, for the most part, ignored, not because they are without interest, but because they are addressed in Chapter 8 (by Wallace Oates and Paul Portney).

2. Charge systems

The conventional wisdom is that European environmental policy has made limited use of pollution taxes, while this approach has been totally ignored in the United States. This is not strictly correct, particularly if one defines charge systems broadly, in which case a significant number of applications around the world can be identified.

---

[13] The distinction between environmental and natural resource policies is somewhat arbitrary. Some policy instruments which are seen to bridge the environmental and natural resource realm, such as removing barriers to water markets, are considered.

For purposes of this review, I identify seven categories of charge systems, but it should be noted at the outset that the categories are neither precisely defined nor mutually exclusive. Hence, the assignment of individual policy instruments to one or another category inevitably involves judgement, if not an arbitrary element. Nevertheless, this set of categories may help readers navigate what would otherwise be a single, very long list of applications. I divide the categories of charges into two primary sets: those for which behavioral impacts are central to their design, implementation, and performance; and those for which anticipated behavioral impacts are secondary.

Within the first set, I distinguish among three categories of charge systems. First, *effluent charges* are those instruments which are closest to the textbook concept of a Pigouvian tax (Section 2.1). Second, *deposit-refund systems* are a special case of Pigouvian taxes in which front-end charges (such as those on some beverage containers) are combined with refunds payable when particular behavior (such as returning an empty container to an approved outlet) is carried out (Section 2.2). Third, *tax differentiation* refers to tax cuts, credits, and subsidies for environmentally desirable behavior (Section 2.7).

The second set of charge systems, those for which behavioral impacts appear to be a secondary consideration, includes four categories of instruments. First, *user charges* provide a mechanism whereby the direct beneficiaries of environmental services finance its provision (Section 2.3). Second, *insurance premium taxes* are levied on particular groups or sectors to finance insurance pools against potential risks associated with the production or use of the taxed product (Section 2.4). Third, *sales taxes* are levied on the sales or value-added of specific goods and services in the name of environmental protection (Section 2.5). Fourth and finally, *administrative charges* are used to raise revenues to help cover the administrative costs of environmental programs (Section 2.6).[14]

2.1. Effluent charges

Most applications of charge systems probably have not had the incentive effects typically associated with a Pigouvian tax, either because of the structure of the systems or because of the low levels at which charges have been set. Nevertheless, a limited number of these systems may have affected behavior.

Within the category of effluent charges, which comes closest to what most economists think of as a pollution tax, member countries of the Organization of Economic Cooperation and Development (OECD) other than the United States have led the way [Blackman and Harrington (1999)].[15] Selected effluent charges are summarized in Table 1, where

[14] For useful surveys of the use of environmentally related taxes in OECD countries, see: OECD (1993d, 1995d, 2001).

[15] Effluent charges have been used more extensively in Europe than in the United States, although – as indicated in the text – it is not clear that the levels have been sufficient to affect behavior in significant ways. For a discussion of the economics and politics surrounding taxation of sulfur dioxide, nitrous oxide, and carbon dioxide in the Scandinavian nations, the Netherlands, France, and Germany, see: Cansier and Krumm (1998) and OECD (1993a, 1995a).

Table 1
Effluent fees

Regulated substance	Country	Rate	Use of revenues
CO	Czech Republic[a]	$22/ton permitted; $33/ton above	State Environmental Fund
	Estonia[b]	$0.27/ton permitted; $1.36/ton above	Estonian Environmental Funds national (50%); county (50%)
	Lithuania[c]	$1.75/ton	Municipal environmental funds (70%); General budget (30%)
	Poland[d]	$22/ton	National, regional and municipal environmental funds
	Russia[e]	$0.02/ton permitted; $0.09/ton above	National and regional environmental funds
	Slovakia[f]	$20/ton	Slovak Environmental Fund
CO_2	Denmark	$42/m^3, diesel, kerosene, gas oil $38/ton, coal $17/ton, LPG $0.03/m^3, natural gas $0.02/kWh, electricity	General budget
	Finland	$38/m^3, leaded and unleaded gasoline $43/m^3, diesel and kerosene $39/ton, coal $0.02/m^3, natural gas $0.003–0.006/kWh, electricity	General budget
	Netherlands	$45/m^3, gas oil and kerosene $54/m^3, LPG $0.05/m^3, natural gas $0.02/kWh, electricity	Corporate and income tax relief
	Norway	$59/m^3, mineral oil $59/ton, coal $0.11/m^3 natural gas (only applied to offshore oil and gas activities)	General budget
	Sweden	$106/m^3 leaded and unleaded gasoline $131/m^3 diesel, kerosene, gas oil $127/ton LPG $135/m^3 heavy fuel oil $114/ton coal $0.03/m^3 natural gas $0.02/kWh electricity	General budget
SO_2	Bulgaria[g]	$0.02/kg	National environmental fund (70%) and polluter's municipality (30%)
	Czech Republic[a]	$30/ton permitted; $45/ton above	State Environmental Fund
	Denmark	All fuels, electricity taxed in proportion to resulting SO_2 emissions, $1.60/kg of SO_2	General budget
	Estonia[b]	$2/ton permitted; $95/ton above	Estonian Environmental Funds national (50%); county (50%)

Table 1
(*Continued*)

Regulated substance	Country	Rate	Use of revenues
SO_2	Finland	$30/m^3 of diesel or gas oil	General budget
	France[h]	$32/ton of direct emissions	Pollution reduction (75%); research (25%)
	Hungary[i]	$2.40/ton	Central Environmental Protection Fund (70%); local government budgets (30%)
	Italy	$62/ton of direct emissions	Reduction of environmental impacts
	Japan	n.a.	Compensation of individuals with chronic breathing problems attributable to pollution
	Lithuania[c]	$46/ton	Municipal environmental funds (70%); general budget (30%)
	Norway[j]	Fuels taxed in proportion to resulting SO_2 emissions, $0.01 per liter of fuel per 0.25% sulfur content	General budget
	Poland[d]	$83/ton	National, regional and municipal environmental funds
	Russia[e]	$1.22/ton permitted; $6.10/ton above	National and regional environmental funds
	Slovakia[f]	$33/ton	Slovak Environmental Fund
	Spain – Galicia	Industrial energy products taxed on sum of SO_2 and NO_x emissions; rate is $35/ton, emissions between 1,001 and 50,000 tons; $39/ton above 50,000 tons	Regional budget
	Sweden	Liquid fuels $3.33/m^3 for each 0.1% by weight of sulfur content; coal and other solid or gaseous fuels $3.70/m^3	General budget
NO_x	Bulgaria[g]	$0.05/kg	National environmental fund (70%) and polluter's municipality (30%)
	Czech Republic[a]	$30/ton permitted; $45/ton above	State Environmental Fund
	Estonia[b]	$4/ton permitted; $216/ton above	Estonian Environmental Funds national (50%); county (50%)
	France	$27/ton, based on direct measurement of emissions	Pollution reduction (75%); research (25%)
	Hungary[i]	$4/ton	Central Environmental Protection Fund (70%); local government budgets (30%)
	Italy	$123/ton of direct emissions	Reduction of environmental impacts
	Lithuania[i]	$67/ton	Municipal environmental funds (70%); General budget (30%)

Table 1
(*Continued*)

Regulated substance	Country	Rate	Use of revenues
NO$_x$	Poland[d]	$83/ton	National, regional and municipal environmental funds
	Russia[e]	$1.02/ton permitted; $5.08/ton above	National and regional environmental funds
	Slovakia[f]	$27/ton	Slovak Environmental Fund
	Sweden	Combustion and incineration plants pay $5/kg of NO$_x$	Redistributed to payees (plants) in proportion to energy produced
Combined industrial air emissions	Latvia[k]	$1.65 to $440/ton, depending on emissions hazard class	National, regional and local general budgets
	China	Varies with pollutants, including SO$_2$, H$_2$S, NO$_x$, HCl, CO, H$_2$SO$_4$, Pb, Hg, dust	Grants, low-interest pollution control loans (80%); local monitoring and administration (20%)
BOD load	Bulgaria	$0.11/kg	National environmental fund (70%); polluter's municipality (30%)
	Colombia	Río Negro basin only, rate n.a.	Wastewater treatment plants (50%); industrial clean technology equipment (30%); research, administration (20%)
	Estonia[b]	BOD$_5$ $77/ton permitted; $386/ton above	Estonian Environmental Funds national (50%); county (50%)
	Lithuania[c]	BOD$_7$ $75/ton	Municipal environmental funds (70%); General budget (30%)
	Malaysia	BOD from palm oil industry; current rates n.a.	n.a.
	Philippines	BOD in Laguna de Bay watershed, rates n.a.	Water quality management, monitoring and enforcement (80%); local government budgets (20%)
	Poland[d]	BOD$_5$ $172 to $1,722/ton, depending on source	National, regional and municipal environmental funds
	South Korea[l]	n.a.	n.a.
TSS	Bulgaria[g]	$0.04/kg	National environmental fund (70%); polluter's municipality (30%)
	Colombia	Río Negro basin only, rate n.a.	Wastewater treatment plants (50%); industrial clean technology equipment (30%); research, administration (20%)
	Estonia[b]	$39/ton permitted; $386/ton above	Estonian Environmental Funds national (50%); county (50%)
	Lithuania[c]	$15/ton	Municipal environmental funds (70%); General budget (30%)

Table 1
(*Continued*)

Regulated substance	Country	Rate	Use of revenues
TSS	Poland[d]	$74/ton	National, regional and municipal environmental funds
	South Korea[l]	n.a.	n.a.
Combined industrial water emissions	China	Varies with pollutants	Grants, low-interest pollution control loans (80%); local monitoring and administration (20%)
	France[m]	Varies by river basin	Water pollution control
	Germany[n]	$42 per "pollution unit"	Water quality management
	Latvia[k]	$1.65 to $27,600/ton, depending on effluent hazard class	National, regional and local general budgets
	Netherlands	Varies by flow and load	Water quality policy
	Slovakia[f]	Varies by effluent load and quantity (not quality) of receiving waters	Slovak Environmental Fund
Nitrogen and phosphorous	Denmark	N $3.10/kg; P $17.30/kg discharged to surface waters	General budget
	Estonia[b]	N $65/ton permitted; $320/ton above P $115/ton permitted; $580/ton above discharged to surface water, ground water or soil	Estonian Environmental Funds national (50%); county (50%)
	Lithuania[c]	N $75/ton; P $260/ton	Municipal environmental funds (70%); General budget (30%)
Landfill, incinerator or hazardous waste	Denmark[o]	$53/ton, landfill waste $41/ton, incinerator waste $393/ton, hazardous waste	General budget
	Estonia[b]	$0.06 to $54/ton permitted; $0.32 to $27,000/ton above for waste dumping or burying, depending on hazard class	Estonian Environmental Funds national (50%); county (50%)
	Finland	$18/ton, landfill waste	n.a.
	Latvia[k]	$0.14/m^3, non-toxic waste disposal $0.83/m^3, toxic waste disposal $28/m^3, highly toxic waste disposal	National, regional and local general budgets
	Netherlands	$16/ton, landfill waste $34/ton, combustible waste disposed of in landfill	General budget
	Poland[d]	$1.60 to $21.50/ton waste disposal, depending on hazard class	National, regional and municipal environmental funds
	United Kingdom	landfill tax, $17/ton on "active" waste; $3/ton on inert waste	General budget

Note: CO is carbon monoxide, SO_2 is sulfur dioxide, and NO_x is nitrogen oxide; BOD is an acronym for biological oxygen demand and TSS is an acronym for total suspended solids. BOD load is the total amount

Table 1

(Continued)

of oxygen that a given amount of effluent will use in biochemical oxidation, during a period of three days at a temperature of 30°C (86°F). Conversion of all currencies to $US made using U.S. Federal Reserve historical bilateral exchange rates for December of the year in which data were gathered, available at http://www.bog.frb.fed.us/releases/H10/hist

[a]Charges from medium and large industrial enterprises in the Czech Republic go to the State Environmental Fund, while charges from small enterprises become part of municipal government budgets. The Czech Republic has established effluent fees for 90 air and 5 water pollutants, though only a few are listed here.

[b]In Estonia, exceeding a permit is not illegal, so long as an enterprise is able to pay the additional effluent fee. Estonia has established effluent fees for 139 air and 8 water pollutants, though only a few are listed here.

[c]Lithuania assesses fines on all air and water pollutants, but rates are available only for those listed here.

[d]Poland's effluent charges are divided among national, regional and municipal environmental funds in specified percentages that vary by substance. For example, NO_x charges are divided between the national (90%) and municipal (10%) funds, while most other air emissions are divided among the national (36%), regional (54%) and municipal (10%) funds. Poland assesses fees on 62 air and 6 water pollutants, though only a few are listed here.

[e]Russia assesses fees on more than 100 air and more than 100 water pollutants, though only a few are listed here.

[f]Slovakia assesses fees on 123 air and five water pollutants, though only a few are listed here.

[g]Bulgaria assesses fees on 16 air and 27 water pollutants, though only a few are listed here.

[h]France taxes sulfur hydrogen and hydrochloric acid emissions at the same rate as sulfur dioxide.

[i]Hungary's air emissions fines vary according to height of emissions and the factor by which permitted levels are exceeded. The charges listed here are "base fines", or those assessed when actual emissions exceed permitted levels by a factor of 1.00–2.00. Hungary has established fines for 150 air and 32 water pollutants, though only a few are listed here.

[j]Gasoline and fuels with sulfur content less than 0.05% (includes most auto diesel used in Norway) are excluded from Norway's SO_2 tax.

[k]Latvia assesses fees on seven air and ten water pollutants, though only a few are listed here.

[l]South Korea's effluent fees are assessed on emissions exceeding 30 percent of maximum allowable limit; penalty fees, assessed on emissions above the allowable maximum, equal the expense of treating actual volume of emitted pollutants. South Korea assesses fees on 10 air pollutants and 15 water pollutants, though only two are listed here.

[m]In 1993, rates ranged from $16/kg of suspended solids in the Loire–Bretagne river basin to $446/kg of soluble salts in the Seine-Normandie basin. See Cadiou and Duc (1994).

[n]In Germany, water pollution units are determined by flow and load; the per unit charge can be reduced by pollution control equipment investment.

[o]Average rate; Danish waste disposal charge depends on type of waste.

Sources: Speck (1998); Gornaja et al. (1997); Brunenieks, Kozlovska and Larson (1997); Semėnienė, Bluffstone and Čekanavičius (1997); Yang, Cao and Wang (1998); Kozeltsev and Markandya (1997); Stepanek (1997); Morris, Tiderenczl and Kovács (1997); Anderson and Fiedor (1997); Owen, Myjavec and Jassikova (1997); Matev and Nivov (1997); Wuppertal Institute (1996); OECD (1997c); World Bank (1997a, 1997b); Panayotou (1998); and World Bank (1999).

I distinguish among ten areas of application: carbon monoxide (CO), carbon dioxide (CO$_2$), sulfur dioxide (SO$_2$), nitrogen oxides (NO$_x$), combined industrial air pollutants, biological oxygen demand (BOD) load, total suspended solids (TSS), combined industrial water emissions, nitrogen and phosphorous, and landfill, incinerator, and hazardous waste discharges.

Several European countries have moved to implement pollution taxes within the framework of ecological or "green tax reform", which seeks a systematic shift of the tax burden away from labor and/or capital and toward the use of environmental resources. As of 1997, environmental taxes in Sweden, Denmark, and Finland were part of a framework green tax reform [Ekins (1999)].

2.1.1. Effluent charges in Western Europe

Seven OECD countries in western Europe have implemented emissions fees to reduce air pollution, but most of the fees are assessed on input proxies, possibly because of monitoring and enforcement costs [Speck (1998)]. Although the effects of direct emissions charges will differ from those of input taxes, both are considered here, following the practice of the OECD (1994a).[16]

As of 1999, six OECD nations levied carbon taxes: Denmark, Finland, Italy, the Netherlands, Norway, and Sweden. Finland's carbon tax, the world's first, was introduced in 1990 [Haugland (1993)]. Italy's carbon tax is a revenue-generating mechanism, part of a broad-ranging attempt to use indirect taxation to compensate for weaknesses in the direct taxation system [Schlegelmilch (1998)]. Carbon taxes in Denmark, Norway, and Sweden are intended to have an incentive effect, in addition to a revenue-generating effect, but it has been difficult to determine the actual impacts of these policies [Blackman and Harrington (1999)].

Claims have been made that the Swedish and Norwegian taxes have reduced carbon emissions [Bohlin (1998), Larsen and Nesbakken (1997)], but in all the Nordic countries, except Finland, a variety of tax exemptions have made effective carbon tax rates significantly lower than nominal rates, thereby increasing skepticism regarding the efficacy of these policies. For example, Sweden's manufacturing tax exemptions and reductions result in effective CO$_2$ tax rates ranging from 19 to 44 percent of nominal rates [Ekins and Speck (1999)]. Danish industry has obtained tax relief on process energy, and power stations are exempt from coal taxes. Norway taxes only 60 percent of domestic CO$_2$ emissions, and only 25 percent of SO$_2$ emissions, when exemptions and reductions are taken into account [Ekins and Speck (1999)].

Norway, Sweden, France, Denmark, Italy, and the Spanish autonomous region of Galicia tax sulfur emissions or the sulfur content of fuels. The Swedish tax seems to have reduced sulfur emissions [Lövgren (1994)], not surprising given that it is very high by international standards [OECD (1996)]. Indeed, Sweden met its national sulfur

---

[16] See also: O'Connor (1994).

emissions targets well ahead of schedule through fuel-switching and emission reductions that have been attributed to the tax [World Bank (1997b)].

France, Italy, Sweden, and Galicia tax nitrogen oxide emissions, but only the Swedish tax has reduced emissions [Blackman and Harrington (1999)]. Energy plants in Sweden with production of 25 GWh or more pay $5/kg on NO_x emissions. The tax is revenue-neutral, with payees (plants) receiving rebates in proportion to energy output.[17] In the first two years of the program, total emissions from monitored plants fell by 40 percent [Blackman and Harrington (1999)], attributed to the emissions fee system [Lövgren (1994), Sterner and Hoglund (1998)], but only about 3 percent of Sweden's domestic NO_x emissions are taxed under the program [Ekins and Speck (1998)].

Effluent charges have also been used in western Europe for water pollution. Since 1970, the Netherlands has assessed effluent fees on heavy metals discharges from large enterprises, and organic discharges from urban and farm households, and small, medium, and large enterprises. The Dutch charges were originally earmarked to finance construction of wastewater treatment facilities, but the high cost of facilities resulted in very high charges, in some cases equal to marginal abatement costs at high levels of cleanup [World Bank (2000)]. By 1990, the charges had reduced total organic discharges by one-half, and industrial organic emissions by 75 percent [World Bank (2000)]. Germany also levies wastewater effluent charges, with revenues earmarked for water pollution control programs [OECD (1993b)]. France has a system of water pollution charges, the revenues from which are reinvested in water infrastructure and pollution control [Cadiou and Duc (1994), OECD (1997b)].

2.1.2. Effluent charges in the transition economies

Some transition economies in Central and Eastern Europe and the former Soviet republics may view air and water pollution charges as means of efficient restructuring of their environmental management and regulatory systems [Bluffstone and Larson (1997)]. In other cases, effluent charge systems were introduced well before the beginnings of the economic transitions in the late 1980's: the former Czechoslovakia introduced charges in the 1960's; Bulgaria, Hungary, and Poland in the 1970's, and parts of the former Soviet Union in the 1980's [Vincent and Farrow (1997)].

Although effluent fees have been implemented throughout the region, Poland is the only country in which the fees may have reduced emissions. Poland restructured its emissions fee system for airborne pollutants in 1991, increasing fees dramatically to twenty times their levels under Communist rule [Anderson and Fiedor (1997)], so that Polish effluent fees are now among the highest in the world. Typically, the Polish fees include a "normal fee" levied on emissions below the regulatory standard, and a penalty fee for emissions thereafter.[18] While fees have been nominally calculated from ambient

---

[17] The program's administrative costs, less than 1 percent of tax revenues, are deducted before the re-distribution.

[18] This is one of an exceptionally small number of non-linear effluent charges. See discussion in Section 6.3.

air quality guidelines and marginal abatement costs, they have been heavily influenced by political factors and revenue requirements [Anderson and Fiedor (1997)]. Fee revenues – on the order of $450 to $500 million annually – flow to national and regional environmental funds.

In other parts of the region, air and water effluent charges have been ineffective for a number of reasons: (1) legislated charges have been significantly eroded by the high inflation that has accompanied economic transition; (2) charges typically have been set below marginal abatement costs [Morris, Tiderenczl and Kovács (1997), Stepanek (1997), Żylicz (1996)]; (3) pollution limits – the point above which emissions are charged at a penalty rate – are typically set too high to influence firm behavior [Brunenieks, Kozlovska and Larson (1997)]; (4) tax rates are often the result of implicit or explicit negotiation between industries and state or regional governments [Gornaja et al. (1997), Kozeltsev and Markandya (1997)]; (5) many countries set upper bounds on pollution charge liabilities; (6) unprofitable enterprises are often exempted [Kozeltsev and Markandya (1997), Owen, Myjavec and Jassikova (1997)]; and (7) regulatory systems are insufficient to support adequate monitoring and enforcement [Gornaja et al. (1997), Kozeltsev and Markandya (1997), Morris, Tiderenczl and Kovács (1997), Bluffstone and Larson (1997)]. While pollution charges rarely induce abatement in Eastern Europe and the former Soviet republics, they do raise revenue for environmental projects, and some argue that they are contributing to the establishment and acceptance of a "polluter pays principle" [Bluffstone and Larson (1997)].

2.1.3. Effluent charges in other countries[19]

A number of other countries have utilized effluent charges, albeit typically at levels too low to induce behavioral changes. For example, China assesses levies on 29 pollutants in wastewater, 13 industrial waste gases, and various forms of industrial solid and radioactive waste [World Bank (1997b)]. Regulated substances include SO_2, NO_x, CO, hydrogen sulfide, dust, mercury, and lead [Yang, Cao and Wang (1998)]. Plants pay a fee for emissions greater than the regulatory standard for each substance, but when more than one pollutant exceeds the standard, plants pay only for the single pollutant which will result in the largest fee. Firms that pay penalty charges, rather than reducing emissions, face a five percent annual charge increase beginning in the third year of noncompliance.

Chinese pollution fees are often lower than the marginal cost of abatement. For example, the World Bank estimates that SO_2 emission charges in Zhengzhou would have to be increased more than fiftyfold to equalize marginal abatement costs and marginal social damages [World Bank (2000)]. Of the fees collected, 80 percent are used for grants

---

[19] The closest that any charge system in the United States comes to operating as a Pigouvian tax may be the unit-charge approach to financing municipal solid waste collection, where households (and businesses) are charged the incremental costs of collection and disposal. I discuss these later within the category of "user charges" for municipal environmental services.

and low-interest loans for pollution control projects, and the remaining 20 percent are dedicated to local administration and monitoring activities [World Bank (1997a)]. These effluent charges appear to have helped reduce both water and air pollution intensity during the period of rapid industrial growth in China since 1979. Each 1 percent increase in the water pollution levy has reduced the intensity of organic water pollution by 0.8 percent; each 1 percent rise in the air pollution levy has reduced the pollution intensity of industrial air emissions by 0.4 percent [Wang and Wheeler (1996, 1999)]. The effluent fees are also a major source of revenue for environmental projects [Sterner (1999), World Bank (2000)]. In 1995, pollution levies were applied to 368,200 Chinese enterprises and raised about $460 million, or 0.6 percent of national income [Wang and Lu (1998)]. Of the fees collected, 80 percent are used for grants and low-interest loans for pollution control projects, and the remaining 20 percent refund local administration and monitoring activities [World Bank (1997a)].

Malaysia was one of the first countries to use effluent charges, having introduced effluent fees, paired with licensing, to control pollution from the palm oil industry as early as 1978 [World Bank (1997b)]. The Philippines instituted environmental fees for wastewater discharge from industrial sources in 1997 [World Bank (1997b)], although the program is active in only one area of the country, Laguna Lake. BOD discharges from affected plants dropped 88 percent between 1997 and 1999 [World Bank (2000)]. South Korea imposes charges for emissions in excess of regulatory limits on ten air pollutants and fifteen water pollutants [OECD (1997c)], and Japan assesses a minor charge on industrial SO_2 emissions [Wuppertal Institute (1996)].

Colombia implemented a pilot program of water effluent charges after experiencing no success in pollution reduction with command and control regulations. Industrial polluters pay effluent fees based on BOD and TSS [World Bank (1999)]. Although emission decreases have been recorded since the program came into existence, it is difficult to separate the effect of the charges from that of voluntary agreements [World Bank (1999)]. The municipality of Quito, Ecuador has implemented a water effluent charge system [Huber, Ruitenbeek and Serôa da Motta (1998)], whereby enterprises discharging above national standards for organic content and TSS pay a per-unit charge equal to the cost of municipal treatment. In addition, Quito assesses fines on mobile air pollution sources, including cars, trucks, and buses in an effort to reduce air pollution in the city's central historical district. The fines are set above the cost of installing low-emissions technology or obtaining a tune-up. Mexico created a system of water effluent fees in 1991 in order to regulate BOD and TSS from municipal and industrial sources. Most municipalities and a large proportion of industrial dischargers do not pay the fees [Serôa da Motta (1998)]. Penalties for non-compliance were established in 1997, but no study has shown whether enforcement has been sufficient to induce abatement, or payment of fees and penalties.

2.2. *Deposit-refund systems*

Policies intended to reflect the social costs of waste disposal (such as waste-end fees, discussed in Section 2.3.2) can have the effect of increasing the experienced cost of legal disposal, and thereby providing unintended incentives for improper (illegal) disposal. For waste that poses significant health or ecological impacts, *ex post* clean up is frequently an especially unattractive option. For these waste products, the prevention of improper disposal is particularly important. One alternative might seem to be a front-end tax on waste precursors, since such a tax would give manufacturers incentives to find safer substitutes and to recover and recycle taxed materials. But substitutes may not be available at reasonable costs, and once wastes are generated, incentives that affect choices of disposal methods are unaffected.

This dilemma can be resolved with a front-end charge (deposit) combined with a refund payable when quantities of the substance in question are turned in for recycling or (proper) disposal. In principle, for economic efficiency, the size of the deposit should be set equal to the marginal social cost of the product being disposed of illegally (at the efficient level of return) minus the real welfare costs of the program's operation, assuming that these costs are proportional to the quantity of returns. As the product changes hands in the production and consumption process (through wholesalers and distributors to consumers), the purchaser of the product pays a deposit to the seller. Deposit-refund systems are most likely to be appropriate when the incidence and the consequences of improper disposal are great [Bohm (1981), Russell (1988), Macauley, Bowes and Palmer (1992)].

The major applications of this approach in the United States have been in the form of ten state-level "bottle bills" for beverage containers (Table 2). A brief examination of these systems provides some insights into the merits *and* the limitations of the approach. In most programs, consumers pay a deposit at the time of purchase which can be recovered by returning the empty container to a redemption center. Typically, the deposit is the same regardless of the type of container.

In some respects, these bills seem to have accomplished their objectives; in Michigan, for example, the return rate of containers one year after the program was implemented was 95 percent [Porter (1983)]; and in Oregon, littering was reduced and long-run savings in waste management costs were achieved [U.S. General Accounting Office (1990)]. But by charging the same amount for each type of container material, these programs do not encourage consumers to choose containers with the lowest product life-cycle costs (including those of disposal).

Analysis of the effectiveness, let alone the cost-effectiveness or efficiency, of beverage container deposit-refund systems has been limited. The few rigorous studies that have been carried out of the benefits and costs of bottle bills have found that social desirability depends critically on the value of the time it takes consumers to return empty containers and the willingness to pay for reduced litter [Porter (1978)]. By requiring consumers to separate containers and deliver them to redemption centers, deposit-refund systems can foster net welfare losses, rather than gains.

Table 2
Deposit-refund systems

Regulated products	Country	Jurisdiction / Size of deposit or description
Specified beverage containers	Australia	South Australia / 3¢ (aluminum cans) to 13¢ (glass bottles)
	Austria	National / 40¢ (reusable plastic bottles)
	Barbados	Local / glass containers
	Belgium	National / beer, soft drink containers
	Bolivia	Local / glass and plastic containers
	Brazil	Regional / glass and aluminum containers
	Canada	Newfoundland / 4¢ deposit, 2¢ return; Nova Scotia / 7¢ deposit, full return on refillables, 4¢ return on non-refillables; Quebec / 4¢ ; British Columbia, Alberta, Yukon / deposit n.a. (specified containers)
	Chile	Local / glass and plastic containers
	Colombia	Local / glass containers
	Czech Republic	National / 9 to 15¢ (glass bottles); 15 to 30¢ (PET bottles)
	Denmark	National / 18¢ to 70¢ (glass bottles)
	Ecuador	Local / glass containers
	Finland	National / 9¢ (small bottle); 46¢ (liter bottle); 18¢ (can)
	Iceland	National / various containers
	Jamaica	Local / glass containers
	Japan[a]	National / $2.40 per case (glass bottles)
	Mexico	Local / glass containers
	Netherlands	National / up to 28¢ (glass bottles); 50¢ (PET bottles)
	Norway	National / glass and PET bottles, up to 28¢
	Sri Lanka	National / 7¢ (glass bottle)
	Sweden	National / 33¢ (glass bottles); 8¢ (cans); 60¢ (PET bottles)
	Switzerland	National / various containers; operated by private sector
	Taiwan[b]	National / 8¢ (PET bottles)
	United States[c]	Connecticut, Delaware, Iowa, Maine, Massachusetts, New York / 5¢ ; Vermont / 5¢ & 15¢ ; Oregon / 3¢ & 5¢ ; Michigan / 5¢ & 10¢ ; California / 2.5¢ & 5¢
	Venezuela	Local / glass containers
Auto batteries	United States	Arizona, Connecticut, Idaho, Minnesota, New York, Rhode Island, Washington, Wisconsin / $5.00; Michigan / $6.00; Arkansas, Maine / $10.00
	Mexico	Old battery must be returned to purchase new battery
Scrap autos	Sweden	National / $160 deposit paid on new car purchase; $185 returned when consumer renders old car being replaced
Small chemical containers	Denmark	National
Tires	South Korea	National / 5¢ to 50¢ , depending on size
Plastic shopping bags	Italy	National / 5¢ per bag
Packaging waste	France	National / Eco-emballages; operated by private sector
	Germany	National / Duales System; operated by private sector
Flourescent light bulbs	Austria	National / $1.20 per bulb
Refrigerators	Austria	National / $10–$100

Table 2

(*Continued*)

Note: Conversion of all currencies to $US made using U.S. Federal Reserve historical bilateral exchange rates for December of the year in which data were gathered, available at http://www.bog.frb.fed.us/releases/H10/hist.

[a]Japan's deposit fee for glass bottles includes approximately 60¢ for the bottles, and 80¢ for the case or container.

[b]Taiwan's deposit-refund system for PET bottles pays 8¢ to consumers bringing bottles to collection locations, and 2¢ for collectors bringing bottles to recycling centers.

[c]Oregon's rate for refillables is 3¢. California's deposit for containers smaller than 24 oz. is 2.5¢, and 5¢ for containers 24 oz. and larger.

Sources: U.S. Environmental Protection Agency (1992); OECD (1993a, 1993c, 1995a, 1995b, 1997a, 1998b, 1998e, 1999a, 1999b); Huber, Ruitenbeek and Serôa da Motta (1998); Steele (1999); and Rhee (1994).

Deposit-refund systems are most likely to be appropriate where: (1) the objective is one of reducing illegal disposal, as opposed to such objectives as general reductions in the waste stream or increased recycling; and (2) there is a significant asymmetry between *ex ante* (legal) and *ex post* (illegal or post-littering) clean-up costs. For these reasons, deposit refund systems may be among the best policy options to address disposal problems associated with containerizable hazardous waste, such as lead in motor vehicle batteries [Sigman (1995)].

As a means of reducing the quantity of lead entering unsecured landfills and other potentially sensitive sites, several U.S. states have enacted deposit-refund programs for lead acid motor vehicle batteries (Table 2).[20] Under these systems, a deposit is collected when manufacturers sell batteries to distributors, retailers, or original equipment manufacturers; likewise, retailers collect deposits from consumers at the time of battery purchase. Consumers can collect their deposits by returning their used batteries to redemption centers; these redemption centers, in turn, redeem their deposits from battery manufacturers. The programs are largely self-enforcing, since participants have incentives to collect deposits on new batteries and obtain refunds on used ones, but a potential problem inherent in the approach is an increase in incentives for battery theft. A deposit of $5–$10 per battery, however, appears to be small enough to avoid much of the theft problem, but large enough to encourage a substantial level of return.

Glass container deposit-refund systems are widely used in other OECD countries, including Australia, Austria, Belgium, Canada, Denmark, Finland, Iceland, the Netherlands, Norway, Portugal, Sweden, Germany, Sri Lanka, and Switzerland [OECD (1993a)]. Non-glass systems include a plastic shopping bag deposit-refund system in Italy, and a small chemicals container system in Denmark. In addition, Austria's deposit-refund system includes fluorescent light bulbs and refrigerators [OECD (1995a)], and

---

[20] Minnesota was the first state to implement deposit refund legislation for car batteries in 1988. By 1991, there were ten states with such legislation: Arizona, Arkansas, Connecticut, Idaho, Maine, Michigan, Minnesota, New York, Rhode Island, and Washington. Deposits range from $5 to $10.

since 1975, Sweden has maintained a deposit-refund system to encourage proper disposal of old vehicles.[21]

Japan's beer bottle deposit-refund system involves a levy paid by wholesale dealers, retail shops, and consumers, and refunded at each distribution stage upon bottle collection. Mexico requires the return of car batteries for deposit refund at the wholesale level [Huber, Ruitenbeek and Serôa da Motta (1998)]. Taiwan has a deposit-refund system for polyethylene terephthalate (PET) soft drink bottles [World Bank (1997b)]; South Korea for beverage containers, tires, batteries, and lubricants [OECD (1997c)]; and the Czech Republic for glass and polyethylene bottles [OECD (1999a)]. Voluntary deposit-refund systems for glass containers have been instituted in Barbados, Bolivia, Brazil, Chile, Colombia, Ecuador, Jamaica, Mexico and Venezuela [Huber, Ruitenbeek and Serôa da Motta (1998)].

2.3. User charges

Environmental user charges are typically structured to require those who directly benefit from a specific environmental service to finance its provision. Thus, I define user charges as those designed to fund environmentally related services, in contrast with effluent charges which I previously defined as those intended to influence behavior. In many cases, the distinction between this category of charge mechanism and effluent charges (or true Pigouvian taxes) is clear. But the distinction is somewhat clouded in the case of those charges that combine the following characteristics: they are directly related to pollutant emission levels (Pigouvian in principle); set too low to influence behavior (not Pigouvian in practice); and have their revenues earmarked for the provision of closely related environmental services. I consider three sub-categories of user charges: transportation; municipal services; and product disposal (Table 3).[22]

2.3.1. Transportation

Motor-vehicle fuels are heavily taxed in many parts of the world, including European nations, but the income from these taxes typically flows to general revenues.[23] Although the levels of such taxes in the United States are set relatively low, they fall more clearly within the user charge category, because revenues are dedicated exclusively to highway construction and maintenance (and now mass transit).[24] Likewise, revenues from U.S.

[21] Over the period of the program's existence, however, inflation has eroded the deposit in real terms so that it is currently less than 10 percent of its original value [Bohm (1999)].

[22] A considerable number of user charges are for parks and recreation, but these fall within the natural resource area and so are considered to be outside of the scope of this chapter. For a discussion of the history of recreation fees on U.S. public lands, see: Reiling and Kotchen (1996).

[23] Exceptions include Austria, Kenya, New Zealand, the United States, and Switzerland, where motor fuel tax revenues are partially or fully dedicated to road construction and other public transportation projects [Ayoo and Jama (1999); Speck (1998)].

[24] In addition, Federal taxes on automobile and truck tires flow to the U.S. Highway Trust Fund.

Table 3
User charges

Country	Item taxed	Rate	Use of revenues
Austria	Motor fuels Annual vehicle use	Varies by fuel type $(kW-24) \cdot \$0.47/month$, plus 20% for cars without catalytic converter	Public transport investments Partially earmarked for public transport subsidies
	Natural gas Electricity	$\$0.05/m^3$ $\$0.009/kWh$	Partially earmarked for energy-saving measures and public transport
	Landfill waste disposal	$5 to $9/ton	Contaminated site cleanup
Belgium	Landfill and incinerator waste Hazardous waste Batteries[a] Disposable beverage containers[a] Disposable razors Disposable cameras[a] Packaging of solvents[a] Packaging of glue[a] Packaging of inks[a] Packaging of pesticides[a] Surplus manure	$4 to $26/ton $11 to $87/ton $0.58/battery $0.44/container $0.29/razor $8.73/camera $0.15/5 liters $0.73/10 liters $0.73/2.5 liters $0.73/5 liters Based on kg of phosphate and nitrogen	National environmental expenditure Regional environmental expenditure Funds manure transport and disposal
Denmark	Batteries Tires	NiCd $0.94 to $5.66 Lead $1.89 to $3.77 $1.26/tire (new or used) $0.63/tire made of recycled material	Funds collection and recycling of old batteries Funds tire collection and recycling
Finland	Tires Lubricant oils and greases Hazardous waste Nuclear power generation	$2.50 to $50/tire $0.05/kg $336/ton $2.40 to $3.20/MWh	Funds tire recovery and recycling[b] Funds treatment of oil wastes Funds waste processing Funds waste processing
France	Lubricant oils, oil products Conventional waste Industrial and hazardous waste Automobile use of bridges to islands Use of inland waterways	$27/ton $7.20/ton, landfill disposal $7.20/ton, treated; $14.40/ton, stored $3.58/vehicle[c] Varies	Funds collection, recycling of used oil and oil products Funds research, treatment and equipment for contaminated site cleanup Funds protection of island environments Finances inland waterways authority
Italy	Lubricant oils	$0.03/kg	Funds collection, reuse and dumping costs
Kenya	Gasoline Diesel	$34/m^3$ $17/m^3$	Finances road maintenance
Netherlands	Surplus manure	$0.13 to $0.26/kg[d]	Funds manure transport, storage and processing

Table 3
(Continued)

Country	Item taxed	Rate	Use of revenues
South Korea	Toxic substance containers	1¢ /container over 500 ml	Funds waste disposal
	Cosmetics containers	0.2¢ to 0.7¢ /container	
	Batteries	0.2¢ /battery (all types)	
	Anti-freeze containers	2¢ /container	
	Flourescent light bulbs	0.6¢ /bulb	
	Chewing gum	0.25% of sale price	
	Disposable diapers	0.1¢ /diaper	
	Commercial operations and tourism within national parks	n.a.	Finances Korea National Park Authority (40%)
Spain	Pollutant spills into coastal waters	Varies with content and quantity of spill	Funds spill cleanup and sea quality improvement
Sweden	Fertilizers	$0.22/kg N for N > 2%; $3.70/g Cd for Cd > 5 g/ton of phosphorous	Finances environmental improvements in agriculture
	Tires	$1.50, automobiles; $37, trucks; $9.30 tractors	Finances recovery and recycling of used tires[e]
	Batteries	Lead, $4.90; NiCd, $5.70 Alkaline and HgO, $2.80	Covers used battery collection and disposal costs
Switzerland	Motorway use (cars and trucks)	Varies by weight, distance	Finances road construction and other road-related expenditures
	Leaded gasoline	$588/m^3	
	Unleaded gasoline	$529/m^3	
	Diesel fuel	$552/m^3	
United States	Motor fuels	$.183/gal	Highway Trust Fund/ Mass Transit Account
	Annual use of heavy vehicles	$100–$500/vehicle	
	Trucks and trailers (excise tax)	12%	
	Auto and truck tires	$0.15/lb (> 40 lbs) $4.50 + $0.30/lb (> 70 lbs) $10.50 + $0.50/lb (> 90 lbs)	
	Noncommercial motorboat fuels	$0.183/gal	Aquatic Resource Trust Fund
	Inland waterways fuels	$0.233/gal	Inland Waterways Trust Fund
	Non-highway recreational fuels and small-engine motor fuels	$0.183/gal gasoline $0.243/gal diesel	National Recreational Trails Trust Fund and Wetlands Account of Aquatic Resources Trust Fund
	Sport fishing equipment	10% (outboard motors, 3%)	Sport Fishing Restoration Account of Aquatic Resources Trust Fund
	Bows and arrows	11%	Federal Aid to Wildlife
	Firearms and ammunition	10%	Program

Note: Conversion of all currencies to $US made using U.S. Federal Reserve historical bilateral exchange rates for December of the year in which data were gathered, available at http://www.bog.frb.fed.us/releases/H10/hist.

Table 3

(*Continued*)

[a]Belgium exempts these products from the tax when organized deposit-refund or collection system exists and minimum recycling or collection targets are achieved.

[b]Finland's tire recycling is managed by a private company. Rates are lower for tires made of recycled materials.

[c]Maximum rate.

[d]The Netherlands' manure charge is based on amount of manure produced per hectare: $.13/kg for amounts between 125 and 200 kg/ha; double that amount for amounts greater than 200 kg/ha.

[e]In Sweden, manufacturers, importers and sellers of tires are required to ensure that used tires are reused, recycled, or disposed of in an environmentally friendly manner.

Source: Barthold (1994); Speck (1998); OECD (1997a); Ayoo and Jama (1999); and Rhee (1994).

noncommercial motor boat fuels are turned over to an Aquatic Resource Trust Fund; revenues from an inland waterways fuels tax are dedicated to the Inland Waterways Trust Fund; revenues from non-highway recreational fuels and small-engine motor fuels taxes are turned over to recreational trusts; and excise taxes on trucks, sport fishing and hunting equipment, and fishing and hunting licenses are similarly dedicated to specific, closely related uses (Table 3).

In European countries, airline traffic taxes are frequently used to finance noise pollution abatement. Aircraft landing charges in Belgium, France, Germany, the Netherlands, and Switzerland resemble Pigouvian taxes, as they relate the charge level to noise levels [McMorran and Nellor (1994)], and in Germany, France, Italy, the Netherlands, Sweden, and Switzerland, revenues from aircraft landing taxes are used to finance noise abatement programs [Speck (1998)].

In the late 1970s, Singapore implemented a comprehensive traffic management program. In order to drive a vehicle through the city center at peak travel periods, drivers must purchase monthly licences [Panayotou (1998), Sterner (1999)]. In Seoul, South Korea, drivers pay congestion surcharges for vehicles carrying fewer than three passengers through particular tunnels [OECD (1997c)]. The Norwegian cities of Oslo, Bergen, and Trondheim charge vehicles for entry into the urban core, but the fees are not differentiated by time of day and have had little incentive effect [Ekins (1999)]. Milan, Italy has introduced a peak-period licensing program which has been credited with a 50 percent reduction in traffic in the urban center [Ekins (1999)].

2.3.2. Municipal environmental services

The closest that any charge system in the United States comes to operating as a Pigouvian tax may be the unit-charge approach to financing municipal solid waste collection, where households (and businesses) are charged the incremental costs of collection and disposal. So called "pay-as-you-throw" policies, where users pay in proportion to the volume of their waste, are now used in well over 4,000 communities in 42 states, reaching an estimated 10 percent of the U.S. population [U.S. Environmental Protection Agency (2001)]. This collective experience provides evidence that unit charges have been somewhat successful in reducing the volume of household waste generated [Efaw and Lanen (1979), McFarland (1972), Skumatz (1990), Stevens (1978), Wertz (1976), Lave and Gruenspecht (1991), Repetto et al. (1992), Jenkins (1993), Fullerton and Kinnaman (1996), Miranda et al. (1994)].[25]

Like many U.S. cities, Switzerland has instituted a pay-as-you-throw system for solid waste disposal, in which ratepayers pay per bag. The system finances waste disposal and seeks to encourage lower volume. The evidence indicates that the volume of municipal solid waste has indeed decreased as a result of the program, but increased illegal disposal may be part of the explanation [OECD (1998e)]. In New Zealand, as many as 25 percent of communities employ volume-based charges for municipal solid waste collection [New Zealand Ministry for the Environment (1997)]. Similarly, Bolivia, Venezuela, Jamaica, and Barbados have adopted volume-based fees for solid waste collection [Huber, Ruitenbeek and Serôa da Motta (1998)].

More broadly, there is significant movement in many developing countries and transition economies toward cost-recovery (full-cost) pricing of environmental services, such as electric power, solid waste collection, drinking water, and wastewater treatment.[26] Full-cost pricing for municipal environmental services is becoming increasingly common in Latin America and the Caribbean [Huber, Ruitenbeek and Serôa da Motta (1998)], but major problems persist. Since 1993, for example, Colombian law has required water charges to incorporate the cost of service and environmental damages, but 90 percent of Colombia's regional governments have declared the law too difficult to implement.

The pace of progress in the transition economies of Eastern Europe and the former Soviet Union is somewhat faster. Decentralization of public services and the lifting of restrictions on tariff increases has reduced municipal reliance on state transfers for environmental services. Between 1989 and 1995, for example, the Hungarian central government's subsidy of public water supplies decreased from 100 percent to 30 percent [World Bank (1997b)]. Drinking water services in cities such as Budapest, Prague, and Zagreb have been privatized, bringing tariffs from minimal levels to ones sufficient to support full operating, and in some cases, capital cost recovery [World Bank (1997b), OECD (1999a)].

---

[25] Volume-based pricing can provide incentives, however, for citizens to compact their waste prior to disposal, so that reductions in quantity of waste (measured by weight, for example) may be significantly less than volume reductions. Also, as the costs of legal disposal increase, incentives for improper (illegal) disposal also increase. Hence, waste-end fees designed to cover the costs of disposal, such as unit curbside charges, can lead to increased incidence of illegal dumping [Fullerton and Kinnaman (1995)].

[26] While the text focuses on progress in environmental service cost-recovery in developing and transition economies, this is not to imply that economically rational tariffs fully characterize conditions in industrialized nations. For example, water metering is not used in many urban areas in Canada, and many Canadian municipal water and wastewater charges are not related to actual volumes consumed or produced [OECD (1995b)]. Likewise, Japan raises less than five percent of the cost of municipal waste collection, treatment, and disposal through user charges [OECD (1994a)].

2.3.3. Product disposal

Product taxes are used in many European countries to reduce the volume of materials in the waste stream. Where the size of such product taxes is insufficient to induce behavioral response, and revenues are used to cover disposal costs, the taxes can be considered user charges. In those cases in which product tax revenues go into general funds, I consider them sales taxes (see Section 2.5 below); product taxes that induce significant behavioral impact are rightly considered pollution taxes.

Thus, the classification of product taxes as user charges is complicated. For example, four EU member states (Belgium, Denmark, Italy, and Sweden) tax batteries. No attempts to measure the behavioral impacts associated with these taxes have been reported. The battery taxes in both Sweden and Denmark are earmarked to cover battery collection and recycling costs, and so these could be considered user charges, provided the taxes do not significantly influence battery purchases. Belgium's battery tax revenues are earmarked for environmental purposes. Italy's battery tax is differentiated according to lead content, but revenues go into general funds and are not used for environmental purposes.

Tire taxes in Denmark, Finland and Sweden can be considered user charges, as revenues are earmarked for tire collection and recycling, and there appear to be no behavioral impacts. France, Finland, and Italy levy lubricant oil taxes, the revenues from which cover disposal expenses. Surplus manure charges in Belgium and the Netherlands might also be considered user charges, as revenues are earmarked for transport, storage, and processing. Finland levies nuclear waste management charges that are earmarked for waste processing [Speck (1998)]. Finally, South Korea imposes waste disposal charges on containers from insecticides and toxic substances, and on butane gas, cosmetics, confectionery packaging, batteries, and antifreeze [OECD (1997c)].

2.4. Insurance premium taxes

In a relatively small number of countries, taxes are levied on industries or groups to fund insurance pools against potential environmental risks associated with the production or use of taxed products (Table 4). Such taxes can have the effect of encouraging firms to internalize environmental risks in their decision making, but, in practice, these taxes have frequently not been targeted at respective risk-creating activities. In the United States, for example, to support the Oil Spill Liability Trust Fund, all petroleum products are taxed, regardless of how they are transported, possibly creating small incentives to use less petroleum, but not to use safer ships or other means of transport. The fund can be used to meet unrecovered claims from oil spills.

An excise tax on specified hazardous chemicals is used to fund (partially) the cleanup of hazardous waste sites through the Superfund program in the United States. The tax functions as an insurance premium to the extent that funds are used for future clean-ups [Barthold (1994)]. The Leaking Underground Storage Trust Fund, established in 1987, is replenished through taxes on all petroleum fuels. Finally, the Black Lung Disability

Table 4
Insurance premium taxes

Country	Item/Action taxed	First enacted/ modified	Rate	Use of revenues
Belgium	Ionizing radiation	1994	n.a.	Fund for Risks of Nuclear Accidents
Finland	Oil imports	1970s	$0.43/ton[a]	Oil Pollution Compensation Fund
United States	Chemical production	1980/1986	$0.22 to $4.88/ton	Superfund (CERCLA)
	Petroleum production	1980/1986	$0.097/barrel crude	
	Corporate income	1986	0.12%[b]	
	Petroleum and petroleum products	1989/1990	$0.05/barrel	Oil Spill Liability Trust Fund
	Petroleum-based fuels, except propane	1986/1990 (expired 1995)	$0.001/gal	Leaking Underground Storage Trust Fund
	Coal production	1977/1987	$1.10/ton underground $0.55/ton surface	Black Lung Disability Trust Fund
	Surface coal mining and reclamation	1977	Varies with specific case	Repayment of performance bonds

Note: Conversion of all currencies to $US made using U.S. Federal Reserve historical bilateral exchange rates for December of the year in which data were gathered, available at http://www.bog.frb.fed.us/releases/H10/hist

[a]Rate is twice as high for tankers without double hulls.
[b]Rate is 0.12% of "alternative minimum taxable income" that exceeds $2 million.

Sources: Barthold (1994); Speck (1998); and OECD (1997a).

Trust Fund was established in 1954 to pay miners who became sick and unable to work because of prolonged exposure to coal dust in mines. Since 1977, it has been financed by excise taxes on coal from underground and surface mines.

Finland maintains an Oil Pollution Compensation Fund, financed by an oil import fee, to cover spill preparedness, clean-up, and damages [OECD (1997a)]. Since 1989, Sweden has had a compulsory insurance system to compensate for damages when polluters cannot be identified [OECD (1996)], managed by private insurance companies and financed by 10,000 "operators of dangerous facilities". France requires operators of quarries and waste storage facilities to post financial guarantees protecting the public from potential non-payment of mitigation expenses [OECD (1997b)], and Belgium requires insurance for waste import and export, and for the operation of oil storage yards. Spain requires pollution liability insurance of companies handling hazardous waste in the chemical industry [OECD (1997d)], and operators of waste and tire disposal sites in the Canadian province of Quebec deposit a required financial guarantee and take out mandatory environmental liability insurance to cover disposal costs and potential damage costs [OECD (1995b)]. Similarly, in the United States, under the 1977 Surface Mining Control and Reclamation Act, the purchase of performance bonds[27] are required before surface coal mining and reclamation permits are issued.

2.5. Sales taxes

Nations around the world have levied sales and value-added taxes, frequently in the name of environmental protection, on diverse goods and services, including motor fuels, other energy products, new automobiles, pesticides, fertilizers, chlorinated solvents, volatile organic compounds, lubricating oils, non-refillable containers, ozone-depleting substances, and new tires (Table 5). I focus on four categories of such taxes: motor fuels; ozone-depleting chemicals; agricultural inputs; and product taxes.

2.5.1. Motor fuels

All EU member states tax motor fuels to raise revenues for general funds. Rates are typically differentiated for leaded and unleaded gasoline, diesel fuel, light heating fuels, and heavy fuel oil, indicating that these taxes may also have environmental functions. Motor fuel taxes in European countries also include value-added taxes, ranging from 12 percent (Luxembourg) to 25 percent (Denmark and Sweden). In Mexico, the fuel tax includes a special surcharge in Mexico City, the revenues from which are used to fund gas station modifications to reduce volatile organic compound emissions [OECD (1998d)].

2.5.2. Ozone-depleting chemicals

It has been argued that only two U.S. national sales taxes have affected behavior in the manner of a Pigouvian tax: the "gas guzzler tax" on new cars, discussed later, and the excise tax on ozone-depleting chemicals [Barthold (1994)], although it is far from clear that the chloroflourocarbon (CFC) tax actually affected business decisions (Table 5). To meet international obligations established under the Montreal Protocol to limit the release of chemicals that deplete stratospheric ozone, the Federal government set up a tradable permit system (discussed below in Section 3.2.1) and levied an excise tax on specific CFCs in 1989. Producers are required to have adequate allowances, and users pay a fee (set proportional to a chemical-specific ozone depleting factor). There is considerable debate regarding which mechanism should be credited with the successful reduction in the use of these substances [Hahn and McGartland (1989), U.S. Congress

---

[27] Although I consider performance bonds under the heading of insurance premium taxes, this instrument can also be considered be the generic form of a deposit-refund system, since the amounts deposited with a performance bond can be refunded only when the affected firm fulfills particular obligations.

Table 5
Sales and value-added taxes

Item/Action taxed	Country	Rates	Use of revenues
Motor fuel, other energy products (excise taxes)	Austria[a]	Gas oil: heating, $81/m^3; industrial, $332/m^3	General budget
	Belgium[b]	Gasoline: leaded, $648/m^3; unleaded, $580/m^3	General budget
		Gas oil: heating, $6/m^3; industrial, $22/m^3	
	China	Gasoline: $3.44/m^3	General budget
		Diesel oil: $1.72/m^3	
	Denmark[c]	Gasoline: leaded, $632/m^3; unleaded, $530/m^3	General budget
		Gas oil: heating, $267/m^3; industrial, $267/m^3	
	Finland[d]	Gasoline: leaded, $709/m^3; unleaded, $620/m^3	General budget
		Gas oil: heating and industrial, $22/m^3	
	France[e]	Gasoline: leaded, $737/m^3; unleaded, $688/m^3	General budget
		Gas oil: heating and industrial, $91/m^3	
	Germany[f]	Gasoline: leaded, $648/m^3; unleaded, $588/m^3	General budget
		Gas oil: heating and industrial, $48/m^3	
	Greece[g]	Gasoline: leaded, $454/m^3; unleaded, $397/m^3	General budget
		Gas oil: heating, $150/m^3; industrial, $275/m^3	
	Ireland[h]	Gasoline: leaded, $242/m^3; unleaded, $198/m^3	General budget
		Gas oil: heating and industrial, $25/m^3	
	Italy[i]	Gasoline: leaded, $672/m^3; unleaded, $618/m^3	General budget
		Gas oil: heating, $452/m^3; industrial, $136/m^3	
	Kenya	Gasoline: premium, $100/m^3; regular, $194/me; diesel, $98/m^3	General budget
	Luxembourg[j]	Gasoline: leaded, $426/m^3; unleaded, $371/m^3	General budget
		Gas oil: heating, $6/m^3; industrial, $20/m^3	
	Netherlands[k]	Gasoline: leaded, $732/m^3; unleaded, $656/m^3	General budget
		Gas oil: heating and industrial, $55/m^3	
		Uranium-235, $17/g used in nuclear power generation	
	Norway[l]	Gasoline: leaded, $575/m^3; unleaded, $542/m^3	General budget
	Portugal[m]	Gasoline: leaded, $591/m^3; unleaded, $555/m^3	General budget
		Gas oil: heating, $117/m^3; industrial, $324/m^3	
	Spain[n]	Gasoline: leaded, $465/m^3; unleaded, $427/m^3	General budget
		Gas oil: heating and industrial, $91/m^3	
	Sweden[o]	Gasoline: leaded, $527/m^3; unleaded, $446/m^3	General budget
		Gas oil: heating and industrial, $92/m^3	
	United Kingdom[p]	Gasoline: leaded, $819/m^3; unleaded, $731/m^3	General budget
		Gas oil: heating and industrial, $49/m^3	
Motor fuels, other energy products (VAT)	Austria	20%	General budget
	Belgium	21%; except coal and other solid fuels (12%)	General budget
	Denmark	25%	General budget
	Finland	22%	General budget
	France	20.6%; 5.5% on fixed charge portion of utility bills	General budget
	Germany	16%	General budget
	Greece	18%; natural gas and coal are exempt	General budget

Table 5
(*Continued*)

Item/Action taxed	Country	Rates	Use of revenues
	Ireland	21% motor fuels; 12.5% other energy products; fuels for public transport are exempt	General budget
	Italy	19%, except coal (9%) and electricity (10%)	General budget
	Kenya	$34/m^3 industrial diesel and fuel oil; $52/m^3 LPG	General budget
	Luxembourg	15% motor fuels, except unleaded gasoline (12%); 12% gas oil, kerosene and coal; 6% LPG	General budget
	Netherlands	17.5%	General budget
	Norway	23%	General budget
	Portugal	17% motor fuels and kerosene; 12% electricity; 5% natural gas	General budget
	Spain	16%	General budget
	Sweden	25%	General budget
	Switzerland	6.5%	General budget
	United Kingdom	17.5%, except domestic heating fuels (5%)	General budget
New automobiles	Austria	[(fuel consumption per 100 km − 3 liters) · 2% of net price]; electric cars are exempt	General budget
	Belgium	$73–$5,800/vehicle, based on engine power	General budget
	China	Sedans, cross-country vehicles and minibuses: 3% to 8%, depending on cylinder volume	General budget
	France	Varies with engine power	Regional budget
	Germany	$21–$30	General budget
	Greece	Varies with cubic capacity; vehicles with anti-pollution technology subject to reduced rate	General budget
	Ireland	13.3%–28%, depending on cubic capacity	General budget
	Italy	$91–$236, depending on type and size of vehicle	General budget
	Netherlands	Varies with vehicle type, weight, and fuel type	General budget
	Norway	Varies with weight, horsepower and piston displacement	General budget
	Portugal	$1.47–$12 per 100 cc	General budget
	Spain	7% of sale price	General budget
	United States	$1,000–$7,700/auto exceeding fuel efficiency maxima	U.S. Treasury
Pesticides	Belgium	$.06/g of specified contents	General budget
	Denmark	3%–37% of retail price, varies by toxicity	General budget
	Finland	2.5% of total annual sales	General budget
Fertilizers	Sweden	$0.16/kg nitrogen; $0.30/kg phosphorous	General budget
Chlorinated solvents	Denmark	$0.31/kg of tetrachlorethylene, trichloroethylene, and dichloromethane	General budget
VOC	Switzerland	$0.73/kg	General budget
Lubricant oils	Denmark	$0.28/liter	General budget
	Sweden	$0.14/liter	General budget
Non-refillable containers	Finland	$0.80/liter	General budget
	Sweden	$0.04–$0.42/container	General budget

Table 5
(*Continued*)

Item/Action taxed	Country	Rates	Use of revenues
Ozone-depleting substances	Australia	$1,225/ton CFCs; $55/ton methyl bromide	General budget
	Denmark	$4.70/kg CFCs or halons	General budget
	United States	$4.35/pound	U.S. Treasury
New tires	United Sates	$0.15–$0.50/pound	U.S. Treasury

Note: VAT is an acronym for value-added tax, and VOC is an acronym for volatile organic compounds. Conversion of all currencies to $US made using U.S. Federal Reserve historical bilateral exchange rates for December of the year in which data were gathered, available at http://www.bog.frb.fed.us/releases/H10/hist.
[a] Austria also assesses excise taxes on heavy fuel oil, LPG and kerosene, at varying rates. Gas oil for cogeneration is taxed at the same rate as domestic heating oil. Austria's motor fuel excise taxes are excluded here because revenues are used for public transport expenses and can therefore be considered user charges. See Table 3.
[b] Belgium also assesses excise taxes on diesel, LPG and kerosene, at varying rates. In addition to excise taxes, most motor fuels and other energy products are subject to an energy tax of $10 to $15/m^3, the revenues from which are earmarked for a social security fund.
[c] Denmark also assesses excise taxes on heavy fuel oil, LPG, kerosene, coal, natural gas, and electricity, at varying rates. Partial rebates are available for gas stations with vapor recovery systems.
[d] Finland also assesses excise taxes on diesel and kerosene, at varying rates.
[e] France also assesses excise taxes on diesel, LPG, kerosene and heavy fuel oil, at varying rates.
[f] Germany also assesses excise taxes on diesel, LPG, kerosene and heavy fuel oil, at varying rates.
[g] Greece also assesses excise taxes on diesel, LPG, kerosene and heavy fuel oil, at varying rates.
[h] Ireland also assesses excise taxes on diesel, LPG, kerosene and heavy fuel oil, at varying rates.
[i] Italy also assesses excise taxes on diesel, LPG, kerosene, heavy fuel oil, natural gas and electricity, at varying rates.
[j] Luxembourg also assesses excise taxes on diesel, LPG, kerosene and heavy fuel oil, at varying rates.
[k] The Netherlands also assesses excise taxes on diesel, LPG, kerosene and heavy fuel oil, at varying rates. All fuels are also subject to a general energy tax, which ranges from $13/m^3 for leaded and unleaded gasoline to $18/m^3 for LPG.
[l] Norway also assesses excise taxes on diesel and electricity, at varying rates, although manufacturing enterprises are exempt from the tax on electricity.
[m] Portugal also assesses excise taxes on diesel, LPG, kerosene, heavy fuel oil and electricity, at varying rates.
[n] Spain also assesses excise taxes on diesel, LPG, kerosene and heavy fuel oil, at varying rates.
[o] Sweden also assesses excise taxes on diesel, LPG, kerosene, heavy fuel oil, coal, natural gas and electricity, at varying rates.
[p] The United Kingdom also assesses excise taxes on diesel, LPG, kerosene and heavy fuel oil, at varying rates.

Sources: Ayoo and Jama (1999); Barthold (1994); Zou and Yuan (1998); Speck (1998); and OECD (1997a, 1998b).

(1989), U.S. Congress, Office of Technology Assessment (1995), Cook (1996)]. Denmark and Australia also tax ozone-depleting chemicals (ODCs), and the Danish ODC tax seems to have affected use [Blackman and Harrington (1999)].

2.5.3. Agricultural inputs

Several states in the United States impose taxes on fertilizers and pesticides, but at levels below those required to affect behavior significantly. The taxes generate revenues that are used to finance environmental programs [Morandi (1992), International Institute for Sustainable Development (1995)]. Likewise, Sweden imposes sales taxes on agro-chemicals, including commercial fertilizers (containing nitrogen and phosphorous) and pesticides [OECD (1996)]. There is evidence that the Swedish taxes have reduced nitrogen use by 10 percent and total pesticide use by 35 percent [Ekins and Speck (1998)]. Denmark and Finland also tax pesticides [Speck (1998), OECD (1999b)].

2.5.4. Product taxes

The U.S. Energy Tax Act of 1978 established a "gas guzzler" tax on the sale of new vehicles that fail to meet statutory fuel efficiency levels, set at 22.5 miles per gallon. The tax ranges from $1,000 to $7,700 per vehicle, based on fuel efficiency; but the tax does not depend on actual performance or on mileage driven. The tax is intended to discourage the production and purchase of fuel inefficient vehicles [U.S. Congress (1978)], but it applies to a relatively small set of luxury cars, and so has had limited effects.[28]

In the European Union, disposable products as diverse as cameras, light bulbs, and razors are taxed, in addition to disposable containers and packaging. Denmark's carrier bag tax, differentiated so that plastic bags are more expensive than paper (though both are taxed), is an example of such a sales tax; revenues go to the general budget. Belgium's disposable camera, disposable razor, and beverage container taxes are earmarked for general environmental purposes [Speck (1998)].

2.6. Administrative charges

These charges raise revenues to help cover the administrative costs of environmental programs (Table 6); the charges are not intended to change behavior. For example, under the National Pollution Discharge Elimination System of the U.S. Clean Water Act, charges by individual states for discharge permits are based in some states on the quantity and type of pollutant discharged. Likewise, the Clean Air Act Amendments of 1990 allow states to tax regulated air pollutants to recover administrative costs of state programs, and allow areas in extreme non-compliance to charge higher rates. Under this structure, the South Coast Air Quality Management District (SCAQMD) in Los Angeles has the highest permit fees in the country [U.S. Congress, Office of Technology Assessment (1995)].

---

[28] Light trucks, which include "sport utility vehicles", are fully exempt from the tax [Bradsher (1997)]. Ontario, Canada has a gas-guzzler tax combined with a rebate for fuel-efficient vehicles, but because the coverage of the tax is very limited and the rates are very low, the overall effect is negligible [Haites (1999)].

Table 6
Administrative charges

Country	Item/Action taxed	First enacted/ Modified	Rate	Use of revenues
Australia	Ozone-depleting substances	n.a.	$6,100 administration fee, $1,200 license fee	Covers cost of licensing and administration
Finland	Pesticides	n.a.	$990 one-time registration charge (new pesticides)	Covers cost of registration
France	Use of inland waterways	n.a.	Varies by waterway and type of craft	Earmarked for financing of inland waterways authority
Malaysia	Palm oil industrial effluent discharges	1978	$2.54 annually per enterprise	Covers license-processing costs
Sweden	Pesticides	1984	Inspection charge, plus 15.5% of wholesale price	Finances administrative costs of biocide registry
United Kingdom	Water pollutant discharges	1992	$840 one-time application charge, annual charge $650 per pollution unit	Finances national water discharge licensing policy
United States	Water pollutant discharges	1972	Varies by substance	State administrative cost of National Pollution Discharge Elimination System, Clean Water Act
	Criteria air pollutants	1990	Varies by implementing state	State administrative cost of state clean air programs under Clean Air Act

Note: Conversion of all currencies to $US made using U.S. Federal Reserve historical bilateral exchange rates for December of the year in which data were gathered, available at http://www.bog.frb.fed.us/releases/H10/hist

Sources: U.S. Congress, Office of Technology Assessment (1995); Speck (1998); and World Bank (1997b).

Sweden has implemented registration charges for pesticides and other chemicals, as well as a CFC charge, which pays for inspections [OECD (1996)]. Belgium levies licensing charges on pesticides, radioactive materials, and hazardous waste import and export, which cover inspection and control costs [OECD (1998c)]. Annual charges for pesticide use increase with pesticide toxicity, and hazardous material license fees are based on an index that accounts for fire, explosion, and toxicity risks. A pesticide registration charge has also been implemented in Finland [Speck (1998)]. Malaysia uses a licensing system to reduce effluents from the palm oil industry. Firms pay a non-refundable annual license processing fee that is reduced for mills that develop pollution-reducing technologies [World Bank (1997b)]. But the effluent fee should not be given excessive credit for Malaysia's significant reductions in water pollutant emissions [Vincent and Ali (1997)]. Canada recovers part or all of its regulatory costs in some sectors through permit fees [OECD (1995b)].

2.7. Tax differentiation

I use the phrase, "tax differentiation", to refer to credits, tax cuts, and subsidies for environmentally desirable behavior (Table 7). These serve as implicit taxes on environmentally undesirable behavior.

A number of U.S. national and state taxes have been implemented in attempts to encourage the use of renewable energy sources, implicitly taking into account externalities associated with fossil fuel energy generation and use. Under the Energy Policy Act of 1992, for example, electricity produced from wind and biomass fuels received a 1.5 cent per kWh credit, and solar and geothermal investments received up to a 10 percent tax credit. Although economists' natural response to energy-related externalities is to advise that fuels or energy use be taxed, there is econometric evidence that energy-efficient technology adoption subsidies may be more effective – in some circumstances – than proportional energy taxes [Jaffe and Stavins (1995)]. In other programs, from 1979 to 1985, employers could provide implicit subsidies to employees for certain commuting expenses, such as free van pools and mass transit passes on a tax-free basis. Likewise, subsidies from utilities to households for energy conservation investments have been excludable from individual income taxes.

European countries have used tax differentiation to reduce vehicle-related emissions by encouraging the switch from leaded to unleaded gasoline (as did New Zealand) and by encouraging clean car sales [Panayotou (1998)]. The drastic reduction in the market share of leaded gasoline in Europe between 1985 and 1995 can be attributed, in part, to the tax differentiation of leaded and unleaded gasoline, and to the tax preferences afforded vehicles with catalytic converters, which require unleaded gasoline [Ekins and Speck (1998)].

Many European countries assess differentiated taxes and fees on vehicles according to cylinder capacity, age, fuel efficiency, and other environmentally relevant grounds [Speck (1998)]. Iceland has differentiated import levies to promote smaller, more fuel efficient cars [OECD (1993c)]. Spain granted rebates on purchases of new cars during 1994 and 1995, provided that old cars were removed from use, a program subsequently replaced by a differential vehicle registration tax [OECD (1997d)]. Austria offers tax incentives for environmental investment enterprises, household energy saving measures, low-noise vehicles, catalytic converters, and electric cars [OECD (1995a)]; and Germany, Sweden, and the Netherlands report significant changes in consumer behavior due to vehicle-related tax differentiation [Panayotou (1998)]. Mexico has reduced its sales tax on new cars and raised fees on older vehicles in an attempt to reduce emissions. A number of other countries have implemented differentiated motor vehicle taxes to discourage vehicle use and fuel consumption, including Côte d'Ivoire (Ivory Coast), Kenya, Australia, Japan, Russia, Italy, Portugal, and Argentina [McMorran and Nellor (1994)].

Subsidized credit and tax or tariff relief for environmentally desirable investments are common in Latin America and the Caribbean [Huber, Ruitenbeek and Serôa da Motta (1998)]. Since 1995, an Argentinian tax exemption has encouraged the switch from

Table 7
Tax differentiation

Item/Action taxed	Country	Provision and differentiated rate
Motor fuels excise tax reductions and exemptions[a]	Belgium	Tax exemptions for motor fuels used in development of environmentally friendly products, rail carriage of passengers and goods
	Denmark	Tax rebate of $0.005/liter for gas stations with vapor recovery, full exemption for public transport
	Norway	Exemption for use of vapor recovery unit
	United States	Reduced rates for natural gas ($0.07/gal); methanol ($0.06/gal); and ethanol ($0.054/gal)
	United Kingdom	Reduction of $33/m^3 for diesel with low sulfur content
Motor fuels VAT reductions and exemptions[a]	Austria	Reduced rate for public transport services (10%)
	Belgium	Reduced rate for public transport services (6%)
	Denmark	Exemption for public transport services
	Finland	Reduced rate for public transport services (6%)
	France	Reduced rate for public transport services (5.5%)
	Germany	Reduced rate for urban public transport (7%)
	Greece	Reduced rate for public transport (8%)
	Ireland	Exemption for public transport
	Italy	Reduced rate for public transport (10%); urban bus/rail transit exempt
	Luxembourg	Reduced rate for public transport (3%)
	Netherlands	Reduced rate for public transport (6%)
	Portugal	Reduced rate for public transport (5%)
	Spain	Reduced rate for public transport (7%)
	Sweden	Reduced rate for public transport (12%)
Income tax credits and deductions	Australia	Deductions for prevention of land degradation
	Austria	Deductions for household energy saving measures, purchase of low-noise trucks (double normal capital deduction); exemption for industrial/commercial environmental investments
	Belgium	Increased deductions for green investments, energy-saving devices
	Colombia	Credits and deductions for reforestation activities
	Denmark	Deductions for environmental improvement equipment on small farms
	Ireland	Deductions for investments in renewable energy (maximum 50% of capital expenditure, investment must be held five years)
	Netherlands	Credit (40–52%) for specified corporate energy investments
	Russia	Credit (100%) for environmental protection equipment investments
	Spain	Deductions (maximum 10% of investment) for investments in environmental protection
	United States	Alcohol fuels: methanol ($0.60/gal) and ethanol ($0.54/gal)
		Business energy: solar (10%) and geothermal (10%)
		Non-conventional fuels: $3.00/Btu-barrel equivalent of oil
		Wind production (1.5¢ /kWh)
		Biomass production (1.5¢ /kWh)
		Electric automobiles (10% credit)

Table 7
(*Continued*)

Item/Action taxed	Country	Provision and differentiated rate
Other income tax provisions	Australia	Accelerated depreciation for water conservation and capital expenditure on environmental impact studies
	Barbados	Income tax rebate for water conservation and solar energy equipment in the tourism sector
	Brazil	Income tax rebates for adoption of clean technology
	Colombia	Income tax rebates for industrial pollution abatement investments
	Ecuador	Income tax relief for investments in mercury recovery in mining
	Finland	Accelerated depreciation (maximum 25% of purchase price for four years) for environmental investments
	France	Accelerated depreciation: 100% in first year for specified energy-saving equipment; lesser percentages for industrial water pollution, air pollution and noise reduction technologies
	Germany	Accelerated depreciation for pollution reduction equipment
	Hungary	Reduced rate for manufacturers of environmental products
	Japan	Capital allowance for solar energy, pollution prevention and recycling equipment; reduced rate for specified facilities for air, water and noise abatement, asbestos emission reduction, oil desulfurization and waste recycling
	Kenya	Capital expenditure for preventing soil erosion or planting permanent crops treated as current expense
	Netherlands	Accelerated depreciation for specified environmental technologies
	Switzerland	Accelerated depreciation for energy-saving and solar energy investments
	Tanzania	Capital expenditure for prevention of soil erosion treated as current expenditure
	United States	Van Pools: tax-free employer provided benefits Mass transit passes Utility rebates: exclusion of subsidies from utilities for energy conservation measures
	Venezuela	Income tax relief for industrial pollution abatement investments
Sales tax and VAT provisions	Australia	Sales tax exemption for recycled paper, solar power equipment and conversion of engines to LPG or natural gas
	Brazil	VAT rebates for adoption of clean technology
	Colombia	VAT rebates for industrial pollution abatement investments
	Denmark	Energy-saving light bulbs exempt from sales tax
	Germany	Reduced energy product excise tax (50%) for hydroelectricity
	Hungary	Reduced VAT rate for cars with catalytic converters
	Portugal	Reduced energy VAT rate of 5% for equipment related to solar or geothermal energy, and for generation of energy from waste
	Sweden	Energy VAT reduction for cogeneration plants (50%), exemption for electricity generated by wind power
	United Kingdom	Reduced VAT rate of 5% on installation of household energy-saving equipment

Table 7

(*Continued*)

Item/Action taxed	Country	Provision and differentiated rate
Tax exempt private activity bonds	United States	Interest exempt from Federal taxation: mass transit, sewage treatment, solid waste disposal, water treatment, high speed rail

Note: Conversion of all currencies to $US made using U.S. Federal Reserve historical bilateral exchange rates for December of the year in which data were gathered, available at http://www.bog.frb.fed.us/releases/H10/hist.
[a]For full motor fuels excise tax and VAT rates in each country, see Table 5. For full rates in the United States and Austria, in which motor fuels taxes are used to finance road investments, see Table 3.

Sources: Barthold (1994); Speck (1998); McMorran and Nellor (1994); and Huber, Ruitenbeek and Serôa da Motta (1998).

diesel and gasoline-powered vehicles to those that use compressed natural gas. Brazil and Colombia offer subsidies for industrial pollution abatement investments, as well as income tax and value-added tax rebates for clean technology adoption. Ecuador offers subsidies and tax relief for mining sector mercury recovery investments. Jamaica offers tax and tariff relief for pollution abatement investments. Mexico offers subsidies for industrial pollution abatement investments, and a set of pollution control equipment is exempt from import taxation. Venezuela offers tax and tariff relief for industrial abatement investments. However, weak enforcement and sporadic monitoring of investments have minimized the effects of these policies World Bank (1997b)].

Many countries include environmentally-friendly provisions within their corporate tax systems [McMorran and Nellor (1994)]. South Korea offers tax deductions for companies involved in environmental conservation, and for investments in anti-pollution facilities and waste recycling [OECD (1997c)]. Japan offers a capital allowance for solar energy equipment, and Germany offers accelerated depreciation for energy-saving and pollution-reducing equipment.

3. Tradeable permit systems

It is well known that over the past decade tradeable permit systems have been adopted for pollution control with increasing frequency in the United States [U.S. Environmental Protection Agency (1992), Tietenberg (1997b)], but it is also true that this market-based environmental instrument has begun to be applied in a number of other countries as well. World wide, these programs are of two basic types: credit programs and cap-and-trade systems. Under credit programs, credits are assigned (created) when a source reduces emissions below the level required by existing, source-specific limits; these credits can enable the same or another firm to meet its control target. Under a cap-and-trade system, an allowable overall level of pollution is established and allocated among firms in the form of permits, which can be freely exchanged among sources. In theory, the allocation can be carried out through free distribution or through sale (for example, auction) by the government.

3.1. Credit programs

There have been several significant applications of the credit program model: the U.S. Environmental Protection Agency's (EPA's) Emissions Trading Program (including a variety of state-level credit programs); the phasedown of leaded gasoline in the United States; U.S. heavy duty motor vehicle engine emissions trading; water quality permit trading; and two Canadian pilot programs (Table 8).[29] Activities implemented jointly (AIJ) under the United Nations Framework Convention on Climate Change (FCCC) are included in Section 3.1.6, even though they are pilot projects and do not generate credits toward greenhouse gas (GHG) commitments for investing nations and firms. There is, as yet, no international agreement in force to provide a framework for international GHG emissions credit programs.

3.1.1. EPA's emissions trading program

Beginning in 1974, EPA experimented with "emissions trading" as part of the Clean Air Act's program for improving local air quality through the control of volatile organic compounds (VOCs), CO, SO_2, particulates, and NO_x. Firms that reduced emissions below the level required by law received "credits" usable against higher emissions elsewhere. Companies could employ the concepts of "netting" or "bubbles" to trade emissions reductions among sources within the firm, so long as total, combined emissions did not exceed an aggregate limit [Tietenberg (1985), Hahn (1989), Foster and Hahn (1995)]. By the mid-1980s, EPA had approved more than 50 bubbles, and states had authorized many more under EPA's framework rules. Estimated compliance cost savings from these bubble programs exceeded $430 million [Korb (1998)].

The "offset" program, which began in 1977, goes further in allowing firms to trade emission credits. Firms wishing to establish new sources in areas that are not in compliance with ambient standards must offset their new emissions by reducing existing emissions. This can be accomplished through internal sources or through agreements with other firms. Finally, under the "banking" program, firms may store earned emission credits for future use. Banking allows for either future internal expansion or the sale of credits to other firms.

EPA codified these programs in its Emissions Trading Program in 1986, but the programs have not been widely used. States are not required to use the programs, and

---

[29] Also, California has used a vehicle retirement program that operates much like a credit system to reduce mobile-source air emissions by removing the oldest and most polluting vehicles from the road [Kling (1994), Alberini, Harrington and McConnell (1995), Tietenberg (1997b)].

Table 8
Tradeable permit systems

Country	Program	Traded commodity	Period of operation	Environmental and economic effects
Canada	ODS Allowance Trading PERT GERT	CFCs and methyl chloroform HCFCs Methyl bromide NO_x, VOCs, CO, CO_2, SO_2 CO_2	1993–1996 1996–present 1995–present 1996–present 1997–present	Low trading volume, except among large methyl bromide allowance holders Pilot program Pilot program
Chile	Santiago Air Emissions Trading	Total suspended particulates emission rights trading among stationary sources	1995–present	Low trading volume; decrease in emissions since 1997 not definitively tied to TP system
European Union	ODS Quota Trading	ODS production quotas under Montreal Protocol	1991–1994	More rapid phaseout of ODS
Singapore	ODS Permit Trading	Permits for use and distribution of ODS	1991–present	Increase in permit prices; environmental benefits unknown
United States	Emissions Trading Program	Criteria air pollutants under the Clean Air Act	1974–present	Performance unaffected; savings = $5–$12 billion
	Leaded Gasoline Phasedown	Rights for lead in gasoline among refineries	1982–1987	More rapid phaseout of leaded gasoline; $250 m annual savings
	Water Quality Trading	Point–nonpoint sources of nitrogen and phosphorous	1984–1986	No trading occurred, because ambient standards not binding
	CFC Trades for Ozone Protection	Production rights for some CFCs, based on depletion potential	1987–present	Environmental targets achieved ahead of schedule; effect of TP system unclear
	Heavy Duty Engine Trading	Averaging, banking, and trading of credits for NO_x and particulate emissions	1992–present	Standards achieved; cost savings unknown
	Acid Rain Reduction	SO_2 emission reduction credits; mainly among electric utilities	1995–present	SO_2 reductions achieved ahead of schedule; savings of $1 billion/year
	RECLAIM Program	SO_2 and NO_x emissions among stationary sources	1994–present	Unknown as of 2000
	N.E. Ozone Transport	Primarily NO_x emissions by large stationary sources	1999–present	Unknown as of 2000

Sources: Hahn and Hester (1989a); Hahn (1989); Schmalensee et al. (1998); Montero and Sánchez (1999); Klaassen (1999); and Haites (1996). "TP" refers to tradeable permits; "ODS" – ozone-depleting substances; "CFCs" – chlorofluorocarbons.

uncertainties about their future course may have made firms reluctant to participate [Liroff (1986)]. Nevertheless, companies such as Armco, DuPont, USX, and 3M have traded emissions credits, and a market for transfers has long since developed [Main (1988)]. Even this limited degree of participation in EPA's trading programs may have saved between $5 billion and $12 billion over the life of the programs [Hahn and Hester (1989b)].

State-level emissions credit programs authorized under the U.S. EPA framework include ones operating in California, Colorado, Georgia, Illinois, Louisiana, and New York. In California, sources that exceed VOC standards for one product can offset excess emissions through over-compliance in other products. Since 1996, Colorado has allowed sources to generate emission reduction credits by reducing production or changing processes and materials. Mobile sources can generate credits by scrapping high-emission vehicles and replacing them with cleaner ones, by fuel switching, or by trip reduction [Bryner (1999)]. In Telluride, Colorado, residents must turn in two existing wood-burning stove or fireplace permits for every new permit.

Georgia allows vehicle fleet operators to earn credits for vehicles that over-comply with Federal clean-fueled fleet regulations, and to bank and trade credits. Illinois instituted a program in 1993 that purchases and scraps pre-1980 automobiles. The program allows "allotment trading units" to be earned by scrapping vehicles (after tailpipe emissions and fuel evaporation have been measured). The trading units can be purchased by stationary sources operating in areas that violate Federal air quality standards. Stationary sources in Louisiana, within areas with current or past ozone pollution problems, can obtain NO_x and VOC allowances by scrapping old vehicles purchased from motorists at fair market value [Bryner (1999)]. New York's New Source Review Offset Program allows new sources to offset emissions with credits generated by all types of emission reductions, including shutdowns of old facilities.

3.1.2. Lead trading

The purpose of the U.S. lead trading program, developed in the 1980s, was to allow gasoline refiners greater flexibility in meeting emission standards at a time when the lead-content of gasoline was reduced to 10 percent of its previous level. In 1982, EPA authorized inter-refinery trading of lead credits, a major purpose of which was to lessen the financial burden on smaller refineries, which were believed to have significantly higher compliance costs. If refiners produced gasoline with a lower lead content than was required, they earned lead credits. Unlike a cap-and-trade program, there was no explicit allocation of permits, but to the degree that firms' production levels were correlated over time, the system implicitly awarded property rights on the basis of historical levels of gasoline production [Hahn (1989)].

In 1985, EPA initiated a program allowing refineries to bank lead credits, and subsequently firms made extensive use of this option. In each year of the program, more than 60 percent of the lead added to gasoline was associated with traded lead credits [Hahn and Hester (1989a)], until the program was terminated at the end of 1987, when the lead phasedown was completed.[30]

---

[30] Under the banking provisions of the program, excess reductions made in 1985 could be banked until the end of 1987, thereby providing an incentive for early reductions to help meet the lower limits that existed

The lead program was clearly successful in meeting its environmental targets, although it may have produced some (temporary) geographic shifts in use patterns [Anderson, Hofmann and Rusin (1990)]. Although the benefits of the trading scheme are more difficult to assess, the level of trading activity and the rate at which refiners reduced their production of leaded gasoline suggest that the program was relatively cost-effective [Kerr and Maré (1997), Nichols (1997)]. The high level of trading between firms far surpassed levels observed in earlier environmental markets.[31] EPA estimated savings from the lead trading program of approximately 20 percent over alternative programs that did not provide for lead banking, a cost savings of about $250 million per year [U.S. Environmental Protection Agency, Office of Policy Analysis (1985)]. The program provided measurable incentives for cost-saving technology diffusion [Kerr and Newell (2000)].

3.1.3. Heavy duty motor vehicle engine emission trading

For nearly a decade, the U.S. Environmental Protection Agency (EPA) has allowed averaging, banking, and trading of credits for NO_x and particulate emissions reductions among eleven heavy-duty truck and bus engine manufacturers. EPA introduced these provisions to facilitate compliance with stricter emissions standards [Haites (1997)]. Emissions reduced below the "standard rate" can be credited to offset emissions for other engines manufactured by the same firm in the same year (averaging), banked to offset emissions for other engines manufactured by the same firm in a future year (banking), or sold to another firm to offset emissions for engines manufactured in the same or a future year (trading).[32]

Manufacturers appear to have used averaging more often than banking, and banking tends to be most common immediately prior to changes in standards; the first inter-firm credit trade occurred in 1997 [Haites (1997)]. EPA has created similar programs for manufacturers of non-road diesel engines, including ones for agricultural and construction equipment, locomotive engines, and certain classes of marine engines.

3.1.4. Water quality permit trading

In contrast with air quality programs, the United States has had very limited experience with tradable permit systems for controlling water pollution. Several experimental, pilot, and new programs are described here.

---

during the later years of the phasedown. The official completion of the phasedown occurred on January 1, 1996, when lead was banned as a fuel additive [Kerr and Newell (2000)].

[31] The program did experience some relatively minor implementation difficulties related to imported leaded fuel. It is not clear that a comparable command-and-control approach would have done better in terms of environmental quality [U.S. General Accounting Office (1986)].

[32] Credits cannot be used to offset emissions above a "maximum rate".

Nonpoint sources, particularly agricultural and urban runoff, may constitute the major, remaining American water pollution problem [Peskin (1986)]. An "experimental program" to protect water quality in Colorado demonstrated how tradable permits could be used to reduce nonpoint-source water pollution. Dillon Reservoir is the major source of water for the city of Denver. Nitrogen and phosphorus loading threatened to turn the reservoir eutrophic, despite the fact that point sources from surrounding communities were controlled to best-available technology standards [U.S. Environmental Protection Agency, Office of Policy Analysis (1984)]. Rapid population growth in Denver, and the resulting increase in urban surface water runoff, further aggravated the problem. In response, state policy makers developed a point-nonpoint-source control program to reduce phosphorus flows, mainly from nonpoint urban and agricultural sources. The program was implemented in 1984 [Kashmanian (1986)]; it allowed publicly owned sewage treatment works to finance the control of nonpoint sources in lieu of upgrading their own treated effluents to drinking water standards [Hahn (1989)].

EPA estimated that the plan could save over $1 million per year [Hahn and Hester (1989a)], due to differences in the marginal costs of control between nonpoint sources and the sewage treatment facilities. However, very limited trading occurred under the program, for a variety of reasons, including: implementation of other regulations that reduced non-point source run off; lower than expected cost for installation of additional treatment facilities; and relatively high regional precipitation that diluted concentrations in the reservoir.

Other states have implemented statewide and local trading programs. In 1981, Wisconsin introduced a discharge trading program to control biological oxygen demand (BOD) on a 45-mile section of the Fox River, which contains the heaviest concentration of paper mills in the world [Svendsen (1998)]. Participants included 15 paper mills and six municipal wastewater treatment plants, but trading activity has been almost nonexistent (one trade), due in part to the fact that paper mills have met permit limits by introducing less water-intensive technologies and recycled wastewater into production processes, rather than trading [Svendsen (1998)]. North Carolina introduced a nitrogen and phosphorous trading system in the Tar-Pamlico River basin in 1989 to control nutrient discharge [OECD (1999c)]. The trading association covers a dozen sewage treatment plants and one industrial discharger. Membership is voluntary, but dischargers that choose not to join are subject to standard individual pollution permits. Members of the trading association can either reduce nutrients internally, trade within the group, or pay a fee of US$56/kg, revenues from which go toward non-point source reductions. Overall discharge of nutrients into the basin was reduced 28 percent between 1989 and 1999, despite an 18 percent increase in average effluent discharge.

Formal rule making for a water quality trading program in Michigan began in January, 2000. The program allows voluntary nutrient trading among and between point and nonpoint sources, consistent with the Clean Water Act and other Federal regulations. A two-year demonstration project for the statewide program, focusing on phosphorous in the Kalamazoo River watershed, was to be completed in June, 2000 [State of Michigan, Department of Environmental Quality (2000)]. The Minnesota Pollution Control

Agency has allowed a producer of malt for brewing to meet the provisions of its National Pollution Discharge Elimination System (NPDES) permit through point-nonpoint water quality trading. The firm, which discharges in the Minnesota River basin, offsets its discharges by paying upstream nonpoint sources to reduce phosphorous discharges, in part by purchasing land easements [Minnesota Pollution Control Agency (1997)].

Overall, by 2001, the U.S. Environmental Protection Agency was actively involved in the development or implementation of 35 effluent trading projects in California, Colorado, Connecticut, the District of Columbia, Florida, Iowa, Idaho, Massachusetts, Maryland, Michigan, Minnesota, North Carolina, New Jersey, Nevada, New York, Ohio, Pennsylvania, Virginia, and Wisconsin [U.S. Environmental Protection Agency (2001)].

3.1.5. Two Canadian pilot programs: PERT and GERT

Canada's Pilot Emission Reduction Trading (PERT) and Greenhouse Gas Emission Reduction Trading (GERT) projects are pilot credit programs. Since 1996, PERT has facilitated the voluntary registry of emission reduction credits in Ontario for industrial emissions reduction greater than required by regulations or voluntary commitments.[33] Ownership of registered credits can be contractually transferred between parties. The initial focus was NO_x and VOC emissions, but in 1997, the program was expanded to include CO_2, SO_2, and CO.

Through 1997, PERT registered 14,000 tons of NO_x, 6,000 tons of SO_2, and more than 1 million tons of CO_2 credits. The volume of registered credits has grown, and there have been a number of purchases of reduction credits. For example, in 1997, the Hartford (Connecticut) Steam Company purchased NO_x reduction credits created by Ontario Hydro and Detroit Edison Company to meet requirements of the Connecticut Department of Environmental Protection [Pilot Emissions Reduction Trading (1999)].

The GERT pilot project began in 1997 and was scheduled to end in December 1999. The project applies to six Canadian provinces: Alberta, British Columbia, Manitoba, Nova Scotia, Saskatchewan, and Quebec. The program's administrators review projects and evaluate trades. Government partners, such as provincial and federal environmental agencies, are included. These partners reserve the right to restrict emissions reductions considered under the pilot. GERT reviews only matched trades, i.e., those with both a buyer and a seller, one of which must be Canadian. Five matched applications were reported through June 1999 [Greenhouse Gas Emissions Trading Pilot Program (1999)]. The Canadian government counts GERT-recognized trades against any subsequent emission commitments [Sonneborn (1999)].

---

[33] PERT reviews but does not approve credits as they are registered. This "buyer beware" approach differs from that of GERT.

3.1.6. Activities implemented jointly under the Framework Convention on Climate Change

Following the 1992 "Earth Summit" in Rio de Janeiro, Brazil, countries that had ratified the Framework Convention on Climate Change (FCCC) met in Berlin in 1995 for the first Conference of the Parties (COP 1). There they decided to establish a pilot phase for "activities implemented jointly" (AIJ), whereby industrialized nations or firms within those nations can finance projects in other countries to reduce net emissions of greenhouse gases and thereby attempt to (partially or fully) meet their own greenhouse gas (GHG) "commitments".[34]

A number of countries have established national AIJ programs, including Japan, Norway, Sweden, Switzerland, and the United States. For example, the U.S. Initiative on Joint Implementation (USIJI), established in 1993, approved 22 projects through 1997, 17 of which were in Latin American countries, including Costa Rica, Honduras, Belize, Bolivia, Mexico, Nicaragua, and Panama [Panayotou (1998)]. Land use and energy appear to be the most common sectors for such programs [World Bank (1997b)].

Specific examples of AIJ projects include: a Norway–Mexico co-financing arrangement for a lighting project in Guadalajara and Monterrey, with additional funding from the World Bank's Global Environmental Facility; and a project switching a district heating plant in Decin, Czech Republic from coal to natural gas, with financing from several U.S. electric utilities [Dudek and Wiener (1996)]. According to one source, 133 AIJ projects had been accepted, approved, and endorsed by designated national authorities for the host and investing countries by September, 1999 [Jepma (1999)]. Limiting attention to those AIJ projects that had been approved by international authorities under the FCCC by mid-1999, the 94 projects included: 62 from the public sector and 32 from private firms; with project lives of one to sixty years; involving CO_2-equivalent reductions of 13 tons to 57 million tons; and average investments of approximately $6 million [Woerdman and Van der Gaast (1999), Dixon (1999)].

These projects cannot really be characterized as true emission credit programs, because the projects are – by definition – pilot programs for which the investing firm or nation receives no actual credit. Furthermore, the likely efficacy of implemented, non-pilot versions of such programs is in doubt due to the fact that they would rely upon hypothetical baselines, i.e., what host nations would have done – in terms of emissions – in the absence of respective investment projects. Nevertheless, AIJ merits mention because it may be a precursor of future attempts to use emission credit and/or cap-and-trade programs for global climate change, whether under the Kyoto Protocol or some other future international agreement.[35]

---

[34] Developing nations, such as Costa Rica, have also established AIJ programs. In any event, this should be distinguished from the more recent use of the phrase "joint implementation", which refers to prospective use of project-level credits among industrialized countries, each of which has targets under the Kyoto Protocol.

[35] For more on the economics of climate change, see the chapter by Charles Kolstad and Michael Toman in this Handbook.

3.2. Cap-and-trade programs

When economists, other scholars, and policy-makers reflect on experiences with market-based instruments for environmental protection, they typically highlight several prominent cap-and-trade systems employed in the United States. A complete list is somewhat longer: CFC trading under the Montreal Protocol to protect the ozone layer; SO_2 allowance trading under the U.S. Clean Air Act Amendments of 1990; NO_x trading, initiated in 1999 to control regional smog in the eastern United States; the Regional Clean Air Markets (RECLAIM) program in the Los Angeles area; the use of auctioned bus licenses and particulates trading in Chile; and other quantity instruments of various degrees of flexibility and cost-effectiveness.

3.2.1. CFC trading

A market in tradable permits was used in the United States to help comply with the Montreal Protocol, an international agreement aimed at slowing the rate of stratospheric ozone depletion. The Protocol called for reductions in the use of CFCs and halons, the primary chemical groups thought to lead to ozone depletion.[36] The market places limitations on both the production and consumption of CFCs by issuing allowances that limit these activities. The Montreal Protocol recognizes the fact that different types of CFCs are likely to have different effects on ozone depletion, and so each CFC is assigned a different weight on the basis of its depletion potential. If a firm wishes to produce a given amount of CFC, it must have an allowance to do so, calculated on this basis [Hahn and McGartland (1989)].

Through mid-1991, there were 34 participants in the market and 80 trades [Feldman (1991)]. However, the overall efficiency of the market is difficult to determine, because no studies were conducted to estimate cost savings. The timetable for the phaseout of CFCs was subsequently accelerated, and a tax on CFCs was introduced, principally as a "windfall-profits tax" to prevent private industry from retaining scarcity rents created by the quantity restrictions [Merrill and Rousso (1990)]. The tax may have become the binding (effective) instrument. Nevertheless, low transaction costs associated with trading in the CFC market suggest that the system was relatively cost-effective.

In similar fashion, production quotas for ozone-depleting substances (ODS) were transferred within and among European Union (EU) countries between 1991 and 1994, until production was nearly phased out. During that period, there were 19 transfers (all but two of which were intrafirm), accounting for 13 percent of the EU's allowable ODS production.

Singapore has operated a tradeable permit system for ODS since 1991. The government records ODS requirements and bid prices for registered end-users and distributors,

---

[36] The Montreal Protocol called for a 50 percent reduction in the production of particular CFCs from 1986 levels by 1998. In addition, the Protocol froze halon production and consumption at 1986 levels beginning in 1992.

and total national ODS consumption (based on the Montreal Protocol) is distributed to registered firms by auction and free allocation. Firms can trade their allocations. Auction rents, captured by the government, have been used to subsidize recycling services and environmentally-friendly technologies [Annex I Expert Group on the United Nations Framework Convention on Climate Change (1997)]. Likewise, New Zealand implemented a CFC import permit system in 1986, whereby CFC permits are distributed by the Ministry of Commerce (based on the Montreal Protocol), and trading is allowed among permit holders.

Canada has also experimented with cap-and-trade systems for ozone-depleting substances since 1993. A system of tradeable permits for CFCs and methylchloroform operated from 1993 to 1996, when production and import of these substances ceased. Producers and importers received allowances for use of CFCs and methylchloroform equivalent to consumption in the base year and were permitted to transfer part or all of their allowances with the approval of the federal government. There were only a very small number of transfers of allowances during the three years of market operation, however [Haites (1996)].

Canada first distributed tradeable allowances for methylbromide in 1995. Due to concerns about the small number of importers (five), allowances were distributed directly to Canada's 133 users of methylbromide. Use and trading of allowances was active among large allowance holders. In addition, Canada has operated an HCFC allowance system since 1996, distributing consumption permits for its maximum allowable use under the Montreal Protocol, but no HCFC transfers were recorded through 1999.

3.2.2. SO₂ allowance trading system

The most important application ever made of a market-based instrument for environmental protection is arguably the tradable permit system in the United States that regulates SO_2 emissions, the primary precursor of acid rain. This system, which was established under Title IV of the U.S. Clean Air Act Amendments of 1990, is intended to reduce sulfur dioxide and nitrogen oxide emissions by 10 million tons and 2 million tons, respectively, from 1980 levels.[37] The first phase of sulfur dioxide emissions reductions was started in 1995, with a second phase of reduction initiated in the year 2000.

In Phase I, individual emissions limits were assigned to the 263 most SO_2-emissions intensive generating units at 110 plants operated by 61 electric utilities, and located largely at coal-fired power plants east of the Mississippi River. After January 1, 1995, these utilities could emit sulfur dioxide only if they had adequate allowances to cover their emissions.[38] During Phase I, the EPA allocated each affected unit, on an annual

---

[37] For a description of the legislation, see: Ferrall (1991).

[38] Under specified conditions, utilities that had installed coal scrubbers to reduce emissions could receive two-year extensions of the Phase I deadline plus additional allowances.

basis, a specified number of allowances related to its share of heat input during the baseline period (1985–1987), plus bonus allowances available under a variety of special provisions.[39] Cost-effectiveness is promoted by permitting allowance holders to transfer their permits among one another and bank them for later use.

Under Phase II of the program, beginning January 1, 2000, almost all electric power generating units were brought within the system. Certain units are exempted to compensate for potential restrictions on growth and to reward units that are already unusually clean. If trading permits represent the carrot of the system, its stick is a penalty of $2,000 per ton of emissions that exceed any year's allowances (and a requirement that such excesses be offset the following year).

A robust market of bilateral SO_2 permit trading has emerged, resulting in cost savings on the order of $1 billion annually, compared with the costs under some command-and-control regulatory alternatives [Carlson et al. (2000)]. Although the program had low levels of trading in its early years [Burtraw (1996)], trading levels increased significantly over time [Schmalensee et al. (1998), Stavins (1998), Burtraw and Mansur (1999), Ellerman et al. (2000)].

Concerns were expressed early on that state regulatory authorities would hamper trading in order to protect their domestic coal industries, and some research indicates that state public utility commission cost-recovery rules have provided poor guidance for compliance activities [Rose (1997), Bohi (1994)]. Other analysis suggests that this has not been a major problem [Bailey (1996)]. Similarly, in contrast to early assertions that the structure of EPA's small permit auction market would cause problems [Cason (1995)], the evidence now indicates that this has had little or no effect on the vastly more important bilateral trading market [Joskow, Schmalensee and Bailey (1998)].

The allowance trading program has apparently had exceptionally positive welfare effects, with benefits being as much as six times greater than costs [Burtraw et al. (1998)]. The large benefits of the program are due mainly to the positive human health impacts of decreased local SO_2 and particulate concentrations, not to the ecological impacts of reduced long-distance transport of acid deposition. This contrasts with what was assumed and understood at the time of the program's enactment in 1990.

Ever since the program's initiation, downwind states, in particular, New York, have been somewhat skeptical about the effects of the trading scheme. This skepticism was translated into specific legislation passed by the New York State legislature and signed by the Governor in May of 2000. The legislation, which is subject to court challenge because of its implicit barrier to interstate commerce, would prevent electric utilities in New York State from selling surplus allowances to sources in upwind states, such as Ohio [Hernandez (2000)]. This legislation was driven by concern that the emissions

---

[39] Utilities that installed scrubbers receive bonus allowances for early clean up. Also, specified utilities in Ohio, Indiana, and Illinois received extra allowances during both phases of the program. All of these extra allowances were essentially compensation intended to benefit Midwestern plants that rely on high-sulfur coal. On the political origins of this aspect of the program, see: Joskow and Schmalensee (1998).

trading program was failing to curb acid deposition in the Adirondacks in northern New York State [Dao (2000)].

The empirical evidence indicates that New York's concern is essentially misplaced. The first question is whether acid deposition has increased in New York State. If the baseline for comparison is the absence of the Clean Air Act Amendments of 1990, then clearly acid deposition is less now than it would have been otherwise. If the baseline for comparison is the original allocation of permits under the 1990 law, but with no subsequent trading, then acid deposition in New York State is approximately unchanged (slightly increased, but within error bounds). But, such comparisons ignore the fact, as emphasized above, that the greatest benefits of the program have been with regard to human health impacts of localized pollution. When such effects are also considered, it becomes clear that the welfare effects of allowance trading on New York State, using *either* baseline, have been positive and significant [Burtraw and Mansur (1999), Swift (2000)]. Thus, the pending New York State ban on upwind trading would increase in-state emissions, increase ambient concentrations of SO_2 and particulates, and hence have net *negative* welfare effects on the State.

3.2.3. RECLAIM program

The South Coast Air Quality Management District, which is responsible for controlling emissions in a four-county area of southern California, launched a tradable permit program in January, 1994, to reduce nitrogen oxide and sulfur dioxide emissions in the Los Angeles area.[40] One prospective analysis predicted 42 percent cost savings, amounting to $58 million annually [Anderson (1997)]. As of June 1996, 353 participants in this Regional Clean Air Incentives Market program, had traded more than 100,000 tons of NO_x and SO_2 emissions, at a value of over $10 million [Brotzman (1996)]. One particularly interesting aspect of the trading program is its zonal nature, whereby trades are not permitted from downwind to upwind sources. In this way, this geographically-differentiated emissions trading program represents one step toward an ambient trading program.

3.2.4. Ozone transport region NO_x budget program in the Northeast

Under U.S. Environmental Protection Agency guidance, twelve Northeastern states and the District of Columbia implemented a regional NO_x cap-and-trade system in 1999 to reduce compliance costs associated with the Ozone Transport Commission (OTC) regulations of the 1990 Amendments to the Clean Air Act.[41] Required reductions are based on targets established by the OTC, which require reduction in emissions by large

---

[40] For a detailed case study of the evolution of the use of economic incentives in the SCAQMD, see [National Academy of Public Administration (1994, Chapter 2)]. Also see: Thompson (1997) and Harrison (1999).

[41] Seven OTC states have also implemented state-level NO_x trading programs: New Jersey, Connecticut, Delaware, New York, Massachusetts, New Hampshire and Maine [Solomon (1999, Section 3.2.5)].

stationary sources. The program, known as the Northeast Ozone Transport Region, includes three geographic zones.[42] Emissions restrictions from 1999–2003 are to be 35 percent of 1990 emissions in the Inner Zone, and 45 percent in the Outer Zone. After 2003, Inner and Outer Zone sources must reduce to 25 percent of 1990 emissions, and Northern Zone sources to 45 percent [Farrell, Carter and Raufer (1999)].

EPA distributes NO_x allowances to each state, and states then allocate allowances to sources in their jurisdictions. Each source receives allowances equal to its restricted percentage of 1990 emissions, and sources must turn in one allowance for each ton of NO_x emitted over the ozone season. Sources may buy, sell, and bank allowances. Potential compliance cost savings of 40–47 percent have been estimated for the period 1999–2003, compared to a base case of continued command-and-control regulation without trading or banking [Farrell, Carter and Raufer (1999)].

NO_x emissions trading may be complicated by existing command-and-control regulations on many sources, the seasonal nature of ozone formation, and the fact that problems tend to result from a few high-ozone episodes and are not continuous [Farrell, Carter and Raufer (1999)]. The potential for "wrong-way" trades, which would trade emissions reductions near the coastal or northern boundary (downwind of a non-attainment area) for reductions to the south or west (upwind), may also complicate the system [Farrell, Carter and Raufer (1999)].

3.2.5. State-level NO_x and VOC emissions trading programs

Many of the states within the Northeast Ozone Transport Region have established in-state trading programs that coordinate with the regional system in order to meet their statewide caps. Delaware implemented trading and banking of NO_x and VOCs among mobile and stationary sources in 1996, with all credits discounted by 10 percent. Credits can be retroactive for reductions as early as 1991, and trading can include sources outside Delaware within the NOTR. Maine instituted a trading program for NO_x and VOCs among stationary sources in 1998. Credits generated within another New England state require a 15 percent "surcharge" – an in-state source needing a 100-ton credit must purchase 115 tons from an out-of-state source. Credits generated within a state outside of New England, but within the NOTR, require a 100 percent surcharge [Bryner (1999)]. New Jersey created the Open Market Emissions Trading program in 1996, which authorizes trading of emissions reductions for NO_x and VOCs. Credits are discounted by 10 percent, and may be purchased from other states in the NOTR.

NO_x emissions trading and banking for stationary and mobile sources in Connecticut began in 1995. Mobile source emissions are discounted 10 percent, and emissions during the summer ozone season cannot be offset by credits generated at other times

---

[42] The Inner Zone includes the Atlantic coast from Northern Virginia to New Hampshire, to varying distances inland. The Outer Zone is adjacent to the Inner Zone, from western Maryland through most of New York State. The Northern Zone includes northern New York and New Hampshire, and all of Vermont and Maine.

of the year [Bryner (1999)]. Massachusetts' program, which covers NO_x VOCs, and CO, began in 1994. Sources of credits include more stringent controls, source reduction, fuel switching, energy conservation, fleet conversion, lawn and garden equipment trade-in, vehicle scrapping, and ride sharing [Bryner (1999)]. New Hampshire's Emissions Reduction Credits Trading Program allows stationary and mobile sources to generate credits for NO_x, VOC, and CO emissions reductions. Credits cannot be banked, and credits from facility shutdowns cannot be traded. Pennsylvania operates the NO_x Allowance Requirements Program, a mandatory cap-and-trade program that covers fossil-fuel-powered electric generating plants during the summer ozone season. Allowances are allocated each summer, and other types of sources may voluntarily opt in.

While not within the NOTR, Michigan and Illinois also have established NO_x emissions trading programs. The Michigan Department of Environmental Quality began a trading program in 1996 which allows emissions averaging (bubbling) and emissions reduction credit trading for most stationary and mobile sources and for all criteria pollutants other than ozone (O_3). Although the U.S. EPA has yet to approve Michigan's program, by mid-1998, 25,000 NO_x credits and 500 VOC credits were registered with the state [Solomon and Gorman (1998), Solomon (1999)]. The area around Chicago in northeast Illinois began a five-month summer season VOC cap-and-trade system in 2000. The program is mandatory for a set of large stationary sources that account for 26 percent of regional emissions.

3.2.6. Gasoline constituent and Tier 2 emission standard trading

The U.S. Clean Air Act Amendments of 1990 imposed more stringent mobile source emissions standards through two routes – requiring automobile manufacturers to reduce tailpipe emissions on new models, and requiring refineries to develop and market reformulated fuels. In 1992, the U.S. Environmental Protection Agency established a trading program for oxygenates in gasoline (to reduce emissions of carbon monoxide during the winter months). Although the trading program could – in theory – increase cost-effectiveness, virtually none of the affected jurisdictions chose to develop trading rules, citing monitoring costs, and the one area that did develop rules experienced no trading.

In 2000, EPA promulgated new standards for NO_x emissions from motor vehicles and for the sulfur content of gasoline. Vehicle manufacturers are permitted to average their NO_x emissions to comply with a corporate average standard, much like under the Corporate Average Fuel Economy Standards, discussed below. In this case, however, trading (and banking) with other manufacturers is also allowed. Similarly, beginning in 2004, refiners and importers must satisfy corporate average gasoline standards on sulfur content. Both banking and inter-refinery trade are to be allowed.

3.2.7. Chilean bus licenses

Since 1991, Chile has had an auctioning system in place for bus licenses to address congestion-related pollution in Santiago [Huber, Ruitenbeek and Serôa da Motta

(1998)]. Deregulation of Santiago's urban public bus system in the late 1970s had resulted in a significant expansion of the system [Hartje, Gauer and Urquiza (1994)], with congestion thereby increasing traffic-related emissions. In 1991, the Chilean Ministry of Transportation began auctioning access rights to buses and taxis in congested areas. Congestion has apparently been reduced by these measures, with emissions reduced proportionately, although actual emission reductions have not been measured [Panayotou (1998)]. Although the system has characteristics of a cap-and-trade system for vehicle congestion, it is not a cap-and-trade system for emissions control *per se*, because in order to bid for a license, a bus must first comply with the prevailing uniform emissions standard (indeed, through specified technology).

3.2.8. *Chilean TSP tradeable permits*

Chile also has implemented a tradeable permit system for total suspended particulates (TSP) from stationary sources in the Santiago area. Initial allocations were based on 1992 emissions, and new sources must offset all incremental emissions. Trading began in 1995. Emissions have decreased, due to the introduction of natural gas as an alternative fuel, but the volume of emissions trading has been low [Montero and Sánchez (1999)]. Regulatory uncertainty, high transaction costs (especially with respect to a lengthy and uncertain approval process), inadequate enforcement, and market concentration may be partly to blame for the low trading volume. An unexpected benefit of the Chilean TSP system was that the offer of free (and potentially valuable) tradeable permits provided a significant incentive to incumbent polluters to identify themselves and report their emissions, in order to claim their permits. Prior to the program's existence, the government authorities had a very limited inventory of sources and emissions.

3.2.9. *Other flexible quantity-based instruments*

Limited regulatory flexibility has been introduced within the context of several conventional quantity-based instruments in various countries, representing – in some cases – movements toward the use of tradeable permit approaches. For this reason, I review in this section such flexible quantity-based instruments.

The U.S. Energy Policy and Conservation Act of 1975 established a program of Corporate Average Fuel Economy (CAFE) standards for automobiles and light trucks. The standards require manufacturers to meet a minimum sales-weighted average fuel efficiency for their fleet of cars sold in the United States. A penalty is charged per car sold per unit of average fuel efficiency below the standard. The program operates like an internal-firm tradeable permit system or "bubble" scheme, since manufacturers can undertake efficiency improvements wherever they are cheapest within their fleets. Firms that do better than the standard can "bank" their surpluses and — in some cases — are permitted to borrow against their future rights.[43]

In an effort to increase flexibility, the U.S. EPA allows air toxics averaging *within individual facilities* when firms are seeking compliance with the 1990 Clean Air Act Amendments. Likewise, EPA permits the use of "bubbling" of water effluent from iron and steel plants under the U.S. Clean Water Act, but imposes tight constraints on its use [U.S. Environmental Protection Agency (2001)].

European national authorities have increased flexibility under a number of existing national and EU emissions standards to create limited quota and trading arrangements, although none have involved inter-firm financial transfers [Klaassen and Nentjes (1997)]. For example, in Denmark, the Ministry of Environment fixes annual emissions ceilings in the power generation industry as a whole, and leaves allocation of the annual ceilings to the country's two power plant consortia. From 1991 to 1997, the United Kingdom allowed intra-firm trading of SO_2 allowances among large combustion plants, as part of its plan for compliance with the EU's Large Combustion Plant Directive, aimed at acid rain control. Inter-firm trading was not allowed, and in the power sector, only part of a firm's annual emissions limitation was tradable [Sorrell (1999), Pototschnig (1994)]. In the Netherlands, electric power producers face emission standards for SO_2 and NO_x, but can comply through cost-sharing arrangements, whereby plants with higher abatement costs are compensated. The system has resulted in intra-firm trading, with estimated savings of $245 million [Klaassen and Nentjes (1997)].

In Germany, the transfer of emission reduction obligations among firms in air quality non-attainment areas is allowed. Since 1974, firms have been allowed to locate new plants in non-attainment areas, provided they replace existing plants in the same area, and the "replaced" plant need not be owned by the same firm. Since 1983, existing plant renovations can also be used to offset new plant emissions in non-attainment areas [Klaassen and Nentjes (1997)]. The cost savings associated with these rules have been very limited, however [Schärer (1994)]. Germany began a pilot project on tradable permits for VOC emissions among small vehicle refinishing shops in 1998 [Schärer (1999)].

From 1991 to 1992, an experimental program was carried out in Chorzów, one of Poland's most polluted municipalities [Żylicz (1999)]. Although emissions trading was not recognized by Polish law at the project's start, the Chorzów pilot project allowed the city's steel mill and power plant to negotiate collective emissions reductions for particulates, sulfur dioxide, carbon monoxide, and hydrocarbons.

---

[43] For reviews of the literature on CAFE standards, with particular attention to the program's costs relative to "equivalent" gasoline taxes, see Crandall et al. (1986) and Goldberg (1997). Light trucks, which are defined by the Federal government to include "sport utility vehicles", face significantly weaker CAFE standards [Bradsher (1997)].

4. Reducing market frictions

In some situations, environmental protection can be fostered by reducing or eliminating frictions in market activity. I consider three types of such market friction reductions: (1) *market creation* for inputs/outputs associated with environmental quality, as with measures that facilitate the voluntary exchange of water rights and thus promote more efficient allocation and use of scarce water supplies; (2) *liability rules* that encourage firms to consider the potential environmental damages of their decisions; and (3) *information programs*, such as energy-efficiency product labeling requirements.

4.1. Market creation for inputs/outputs associated with environmental quality

Two examples of using market creation as an instrument of environmental policy stand out: measures that facilitate the voluntary exchange of water rights and thus promote more efficient allocation and use of scarce water supplies; and particular policies that facilitate the restructuring of electricity generation and transmission.

First, the western United States has long been plagued by inefficient use and allocation of its scarce water supplies, largely because users do not have incentives to take actions consistent with economic and environmental values. For more than a decade, economists have noted that Federal and state water policies have been aggravating, not abating, these problems [Anderson (1983), Frederick (1986), El-Ashry and Gibbons (1986), Wahl (1989)]. As recently as 1990, in the Central Valley of California, farmers were paying as little as $10 for water to irrigate an acre of cotton, while just a few hundred miles away in Los Angeles, local authorities were paying up to $600 for the same quantity of water. This dramatic disparity provided evidence that increasing urban demands for water could be met at relatively low cost to agriculture or the environment (i.e., without constructing new, environmentally-disruptive dams and reservoirs). Subsequent reforms allowed markets in water to develop, so that voluntary exchanges could take place. For example, an agreement was reached to transfer 100,000 acre-feet of water per year from the farmers of the Imperial Irrigation District (IID) in southern California to the Metropolitan Water District (MWD) in the Los Angeles area.[44] Subsequently, policy reforms spread throughout the west, and transactions soon emerged elsewhere in California, and in Colorado, New Mexico, Arizona, Nevada, and Utah [MacDonnell (1990)].

In Colorado, water-rights trading has continued to develop [OECD (1997e)]. Water rights holders in one district, the Colorado River Basin, send, on average, 5 to 15 applications per month for water transfers to the district's Water Court, which reviews

---

[44] In March of 1983, the Environmental Defense Fund (EDF) published a proposal calling for MWD to finance the modernization of IID's water system in exchange for use of conserved water [Stavins (1983)]. In November, 1988, after five years of negotiation, the two water giants agreed on a $230 million water conservation and transfer arrangement, much like EDF's original proposal to trade conservation investments for water [Morris (1988)].

all transfers. Prices depend on the characteristics of the region and the particular water right: rights near Grand Junction trade for approximately $0.06 per cubic meter, while rights near rapidly-developing Summit City trade for $65 per cubic meter [OECD (1997e)]. Quantities traded range from 300 to 54,000 cubic meters per year. The Colorado market includes 22,000 water rights located in 11,000 diversion structures. All public and private parties, including government agencies, are treated alike in proposed transfer evaluations. For example, the state government must purchase rights to promote ecological uses, like wetlands and in-stream flows.

In February, 2000, Azurix, formerly a division of Enron Corporation, launched an Internet exchange for buying, selling, storing and transporting water in the western U.S., but it is too early to assess whether or how this system will enhance water market activity [Azurix (2000)].[45] In Chile, water rights trading was reintroduced in 1981, having existed from the 1920s through the 1960s, but prohibited in 1969 when water became state property [Huber, Ruitenbeek and Serôa da Motta (1998)]. Transactions are relatively rare, however. Australia has permitted water trading in parts of the country since 1982 [OECD (1998b)].

A second example of "market creation" is the worldwide revolution in electricity restructuring that is motivated by economic concerns[46] but possibly bringing significant environmental impacts. For many years, utilities in the United States – closely overseen by state public utility commissions (PUCs) – have provided electricity within exclusive service areas. The utilities were granted these monopoly markets and guaranteed a rate of return on their investments, conditional upon their setting reasonable rates and meeting various social objectives, such as universal access. The Energy Policy Act of 1992 allowed independent electricity generating companies to sell power directly to utilities, and in 1996, the Federal Energy Regulatory Commission (FERC) required utilities with transmission lines to transmit power for other parties at reasonable rates. The purpose of these regulatory changes was to encourage competition at the wholesale (electricity generation) level, but many states moved to facilitate competition at the retail level as well, so that consumers can contract directly for their electricity supplies. Legislation has been introduced in the U.S. Congress to establish guidelines for retail competition throughout the nation [Kriz (1996)].

These changes have environmental implications. First, as electricity prices fall in the new competitive environment, electricity consumption is expected to increase. This might be expected to increase pollutant emissions, but to whatever degree electricity substitutes for other, more polluting forms of energy, the overall effect may be environmentally beneficial. Second, deregulation will unquestionably make it easier for new

---

[45] The exchange is located at http://www.water2water.com

[46] The primary arguments for restructuring are: (1) the electricity industry is no longer a natural monopoly, since small generation technologies are now competitive with large centralized production; (2) consumers will benefit from buying cheaper electricity from more efficient producers, who currently face significant barriers to entry; and (3) the old system with cost-of-service pricing provides poor incentives for utilities to reduce costs [Brennan et al. (1996)].

firms and sources to enter markets. Since new power plants tend to be both more efficient and less polluting (relying more on natural gas), environmental impacts may decrease.[47] Third, more flexible and robust markets for electricity can be expected to increase the effectiveness of various market-based incentives for pollution control, such as the SO_2 allowance trading system.[48]

4.2. Liability rules

Liability rules can have the effect of providing strong incentives for firms to consider the potential environmental damages of their decisions.[49] In theory, a liability rule can be cost effective as a policy instrument, because technologies or practices are not specified. For example, taxing hazardous materials or their disposal creates incentives for firms to reduce their use of those materials, but does *not* provide overall incentives for firm to reduce societal *risks* from those materials. An appropriately designed liability rule can do just that [Revesz (1997)]. On the other hand, transaction costs associated with litigation may make liability rules appropriate only for acute hazards. It is in these situations, in fact, that this approach has been most frequently employed, particularly in the case of liability for toxic waste sites and for the spill of hazardous materials.

The U.S. Comprehensive Environmental Response, Compensation, and Liability Act (CERCLA) of 1980 established retroactive liability for companies that are found responsible for the existence of a site requiring clean up.[50] Governments can collect cleanup costs and damages from waste producers, waste transporters, handlers, and current and past owners and operators of a site.[51] Similarly, the Oil Pollution Act makes firms liable for cleanup costs, natural resource damages, and third party damages caused by oil spills onto surface waters; and the Clean Water Act makes responsible parties liable for cleanup costs for spills of hazardous substances.

---

[47] There is considerable debate on this point, since – in the short run – more electricity may be generated from old surplus capacity coal plants in the Midwest, increasing pollutant emissions. In any event, in the long run, competition will encourage a more rapid turnover of the capital stock [Palmer and Burtraw (1997)].

[48] Environmental advocates, however, are very concerned that state PUCs will have much less influence than previously over the industry. In the past, PUCs encouraged "demand side management" and supported the use of renewable forms of electricity generation through the investment approval process or by requiring full-cost pricing for generation. Several policies have been proposed to provide these functions in the new, more competitive environment: for example, a system of tradable "renewable energy credits", wherein each generator would need to hold credits for a certain percentage of their generation; and a tax on the transmission of electricity, used to subsidize renewable generation.

[49] These incentives are frequently neither simple nor direct, because firms and individuals may choose to reduce their exposure to liability by taking out insurance. In this regard, see the earlier discussion in this chapter of "Insurance Premium Taxes".

[50] Retroactive liability provisions can of course provide incentive effects only for future actions which might be subject to liability rules.

[51] For economic analyses of the Superfund program, see, for example: Hamilton (1993), Gupta, Van Houtven and Cropper (1996), and Hamilton and Viscusi (1999).

The Nordic countries have strict environmental liability rules. Sweden has held polluters strictly liable for full damage compensation since 1986 [OECD (1996)]; and Norway and Finland enforce strict liability for environmental damage [OECD (1997a)]. Germany, Belgium, France, and the Netherlands enforce strict liability for a variety of polluting activities [OECD (1995c, 1997b, 1998c)]. In the emerging market economies of central and eastern Europe, environmental liability rules have played particularly important roles in the process of economic transition [Panayotou, Bluffstone and Balaban (1994)].

Among developing countries, the nation of Trinidad and Tobago has established a voluntary policy of full compensation for environmental damages, but has not legislated mandatory liability [Huber, Ruitenbeek and Serôa da Motta (1998)]. Mexico has established strict liability of parties who degrade the environment [OECD (1998d)], but in Latin American and Caribbean countries, as in many developing nations, lack of resources among executive and judiciary institutions makes enforcement of these policies relatively uncommon.

4.3. Information programs

Since well-functioning markets depend, in part, on the existence of well-informed producers and consumers, information programs can – in theory – help foster market-oriented solutions to environmental problems.[52]

4.3.1. Product labeling requirements

One approach to government improving the set of information available to consumers is a product labeling requirement (Table 9). The U.S. Energy Policy and Conservation Act of 1975 specifies that certain appliances and equipment (including air conditioners, washing machines, and water heaters) carry labels with information on products' energy efficiency and estimated annual energy costs [U.S. Congress, Office of Technology Assessment (1992)]. More recently, EPA and the U.S. Department of Energy (DOE) developed the Energy Star program, in which energy efficient products can display an *Energy Star* label. The label does not provide specific information on the product, but signals to consumers that the product is, in general, "energy efficient". This program is

---

[52] For a comprehensive review of information programs and their apparent efficacy, see: Tietenberg (1997a). For an overview of international experience with "eco-labels", see: Morris and Scarlett (1996). A number of studies have measured statistically significant reactions of stock values to positive and negative environmental news in the U.S. and Canadian markets [Muoghalu, Robison and Glascock (1990), Lanoie and Laplante (1994), Klassen and McLaughlin (1996), Hamilton (1995), Laplante, Lanoie and Roy (1997)]. Recent work at the World Bank indicates that the same may be true in developing countries [Dasgupta, Laplante and Mamingi (1997)]. The International Standards Organization's (ISO) latest benchmark, ISO 14001, was issued in draft form in 1996 and includes new standards for environmental management systems. In order to obtain ISO 14001 certification, firms must commit to environmental performance targets, among other things. More than 8,000 plants worldwide had obtained certification through 1999 [World Bank (2000)].

Table 9
Information programs

Country	Information program	Year of implementation
Australia	Energy Efficiency Labeling	late 1980s
Canada	Environmental Choice Label	n.a.
China	National Environmental Protection Agency Labeling	1994
EU Members	EU Eco-Label	1993
Nordic Countries	Nordic Swan Label	1989
France	NF Environment Label	n.a.
Germany	Blue Eco-Angel Label	1977
Hungary	Eco-Label	1995
Indonesia	PROPER industrial environmental performance labeling	1995
	Tropical hardwood labeling	n.a.
Japan	Eco-mark	1989
Philippines	Eco-watch industrial environmental performance labeling	1997
Sweden	Good Environmental Choices Label	1990
Taiwan	Green Mark	1993
Thailand	Thai Green Label	1994
United States	Energy Efficiency Product Labeling	1975
	NJ Hazardous Chemical Emissions	1984
	Toxic Release Inventory	1986
	CA Hazardous Chemical Emissions	1987
	CA Proposition 65	1988
	Energy Star	1993

Sources: World Bank (1997a, 1997b); TerraChoice Environmental Services Inc. (1999); China Council Working Group on Trade and Environment (1996); European Union (1999); OECD (1997b); Federal Republic of Germany (1998); Sterner (1999); and Thailand Environment Institute (1999).

much broader in its coverage than the appliance labeling program; by 1997, over 13,000 product models carried the *Energy Star* label [U.S. Department of State (1997)]. There has been little economic analysis of the efficacy of such programs, but limited econometric evidence suggests that product labeling (specifically appliance efficiency labels) can have significant impacts on efficiency improvements, essentially by making consumers (and therefore producers) more sensitive to energy price changes [Newell, Jaffe and Stavins (1999)].

The European Union established an "Eco-label" in 1993; it was initially intended to replace proliferating (and possibly trade-restricting) national labels in Europe, but the European Parliament voted in 1998 to continue to allow national labels. By 1999, the Eco-label had been applied to 200 products, including detergents, light bulbs, linens and t-shirts, appliances, paper, mattresses, and paints.

The EU Eco-label has not supplanted older and more extensive European national systems. The German "Eco-Angel" label program, the world's first, began in 1977. More than 4,200 products in dozens of sectors have received the label, including almost 600 foreign products. Hungary's eco-label, introduced in 1995, borrows its issuance guidelines from the German Eco-Angel program. The Nordic Swan has been applied in

Norway, Sweden, Finland, and Iceland since 1989, and now covers 1,000 products. The market share of eco-labeled laundry detergents in Sweden increased from zero in 1990 to 80 percent by 1997, but analysts see no major improvement in environmental quality as a result of the switch to eco-labeled detergents [Sterner (1999)]. The French "NF Environnement" label has been granted for paint products and garbage bags [OECD (1997b)], and Spain's environmental label, administered by a private, non-profit organization, has been applied to ten classes of consumer products. The Czech Republic uses eco-labels on the basis of product life cycle analysis tests (paid for by applicants), and has issued 262 labels in 21 chiefly industrial product categories [OECD (1999a)].

Canada awards an "environmental choice" label on licensed products including appliances, automotive products, cleansers, office products, paints, paper products, printing services, plastic products, film, and other items. The program, operated in the private sector through an exclusive license agreement, has granted labels to 1,400 products. Environmental labeling programs also exist in several Asian nations, including: Japan (initiated in 1989); Taiwan (1993); China (1994); Thailand (1994); and Indonesia (1997). Australian energy efficiency labels include technical information on energy consumption and a simple rating system [World Bank (1997b)].

4.3.2. Reporting requirements

A second type of government information program is a reporting requirement. The first such program was New Jersey's Community Right-to-Know Act, established in the United States in 1984. Two years later, a similar program was established at the national level. The U.S. Toxics Release Inventory (TRI), initiated under the Emergency Planning and Community Right-to-Know Act (EPCRA), requires firms to report to local emergency planning agencies information on use, storage, and release of hazardous chemicals. Such information reporting serves compliance and enforcement purposes, but may also increase public awareness of firms' actions, which may be linked with environmental risks.[53] This public scrutiny can encourage firms to alter their behavior, although the evidence is mixed [U.S. General Accounting Office (1992), Hamilton (1995), Singh (1995), Bui and Mayer (1997), Konar and Cohen (1997), Ananathanarayanan (1998), and Hamilton and Viscusi (1999)]. In 1989, the Commonwealth of Massachusetts instituted its Toxics Use Reduction Act, which is similar to EPCRA, but includes several additional business categories (SIC codes).

The Safe Drinking Water Act and Toxic Enforcement Act were adopted in California as a ballot initiative ("Proposition 65") in 1986. The law covers consumer products and facility discharges, and requires firms to provide a "clear and reasonable warning" if they expose populations to certain chemicals. A year later, California enacted its Air

---

[53] A non-governmental advocacy group, Environmental Defense (formerly the Environmental Defense Fund), has established an Internet site that provides TRI information in an accessible form: http://www. scorecard.org/

Toxics Hot Spots Information and Assessment Act, which sets up an emissions reporting system to track emissions of over 700 toxic substances. The law requires the identification and assessment of localized risks of air contaminants and provides information to the public about the possible impact of those emissions on public health.

One other U.S. example of environmental reporting requirements is provided by the Drinking Water Consumer Confidence Reports required by EPA since 1999. Under this program, all suppliers of drinking water in the United States must provide households with information on the quality of their drinking water, including specified information regarding water sources and actual and potential contamination.

Indonesia introduced the Program for Pollution Control, Evaluation and Rating with the help of the World Bank (1997b) in 1995. Plants are assigned ratings based on environmental performance, and plants with the lowest ratings were notified privately and given six months to improve performance, after which information was released to the public. The administrative costs of the program have been kept at relatively low levels [Tietenberg and Wheeler (1998)] – on the order of $1 per day per plant – for 187 plants over the first 18 months, and the process resulted in a 40 percent reduction in BOD emissions. The Philippines has instituted EcoWatch, a similar system of public disclosure of plant environmental performance, with rating results announced in the news media [World Bank (1997b)]. Mexico and Colombia are launching information programs based on Indonesia's system [Tietenberg and Wheeler (1998)].

The Scandinavian countries have focused considerable attention on environmental information dissemination [OECD (1996, 1997a)]. The Swedish national environmental regulatory agency regularly produces and circulates information to educators, public authorities, environmental managers, business leaders, and the general public [OECD (1996)], and the Danish Ministry of the Environment and Energy publishes annual environmental indicators [OECD (1999b)]. In addition, Belgium has developed regional pollution release and transfer registers that are available to the public, and Austria issues a comprehensive set of environmental data every three years [OECD (1995a, 1998c)]. But, other than the U.S. and Indonesian studies cited above, there have been no analyses of the effectiveness (or complete costs) of these various policy instruments.

5. Reducing government subsidies

A final category of market-based instruments is government subsidy reduction. Since subsidies are the mirror image of taxes, they can – in theory – provide incentives to address environmental problems. But, in practice, a variety of subsidies are believed to promote economically inefficient and environmentally unsound practices, despite the fact that governments frequently have implemented these subsidies in order to achieve specific goals, such as support of infant industries or income redistribution. Thus, in this section, I consider cases in which direct or indirect subsidies with adverse environmental impacts have been reduced or eliminated (or in which serious consideration has been given to doing so).

According to the World Bank (1997b), subsidies to energy, road transportation, water use, and agriculture in developing and transition economies totaled over $240 billion per year in the 1990s, representing a substantial reduction over the 1980s. A significant increase in energy prices toward efficient levels in transition economies is one important change underlying this trend. A second factor has been reduced protection of inefficient (and ecologically harmful) domestic industries, as a result of greater acceptance of free trade [Fischer and Toman (1998)].

China has reduced energy subsidies drastically since the mid 1980s [World Bank (1997b)]. For example, subsidy rates for coal, which fueled more than 70 percent of China's energy production as of 1994, fell from 61 percent in 1984 to 11 percent in 1995. Through development of private coal mining and removal of price controls, nearly 80 percent of China's coal was sold at unsubsidized international prices by 1995. Many state-owned enterprises, however, face soft-budget constraints, and so higher energy prices have not necessarily led to efficiency improvements, since these firms are insulated from market forces by the central government [Fisher-Vanden (1999)].

Bangladesh and Indonesia have reduced pesticide and fertilizer subsidies significantly. In the late 1970s, fertilizer subsidies accounted for fully four percent of the national budget of Bangladesh [World Bank (1997b)]; the government began reducing subsidies in 1978, and completely deregulated retail fertilizer prices in 1983. Direct subsidies for pesticides in Indonesia, which in the early 1980s were as high as 85 percent, were phased out in 1986–1989 [World Bank (1997b)]; domestic pesticide production was reduced by one-half between 1985 and 1990, and imports fell to one-third the level of the mid-1980s.

Ecuador has completely phased out subsidies on agricultural inputs (pesticide and fertilizer), fuel oil, and motor fuels, with the exception of diesel [Huber, Ruitenbeek and Serôa da Motta (1998)]. Likewise, India, Mexico, South Africa, Saudi Arabia, Brazil, and Jamaica cut fuel subsidies significantly in the mid-1990s [Fischer and Toman (1998), Huber, Ruitenbeek and Serôa da Motta (1998)]. In 1985, New Zealand's removal of agricultural subsidies apparently led to significant abandonment of marginal lands and consequent reductions in land degradation [New Zealand Ministry for the Environment (1997)].

Despite these trends, significant subsidies (of environmental consequence) are common in many parts of the world, particularly on energy production and use. For example, many EU countries, including Germany, the United Kingdom, Spain, and France, continue to subsidize coal production [Ekins and Speck (1999)]. But assessing the magnitude, let alone the effects, of these subsidies is difficult, a point that is illustrated by the case of the United States. Because of concerns about global climate change, increased attention has been given to Federal subsidies and other programs that promote the use of fossil fuels. An EPA study indicates that eliminating these subsidies would have a significant effect on reducing carbon dioxide (CO_2) emissions [Shelby et al. (1997)]. The Federal government is involved in the energy sector through the tax system and through a range of individual agency programs. One study indicates that these activities together cost the government $17 billion annually [Alliance to Save Energy (1993)].

A substantial share of these U.S. subsidies and programs were enacted during the "oil crises" to encourage the development of domestic energy sources and reduce reliance on imported petroleum. They favor energy supply over energy efficiency.[54] Although there is an economic argument for government policies that encourage new technologies that have particularly high risk or long term payoffs, mature and conventional technologies currently receive nearly 90 percent of the subsidies. Furthermore, within fossil fuels, the most environmentally benign fuel – natural gas – receives only about 20 percent of the subsidies. On the other hand, it should also be recognized that Federal user charges (Table 3) and insurance premium taxes (Table 4) include significant levies on fossil fuels, and that Federal tax differentiation has tended to favor renewable energy sources and non-conventional fossil fuels (Table 7).

6. Lessons that emerge from experience

In this chapter, I have defined "market-based instruments" broadly and thereby cast a large net for this review of applications of this relatively new set of policy approaches. As a consequence, the review is extensive, but this should not leave the reader with the impression that market-based instruments have replaced, or have come anywhere close to replacing, the conventional, command-and-control approach to environmental protection. Further, even when and where these approaches have been used in their purest form and with some success, such as in the case of tradeable-permit systems in the United States, they have not always performed as anticipated. In this part of the chapter, therefore, I ask what lessons can be learned from our experiences. In particular, I consider normative lessons for: design and implementation of market-based instruments; analysis of prospective and adopted systems; and identification of new applications.[55]

6.1. Lessons for design and implementation

The performance to date of market-based instruments for environmental protection provides valuable evidence for environmentalists and others that market-based instruments can achieve major cost savings while accomplishing their environmental objectives. The performance of these systems also offers lessons about the importance of flexibility, simplicity, the role of monitoring and enforcement, and the capabilities of the private

---

[54] The Alliance to Save Energy study (1993) claims that end-use efficiency receives $1 from a wide variety of implicit and explicit Federal subsidies for every $35 received by energy supply.

[55] The lessons reviewed here are normative lessons. There is another set which could be characterized as positive (political economy) lessons: Why has the command-and-control approach dominated environmental policy? Why has there been a relatively recent upsurge in attention given by policy makers to market-based instruments? I have addressed these and related questions elsewhere [Hahn and Stavins (1991), Keohane, Revesz and Stavins (1998), Stavins (1998)], but I do not consider such questions in this chapter, because they fall within the scope of Chapter 8 (by Wallace Oates and Paul Portney).

sector to make markets of this sort work. Most of the references in this section are to U.S. programs, simply because those programs have been the subject of more analyses, particularly economic analyses, than have programs in other countries. Similar lessons have been reported for other parts of the world, however [Bluffstone and Larson (1997), World Bank (1997b), OECD (1997e, 1999c)].

In regard to flexibility, it is important that market-based instruments should be designed to allow for a broad set of compliance alternatives, in terms of both timing and technological options. For example, allowing flexible timing and intertemporal trading of permits – that is, banking allowances for future use – played an important role in the SO_2 allowance trading program's performance [Ellerman et al. (1997)], much as it did in the U.S. lead rights trading program a decade earlier [Kerr and Maré (1997)]. One of the most significant benefits of using market-based instruments is simply that technology standards are thereby avoided.[56] Less flexible systems would not have led to the technological change that may have been induced by market-based instruments [Burtraw (1996), Ellerman and Montero (1998), Bohi and Burtraw (1997)], nor the induced process innovations that have resulted [Doucet and T. Strauss (1994)].

In regard to simplicity, transparent formulae – whether for permit allocation or tax computation – are difficult to contest or manipulate. Rules should be clearly defined up front, without ambiguity. For example, prior government approval of individual trades may increase uncertainty and transaction costs, thereby discouraging trading; these negative effects should be balanced against any anticipated benefits due to prior government approval. Such requirements hampered EPA's Emissions Trading Program in the 1970s, while the lack of such requirements was an important factor in the success of lead trading [Hahn and Hester (1989a)]. In the case of SO_2 trading, the absence of requirements for prior approval has reduced uncertainty for utilities and administrative costs for government, and contributed to low transactions costs [Rico (1995)].

Experience also argues for using absolute baselines, not relative ones, as the point of departure for credit programs. The problem is that without a specified baseline, reductions must be credited relative to an unobservable hypothetical – what the source would have emitted in the absence of the regulation. A hybrid system – where a cap-and-trade program is combined with voluntary "opt-in provisions" – creates the possibility for "paper trades", where a regulated source is credited for an emissions reduction (by an unregulated source) that would have taken place in any event [Montero (1999)]. The result is a decrease in aggregate costs among regulated sources, but this is partly due to an unintentional increase in the total emissions cap. As was experienced with EPA's Emissions Trading Program, relative baselines create significant transaction costs by essentially requiring prior approval of trades as the authority investigates the claimed counterfactual from which reductions are calculated and credits generated [Nichols, Farr and Hester (1996)].

---

[56] This is also true, of course, of other performance-based approaches.

Experiences with market-based instruments also provide a powerful reminder of the importance of monitoring and enforcement. These instruments, whether price or quantity based, do not eliminate the need for such activities, although they may change their character. In the many programs reviewed in this chapter where monitoring and/or enforcement have been deficient, the results have been ineffective policies. One counterexample is provided by the U.S. SO_2 allowance trading program, which includes (costly) continuous emissions monitoring of all sources [Burtraw (1996)]. On the enforcement side, the Act's stiff penalties (much greater than the marginal cost of abatement) have provided sufficient incentives for the very high degree of compliance that has been achieved [Stavins (1998)].

In nearly every case of implemented cap-and-trade programs, permits have been allocated freely to participants. The same characteristic that makes such allocation attractive in positive political economy terms – the conveyance of scarcity rents to the private sector – makes free allocation problematic in normative, efficiency terms [Fullerton and Metcalf (1997)]. It has been estimated that the costs of SO_2 allowance trading would be 25 percent less if permits were auctioned rather than freely allocated, because auctioning yields revenues that can be used to finance reductions in pre-existing distortionary taxes [Goulder, Parry and Burtraw (1997)].[57] Furthermore, in the presence of some forms of transaction costs, the post-trading equilibrium – and hence aggregate abatement costs – are sensitive to the initial permit allocation [Stavins (1995)]. For both reasons, a successful attempt to establish a politically viable program through a specific initial permit allocation can result in a program that is significantly more costly than anticipated.

Improvements in instrument design will not solve all problems. One potentially important cause of the mixed performance of implemented market-based instruments is that many firms are simply not well equipped internally to make the decisions necessary to fully utilize these instruments. Since market-based instruments have been used on a limited basis only, and firms are not certain that these instruments will be a lasting component on the regulatory landscape, most companies have chosen not to reorganize their internal structure to fully exploit the cost savings these instruments offer. Rather, most firms continue to have organizations that are experienced in minimizing the costs of complying with command-and-control regulations, not in making the strategic decisions allowed by market-based instruments.[58]

---

[57] Although the positive political economy of instrument choice is outside the scope of this chapter, it should be recognized that the European experience with environmental taxes clearly illustrates that if tax revenues (or tradeable-permit auction revenues) are used to reduce distortionary taxes, those same revenues cannot generally be used to encourage acceptance of the program. The choice in Europe has been to dedicate environmental tax revenues to the environmental resources degraded by the taxed activity.

[58] There are some exceptions. Enron, for example, has attempted to use market-based instruments for its strategic benefit by becoming a leader in creating new markets for trading acid rain permits. Other firms have appointed environmental, health, and safety leaders who are familiar with a wide range of policy instruments, not solely command-and-control approaches, and who bring a strategic focus to their company's pollution-control efforts [Hockenstein, Stavins and Whitehead (1997)].

The focus of environmental, health, and safety departments in private firms has been primarily on problem avoidance and risk management, rather than on the creation of opportunities made possible by market-based instruments. This focus has developed because of the strict rules companies have faced under command-and-control regulation, in response to which companies have built skills and developed processes that comply with regulations, but do not help them benefit competitively from environmental decisions [Reinhardt (2000)]. Absent significant changes in structure and personnel, the full potential of market-based instruments will not be realized.

6.2. Lessons for analysis

When assessing market-based environmental programs, economists need to employ some measure by which the gains of moving from conventional standards to an economic-incentive scheme can be estimated. When comparing policies with the same anticipated environmental outcomes, aggregate cost savings may be the best yardstick for measuring success of individual instruments. The challenge for analysts is to make fair comparisons among policy instruments: either idealized versions of both market-based systems and likely alternatives; or realistic versions of both [Hahn and Stavins (1992)].

It is not enough to analyze static cost savings. For example, the savings due to banking allowances should also be modeled (unless this is not permitted in practice). It can likewise be important to allow for the effects of alternative instruments on technology innovation and diffusion [Milliman and Prince (1989), Jaffe and Stavins (1995), Doucet and T. Strauss (1994)], especially when programs impose significant costs over long time horizons [Newell, Jaffe and Stavins (1999)]. More generally, it is important to consider the effects of the pre-existing regulatory environment. For example, the level of pre-existing factor taxes can affect the total costs of regulation [Goulder, Parry and Burtraw (1997)], as indicated above.

6.3. Lessons for identifying new applications

Market-based policy instruments are now considered for nearly every environmental problem that is raised, ranging from endangered species preservation[59] to what may be the greatest of environmental problems, the greenhouse effect and global climate change.[60] Experiences with market-based instruments offer some guidance to the conditions under such approaches are likely to work well, and when they may face greater difficulties.

---

[59] See, for example Goldstein (1991) and Bean (1997).
[60] See, for example: Fisher et al. (1996), Hahn and Stavins (1995), Schmalensee (1996), and Stavins (1997). More broadly, see Ayres (2000).

First, where the cost of abating pollution differs widely among sources, a market-based system is likely to have greater gains, relative to conventional, command-and-control regulations [Newell and Stavins (1999)]. For example, it was clear early on that SO_2 abatement cost heterogeneity was great, because of differences in ages of plants and their proximity to sources of low-sulfur coal. But where abatement costs are more uniform across sources, the political costs of enacting an allowance trading approach are less likely to be justifiable.

Second, the greater is the degree of mixing of pollutants in the receiving airshed or watershed, the more attractive will a market-based system be, relative to a conventional uniform standard. This is because taxes or tradeable permits, for example, can lead to localized "hot spots" with relatively high levels of ambient pollution. Most applications of market-based instruments have not addressed the hot-spot or hot-time issues, differences in damages associated with emissions from different geographical points or at different times. This is a significant distributional issue, and it can also become an efficiency issue if damages are non-linearly related to pollutant concentrations. These issues can, in principle, be addressed by appropriate differentiation in taxes or permit prices.[61]

Third, the efficiency of price-based (tax) systems compared with quantity-based (tradeable permit) systems depends on the pattern of costs and benefits. If uncertainty about marginal abatement costs is significant, and if marginal abatement costs are quite flat and marginal benefits of abatement fall relatively quickly, then a quantity instrument will be more efficient than a price instrument [Weitzman (1974)]. Furthermore, when there is also uncertainty about marginal benefits, and marginal benefits are positively correlated with marginal costs (which, it turns out, is not uncommon), then there is an additional argument in favor of the relative efficiency of quantity instruments [Stavins (1996)]. On the other hand, the regulation of stock pollutants will often favor price instruments when the optimal stock level rises over time [Newell and Pizer (2000)]. It should also be recognized that despite the theoretical efficiency advantages of hybrid systems – non-linear taxes, or quotas combined with taxes – in the presence of uncertainty [Roberts and Spence (1976), Kaplow and Shavell (1997)],[62] virtually no such hybrid systems have been adopted.

Fourth, the long-term cost-effectiveness of tax systems versus tradeable permit systems is affected by their relative responsiveness to change. This arises in at least three dimensions. In the presence of rapid rates of economic growth (important in the case of some developing countries), a fixed tax leads to an increase in aggregate emissions, whereas with a fixed supply of permits there is no change in aggregate emissions (but an increase in permit prices). In the context of general price inflation, a unit (but not an

[61] Neither problem arose, however, in the case of the U.S. SO_2 allowance trading program, because dirtier plants had lower marginal abatement costs, and hence made the largest emissions reductions.

[62] In addition to the efficiency advantages of non-linear taxes, they also have the attribute of reducing the total (although not the marginal) tax burden of the regulated sector, relative to an ordinary linear tax, which is potentially important in a political economy context.

ad valorem) tax decreases in real terms and so emissions levels increase; whereas with a permit system, there is no change in aggregate emissions. In the presence of exogenous technological change in pollution abatement, a tax system leads to an increase in control levels, i.e., a decrease in aggregate emissions, while a permit system maintains emissions, with a fall in permit prices [Stavins and Whitehead (1992)].

Fifth, tradeable permits will work best when transaction costs are low, and experience demonstrates that if properly designed, private markets will tend to render transaction costs minimal. Sixth, a potential advantage of freely-allocated tradeable permit systems over other policy instruments is associated with the incentive they provide for pollution sources to identify themselves and report their emissions (in order to claim their permits). This was illustrated by Chile's experience with its TSP system, and could be a significant factor in countries where monitoring costs are relatively high and/or self-reporting requirements are ineffective.

Seventh and finally, considerations of political feasibility point to the wisdom (more likely success) of proposing market-based instruments when they can be used to facilitate a cost-effective, aggregate emissions reduction (as in the case of the U.S. SO$_2$ allowance trading program in 1990), as opposed to a cost-effective reallocation of the status quo burden (as in the case of the earlier U.S. EPA Emissions Trading Program). Policy instruments that appear impeccable from the vantage point of research institutions, but consistently prove infeasible in real-world political institutions, can hardly be considered "optimal".

6.4. Conclusion

Given that most experience with market-based instruments has been generated very recently, one should be cautious when drawing conclusions about lessons to be learned. A number of important questions remain. For example, little is known empirically about the impact of these instruments on technological change. Also, much more empirical research is needed on how the pre-existing regulatory environment affects performance, including costs. Moreover, the successes with tradeable permits have involved air pollution: acid rain, leaded gasoline, and chloroflourocarbons. Experience (and success) with water pollution is much more limited [Hahn (1989)], and in other areas, there has been no experience at all. Even for air pollution problems, the tremendous differences between SO$_2$ and acid rain, on the one hand, and the combustion of fossil fuels and global climate change, on the other, indicate that any rush to judgement regarding global climate policy instruments is unwarranted.

Despite these and other uncertainties, market-based instruments for environmental protection now enjoy proven successes in reducing pollution at low cost. Such cost effectiveness is the primary focus of economists when evaluating these public policies, but the political system gives greater weight to distributional concerns. Indeed, individual constituencies, each fighting for its own version of distributional equity, frequently negate efficiency and cost effectiveness.

There are sound reasons why the political world has been slow to embrace the use of market-based instruments for environmental protection, including the ways economists have packaged and promoted their ideas in the past: failing to separate means (cost-effective instruments) from ends (efficiency); and treating environmental problems as little more than "externalities calling for corrective taxes". Much of the resistance has also been due, of course, to the very nature of the political process and the incentives it provides to both politicians and interest groups to favor command-and-control methods instead of market-based approaches.[63]

But, despite this history, market-based instruments have moved center stage, and policy debates look very different from the time when these ideas were characterized as "licenses to pollute" or dismissed as completely impractical. Of course, no single policy instrument – whether market-based or conventional – will be appropriate for all environmental problems. Which instrument is best in any given situation depends upon characteristics of the specific environmental problem, and the social, political, and economic context in which the instrument is to be implemented.

Acknowledgements

Sheila Olmstead provided exceptionally valuable research assistance, contributing greatly to the quality of the final product. Helpful comments on previous versions were provided by Scott Barrett, Peter Bohm, David Dreisen, Denny Ellerman, Karen Fisher-Vanden, Robert Hahn, Erik Haites, Suzi Kerr, Juan-Pablo Montero, Wallace Oates, William Pizer, Ronaldo Serôa da Motta, Thomas Sterner, Tom Tietenberg, Jeffrey Vincent, and Tomasz Żylicz. The author alone is responsible for any remaining errors.

References

Alberini, A., W. Harrington and V. McConnell (1995), "Determinants of participation in accelerated vehicle retirement programs", RAND Journal of Economics 26:93–112.

Alliance to Save Energy (1993), Federal Energy Subsidies: Energy, Environmental and Fiscal Impacts (Lexington, MA).

Ananathanarayanan, A. (1998), "Is there a green link? A panel data value event study of the relationship between capital markets and toxic releases", Working Paper (Rutgers University).

Anderson, G., and B. Fiedor (1997), "Environmental charges in Poland", in: R. Bluffstone and B.A. Larson, eds., Controlling Pollution in Transition Economies (Edward Elgar, Cheltenham, UK) 187–208.

Anderson, G., and T. Żylicz (1999), "The role of Polish environmental funds: too generous or too restrictive?" Environment and Development Economics 4:413–448.

Anderson, R. (1997), The U.S. Experience with Economic Incentives in Environmental Pollution Control Policy (Environmental Law Institute, Washington, DC).

Anderson, R.C., L.A. Hofmann and M. Rusin (1990), The Use of Economic Incentive Mechanisms in Environmental Management, Research Paper 51 (American Petroleum Institute, Washington, DC).

---

[63] See Keohane, Revesz and Stavins (1998).

Anderson, T.L. (1983), Water Crisis: Ending the Policy Drought (Cato Institute, Washington, DC).

Anderson, T.L., and D.R. Leal (1991), Free Market Environmentalism (Westview Press, Boulder, CO).

Annex I Expert Group on the United Nations Framework Convention on Climate Change (1997), "International greenhouse gas emission trading", Working Paper No. 9 (OECD, Paris).

Ayoo, C., and M.A. Jama (1999), "Environmental taxation in Kenya", in: T. Sterner, ed., The Market and the Environment (Edward Elgar, Cheltenham, UK) 301–318.

Ayres, R.E. (2000), "Expanding the use of environmental trading programs into new areas of environmental regulation", Pace Environmental Law Review 18(1):87–118.

Azurix Corporation (2000), "Azurix Corp. Introduces Water2Water.com, the Internet-based, Ariba-powered marketplace for buyers and sellers of water", Press Release, available at http://www.azurix.com

Bailey, E.M. (1996), "Allowance trading activity and state regulatory rulings: evidence from the U.S. Acid Rain Program", MIT-CEEPR 96-002 WP (Center for Energy and Environmental Policy Research, Massachusetts Institute of Technology, Cambridge, MA).

Barthold, T.A. (1994), "Issues in the design of environmental excise taxes", Journal of Economic Perspectives 8(1):133–151.

Batstone, C.J., and B.M.H. Sharp (1999), "New Zealand's quota management system: the first ten years", Marine Policy 23(2):177–190.

Baumol, W.J., and W.E. Oates (1988), The Theory of Environmental Policy, 2nd edn. (Cambridge University Press, New York).

Bean, M.J. (1997), "Shelter from the storm: endangered species and landowners alike deserve a safe harbor", The New Democrat (March/April):20–21.

Bellandi, R.L., and R.B. Hennigan (1980), "The why and how of transferable development rights", Real Estate Review 7:60–64.

Blackman, A., and W. Harrington (1999), "The use of economic incentives in developing countries: lessons from the international experience with industrial air pollution", Discussion Paper No. 99-39 (Resources for the Future, Washington, DC).

Bluffstone, R., and B.A. Larson (eds.) (1997), Controlling Pollution in Transition Economies (Edward Elgar, Cheltenham, UK).

Bohi, D. (1994), "Utilities and state regulators are failing to take advantage of emissions allowance trading", The Electricity Journal 7:20–27.

Bohi, D., and D. Burtraw (1997), "SO_2 allowance trading: how do expectations and experience measure up?" The Electricity Journal 10(7):67–75.

Bohlin, F. (1998), "The Swedish carbon dioxide tax: effects on biofuel use and carbon dioxide emissions", Biomass and Bioenergy 15(4/5):213–291.

Bohm, P. (1981), Deposit-Refund Systems: Theory and Applications to Environmental, Conservation, and Consumer Policy (Resources for the Future/Johns Hopkins University Press, Baltimore, MD).

Bohm, P. (1999), Personal communication.

Bowes, M.D., and J.V. Krutilla (1989), Multiple-Use Management: The Economics of Public Forestlands (Resources for the Future, Washington, DC).

Bradsher, K. (1997), "Light trucks increase profits but foul air more than cars", New York Times (30 November):A1, A38–A39.

Brennan, T.J., K.L. Palmer, R.J. Kopp, A.J. Krupnick, V. Stagliano and D. Burtraw (1996), A Shock to the System: Restructuring America's Electricity Industry (Resources for the Future, Washington, DC).

Bressers, H.Th. A., and J. Schuddeboom (1994), "A survey of effluent charges and other economic instruments in Dutch environmental policy", in: Organization for Economic Cooperation and Development, Applying Economic Instruments to Environmental Policies in OECD and Dynamic Non-Member Countries (OECD, Paris) 153–172.

Brotzman, T. (1996), "Opening the floor to emissions trading", Chemical Marketing Reporter (27 May):SR8.

Brunenieks, J., A. Kozlovska and B.A. Larson (1997), "Implementing pollution permits and charges in Latvia", in: R. Bluffstone and B.A. Larson, eds., Controlling Pollution in Transition Economies (Edward Elgar, Cheltenham, UK) 75–101.

Bryner, G.C. (1999), "New tools for improving government regulation: an assessment of emissions trading and other market-based regulatory tools", Report (PricewaterhouseCoopers Endowment for The Business of Government, Arlington, VA).

Bui, L.T.M., and C.J. Mayer (1997), "Public disclosure of private information as a means of regulation: Evidence from the toxic release inventory in Massachusetts", Mimeo.

Burtraw, D. (1996), "The SO_2 emissions trading program: cost savings without allowance trades", Contemporary Economic Policy 14:79–94.

Burtraw, D., A.J. Krupnick, E. Mansur, D. Austin and D. Farrell (1998), "The costs and benefits of reducing air pollution related to acid rain", Contemporary Economic Policy 16:379–400.

Burtraw, D., and E. Mansur (1999), "The environmental effects of SO_2 trading and banking", Environmental Science and Technology 33(20):3489–3494.

Cadiou, A., and N.T. Duc (1994), "The use of pollution charges in water management in France", in: Organization for Economic Cooperation and Development, Applying Economic Instruments to Environmental Policies in OECD and Dynamic Non-Member Countries (OECD, Paris):131–151.

Cansier, D., and R. Krumm (1998), "Air pollution taxation: an empirical survey", Ecological Economics, forthcoming.

Carlson, C., D. Burtraw, M. Cropper and K. Palmer (2000), "Sulfur dioxide control by electric utilities: What ere the gains from trade?" Discussion Paper No. 98-44-REV (Resources for the Future, Washington, DC).

Cason, T.N. (1995), "An experimental investigation of the seller incentives in EPA's Emission Trading Auction", American Economic Review 85:905–922.

China Council Working Group on Trade and Environment (1996), "Report of the Working Group on trade and environment of CCICED", International Institute for Sustainable Development World Wide Web site, available at http://iisd1.iisd.ca/trade/cciced/wgreport96.htm

Clean Air Act Amendments of 1990 (1990), Public Law No. 101-549, 104 Statute 2399.

Coase, R. (1960), "The problem of social cost", Journal of Law and Economics 3:1–44.

Cook, E. (ed.) (1996), Ozone Protection in the United States: Elements of Success (World Resources Institute, Washington, DC).

Crandall, R.W., H.K. Gruenspecht, T.E. Keeler and L.B. Lave (1986), Regulating the Automobile (The Brookings Institute, Washington, DC).

Crocker, T.D. (1966), "The structuring of atmospheric pollution control systems", in: H. Wolozin, ed., The Economics of Air Pollution (Norton, New York).

Dales, J. (1968), Pollution, Property and Prices (University Press, Toronto).

Dasgupta, S., B. Laplante and N. Mamingi (1997), "Capital market responses to environmental performance in developing countries", World Bank Development Research Group Working Paper No. 1909, October.

Dao, J. (2000), "Acid Rain Law found to fail in Adirondacks", New York Times, March 27.

Dixon, R.K. (1999), The U.N. Framework Convention on Climate Change Activities Implemented Jointly (AIJ) Pilot: Experiences and Lessons Learned (Kluwer Academic, The Netherlands).

Doucet, J., and T. Strauss (1994), "On the bundling of coal and sulphur dioxide emissions allowances", Energy Policy 22(9):764–770.

Downing, P.B., and L.J. White (1986), "Innovation in pollution control", Journal of Environmental Economics and Management 13:18–27.

Dudek, D.J., and J.B. Wiener (1996), "Joint implementation, transaction costs, and climate change" (OECD, Paris).

Efaw, F., and W.N. Lanen (1979), "Impact of user charges on management of household solid waste", Report prepared for the U.S. Environmental Protection Agency under Contract No. 68-3-2634 (Mathtech, Inc., Princeton, NJ).

Ekins, P., and S. Speck (1998), "Evaluation of the environmental effects of environmental taxes", Mimeo (Forum for the Future and Keele University).

Ekins, P., and S. Speck (1999), "Competitiveness and exemptions from environmental taxes in Europe", Environmental and Resource Economics 13:369–396.

Ekins, P. (1999), "European environmental taxes and charges: recent experience, issues and trends", Ecological Economics 31:39–62.

El-Ashry, M.T., and D.C. Gibbons (1986), Troubled Waters: New Policies for Managing Water in the American West (World Resources Institute, Washington, DC).

Ellerman, D., P. Joskow, R. Schmalensee, J. Montero and E. Bailey (2000), Markets for Clean Air: The U.S. Acid Rain Program (Cambridge University Press, New York).

Ellerman, D., and J. Montero (1998), "The declining trend in sulfur dioxide emissions: implications for allowance prices", Journal of Environmental Economics and Management 36:26–45.

Ellerman, D., R. Schmalensee, P. Joskow, J. Montero and E. Bailey (1997), Emissions Trading Under the U.S. Acid Rain Program: Evaluation of Compliance Costs and Allowance Market Performance (MIT Center for Energy and Environmental Policy Research, Cambridge, MA).

Environment and Development Foundation (1999), Taiwan Greenmark Program World Wide Web site, available at http://www.greenmark.itri.org.tw/eng/english/htm

European Commission, Directorate-General for Environment, Nuclear Safety and Civil Protection (1998), Database on Environmental Taxes in the European Union Member States, plus Norway and Switzerland: Evaluation of Environmental Effects of Environmental Taxes (Forum for the Future, London).

European Union (1999), Eco-label World Wide Web site, available at http://europa.eu.int/comm/ dg11/ eco-label/.

Farrell, A., R. Carter and R. Raufer (1999), "The NO_x budget: market-based control of tropospheric ozone in the northeastern United States", Resource and Energy Economics 21:103–124.

Federal Republic of Germany (1998), Blue Eco-Angel World Wide Web site, available at http://www.blauer-engel.de

Feldman, R.D. (1991), U.S. Environmental Protection Agency (7 January).

Ferrall, B.L. (1991), "The Clean Air Act amendments of 1990 and the use of market forces to control sulfur dioxide emissions", Harvard Journal on Legislation 28:235–252.

Field, B.C., and J.M. Conrad (1975), "Economic issues in programs of transferable development rights", Land Economics 51:331–340.

Fischer, C. and M. Toman (1998), "Environmentally- and economically-damaging subsidies: concepts and illustrations", Climate Issue Brief No. 14 (Resources for the Future, Washington, DC).

Fisher, B., S. Barrett, P. Bohm, M. Kuroda, J. Mubazi, A. Shah and R. Stavins (1996), "Policy instruments to combat climate change", in: J.P. Bruce, H. Lee and E.F. Haites, eds., Climate Change 1995: Economic and Social Dimensions of Climate Change (Cambridge University Press, New York) 397–439.

Fisher-Vanden, K. (1999), "Structural change and technological diffusion in transition economies: implications for energy use and carbon emissions in China", Ph.D. Dissertation (Harvard University, Cambridge, MA), unpublished.

Foster, V. and R.W. Hahn (1995), "Designing more efficient markets: lessons from Los Angeles smog control", Journal of Law and Economics 38:19–48.

Frederick, K.D. (ed.) (1986), Scarce Water and Institutional Change (Resources for the Future, Washington, DC).

Fullerton, D., and T.C. Kinnaman (1995), "Garbage, recycling and illicit burning or dumping", Journal of Environmental Economics and Management 29:78–92.

Fullerton, D., and T.C. Kinnaman (1996), "Household responses to pricing garbage by the bag", American Economic Review 86:971–984.

Fullerton, D., and G. Metcalf (1997), "Environmental controls, scarcity rents, and pre-existing distortions", Working Paper No. 6091 (NBER).

Fulton, W. (1996), "The big green bazaar", Governing Magazine (June) 38.

Goldberg, P.K. (1997), "The effects of the corporate average fuel efficiency standards", Working Paper (Department of Economics, Princeton University, Princeton, NJ).

Goldstein, J.B. (1991), "The prospects for using market incentives to conserve biological diversity", Environmental Law 21.

Gornaja, L., E. Kraav, B.A. Larson and K. Türk (1997), "Estonia's mixed system of pollution permits, standards and charges", in: R. Bluffstone and B.A. Larson, eds., Controlling Pollution in Transition Economies (Edward Elgar, Cheltenham, UK) 46–74.

Goulder, L., I. Parry and D. Burtraw (1997), "Revenue-raising vs. other approaches to environmental protection: the critical significance of pre-existing tax distortions", RAND Journal of Economics.

Greenhouse Gas Emissions trading pilot program (1999), "Offers and trades: project applications", GERT World Wide Web site, available at http://www.gert.org.

Gupta, S., G. Van Houtven and M. Cropper (1996), "Paying for permanence: an economic analysis of EPA's cleanup decisions at Superfund sites", RAND Journal of Economics 27(3):563–582.

Hahn, R.W. (1989), "Environmental problems: how the patient followed the doctor's orders", Journal of Economic Perspectives 3:95–114.

Hahn, R.W. (1990), "Regulatory constraints on environmental markets", Journal of Public Economics 42:149–175.

Hahn, R.W., and G.L. Hester (1989a), "Marketable permits: lessons for theory and practice", Ecology Law Quarterly 16:361–406.

Hahn, R.W., and G.L. Hester (1989b), "Where did all the markets go? An analysis of EPA's Emissions Trading Program", Yale Journal of Regulation 6:109–153.

Hahn, R.W., and A.M. McGartland (1989), "Political economy of instrumental choice: an examination of the U.S. role in implementing the Montreal Protocol", Northwestern University Law Review 83:592–611.

Hahn, R., and R. Noll (1982), "Designing a market for tradeable permits", in: W. Magat, ed., Reform of Environmental Regulation (Ballinger, Cambridge, MA).

Hahn, R.W., and R.N. Stavins (1991), "Incentive-based environmental regulation: a new era from an old idea?" Ecology Law Quarterly 18:1–42.

Hahn, R.W., and R.N. Stavins (1992), "Economic incentives for environmental protection: integrating theory and practice", American Economic Review 82(May):464–468.

Hahn, R.W., and R.N. Stavins (1995), "Trading in greenhouse permits: a critical examination of design and implementation issues", in: H. Lee, ed., Shaping National Responses to Climate Change: A Post-Rio Policy Guide (Island Press, Cambridge, MA):177–217.

Haites, E. (1996), "Trading for ozone-depleting substances", Mimeo (Margaree Consultants, Toronto, Canada).

Haites, E. (1997), "Heavy-duty vehicle engine emissions", Mimeo (Margaree Consultants, Toronto, Canada).

Haites, E. (1999), Personal communication.

Hamilton, J.T. (1993), "Politics and social costs: estimating the impact of collective action on hazardous waste facilities", RAND Journal of Economics 24:101–125.

Hamilton, J. (1995), "Pollution as news: media and stock market reactions to the toxics release inventory data", Journal of Environmental Economics and Management 28:98–113.

Hamilton, J., and K. Viscusi (1999), Calculating Risks? The Spatial and Political Dimensions of Hazardous Waste Policy (MIT Press, Cambridge, MA).

Harrison, D., Jr. (1999), "Turning theory into practice for emissions trading in the Los Angeles air basin", in: S. Sorrell and J. Skea, eds., Pollution for Sale: Emissions Trading and Joint Implementation (Edward Elgar, London).

Hartje, V., K. Gauer and A. Urquiza (1994), "The use of economic instruments in the environmental policy of Chile", in: Proceedings of the International Conference on Market-Based Instruments of Environmental Management in Developing Countries (German Foundation for International Development, Berlin) 61–80.

Haugland, T. (1993), "A comparison of carbon taxes in selected OECD countries", Environment Monographs, No. 78 (OECD, Paris).

Helfand, G.E. (1991), "Standards versus standards: the effects of different pollution restrictions", American Economic Review 81:622–634.

Hernandez, R. (2000), "Albany battles acid rain fed by other states", New York Times, May 2.

Hockenstein, J.B., R.N. Stavins and B.W. Whitehead (1997), "Creating the next generation of market-based environmental tools", Environment 39(4):12–20, 30–33.

Huber, R.M., J. Ruitenbeek and R. Serôa da Motta (1998), "Market based instruments for environmental policymaking in Latin American and the Caribbean: lessons from eleven countries", Discussion Paper No. 381 (World Bank, Washington, DC).

Hyde, W.F. (1981), "Timber economics in the rockies: efficiency and management options", Land Economics 57:630–637.

Internal Revenue Code (1954), Section 9501.

International Institute for Sustainable Development (1995), Green Budget Reform: An International Casebook on Leading Practices (EarthScan, London).

Jaffe, A.B., S.R. Peterson, P.R. Portney and R.N. Stavins (1995), "Environmental regulation and the competitiveness of U.S. manufacturing: what does the evidence tells us?" Journal of Economic Literature 33:132–163.

Jaffe, A.B., and R.N. Stavins (1995), "Dynamic incentives of environmental regulation: the effects of alternative policy instruments on technology diffusion", Journal of Environmental Economics and Management 29:S43–S63.

Jenkins, R.R. (1993), The Economics of Solid Waste Reduction: The Impact of User Fees (Edward Elgar, Cheltenham, UK).

Jepma, C. (1999), "Planned and ongoing AIJ pilot projects", Joint Implementation Quarterly 5(3):14.

Johnson, S.L., and D.M. Pekelney (1996), "Economic assessment of the regional clean air incentives market: a new emissions trading program for Los Angeles", Land Economics 72:277–297.

Johnston, J.L. (1994), "Pollution trading in La La Land", Regulation (3):44–54.

Jorgenson, D., and P. Wilcoxen (1994), "The economic effects of a carbon tax", Paper presented to the IPCC Workshop on Policy Instruments and their Implications, Tsukuba, Japan (17–20 January).

Joskow, P.L., and R. Schmalensee (1998), "The political economy of market-based environmental policy: the U.S. Acid Rain Program", Journal of Law and Economics 41:81–135.

Joskow, P.L., R. Schmalensee and E.M. Bailey (1998), "Auction design and the market for sulfur dioxide emissions", American Economic Review, forthcoming.

Jung, C., K. Krutilla and R. Boyd (1996), "Incentives for advanced pollution abatement technology at the industry level: an evaluation of policy alternatives", Journal of Environmental Economics and Management 30:95–111.

Kaplow, L., and S. Shavell (1997), "On the superiority of corrective taxes to quantity regulation", NBER Working Paper No. 6251 (National Bureau of Economic Research, Cambridge, MA).

Kashmanian, R. (1986), "Beyond categorical limits: the case for pollution reduction through trading", Unpublished paper presented at the 59th Annual Conference of the Water Pollution Control Federation.

Kelman, S. (1981), What Price Incentives?: Economists and the Environment (Auburn House, Boston, MA).

Keohane, N.O., R.L. Revesz and R.N. Stavins (1998), "The positive political economy of instrument choice in environmental policy", in: P. Portney and R. Schwab, eds., Environmental Economics and Public Policy (Edward Elgar, London).

Kerr, S., and D. Maré (1997), "Efficient regulation through tradeable permit markets: the United States lead phasedown", Working Paper 96-06 (January) (Department of Agricultural and Resource Economics, University of Maryland, College Park, MD).

Kerr, S., and R. Newell (2000), "Policy-induced technology adoption: evidence from the U.S. lead phasedown", Draft Manuscript (Resources for the Future, Washington, DC).

Klaassen, G. (1999), "Emissions trading in the European Union: practice and prospects", in: S. Sorrell and J. Skea, eds., Pollution for Sale (Edward Elgar, Cheltenham, UK) 83–100.

Klaassen, G., and A. Nentjes (1997), "Creating markets for air pollution control in Europe and the USA", Environmental and Resource Economics 10:125–146.

Klassen, R., and C. McLaughlin (1996), "The impact of environmental management on firm performance", Management Science 42(8):1199–1214.

Klier, T.H., R.H. Mattoon and M.A. Prager (1997), "A mixed bag: assessment of market performance and firm trading behaviour in the NO_x RECLAIM Programme", Journal of Environmental Planning and Management 40(6):751–774.

Kling, C.L. (1994), "Emission trading vs. rigid regulations in the control of vehicle emissions", Land Economics 70:174–188.

Konar, S., and M.A. Cohen (1997), "Information as regulation: the effect of community right to know laws on toxic emissions", Journal of Environmental Economics and Management 32:109–124.

Korb, B. (1998), "U.S. experience with economic incentives for emissions trading", in: Applying Market-Based Instruments to Environmental Policies in China and OECD Countries (OECD, Paris): 99–115.

Kozeltsev, M., and A. Markandya (1997), "Pollution charges in Russia: the experience of 1990–1995", in: R. Bluffstone and B.A. Larson, eds., Controlling Pollution in Transition Economies (Edward Elgar, Cheltenham, UK) 128–143.

Kramer, R.A., and K.M. Banholzer (1999), "Tradable permits in water resource management and water pollution control", in: Implementing Domestic Tradable Permits for Environmental Protection (OECD, Paris).

Kriz, M. (1996), "A jolt to the system", National Journal (3 August):1631–1636.

Krupp, F. (1986), "New environmentalism factors in economic needs", Wall Street Journal (20 November):34.

Lanoie, P., and B. Laplante (1994), "The market response to environmental incidents in Canada: a theoretical and empirical analysis", Southern Economic Journal 60:657–672.

Laplante, B., P. Lanoie and M. Roy (1997), "Can capital markets create incentives for pollution control?" Working Paper No. 1753, April (World Bank Policy Research Department).

Larsen, B.N., and R. Nesbakken (1997), "Norwegian emissions of CO_2, 1987–1994", Environmental and Resource Economics 9:275–290.

Lave, L., and H. Gruenspecht (1991), "Increasing the efficiency and effectiveness of environmental decisions: benefit-cost analysis and effluent fees", Journal of Air and Waste Management 41:680–690.

Lents, J. (1998), "The RECLAIM program at three years", Working Paper (April).

Liroff, R.A. (1986), Reforming Air Pollution Regulations: The Toil and Trouble of EPA's Bubble (Conservation Foundation, Washington, DC).

Lövgren, K. (1994), "Economic instruments for air pollution control in Sweden", in: G. Klaassen and F.R. Forsund, eds., Economic Instruments for Air Pollution Control (Kluwer Academic, Boston, MA) 107–121.

Lowry, R.C. (1993), "The political economy of environmental citizen groups", Ph.D. Thesis (Harvard University, Cambridge, MA), unpublished.

Macauley, M.K., M.D. Bowes and K.L. Palmer (1992), Using Economic Incentives to Regulate Toxic Substances (Resources for the Future, Washington, DC).

MacDonnell, L.J. (1990), The Water Transfer Process As a Management Option For Meeting Changing Water Demands, Vol. 1, submitted to the U.S. Geological Survey, Washington, DC (April).

Main, J. (1988), "Here comes the big new cleanup", Fortune (November) 102–118.

Malueg, D.A. (1989), "Emission credit trading and the incentive to adopt new pollution abatement technology", Journal of Environmental Economics and Management 16:52–57.

McFarland, J.M. (1972), "Economics of solid waste management", in: Comprehensive Studies of Solid Waste Management, Final Report, Report No. 72-3 (Sanitary Engineering Research Laboratory, College of Engineering and School of Public Health, University of California, Berkeley, CA) 41–106.

McMorran, R.T., and D.C.L. Nellor (1994), "Tax policy and the environment: theory and practice", Working Paper No. 94/106 (International Monetary Fund Fiscal Affairs Department, Washington, DC).

Menell, P. (1990), "Beyond the Throwaway Society: an incentive approach to regulating municipal solid waste", Ecology Law Quarterly 17:655–739.

Merrill, P.R., and A.S. Rousso (1990), "Federal environmental taxation", presented at the Eighty-third Annual Conference of the National Tax Association, San Francisco, CA (November 13).

Milliman, S.R., and R. Prince (1989), "Firm incentives to promote technological change in pollution control", Journal of Environmental Economics and Management 17:247–265.

Mills, D.E. (1980), "Transferable development rights markets", Journal of Urban Economics 7:63–74.

Minnesota Pollution Control Agency (1997), "Case study: Minnesota-pollutant trading at Rah Malting Co.", Environmental Regulatory Innovations Symposium (November 5–7), available at http://www.pca.state.mn.us/hot/es-mn-r.html

Miranda, M.L., J.W. Everett, D. Blume and B.A. Roy, Jr. (1994), "Market based incentives and residential municipal solid waste", Journal of Policy Analysis and Management 13:681–698.

Montero, J.P. (1999), "Voluntary compliance with market-based environmental policy: evidence from the U.S. Acid Rain Program", Journal of Political Economy 107:998–1033.

Montero, J.P., and J.M. Sánchez (1999), "A market-based environmental policy experiment in Chile", Mimeo (Department of Industrial Engineering, Catholic University of Chile, Chile).

Montgomery, D. (1972), "Markets in licenses and efficient pollution control programs", Journal of Economic Theory 5:395–418.

Morandi, L. (1992), "An outside perspective on Iowa's 1987 Groundwater Protection Act" (National Conference of State Legislatures).

Morris, G.E., J. Tiderenczl and P. Kovács (1997), "Environmental emission charges and air quality protection in Hungary: recent practice and future prospects", in: R. Bluffstone and B.A. Larson, eds., Controlling Pollution in Transition Economies (Edward Elgar, Cheltenham, UK) 156–186.

Morris, J., and L. Scarlett (1996), "Buying green: consumers, product labels and the environment", Policy Study No. 202 (The Reason Foundation, Los Angeles, CA).

Morris, W. (1988), "IID approves state's first water swap with MWD", Imperial Valley Press (9 November).

Muoghalu, M., D. Robison and J. Glascock (1990), "Hazardous waste lawsuits, stockholder returns, and deterrence", Southern Economic Journal 57:357–370.

National Academy of Public Administration (1994), The Environment Goes to Market: The Implementation of Economic Incentives for Pollution Control (July).

New Zealand Ministry for the Environment (1997), The State of New Zealand's Environment 1997 (Ministry for the Environment, Wellington, New Zealand).

Newell, R.G., A.B. Jaffe and R.N. Stavins (1999), "The induced innovation hypothesis and energy-saving technological change", Quarterly Journal of Economics 114(3):941–975.

Newell, R.G., and W. Pizer (2000), "Regulating stock externalities under uncertainty", Discussion Paper 98-02 (revised) (Resources for the Future, Washington, DC).

Newell, R., and R.N. Stavins (1999), "Abatement cost heterogeneity and potential gains from market-based instruments", Working Paper (John F. Kennedy School of Government, Harvard University, Cambridge, MA).

Nichols, A.L. (1997), "Lead in gasoline", in: R.D. Morgenstern, ed., Economic Analyses at EPA: Assessing Regulatory Impact (Resources for the Future, Washington, DC) 49–86.

Nichols, A., J. Farr and G. Hester (1996), "Trading and the Timing of Emissions: Evidence from the Ozone Transport Region" (National Economic Research Associates, Cambridge, MA).

Oates, W.E., P.R. Portney and A.M. McGartland (1989), "The net benefits of incentive-based regulation: a case study of environmental standard setting", American Economic Review 79:1233–1243.

O'Connor, D. (1994), "The use of economic instruments in environmental management: the East Asian experience", in: Applying Economic Instruments to Environmental Policies in OECD and Dynamic Non-Member Countries (OECD, Paris) 33–58.

OECD, Organization for Economic Cooperation and Development (1989), Economic Instruments for Environmental Protection (OECD, Paris).

OECD (1991), Environmental Policy: How to Apply Economic Instruments (OECD, Paris).

OECD (1993a), Applying Economic Instruments to Packaging Waste: Practical Issues for Product Charges and Deposit-Refund Systems, Environment Monographs No. 82 (OECD, Paris).

OECD (1993b), Environmental Performance Reviews: Germany (OECD, Paris).

OECD (1993c), Environmental Performance Reviews: Iceland (OECD, Paris).

OECD (1993d), Taxation and the Environment, Complementary Policies (OECD, Paris).

OECD (1994a), Applying Economic Instruments to Environmental Policies in OECD and Dynamic Non-Member Countries (OECD, Paris).

OECD (1994b), The Distributive Effects of Economic Instruments for Environmental Policy (OECD, Paris).

OECD (1994c), Evaluating Economic Incentives for Environmental Policy (OECD, Paris).

OECD (1994d), Managing the Environment – The Role of Economic Instruments (OECD, Paris).

OECD (1994e), Public Policies for the Protection of Soil Resources, Environment Monographs, No. 89 (OECD, Paris).

OECD (1995a), Environmental Performance Reviews: Austria (OECD, Paris).

OECD (1995b), Environmental Performance Reviews: Canada (OECD, Paris).

OECD (1995c), Environmental Performance Reviews: Netherlands (OECD, Paris).

OECD (1995d), Environmental Taxation in OECD Countries (OECD, Paris).

OECD (1996), Environmental Performance Reviews: Sweden (OECD, Paris).

OECD (1997a), Environmental Performance Reviews: Finland (OECD, Paris).

OECD (1997b), Environmental Performance Reviews: France (OECD, Paris).

OECD (1997c), Environmental Performance Reviews: Korea (OECD, Paris).

OECD (1997d), Environmental Performance Reviews: Spain (OECD, Paris).

OECD (1997e), Putting Markets to Work: The Design and Use of Marketable Permits and Obligations, Public Management Occasional Paper No. 19 (OECD, Paris).

OECD (1998a), Applying Market-Based Instruments to Environmental Policies in China and OECD Countries (OECD, Paris).

OECD (1998b), Environmental Performance Reviews: Australia (OECD, Paris).

OECD (1998c), Environmental Performance Reviews: Belgium (OECD, Paris).

OECD (1998d), Environmental Performance Reviews: Mexico (OECD, Paris).

OECD (1998e), Environmental Performance Reviews: Switzerland (OECD, Paris).

OECD (1999a), Environmental Performance Reviews: Czech Republic (OECD, Paris).

OECD (1999b), Environmental Performance Reviews: Denmark (OECD, Paris).

OECD (1999c), Implementing Domestic Tradable Permits for Environmental Protection (OECD, Paris).

OECD (2001), Environmentally Related Taxes in OECD Countries – Issues and Strategies (OECD, Paris).

OECD Centre for Co-operation with the European Economies in Transition (1994), "Taxation and environment in European economies in transition" (OECD, Paris).

OECD Centre for Co-operation with the European Economies in Transition (1995), "Environmental funds in economies in transition" (OECD, Paris).

Opschoor, J.B. (1994), "Developments in the use of economic instruments in OECD countries", in: G. Klaassen and F.R. Forsund, eds., Economic Instruments for Air Pollution Control (Kluwer Academic, Boston, MA).

Owen, T.H., J. Myjavec and D. Jassikova (1997), "Implementation of pollution charge systems in a transition economy: the case of Slovakia", in: R. Bluffstone and B.A. Larson, eds., Controlling Pollution in Transition Economies (Edward Elgar, Cheltenham, UK) 209–220.

Palmer, K., and D. Burtraw (1997), "Electricity restructuring and regional air pollution", Resource and Energy Economics 19:139–174.

Panayotou, T. (1998), Instruments of Change: Motivating and Financing Sustainable Development (Earthscan Publications for UNEP, London).

Panayotou, T., R. Bluffstone and V. Balaban (1994), "Lemons and liabilities: privatization, foreign investment, and environmental liability in Central and Eastern Europe", Environmental Impact Assessment 14:157–168.

Peskin, H.M. (1986), "Nonpoint pollution and national responsibility", Resources (Spring):10, 11, 17.

Peszko, G., and T. Żylicz (1998), "Environmental financing in European economies in transition", Environmental and Resource Economics 11(3–4):521–538.

Pigou, A.C. (1920), The Economics of Welfare (Macmillan & Co, London).

Pilot Emissions Reduction Trading (PERT) Project (1999), "Emission trading in Canada: the PERT experience", Draft Discussion Paper (Toronto, ON), available at http://www.pert.org.

Porter, R. (1978), "A social benefit-cost analysis of mandatory deposits on beverage containers", Journal of Environmental Economics and Management 5:351–375.

Porter, R. (1983), "Michigan's experience with mandatory deposits on beverage containers", Land Economics 59:177–194.

Pototschnig, A. (1994), "Economic instruments for the control of acid rain in the UK", in: O. Klaassen and F.R. Førsund, eds., Economic Instruments for Air Pollution Control (Kluwer, Dordrecht) 22–45.

Reiling, S.D., and M.J. Kotchen (1996), "Lessons learned from past research on recreation fees", in: A.L. Lundgren, ed., Recreation Fees in the National Park Service: Issues, Policies and Guidelines for Future Action, Minnesota Extension Service Pub. No. BU-6767 (Cooperative Park Studies Unit, Department of Forest Resources, University of Minnesota, St. Paul, MN).

Reinhardt, F.L. (2000), Down to Earth: Applying Business Principles to Environmental Management (Harvard Business School Press, Boston, MA).

Repetto, R., R. Dower, R. Jenkins and J. Geoghegan (1992), Green Fees: How a Tax Shift Can Work for the Environment and the Economy (World Resources Institute, Washington, DC).

Repetto, R., and M. Gillis (eds.) (1988), Public Policies and the Misuse of Forest Resources (Cambridge University Press, New York).

Revesz, R.L. (1997), Foundations in Environmental Law and Policy (Oxford University Press, New York).

Rhee, H.S. (1994), "The use of economic instruments in environmental protection in Korea", in: Organization for Economic Cooperation and Development, Applying Economic Instruments to Environmental Policies in OECD and Dynamic Non-Member Countries (OECD, Paris) 97–104.

Rico, R. (1995), "The U.S. allowance trading system for sulfur dioxide: an update of market experience", Environmental and Resource Economics 5(2):115–129.

Roberts, M.J., and M. Spence (1976), "Effluent charges and licenses under uncertainty", Journal of Public Economics, 5(3-4):193–208.

Rose, K. (1997), "Implementing an emissions trading program in an economically regulated industry: Lessons from the SO_2 trading program", in: R.F. Kosobud and J.M. Zimmerman, eds., Market Based Approaches to Environmental Policy: Regulatory Innovations to the Fore (Van Nostrand/Reinhold, New York).

Russell, C.S. (1988), "Economic incentives in the management of hazardous wastes", Columbia Journal of Environmental Law 13:257–274.

Schärer, B. (1994), "Economic incentives in air pollution control: the case of Germany", European Environment 4(3):3–8.

Schärer, B. (1999), "Tradable emission permits in German clean air policy: considerations on the efficiency of environmental policy instruments", in: S. Sorrell and J. Skea, eds., Pollution for Sale (Edward Elgar, Cheltenham, UK) 141–153.

Schlegelmilch, K. (1998), "Energy taxation in the EU and some Member states: looking for opportunities ahead" (Wuppertal Institute for Climate, Environment and Energy on behalf of the Heinrich-Böll-Foundation, Brussels).

Schmalensee, R. (1996), "Greenhouse policy architecture and institutions", paper prepared for National Bureau of Economic Research Conference, "Economics and policy issues in global warming: an assessment of the intergovernmental panel report", Snowmass, CO (July 23–24).

Schmalensee, R., P.L. Joskow, A.D. Ellerman, J.P. Montero and E.M. Bailey (1998), "An interim evaluation of sulfur dioxide emissions trading", Journal of Economic Perspectives 12 (3):53–68.

Scodari, P., L. Shabman and D. White (1995), "Commercial wetland mitigation credit markets: theory and practice", IWR Report 95-WMB-7 (Institute for Water Resources, Water Resources Support Center, US Army Corps of Engineers, Alexandria, VA).

Seligman, D.A. (1994), Air Pollution Emissions Trading: Opportunity or Scam? A Guide for Activists (Sierra Club, San Francisco, CA).

Senator Richard Bryan, Statements before Senate, Oct. 28, 1997; Conference Report on Interior Appropriations Act.

Serôa da Motta, R. (1998), "Application of economic instruments for environmental management in Latin America: from theoretical to practical constraints", paper presented at the Organization of American States Meeting on Sustainable Development in Latin America and the Caribbean: Policies, Programs and Financing, Washington, DC (30 October).

Shelby, M., R. Shackleton, M. Shealy and A. Cristofaro (1997), The Climate Change Implications of Eliminating U.S. Energy (and Related) Subsidies (U.S. Environmental Protection Agency, Washington, DC).

Sigman, H.A. (1995), "A comparison of public policies for lead recycling", RAND Journal of Economics 26(3):452–478.

Singh, M. (1995), "Pollution control through the provision of information: assessing the impact of the toxics release inventory", Ph.D. Thesis (Boston University, Boston, MA), unpublished.

Skumatz, L.A. (1990), "Volume-based rates in solid waste: Seattle's experience", Report for the Seattle Solid Waste Utility (Seattle Solid Waste Utility, Seattle, WA).

Smith, R. (2000), "Azurix is launching online exchange for buying and selling water in the West", The Wall Street Journal (9 February).

Solomon, B.D. (1999), "New directions in emissions trading: the potential contribution of new institutional economics", Ecological Economics 30:371–387.

Solomon, B.D., and H.S. Gorman (1998), "State-level air emissions trading: the Michigan and Illinois models", Journal of the Air & Waste Management Association 48(12):1156–1165.

Sonneborn, C.L. (1999), "An overview of greenhouse gas emissions trading pilot schemes and activities", Ecological Economics 31:1–10.

Sorrell, S. (1999), "Why sulphur trading failed in the UK", in: S. Sorrell and J. Skea, eds., Pollution for Sale: Emissions Trading and Joint Implementation (Edward Elgar, Cheltenham, UK) 170–207.

Sorrell, S., and J. Skea (1999), Pollution for Sale: Emissions Trading and Joint Implementation (Edward Elgar, Cheltenham, UK).

Speck, S. (1998), "A database of environmental taxes and charges", The Eco-Tax Database of Forum for the Future at Keele University, in: European Commission, Database on Environmental Taxes in the European Union Member States, plus Norway and Switzerland (Forum for the Future, London) 37–153.

State of Michigan, Department of Environmental Quality (2000), Water Quality Trading Program World Wide Web site, available at http://www.deq.state.mi.us/swq/trading/tem5x.htm.

Stavins, R.N. (1983), Trading Conservation Investments for Water (Environmental Defense Fund, Berkeley, CA).

Stavins, R.N. (ed.) (1988), Project 88: Harnessing Market Forces to Protect Our Environment, sponsored by Senator Timothy E. Wirth, Colorado, and Senator John Heinz, Pennsylvania (Washington, DC).

Stavins, R.N. (ed.) (1991), Project 88 – Round II Incentives for Action: Designing Market-Based Environmental Strategies, sponsored by Senator Timothy E. Wirth, Colorado, and Senator John Heinz, Pennsylvania (Washington, DC).

Stavins, R.N. (1995),"Transaction costs and tradable permits", Journal of Environmental Economics and Management 29:133–148.

Stavins, R.N. (1996), "Correlated uncertainty and policy instrument choice", Journal of Environmental Economics and Management 30:218–232.

Stavins, R.N. (1997), "Policy instruments for climate change: how can national governments address a global problem?" The University of Chicago Legal Forum:293–329.

Stavins, R.N. (1998), "What have we learned from the grand policy experiment: lessons from SO_2 allowance trading", Journal of Economic Perspectives 12(3):69–88.

Stavins, R.N. (2000), "Market-based environmental policies", in: P.R. Portney and R.N. Stavins, eds., Public Policies for Environmental Protection (Resources for the Future, Washington, DC).

Stavins, R.N., and B.W. Whitehead (1992), "Pollution charges for environmental protection: a policy link between energy and environment", Annual Review of Energy and the Environment 17:187–210.

Steele, P. (1999), "Market-based instruments for environmental policy in developing countries – from theory to practice with a case study of Sri Lanka's recent experience", in: T. Sterner, ed., The Market and the Environment (Edward Elgar, Cheltenham, UK) 274–300.

Stepanek, Z. (1997), "Integration of pollution charge systems with strict performance standards: the experience of the Czech Republic", in: R. Bluffstone and B.A. Larson, eds., Controlling Pollution in Transition Economies (Edward Elgar, Cheltenham, UK) 144–155.

Sterner, T. (Draft, 1999), "The selection and design of policy instruments", Draft Manuscript.

Sterner, T. (ed.) (1999), The Market and the Environment (Edward Elgar, Cheltenham, UK).

Sterner, T., and L. Hoglund (1998), "Refunded emission payments: a hybrid instrument with some attractive properties", Working Paper (Resources for the Future, Washington, DC).

Stevens, B.J. (1978), "Scale, market structure, and the cost of refuse collection", The Review of Economics and Statistics 40:438–448.

Svendsen, G.T. (1998), Public Choice and Environmental Regulation: Tradable Permit Systems in the United States and CO_2 Taxation in Europe (Edward Elgar, Cheltenham, UK).

Swift, B. (2000), "Allowance trading and SO_2 hot spots – good news from the Acid Rain Program", Environment Reporter, May 12.

TerraChoice Environmental Services Inc. (1999), Environmental Choice Program World Wide Web site, available at http://www.environmentalchoice.com

Thailand Environment Institute (1999), Thai Green Label World Wide Web site, available at http://www.tei.or.th/bep/GreenLabel.cfm

Thompson, D.B. (1997), "The political economy of the RECLAIM emissions market for Southern California", Working Paper (March) (University of Virginia, VA).

Tietenberg, T. (1980), "Transferable discharge permits and the control of stationary source air pollution: A survey and synthesis", Land Economics 56:391–416.

Tietenberg, T. (1985), Emissions Trading: An Exercise in Reforming Pollution Policy (Resources for the Future, Washington, DC).

Tietenberg, T. (1995), "Tradeable permits for pollution control when emission location matters: what have we learned?" Environmental and Resource Economics 5:95–113.

Tietenberg, T. (1997a), "Information strategies for pollution control", Paper presented at the Eighth Annual Conference, European Association of Environmental and Resource Economists, Tilburg, The Netherlands (June 26–28).

Tietenberg, T. (1997b), "Tradeable permits and the control of air pollution in the United States", Paper prepared for the 10th Anniversary Jubilee edition of Zeitschrift Fürangewandte Umweltforschung.

Tietenberg, T., and D. Wheeler (1998), "Empowering the community: information strategies for pollution control", in: Frontiers of Environmental Economics Conference (Airlie House, VA).

Tripp, J.T.B., and D.J. Dudek (1989), "Institutional guidelines for designing successful transferable rights programs", Yale Journal of Regulation 6:369–391.

U.S. Congress (1951), "Independent Offices Appropriations Act of 1951", Ch. 375, §501, 654 Stat. 290. 31 U.S.C. §9701 (31 August).

U.S. Congress (1978), "Gas Guzzler Tax", 26 U.S.C. Sec. 4064.

U.S. Congress (1986), "Job Training Partnership Act Amendments of 1986", 99th Congress, 1st Session, P.L. 99-499, Sec. 522(a) (16 October).

U.S. Congress (1989), "Omnibus Budget Reconciliation Act of 1989", 101st Congress, 1st Session, P.L. 101-239 (19 December).

U.S. Congress (1989), "Omnibus Budget Reconciliation Act of 1989", 101st Congress, 1st Session, P.L. 101-239, Sec. 7506, "Excise Tax on the Sale of Chemicals Which Deplete the Ozone Layer and of Products Containing Such Chemicals" (19 December).

U.S. Congress (1995), "Appliance Labeling Rule", 16 C.F.R., Ch. 1, Federal Trade Commission, Part 305 (16 March).

U.S. Congress (1997), "The Electric Consumers' Power to Choose Act", 105th Congress, 1st Session, H.R. 655 (10 February).

U.S. Congress, Office of Technology Assessment (1992), Building Energy Efficiency (Washington, DC).

U.S. Congress, Office of Technology Assessment (1995), Environmental Policy Tools: A Users Guide (Washington, DC).

U.S. Congressional Budget Office (1990), Reducing the Deficit: Spending and Revenue Options (Washington, DC).

U.S. Department of State (1997), U.S. Climate Action Report, Publication 10496 (Washington, DC).

U.S. Environmental Protection Agency (1982), "Regulation of fuel and fuel additives – final rule", 24 F.R. 49322 (29 October).

U.S. Environmental Protection Agency (1986), "Emissions trading policy statement – final policy statement", 51 F.R. 43814 (4 December).

U.S. Environmental Protection Agency (1990), Environmental Investments: The Cost of a Clean Environment, Report of the Administrator to Congress (U.S. EPA, Washington, DC).

U.S. Environmental Protection Agency (1991), Economic Incentives, Options for Environmental Protection, Document P-2001 (EPA, Washington, DC).

U.S. Environmental Protection Agency (1992), The United States Experience with Economic Incentives to Control Environmental Pollution, EPA-230-R-92-001 (Washington, DC).

U.S. Environmental Protection Agency (2001), The United States Experience with Economic Incentives for Protecting the Environment, EPA-240-R-01-001 (Washington, DC).

U.S. Environmental Protection Agency (1998), "EPA proposes emissions trading program to help protect eastern U.S. from smog", Press Release (29 April).

U.S. Environmental Protection Agency, Office of Policy Analysis (1984), "Case studies on the trading of effluent loads, Dillon Reservoir", Final Report.

U.S. Environmental Protection Agency, Office of Policy Analysis (1985), Costs and Benefits of Reducing Lead in Gasoline, Final Regulatory Impact Analysis (Washington, DC).

U.S. Federal Energy Regulatory Commission (1996), "Promoting wholesale competition through open access non-discriminatory transmission services by public utilities; recovery of stranded costs by public utilities and transmitting utilities", Order 888 (24 April).

U.S. General Accounting Office (1986), Vehicle Emissions: EPA Program to Assist Leaded-Gasoline Producers Needs Prompt Improvement, GAO/RCED-86-182 (U.S. GAO, Washington, DC).

U.S. General Accounting Office (1990), Solid Waste: Trade-offs Involved in Beverage Container Deposit Legislation, GAO/RCED-91-25 (U.S. GAO, Washington, DC).

U.S. General Accounting Office (1992), Toxic Chemicals: EPA's Toxics Release Inventory Is Useful But Could Be Improved (U.S. GAO, Washington, DC).

Vincent, J., R. Ali et al. (1997), Environment and Development in a Resource-Rich Economy (Harvard University Press, Cambridge, MA).

Vincent, J., and S. Farrow (1997), "A survey of pollution charge systems and key issues in policy design", in: R. Bluffstone and B.A. Larson, eds., Controlling Pollution in Transition Economies (Edward Elgar, Cheltenham, UK).

Voigt, P.C., and L.E. Danielson (1996), "Wetlands mitigation banking systems: a means of compensating for wetlands impacts", Applied Resource Economics and Policy Group Working Paper AREP96-2 (Department of Agricultural and Resource Economics, North Carolina State University, NC).

Wahl, R.W. (1989), Markets for Federal Water: Subsidies, Property Rights, and the Bureau of Reclamation (Resources for the Future, Washington, DC).

Wang, H., and D. Wheeler (1996), "Pricing industrial pollution in China: an econometric analysis of the levy system" (World Bank Policy Research Department, Washington, DC).

Wang, H., and D. Wheeler (1999), "China's pollution levy: an analysis of industry's response", presented to the Association of Environmental and Resource Economists Workshop, Market-Based Instruments for Environmental Protection, John F. Kennedy School of Government, Harvard University, July 18–20.

Wang, J., and X. Lu (1998), "Economic policies for environmental protection in China: practice and perspectives", in: Organization for Economic Cooperation and Development, Applying Market-Based Instruments to Environmental Policies in China and OECD Countries (OECD, Paris): 15–29.

Weitzman, M. (1974),"Prices vs. quantities", Review of Economic Studies 41:477–491.

Wertz, K.L. (1976), "Economic factors influencing households' Production of Refuse", Journal of Environmental Economics and Management 2:263–272.

Woerdman, E., and W. Van der Gaast (1999), "A review of the cost-effectiveness of activities implemented jointly: projecting JI and CDM credit prices", Working Paper (September).

World Bank (1997a), China 2020: Clear Water, Blue Skies. China's Environment in the New Century (World Bank, Washington, DC) Chapter 5.

World Bank (1997b), Five Years after Rio: Innovations in Environmental Policy, Environmentally Sustainable Development, Studies and Monographs Series No. 18 (World Bank, Washington, DC).

World Bank (1999), "Using market-based instruments in the developing world: the case of pollution charges in Colombia", New Ideas in Pollution Regulation World Wide Web site, available at http://www.worldbank.org/nipr/lacsem/columpres/index.htm

World Bank (2000), Greening Industry: New Roles for Communities, Markets and Governments (Oxford University Press for the World Bank, New York).

Wuppertal Institute (1996), "Japan back as world-beater – this time with green taxes?" Wuppertal Bulletin on Ecological Tax Reform 2(3):1–2.

Yang, J., D. Cao and D. Wang (1998), "The air pollution charge system in China: practice and reform", in: Organization for Economic Cooperation and Development, Applying Market-Based Instruments to Environmental Policies in China and OECD Countries (OECD, Paris): 67–86.

Zou, J., and H. Yuan (1998), "Air pollution, energy and fiscal policies in China: a review", in: Organization for Economic Cooperation and Development, Applying Market-Based Instruments to Environmental Policies in China and OECD Countries (OECD, Paris): 119–137.

Żylicz, T. (1996), "Economic instruments in Poland's environmental policy", paper prepared for Central and Eastern European Environmental Economics and Policy Project (Harvard Institute for International Development).

Żylicz, T. (1999), "Towards tradability of pollution permits in Poland", in: S. Sorrell and J. Skea, Pollution for Sale (Edward Elgar, Cheltenham, UK) 124–138.

Żylicz, T. (2000), Personal communication.

World Bank (1997a). *Five Years after Rio: Innovations in Environmental Policy.* Environmentally Sustainable Studies and Monographs Series No. 18. World Bank, Washington, DC.

World Bank (1997b). *Using market-based instruments in the developing world.* In *Expanding Environmental Policy Instruments to Developing Countries.* World Bank, Washington, DC.

World Bank (1998). *Greening Industry: New Roles for Communities, Markets and Governments.* Oxford University Press, New York.

Chapter 10

EXPERIMENTAL EVALUATIONS OF POLICY INSTRUMENTS

PETER BOHM

Department of Economics, Stockholm University, 106 91 Stockholm, Sweden

Contents

Handbook of Environmental Economics, Volume 1, Edited by K.-G. Mäler and J.R. Vincent

Abstract

Experimental methods have recently been used to evaluate environmental policy instruments, in particular – and most suitably, it seems – emissions trading programs of various designs. Some studies have focused on domestic emissions trading programs, while others have focused on international programs, in particular ones related to greenhouse gases. Much emphasis has been put on investigating the implications of market power in emissions trading. Other topics of the experimental studies reviewed here include the relative merits of different policy instruments (permits, taxes, standards), and the possibility of eliminating the need for conventional environmental policy through application of the Coase theorem.

Keywords

experimental economics, environmental policy, emissions trading, market power, climate change policy

JEL classification: C9, D44, Q25, Q28

Introduction

Both environmental and experimental economics are young sub-disciplines of economic science. The intersection of the two is even younger. This explains why almost none of the sources reviewed in this chapter is from more than ten years back.

Experimental methods have been used in economics, first of all, to test propositions in economic theory concerning areas such as game theory, bargaining, decisions under uncertainty and, in particular, behavior under various market institutions.[1] Second, experimental economics can be seen to provide an approach to empirical insights as a substitute or complement to certain forms of analyses of traditional data. Third, experimental economics has been used, and increasingly so, to investigate options to attain specific policy targets, which is directly relevant to the topic of this chapter.[2]

As will become obvious below, most experimental analyses of environmental policy refer to a fairly new policy instrument, emissions trading – originally proposed by Crocker (1966) and Dales (1968).[3] By contrast, environmental taxation dates back to Pigou in the 1910s. Direct regulation or "command and control" emerged as the dominant real-world practice of environmental policy, especially in the 1960s to 1980s. It is perhaps fair to say that many of the actual or potential designs of command and control have been ill suited for experimental testing, given this method's reliance on simple and straight-forward incentive systems. A reason why environmental taxes have been little in focus for experimental analysis is that their effects are unlikely to deviate much from what is known about the effects of excise taxes, in general. Therefore, emissions trading seems to have emerged as the instrument both most suitable and most interesting for experimental testing. This contributes to explaining why the chapter and the literature on which it is based has emissions trading in focus.

Before entering into the topic area of this chapter, the reader should be forewarned that the present author is not convinced that the methodology in dominant use in experimental economics always has been the appropriate one or one that is known to generate relevant and reliable results. These doubts (elaborated in Bohm (2002)) relate in particular to (a) the use of 'low' – instead of *clearly* significant – incentive levels (discussed further in the next section), (b) the attempts to obtain generality by couching experiments in a 'context-free' style (discussed in the concluding section), and (c) the practice of repeating exactly the same setting, with the same subjects and the same (induced) values, over a number of periods, one right after the other. Such repetitions

---

[1]   As a case in point, relevant for the subject matter of this chapter, mechanisms central to theories of the overuse of common-pool resources and of market neglect of (other) externalities have been tested in laboratory experiments. See Davis and Holt (1993).

[2]   For an early general review of the use of experimental methods for policy evaluations and some applications, see Plott (1987). Bjornstad, Elliott and Hale (1999) is a more recent review, focused on tradable permits.

[3]   See Chapter 9 (by Robert Stavins) for a review of empirical experience with emissions trading and other market-based environmental policy instruments.

are frequently made in the experiments to be reviewed here, with the primary purpose of making subjects more familiar with the task they are asked to perform. However, there is the risk that identical repetitions, in particular in large numbers, significantly increase the artificiality of the experimental exercise in a way that tends to confuse the subjects' perception of this exercise. Or, the repetitions may simply make the subjects bored enough to stop thinking and/or cause them to change their behavior just to avoid being bored or to do some private, uncalled for, experiments of their own.[4]

Although an attempt is made below not to exclude reporting any work just because of this author's personal doubts about some aspects of the methodology that has been in use, such doubts may still color the report. Therefore, the reader may want to check also other overviews related to the kind of experimental work reported here, such as Shogren and Hurley (1999) or the chapter by Jason Shogren in this Handbook, which describes how experiments are run in economics but focuses on applications related to environmental valuation. For an overview of methods and achieved results in experimental economics in general, see Hey (1991), Davis and Holt (1993), and Kagel and Roth (1995).

Section 1 deals with the Coase theorem and the possibility to reduce the need for conventional environmental policy. Section 2 reports a study of the relative merits of different policy instruments. Broad issues related to domestic emissions trading are discussed in Section 3, while studies of market power in such markets are reviewed in Section 4. International emissions trading is the topic in Section 5. The final section offers some concluding remarks.

1. Property-rights allocation as an instrument to eliminate the need for environmental policy

Externalities such as environmental effects would cease to exist if the polluter or the party harmed had enforceable rights to the property affected. The much-cited theorem by Coase (1960) holds that, under certain circumstances, efficiency is attained regardless of the choice of party to whom the rights are allocated. These circumstances are characterized by, in particular, zero transaction (= negotiation, monitoring, enforcement, etc.) costs, and common knowledge of the parties' payoffs. An interpretation of the theorem is that, whenever these circumstances obtain, no fine-tuned policy intervention is required. Allocating the property rights is 'all' a government needs to do.

The Coase theorem has attracted a lot of attention among experimental economists. Most of the experimental work has focused on the two-party case, in which the assumption of zero transaction costs seems most likely to hold. In these experiments, subjects are asked to perform in an abstract, i.e., context-free, situation. For each subject pair, one subject has been randomly selected to be the controller, i.e., the party who has the

---

[4] For a critique of the use of stationary repetition, see Loewenstein (1999).

relevant property right. This subject is free to choose an activity level that not only determines his/her payoff but also the payoff for the other subject. In the case of negative environmental effects, the latter receives lower payoffs, the higher the levels chosen by the controller. The dependent subject could try to influence this choice by paying the controller subject a lump-sum compensation for moving to a less detrimental activity level. As long as the joint payoffs are not maximized, there exist combinations of activity levels and compensations that are mutually advantageous.

In the first tests of this type, Hoffman and Spitzer (1982) found that the two parties predominantly agreed on efficient (= joint-payoff-maximizing) activity levels but not on sharing the joint payoff in a mutually advantageous fashion. Instead, quite often they split the joint payoff equally between them, which gives the controller a lower payoff than in the case of no agreement and an individually profit-maximizing activity level. For this behavior to be generally rational, the controller must value fairness *per se*.[5] However, as shown by Harrison and McKee (1985), another reason for this behavior might be that the incentives for finding individually rational agreements were too small. Increasing the incentive levels, they found (i) a significant reduction of the occurrence of equal splits and (ii) a predominant number of outcomes where the controller received at least as much as in the case of no agreement.

Harrison and McKee's findings illustrate the risk of using small incentives to save on experimental costs, thus generating results that may not represent the important real-world cases.[6] Setting out using small incentives and generating surprising results, which motivates another round of experiments, now with higher incentive levels, implies at least a waste of aggregate experimental resources, which is especially conspicuous if the theoretical predictions then are restored.

The bottom line of the results of these tests, which include extensions by Hoffman and Spitzer (1985), Shogren and Kask (1992) and Shogren (1992, 1998), is that predicted bargaining efficiency and assumed individually rational behavior of the Coase theorem receive considerable support. Furthermore, Shogren (1998) and Rhoads and Shogren (1999) explicitly test the effect of varying the size of transaction costs, finding that efficiency declines significantly with increases in such costs.

A remaining issue concerns the extent of the environmental-policy relevance of the Coase theorem. In many cases, transaction costs are likely to be significant, at least when there are more than two parties involved. A particular case where this becomes apparent is that of latent environmental effects, where the existence of a polluting firm never made it worthwhile for other agents to establish any consumption or production activities in the area. If there were only one potential firm that would want to start negotiating with the existing firm, the original Coasian set-up may apply and lead to a deal where the potential firm could be established by compensating the existing firm for reducing

---

[5] Letting subjects earn, or believe they earned, the right to be the controller has been shown to reduce the role of fairness considerations. This design was used by Hoffman and Spitzer (1985) and turned out to increase the controllers' payoffs.

[6] See Harrison (1989, 1992, 1994) for a discussion of the consequences of small incentive levels.

its externality-generating activity. In a more likely case, there are a large number of potential firms or other agents (possibly mutually exclusive) and the negotiation costs may well be prohibitively high.

When the Coasian solution fails or property rights are not allocated to any party because this solution is deemed to fail, the government may intervene with market-based instruments or direct regulation, possibly on the basis of imperfect best estimates of the harm caused by the polluter(s). In addition to the many cases with multiple polluters and pollutees, where negotiation costs would be too high, there are those in which the government does not want to, or simply cannot, get political support for any particular allocation of the environmental property rights. In such cases, the rights may be allocated to the government itself as a representative for the community at large, e.g., by introducing an environmental tax. The rest of the chapter concerns experimental studies that have contributed to identifying the characteristics of such conventional instruments of environmental policy, in particular tradable permits or quotas.

2. Comparing environmental policy options

In an early experimental study, Plott (1983) compared versions of the three major instruments in environmental policy: taxes, standards, and tradable permits. A laboratory market was created where the aggregate trade volume produced negative external effects on the profits of all traders. Seller and buyer subjects who were given cost/redemption values could submit bids, asks, and acceptances in an oral double auction.[7] That is, the rate of seller profits was given by actual selling prices *minus* the induced costs for the trades accomplished, and the rate of buyer profits was given by redemption values *minus* actual buyer prices. The optimal tax rate (= marginal external cost in the social optimum) as well as the socially optimal output level were taken to be known and therefore imposed by the regulator. In the standards case, firms were assumed to be given permits up to that level on a first-come first-served basis. The same data were used in two market sessions with at least six periods for each of the three policy cases.

Plott found that "the traditional models found in the economics literature were amazingly correct". At the end of the six periods, the efficiency percentages (= actual social surplus, which could be negative, relative to optimum social surplus) ranged from −111.9 to 36.1 for the 'no policy' case and for the three policy cases:

- from 65.5 to 98.0 for the tax policy;
- from −0.4 to 59.1 for the standards policy;
- from 88.4 to 99.6 for the tradable permit policy.

---

[7] Double auction is the trading institution used, e.g., on a running stock exchange such as the NYSE. In auctions of this type traders sequentially enter bids and asks, stating price as well as quantity. Transactions are carried out if and when such bids or asks are accepted by other traders.

Efficiency essentially increased over time for the two market based instruments, but behaved erratically in the standards case. One reason for the latter result is likely the fact that transactions in that case were accepted on a first-come first-served basis. This created a rush for subjects to have their bids or asks accepted. No similar hurry seems to have been imposed in the other cases. Moreover, the design chosen for the standards case may have failed to represent equally well all direct regulatory instruments used in practice, due to the fact that this set of instruments is large and so is the number of possible designs for each option. Furthermore, the subject pools differed between the tradable permit case (often experienced subjects from Caltech) and the other test cases (inexperienced students from less prominent schools); this difference may explain at least part of the success of the tradable permit policy, even though it was the most complicated case with trade in both a permit and a 'product' market.

In spite of such caveats, the experiment produced a significant result, which was not contradicted in a similar test by Harrison et al. (1987) and does not seem to have been challenged by other tests. So, as Plott concluded, "those who wish to offer competing theories about market behavior in externality situations must reconcile their ideas with these experiences".

3. Testing designs of emissions trading

Standard theory has it that the tradable-permit instrument is efficient in allocating a given cap on emissions (sum of permits), if the market for permits and downstream markets are competitive and if transaction costs are insignificant [see Chapter 6 (by Gloria Helfand, Peter Berck, and Tim Maull)]. Some experimental support for this proposition was reported in the preceding section.

Experimental economics has also been used to test the properties of various designs of permit markets. One set of experimental studies has investigated aspects of different revenue-neutral auction designs; see, e.g., Franciosi et al. (1993) and Ledyard and Szakaly-Moore (1994).[8] Another set of experimental studies, to which we now turn, has focused on the incentive effects of an auction design introduced by the US Environmental Protection Agency (EPA). These studies are of particular interest since they give an example of experimental research that provides policy-relevant support to theoretical arguments which alone may fail to convince policy makers.

3.1. The EPA auction mechanism

The 1990 US Clean Air Act stated that sulfur-dioxide emission permits/allowances "shall be sold on the basis of bid price, starting with the highest-price bid and continuing until all allowances for sale at such auction have been allocated". The US EPA

---

[8] For a brief summary of these experiments, see Muller and Mestelman (1998). That paper also reports an early experimental study by Hahn (1988) on revenue-neutral permit auctions.

interpreted this to say that the seller with the lowest ask should receive the highest bid price and so on down to a market-clearing transaction volume. Thus, sellers were ranked from the lowest ask upwards and were matched with buyers, who were ranked from the highest bid downwards and had to pay a price equal to their bids. The reason for the decision by the US Congress is said to have been to let the initial permit holders get a maximum compensation for giving up their (originally grandfathered) permits, possibly because previous experience with tradable permit systems indicated that permit holders were unwilling to give up excess volumes of their permits.

Simple theoretical analysis indicates that bidders would hesitate to state bids equal to their maximum willingness to pay in this kind of discriminative auction and that sellers would have an incentive to compete for the high prices by understating their asks [Cason (1995)]. Then, the prediction is that prices for marginal transactions would fall below that of an efficient non-discriminative auction and send the wrong price signals for R&D and long-term investment decisions. Laboratory experiments could help to establish for policy makers and others whether or not this outcome is more than just a theoretical proposition. It may be argued that more sophisticated theoretical counter-propositions are conceivable, saying, e.g., that buyers would not dare to significantly understate their demand, since they would then risk being eliminated from the market. Another possibility is that sellers might hesitate to understate their asks because of the risk of being matched with buyer bids below their true willingness-to-accept. Another way of stating these reasons for undertaking a controlled test of the incentives of the EPA auction rules is that the incentives may be difficult to grasp and therefore that their effects are unclear.

Cason and Plott (1996) compared experimentally the annual EPA call auction design with that of the more commonly observed uniform-price call auction for an abstract commodity.[9] Buyer values and seller costs were induced for two different market environments. Trading was repeated over a number of periods. The main results were that, in the uniform-price auction,

- the revealed values as well as costs were higher, and the price was closer to the competitive equilibrium price and higher than the marginal market clearing price of the EPA auction;
- seller profits were at least as high; and
- prices responded more rapidly to unexpected parameter changes.

The first result was expected by standard theory, but the last two added new insights. Particularly, noteworthy is that the suggested ambition of the EPA auction design – to maximize seller revenues – clearly seems to have failed to be attained. But this result may be sensitive to the level of information that would exist on this kind of market in the real world. If sellers and buyers had pretty good estimates of what the marginal price could be, those who had a true willingness-to-accept or willingness-to-pay close to that

---

[9]   Cason (1995) reports a test of an inverted version of the EPA auction, where buyers face the same incentives as the sellers in that auction. The results are similar, *mutatis mutandis*, to those of the Cason and Plott study.

level may not feel inclined to distort their bids and asks even in the EPA auction. If so, the misrepresentation incentives would concern intramarginal units only, thus moving demand and supply downwards with the competitive equilibrium quantity and price combination as a pivotal point.

As it turned out, this experimental work had limited relevance for actual sulfur-dioxide emissions trading. Cason and Plott (1996) implicitly assumed that no permit trade occurred outside the EPA auction. After some time, a large private continuous outside market had developed with publicized prices [see Joskow, Schmalensee and Bailey (1998)]. This turned the EPA auctions into a common value auction, as no buyer (seller) in the auctions could be expected to enter bids (asks) higher (lower) than the current prices on the outside market. In fact, once the transaction prices on this market were made commonly known, auction prices never deviated from these prices. Furthermore, Joskow et al. report that the few private seller asks there were in the auction often turned out to be higher rather than lower than the market-clearing prices; the suggested reason was that these asks were made for strategic or demonstrative purposes.

There are some indications that market-clearing auction prices indeed were as predicted by Cason and Plott during the first two EPA auctions, i.e., before the outside market had been firmly established and its prices made public. But there were few privately owned permits even offered, and still fewer sold, in these two auctions. (In the last auction reported in Joskow, Schmalensee and Bailey, privately offered permits had disappeared, leaving the auction entirely to the small permit volume offered by the government.)

Thus, the EPA auction might have been designed in *any* fashion and still not have had any significant effect on trading efficiency and price signaling. What the Cason and Plott work nevertheless indicates is that the policy-makers' actual choice of auction design for permit trading would have been inefficient in the absence of an outside market; if so, it would have created a difficult, possibly counterintuitive and counterproductive, market institution that traders would have tried fast to dissociate themselves from.

3.2. Electric bulletin board trading vs. double auctions

Cason and Gangadharan (1998) analyzed experimentally a Los Angeles tradable emissions program (RECLAIM) that uses an electronic bulletin board institution, in which firms seeking to buy or sell permits post proposed terms of trade. Transactions were executed following bilateral negotiations.

The experimental results of three laboratory sessions indicated that mean transaction prices were roughly the same as in a continuous double auction institution used in another three sessions. However, efficiency was much lower in bulletin board trading (43–66 percent) than in double auction trading (90–93 percent). One reason seems to have been that there were significant procedural differences between the institutions; the bulletin board trading was drawn out over a long period of time (six weeks) and attracted a low participation rate among the subjects enrolled for this test. Another reason, suggested in Hizen and Saijo (2001), could be that each subject was given the role of either seller or buyer and hence could not act as a general trader. Cason and Gangadha-ran also admit that the subjects' earnings were small which may have failed to produce significant trading incentives. Still, they stress that their results do not indicate that the bulletin board system would lead to any "highly inaccurate transaction prices".

3.3. Implications of futures markets for permits and permit banking

Experience with the use of permit systems in the US suggests that the success of such systems depends heavily on their design with respect to market institutions and admin-istrative details concerning trade approval, permit hoarding, constraints on trading, etc. A particular reason why tradable-permit systems have not always been as successful as predicted by theory is that, in several cases, the designs used have led to significant transaction costs [Stavins (1995)].

In some permit systems, permits have an 'eternal' life allowing certain emission vol-umes per period, although their values in terms of per-period emissions permitted may be adjusted over time. In other permit systems, permits are time-limited and allow a given amount of emissions in a specific period. A permit system that includes futures markets as well, i.e., where trading is feasible also for permits that can be used only in a future period, may be expected to have properties quite different from those of systems where trading is limited to permits for emissions during a current permit period. In the case of an existing demand and supply of permits for the following permit period or several subsequent periods, a pretty safe guess is that the availability of futures mar-kets would be better in all relevant normative dimensions than if no such markets were available.

A Canadian tradable-permit proposal for nitrogen-oxide emissions has been evalu-ated with respect to the inclusion of a particular futures market. With complete and per-fect contingent future markets, the analysis would be straightforward. But the proposed Canadian futures market is special in the sense that the cap on emissions in each future period is not known in advance, so only *shares* of the permits that eventually will be issued can be traded along with permits with known denomination in emissions for the current period. Muller and Mestelman (1994) and Mestelman and Muller (1998) com-pare this system with some earlier tests of US sulfur-dioxide permit trading for a current period only. Taken at face value, the results indicate that the addition of a permit-share futures market increases efficiency. However, since they note several possibly signifi-cant differences between the tests with respect to the training of the subjects and the market institution used, it becomes difficult to evaluate these results.

In Godby et al. (1997), double-auction trading in permit shares is tested when con-trol over discharges in the permit period is uncertain. In the test, the net emissions are taken to be stochastic. A reconciliation market is introduced for traders who *ex post* find themselves short of permits. Comparing efficiency with and without the assumed form of uncertainty did not reveal any difference. But it should be noted that prices in the reconciliation period – relevant only in the uncertainty case – were found to be highly variable.

There are several ways to counteract tendencies for high price volatility at the end of a permit period, e.g., letting permit periods overlap or allowing banking and (limited) borrowing of permits to/from the next period.[10] Experimental methods have been used by Godby, Mestelman, and Muller (1999) to study the role of permit banking in connection with their special model of trading in permits as well as shares of future permits. In the uncertainty environment analyzed in Godby et al. (1997), banking is found to greatly reduce price instability in the reconciliation period, while the effect on efficiency is less clear-cut. This may be due to the increased complexity introduced by the banking institution.[11]

4. Market power and domestic emissions trading

Standard theory of market behavior states that dominant sellers or buyers would use their market power to increase their profits by withholding supply or demand. As a result, (a) transactions fall short of the competitive volume and (b) the market-power agent obtains prices that are more favorable than prices at the competitive level. Experiments confirm that markets where price *plus* quantity offers are posted by a monopoly or by colluding sellers have such effects [for an overview see Davis and Holt (1993, Section 4.5)].

Further experiments have been made to test to what extent these results hold also in permit markets. A couple of reasons can be given why they should not. First, the scope for market power in this particular type of market is limited by the very nature of permit trading. For comparison, take a market where a producer has a monopoly position for a particular product and can maintain that position at any price, at least in the short run. By contrast, a permit market is much like a stock market and typically such that many traders can take either a seller or a buyer position, the choice of which depends on the prices established on the market. Thus, for instance, if a permit trader is the only seller for prices below a certain limit, but now tries to charge a price above that limit, some of those who were buyers at lower prices may now become sellers.

Another reason for permit markets being special has to do with the fact that the trade object has a value (apart from speculation in gains from arbitrage over time) only for the purpose of demonstrating compliance at the end of the period, while trade can take place at any time over the entire period. The question here concerns the dates at which a market-power agent – say, a monopolist who is initial holder of all permits, grandfathered or acquired in an initial government auction – wants to offer his permits for

---

[10] Borrowing of permits does not seem to have been much considered in the literature, possibly reflecting that borrowing is disliked for political reasons, in particular, with respect to mandatory domestic tradable-permit systems. However, in the context of international greenhouse-gas emissions trading, allowing limited borrowing would give traders additional flexibility and make poor, risk-averse countries more interested in joining, hence increasing the cost-effectiveness of such trading.

[11] See Muller and Mestelman (1998) for a comprehensive overview of the experiments reported in this subsection.

sale. (We disregard banking and the existence of futures markets here.) Assume the market institution to be one of double auction; this is particularly likely to be relevant for highly liquid permit markets, such as those expected for carbon permits.[12] During the early stages of a trading period, the monopolist may withhold supply to try to earn as high profits as possible from those who wish to buy at that point. Towards the end of the period, when all parties are out to finalize permit holdings to cover emissions, the monopolist can benefit from any trade where price exceeds his opportunity cost. This would make final transactions approach the competitive price and quantity levels at that point in time.

4.1. Market power and double auctions

Experimental tests of stylized double auctions have revealed [Smith and Williams (1989); see also the overview in Godby et al. (1999)] that, even with only one trader on one side of the market and competition among several traders on the other,[13] end-period prices approach the competitive price and hence, the aggregate trade volume approaches the competitive level. Thus, if these results are confirmed, what would remain from standard market-power theory is essentially that early prices may favor the agent with market power. However, some of the experiments have been able to show that repeated exposure to this kind of environment may make the competitive side of the market unwilling to trade at prices that are expected to deviate significantly and unfavorably from the competitive price level. If this learning process could be counted on, both of the above-mentioned effects of market power would be eliminated. If not, at least the tendency towards market inefficiency would be removed (see further below).

At McMaster University, a group of experimentalists have investigated the permit markets' susceptibility to market power in double auctions. Muller et al. (2003) summarize part of this work, ending up being very critical of the results cited in the preceding paragraph and of their relevance for emissions trading, in particular. They claim that double auctions do not provide an effective constraint on market power. This is supported by tests in which one group of subjects move around, first trading on a competitive market, then on a monopoly or monopsony market, and finally on a competitive market; other subjects are exposed to the opposite crossover design. Each market form is tested over a number of periods. The test results imply that monopoly and monopsony subjects are clearly able to manipulate *mean* within-period prices to their favor and that the successful application of this strategy is not eliminated by the learning provided across periods.

Thus, these tests do not confirm the Smith and Williams result, where repeated trading with the same subjects and the same induced values eventually, i.e., in late periods, leads *mean* within-period prices to approach the competitive price level. But

---

[12] See the chapter in this Handbook on climate change by Charles Kolstad and Michael Toman.

[13] This was explicitly tested for the monopoly case only. See Carlén (forthcoming) for a test with similar results for a monopsony case.

What they do confirm is the important prediction that – regardless of repetition – *final* within-period prices and efficiency in double auctions tend to approach the competitive level, even if early within-period prices do not. The tests show, first, that efficiency in monopoly/monospony sessions is essentially as high (0.89–0.92) as in the competitive sessions. Second, *final* prices in each period, e.g., either an early period or a late one, approach the competitive price, although *mean* prices for the monopoly (monopsony) within each period remain above (below) that level. Moreover, prices tend to converge from below in monopsony cases and from above in monopoly cases.[14]

To sum up, the experimental evidence seems to imply that market power in double auctions manifests itself not as making the market inefficient, but as providing agents with more than their competitive share of the joint profits or social surplus. Experimental studies concerning market power in the context of *international* emissions trading are presented later, in Section 5.3.

4.2. Market power in vertical markets

Theoretical work by Misolek and Elder (1989) shows how market power can be used by a large firm that competes with a given set of small firms in an emissions permit market as well as in a market for a commodity, the production of which requires such permits. By hoarding permits the large firm may abstain from maximizing profits in the permit market, but by doing so it could increase the costs for its competitors in the product market and hence maximize its total profits. The increased distortion in the product market that follows from this so-called exclusionary manipulation (EM) may outweigh the aggregate trade gains achieved on the permit market. If so, tradable permits would emerge as a less efficient system than non-tradable permits, other things equal.

Godby (2000) tests EM experimentally for the case where permit trading is governed by double-auction rules and where permit and product markets are taken to operate sequentially. In the EM sessions the product was sold on a uniform-price call market. In the control sessions the product price was held fixed in order to eliminate the vertical relation between the two markets. The results show substantially lower efficiency in sessions where EM is possible. This seems to indicate that market power in the form of EM causes efficiency losses even when emissions trading is governed by double-auction rules.

Moreover, in a number of the Godby (2000) sessions, the aggregate trade gains were negative, implying that command and control would be Pareto superior. However, the robustness of these results must be checked, in particular with respect to two assumptions made. A special characteristic of this study that may have contributed to the results

---

[14] It should be noted that differences in convergence patterns that have been observed in *competitive* experimental markets seem to reflect differences in relative aggregate consumer and producer surpluses [Smith and Williams (1982)]. Thus, if the aggregate consumer surplus exceeded the aggregate producer surplus, prices tended to converge from above, and *vice versa*. What remains of such influences when the market is not fully competitive is not taken into consideration here.

obtained is that the subject with market power was given complete information about the cost-schedules of its competitors, while subjects representing small firms only were informed about their own costs. Is this high degree of asymmetric information a generally valid reflection of real-world conditions? Furthermore, the cost schedules for subjects representing the small firms were changed during the experiment but were not changed for the subject representing the firm with market power. Thus, the small firms subjects may well have expected that the aggregate market conditions also would change during the experiment.

In any event, this evidence of exclusionary manipulation may apply to the case where the dominant seller and the fringe of small competitors is replaced by a situation where some firms have been grandfathered permits and others – new firms or rapidly expanding firms – demand permits that must come from their competitors. It has long been suspected that, in certain permit markets, firms do not offer their surplus permits for sale in order to avoid helping competitors enter the market or expand their operations. This is another interpretation of the market responses that have been observed in some of the experiments. But, if, contrary to the experimental design employed, a dominant firm or tacitly colluding firms do not have any clear vision of their (new) competitors' cost functions and strategy choices, they would not be in a position to fine-tune their permit sales policy; instead, the party who wants to use its market power may simply decide to refrain from offering its permit surplus at 'any' current bid price.

In the particular case of domestic carbon emissions trading across sectors, which is under consideration in a number of countries, exclusionary manipulation is hardly relevant. If permit liability is placed at the upstream level, i.e., on producers and importers of fossil fuels, market power in the permit market may be conceivable in some countries. But even so, given the many different uses of fossil fuels downstream, it seems unusual that a firm with market power in the permit market would also have market power in any product market with the same set of competitors. The same is true, when we now turn to discuss experiments concerning *international* carbon emissions trading.

5. Experiments with international emissions trading

Experimental economics has been used to test the implications of international emissions trading, a policy instrument proposed for combating, among other things, the risk of global warming. This risk derives from the emissions of so-called greenhouse gases, primarily carbon dioxide.[15] The experimental work started before any clear indications existed that an international treaty to control these emissions would include a system of tradable carbon emissions quotas.[16] In late 1996, about a year before the Kyoto Proto-

---

[15] Chapter 1 (by Bert Bolin) reviews the science of climate change, while the chapter by Charles Kolstad and Michael Toman reviews the economics.

[16] In line with the linguistic convention used in the International Panel on Climate Change (1995), the term tradable quotas is here used for emissions trading among countries while the term tradable permits are reserved for domestic emissions trading.

col to the Framework Convention of Climate Change, UN FCCC (1997), was signed, the US government proposed that international emissions trading should be part of such a treaty. However, up to that point in time, governments in most other countries had been clearly opposed to using this instrument. This is the background to two of the tests reported below.

5.1.   Gains from international emissions trading

In 1996, an experiment was carried out to test the performance of a tradable carbon emissions quota market [Bohm (1999)]. A particular purpose of this test was to use a design that could be more successful in attracting the interest of policy makers than the standard laboratory test with student subjects might be. Thus, an attempt was made to make this test as field-like as possible.

The governments of Denmark, Finland, Norway and Sweden had been generally in favor of international commitments to reduce carbon emissions. Since the beginning of the 1990s, they had all undertaken unilateral measures to cut back carbon emissions by introducing carbon-dioxide taxes. In 1996, their Energy Ministries agreed to let an experiment be carried out in which it was assumed that their respective countries' governments had accepted (a) to stay within their 1990 carbon dioxide emission levels for the year 2000 (a recommended UN FCCC target from 1992) and (b) to trade emission reductions with respect to that year among themselves. Given the countries' existing use of carbon dioxide taxes, the basic instrument for adjusting the emissions in each of the four countries was taken to be to change their respective carbon dioxide tax rates. Thus, the test referred to a case of government emissions trading and not to a case where the participating countries had allocated their national quotas to domestic permit-liable entities to whom all trade, international as well as domestic, would be delegated.

In the experiment, the four governments were represented by negotiating teams of relevant public officials and experts, appointed by the countries' energy ministries. Since the ministries deemed it likely that, in a case of real trade negotiations, their governments would prefer to negotiate bilaterally, this trade design (with communication by fax) was used in the experiment. The incentive mechanism used was that the negotiating teams, prior to the negotiations, would deposit their trade-relevant *social* emission reduction cost functions so that, after the negotiations were completed, an appointed group of international experts could evaluate the attained share of their respective feasible competitive trade gains and then publish this evaluation of the teams' performance. Experts from the four countries had for some years up to 1996 exchanged information about their countries' technical abatement options and marginal *technical* abatement costs for the year 2000, i.e., the costs incurred before political considerations of employment and income distribution effects are taken into account. Hence, the trader teams had some cost information about one another, to an extent similar to what countries considering to participate in real-world emissions trading can be expected to have collected before they would feel ready to engage in any such trading.

As it turned out, the four countries were almost equally capable of reaping the profits that perfectly competitive behavior would have implied. The results of the test revealed a trade volume of more than one third of the countries' required aggregate emission reductions. This trade reduced aggregate costs by almost 50 percent. The efficiency of the trade amounted to 97 percent of the maximum aggregate trade gains.

Since this was a one-shot test, where the results may have been influenced by chance events, it may be noted that a preceding set of eleven pilot tests, conducted with Ph.D. students in economics and using monetary rewards, gave quite similar results, 87–99 percent efficiency. See Bohm and Carlén (1999).

5.2. Comparing bilateral trading and double auction

Hizen and Saijo (2001) report a set of greenhouse-gas (GHG) emissions trading experiments investigating the effect of disclosure vs closure of the trade-relevant (social) marginal abatement cost curves as well as disclosure vs closure of the contracted prices. This was done by comparing the results of bilateral trading and double auction trading (given that, by definition in the latter case, all bids, asks, and prices are common knowledge). Student subjects were used to represent the EU, Japan, Poland, Russia, Ukraine, and the US, although the subjects were not informed about the identity of their representation. Trading was supported by the provision of monetary incentives, averaging some $30 (1998 US dollars) for sessions of less than 60 minutes.

The main findings were:

(1) Efficiency was high, mostly 99 percent, for both bilateral trading and double auctions, which means that marginal abatement costs were equalized in both institutions. This is in line with the bilateral trading result in Bohm (1999), but it is still a striking result, given the conventional wisdom that bilateral trading is expected to be less efficient than double auctions.

(2) Contracted prices roughly converged to the competitive equilibrium price in the double auction experiment, but not in the case of bilateral trading. Also, price variance as well as the number of bids and asks were smaller in the double auction institution.

(3) Efficiency was high regardless of whether all relevant information was private or common. Thus, information about abatement costs and/or contracted prices did not improve efficiency, nor was it found to influence the extent to which prices converged to the competitive level.

In another experiment, Soeberg (2000) studied double auctions among seven countries which traded emissions during five periods. The main purpose was to investigate (a) to what extent quota prices are influenced by an *expected* market-clearing price, assumed to be commonly known, and (b) to what extent quota prices converge to the *true* equilibrium price, which is affected by the traders' uncertain marginal abatement costs. Soeberg found that the expected price emerges as a focal point on which the bulk of quota prices are keyed in spite of the uncertainty of quota demand and supply. The efficiency attained in Soeberg's emissions trading experiment agrees with those in Bohm and Carlén (1999), Bohm (1999), and Hizen and Saijo (2001), which all suggest that efficiency would likely be high in international emissions trading even with only a limited number of (government) traders.

5.3. Market power in international emissions trading

The role of market power has been investigated in two studies with special reference to international GHG emissions trading. Both relate to a case where the US would be a dominant buyer. This case has been placed in focus, e.g., by countries that have criticized the use of an international tradable quotas for being a policy advocated by the US and one that would benefit that country.

In the tests by Hizen and Saijo, just referred to, the assumed marginal abatement cost curves used for the six traders were such that one country emerged as being able to exercise market power, the US. At prices around the competitive level, US demand was approximately equal to the demand by the other two net buyers, the EU and Japan. Although Russia was at least equally dominant on the seller side, the stepwise rising marginal abatement cost curve happened to be such that subjects in that role found themselves unable to exercise market power in the sense of a meaningful withholding of supply.

Although the authors caution the reader that their test may not be ideally suited for investigating the role of market power, some observations in that regard could be made. As it turned out, subjects representing the buyer with market power did not withhold demand in the bilateral trading experiment. In the double auction case, such subjects earned larger profits than they would have at the uniform price of a competitive equilibrium. Then, recalling the earlier discussion of market power in double-auction institutions and the fact that efficiency in the Hizen and Saijo test was close to 100 percent, we observe once again an example of double auctions where a dominant (although now not monopsonistic) buyer succeeds in trading early at prices below the competitive level while ending up trading at that level.

Carlén (forthcoming) tested the effects that large countries may have on the outcome of international carbon emissions trading when the trading mechanism is a double auction. The test environment mimicked a case in which the US, Japan, and ten EU countries engage in such trading while approaching the end of a Kyoto-like commitment period. This is the crucial period for finalizing the participating countries' net emissions trading. It is also the time at which uncertainty and information asymmetries are likely to be small. The assumed market structure at the end of the commitment period was such that, in a perfectly competitive market, the US would emerge as a clearly dominant buyer and purchase as much as 90 percent of the available supply. Thus, Carlén used a test case with an *a priori* much stronger potential for market power than in the Hizen and Saijo experiment.

A crucial assumption in Carlén's experiment was that, given the likely significant values at stake in real-world international trading of this kind, the participating countries would have strong incentives to gather information about the other countries' marginal abatement costs even before any trade starts. As already indicated, approximately reliable information is particularly likely to be available at the time when the commitment period draws to a close, and may be taken to be roughly common to all the participating countries at that stage. (In the experiment, this situation was implemented so that all traders obtained information about the *expected* marginal abatement cost functions for all countries, but were told that *actual* costs could deviate from these data, to an extent known only by the individual trader concerned.) In particular, this means that they would have more or less common expectations of a competitive price level for the market. Given the incentives that experienced professional traders are likely to have in a double-auction market,[17] final prices could be expected to approach the vicinity of that commonly known price level. Given that prediction, the hypothesis in Carlén was that, in this particular case of market power on the buyer side of the market, sellers would hesitate to accept low bids in early trading and would force buyers to accept prices close to the competitive level as the deadline moved closer.

Incentive payoffs to subjects as a percent of their trade gains achieved ranged from $0.3 to $250 for the three-hour test. The efficiency attained ranged between 78 and 99 percent in the first trading period. It increased to 96–100 percent in the second trading period, where *new* cost data and a different competitive price now were relevant. Four out of eight price paths were flat and close to the competitive price level; three of these four occurred in the second period. The other four price paths revealed a price convergence from below, thus benefiting early buyers. During the first period the buyer with market power obtained profit shares clearly above the share (s)he would have obtained had the market been competitive, 45–107 percent as compared to 31 percent. In the second period, this difference was smaller, 43–64 percent as compared to 42 percent.[18]

The results of these two tests seem to suggest that double auctions – an emissions exchange – would produce results close to efficiency in environments structured to reflect international emissions trading. However, dominant traders, here buyers, attained profits above the competitive level. Still, in the study with assumed real-world-like common expectations of final prices (i.e., the one by Carlén), there was a tendency for subjects in a second trading period to trade constantly at prices close to the competitive price level.

5.4. Attracting countries to participate in international emissions trading

An important difference between domestic tradable permit systems and international (GHG) tradable quota systems is that participation is mandatory only in the former

---

[17] It is often argued that government emissions trading would be run by incompetent bureaucrats and therefore bear no resemblance to professional profit-motivated trading activities. Although such government failures are possible, the reason remains unclear why a government in an open democratic society would not hire professional traders to join its trading team.

[18] To avoid giving market power a poor representation, the US trader subjects were selected from a group of experienced and previously highly successful Ph.D. student traders.

case. A country may find that its expected net costs of accepting a tradable-quota treaty proposal may be too high and therefore abstain from joining. In the context of treaties like the Kyoto Protocol, a distinction needs to be made between rich and poor countries; rich countries will have to accept that strictly positive costs will arise (although they may find them too high), while poor countries can be assumed to be unwilling to accept any (early) commitment to a treaty that would imply a positive cost to them.

In the first half of the 1990s, as already pointed out, there was a broad consensus among spokesmen for a large number of countries that treaties involving international emissions trading would not be acceptable. There were a number of indications that part of the reason for this was a poor understanding of the properties of emissions trading. In early 1996, a study was undertaken that tried to investigate whether governments would remain as negative towards international climate-change policy in the form of a tradable-quota treaty when more information about its properties was provided [Bohm (1999)]. If the study design was sufficiently convincing to attract the attention of real-world policy negotiators, the results could contribute to raising their awareness of the potential advantages of this policy option.

For such an inquiry to have a chance to be informative, the proposed treaty would have to be credible and possible to interpret as 'fair' in the sense stated above. Specifically, it was taken as given that rich countries would share the treaty costs in proportion to their GDP, while poor countries would be kept fully compensated. Moreover, subjects would have to be informed in some considerable detail about the relevant policy issues and policy options and about the proposed treaty.

Before proceeding, it should be acknowledged that this kind of test bore little resemblance to traditional work in experimental economics. One reason is that the use of a standard experimental design, although conceivable, would hardly be successful, in particular because of a shortage of suitable subjects. But the endeavor to identify an approachable competent subject pool and a relevant incentive instrument, to which we now turn, had crucial similarities with the challenges of ambitious economic experiments.

Here, as in Bohm (1999), the relevance of the results would turn on the qualifications of the subjects and the appropriateness of the incentives presented to them. The subjects chosen were high-level diplomats in the service of one country (Sweden), each of whom had recently been stationed in one of a set of selected countries, which they had now left, and which they would now be asked to represent. Crucial for this test was that it was sanctioned by the Swedish State Department, where a chief official (No. 3 in command) asked the 29 participating diplomats (24 of whom were ambassadors) to respond to the questions posed to them. The subjects' identities would not be revealed to anyone outside the Department. But since the responses were available for scrutiny by the chief official and others inside the Department, the responses could be peer reviewed. This was taken to provide an incentive for the subjects to consider the questions carefully before responding.

Each participant was asked whether 'his/her' government, as (s)he perceived the government's policies 'today' (1996), would accept or reject the treaty proposed at a realistic date 'tomorrow' (2005). As it turned out, 17 of the 29 respondents – 8 of the 12 'representatives' for rich countries and 9 of the 17 'representatives' for non-rich countries – said they believed that their countries' governments would accept the proposal. These indications, taken at face value, differed substantially from the impression given by the opposition to international emissions trading at the time. Given that the incentives for subjects to carefully consider the issues before responding and noting that experienced diplomats would hardly say "yes", if not convinced that "yes" was an appropriate answer, a possible interpretation of the results was that the tradable-quota solution should not be excluded as an international climate change policy. This interpretation may seem to have been confirmed some 18 months later when a version of such a treaty (here, without any developing countries committed to tradable emissions quotas) was signed by a large number of countries at Kyoto.

In another study, Bohm and Carlén (2002) analyzed cost-effective ways to attract poor countries to participate in international emissions trading, e.g., by enrolling them as Annex B countries in the Kyoto Protocol. Again, it was assumed that poor countries needed to be fully compensated. Moreover, it was assumed that the industrial countries, already included in Annex B, would need to reduce their emissions quotas to avoid benefiting from the quota-price reduction caused by the participation of poor countries. An experiment was conducted with graduate students in economics acting in the make-believe role of consultants to a developing country, earning a share of the overall gains from the country's participation in emissions trading. Subjects were asked to state their minimum compensations in terms of (a) an emissions quota and (b) a financial transfer in addition to a given small quota. The incentives of a Becker–DeGroot–Marschak mechanism were provided for demand revelation.

The results indicated that, if the two options were designed to be equally costly for the industrial countries, risk-averse new trader countries would prefer the financial transfer option. The implied increase in cost-effectiveness could be shared between the two groups of countries, of course. Even so, there is the political problem that actual use of financial transfers might be blocked by industrial countries' possible aversion to commitments involving financial transfers instead of only emissions quotas.

6. Concluding remarks

Experimental economics still has several methodological issues to come to grips with, such as the relevance of alternative subject pools and the incentive levels appropriate for the study purpose. In that respect, experimental analyses of environmental policy issues probably do not differ from those of other economic topics. However, in spite of such unsettled methodological problems, there is no doubt that experimental studies of the type reviewed here have contributed to a better understanding of the implications of various environmental policy instruments. An example that has dominated the review is the contribution of experiments to a significantly increased insight into emissions trading and the importance of the design selected for such trading – an example of the oft-quoted primary contribution of general experimental economics that "institutions matter".

We have seen that experiments indicate that permit futures and permit banking increase efficiency and reduce price volatility, respectively. Market power on emissions exchanges for permit or quota trading has in a large set of experimental studies turned out not to cause losses of efficiency but it has not, in general, avoided the generation of profits to market-power agents in excess of those under competition. Moreover, experiments or experiment-like studies have indicated that international GHG emissions trading promotes cost-effectiveness.

We have also seen that a number of theoretical propositions have been given support by experimental evidence. This was true, *inter alia,* with respect to the Coase theorem, the cost-effectiveness of market-based instruments, and the critique of the US EPA auction design.

As much as the reverse is true, in certain respects studies of issues concerning environmental policy would seem to be equally relevant for other topics in economics. This may be seen to have manifested itself by the explicit methodological choice in many experiments to avoid the exposure of subjects to specific contexts in favor of a more generally defined, 'context-free' issue. For example, several 'emissions trading' experiments have been presented as trading in an unspecified commodity. The laudable ambition behind this choice is to avoid having the subjects' decisions distorted by idiosyncrasies that a specific issue may evoke and to allow them to focus on the relevant principles involved.

The context of 'the environment' or environmental policy provides a pertinent illustration of this type of consideration. If subjects turn out to question the relevance of environmental concerns, their participation in an experiment with explicit reference to that context may not appropriately serve the objective of the test. Or if the subjects were recently exposed to the news of an environmental disaster, their responses may not be relevant for the purpose of the experiment in which they participate.

However, there is another side of this coin. What happens to the subjects' behavior when they are exposed to something 'context-free' that may seem nonsensical or for which they explicitly or implicitly try to imagine a real-world illustration (possibly false or even counterproductive), or behind which they feel there must lie an issue that is intentionally being concealed from them? We do not know to what extent such noise exists, and we know even less what effect it might have or whether its distortionary effects are more serious than those of the noise discussed in the preceding paragraph.[19]

A crucial aspect of all experimental activity is that results are preliminary unless sufficient replication has been achieved. Little of that has yet been accomplished in experimental economics. One reason may be, of course, that it becomes boring to engage in replications of earlier studies given all the alternative new topics that could be tested instead. But, to check the robustness of experimental results, it may matter less if the

---

[19] For a discussion of this issue, see Loewenstein (1999), Loomes (1999) and Bohm (2002).

same abstract issue is replicated or if a large number of tests that are similar in principle, but different in context, are carried out. The topic area of the experiments presented in this chapter may be suitable for a closer investigation of this methodological issue. To design and run a large number of experiments investigating policy choices for different specific real-world environmental policy contexts could lay bare general principles as well as provide information about specific current policy issues. This would have additional benefits in that it would provide tentative results to policy makers who have an interest only in the applications.

Acknowledgements

Helpful comments on earlier versions of this chapter by Björn Carlén, Gloria Helfand, Charlie Plott, Jay Shogren, and Jeffrey Vincent are gratefully acknowledged.

References

Bjornstad, D.J., S.R. Elliott and D.R. Hale (1999), "Understanding experimental economics and policy analysis in a federal agency: the case of marketable emissions trading", in: C.A. Holt and R.M. Isaac, eds., Research in Experimental Economics, Vol. 7, Emissions Permit Experiments (JAI Press, Greenwich, CT) 163–180.
Bohm, P. (1997), "Are tradable carbon emission quotas internationally acceptable? An inquiry with diplomats as country representatives", Nord 1997:8 (Nordic Council of Ministers, Copenhagen).
Bohm, P. (1998), "Determinants of the benefits of international carbon emissions trading: theory and experimental evidence", in: B. Fisher, ed., Emissions Trading, Proceedings of the Conference on Greenhouse Gas Emissions Trading (ABARE, Canberra).
Bohm, P. (1999), "An emission quota trade experiment among four Nordic countries", in: S. Sorrell and J. Skea, eds., Pollution for Sale: Emissions Trading and Joint Implementation (Edward Elgar, Cheltenham) 299–321, abbreviated version of "Joint implementation as emission quota trade: an experiment among four nordic countries", Nord 1997:4 (Nordic Council of Ministers, Copenhagen).
Bohm, P. (2002), "Pitfalls in experimental economics", in: F. Andersson and H. Holm, eds., Experimental Economics: Financial Markets, Auctions, and Decision Making (Kluwer, Dordrecht, 2002) Chapter 10.
Bohm, P., and B. Carlén (1999), "Emission quota trade among the few: laboratory evidence of joint implementation among committed countries", Resource and Energy Economics 21(1):43–66.
Bohm, P., and B. Carlén (2002), "A cost-effective approach to attracting low-income countries to international emissions trading: Theory and experiments", Environmental and Resource Economics 23:187–211.
Carlén, B. (forthcoming), "Market power in international carbon emissions trading: a labaratory test", Energy Journal.
Cason, T.N. (1995), "An experimental investigation of the seller incentives in EPA's emission trading auction", American Economic Review 85(4):905–922.
Cason, T.N., and C.R. Plott (1996), "EPA's new emissions trading mechanism: a laboratory evaluation", Journal of Environmental Economics and Management 30(2):133–160.
Cason, T.N., and L. Gangadharan (1998), "An electronic bulletin board trading for emission permits", Journal of Regulatory Economics 14(1):55–73.
Coase, R. (1960), "The problem of social cost", Journal of Law and Economics 3:1–31.
Crocker, T. (1966). "The structuring of atmospheric pollution control systems", in: H. Wolozin, ed., The Economics of Air Pollution (Norton, New York) 61–86.

Davis, D.D., and C.A. Holt (1993), *Experimental Economics* (Princeton University Press, Princeton, NJ).

Dales, J.H. (1968), Pollution, Property and Prices (University of Toronto, Toronto).

Franciosi, R., R.M. Isaac, D.E. Pingry and S.S. Reynolds (1993), "An experimental investigation of the Hahn–Noll revenue neutral auction for emissions licenses", Journal of Environmental Economics and Management 24(1):1–24.

Godby, R.W. (1999), "Market power in emissions permit double auctions", in: C.A. Holt and R.M. Isaac, eds., Research in Experimental Economics, Vol. 7, Emissions Permit Experiments (JAI Press, Greenwich, CT) 121–162.

Godby, R.W. (2000), "Market power and emission trading: theory and laboratory results", Pacific Economic Review 5(3):349–363.

Godby, R.W., S. Mestelman and R.A. Muller (1999), "Experimental tests of market power in emission trading markets", in: E. Petrakis, E. Sartzetakis and A. Xepapadeas, eds., Environmental Regulation and Market Structure (Edward Elgar, Cheltenham) 67–94.

Godby, R.W., S. Mestelman, R.A. Muller and J.D. Welland (1997), "Emission trading with shares and coupons when control over discharges is uncertain", Journal of Environmental Economics and Management 32:359–381.

Hahn, R.W. (1988), "Promoting efficiency and equity through institutional design", Policy Sciences 21:41–66.

Harrison, G.W. (1989), "Theory and misbehavior of first-price auctions", American Economic Review 79(4):749–762.

Harrison, G.W. (1992), "Theory and misbehavior of first-price auctions: reply", American Economic Review 82(5):1426–1443.

Harrison, G.W. (1994), "Expected utility theory and the experimentalists", Empirical Economics 19(2):43–74.

Harrison, G.W., and M. McKee (1985), "Experimental evaluation of the Coase theorem", Journal of Law and Economics 28(3):653–670.

Harrison, G.W., E. Hoffman, E.E. Rutström and M.L. Spitzer (1987), "Coasian solutions to the externality problem in experimental markets", Economic Journal 97:388–402.

Hey, J.D. (1991), Experiments in Economics (Blackwell, Oxford).

Hizen, Y., and T. Saijo (2001), "Designing GHG emissions trading institutions in the Kyoto Protocol: an experimental approach", Environmental Modelling and Software 16(6):533–543.

Hoffman, E., and M.L. Spitzer (1982), "The Coase theorem: some experimental tests", Journal of Law and Economics 25:73–98.

Hoffman, E., and M.L. Spitzer (1985), "Entitlements, rights and fairness: An experimental examination of subjects' concepts of distributive justice", Journal of Legal Studies 14:259–297.

Joskow, P.L., R. Schmalensee and E.M. Bailey (1998), "The market for sulphur dioxide emissions" 88(4):669–685.

International Panel on Climate Change (1995), Second Assessment Report (Working Group III).

Kagel, J., and A. Roth (eds.) (1995), Handbook of Experimental Economics (Princeton University Press, Princeton, NJ).

Kunreuther, H., P. Kleindofer and P.J. Knez (1987), "A compensation mechanism for siting noxious facilities", Journal of Environmental Economics and Management 14:371–383.

Ledyard, J.O., and K. Szakaly-Moore (1994), "Designing organizations for trading pollution rights", Journal of Economic Behavior and Organization 25:167–196.

Loewenstein, G. (1999), "Experimental economics from the Vantage-point of behavioural economics", Economic Journal 109(453):25–34.

Loomes, G. (1999), "Experimental economics: introduction", Economic Journal 109(453):1–4.

Mestelman, S., R. Moir and R.A. Muller (1999), "A laboratory test of Canadian proposals for an emission trading program" in: C.A. Holt and R.M. Isaac, eds., Research in Experimental Economics, Vol. 7 (JAI Press, Greenwich, CT) 45–92.

Mestelman, S., and R.A. Muller (1998), "The choice of instruments for pollution emission permit trading: designing a laboratory environment", in: E.T. Loehman and D.M. Kilgour, eds., Designing Institutions for

Environmental and Resource Management, New Horizons in Environmental Economics (Edward Elgar, Cheltenham, UK, and Northampton, MA).

Misolek, W.S., and H.W. Elder (1989), "Exclusionary manipulation of markets for pollution rights", Journal of Environmental Economics and Management 16:156–166.

Muller, R.A., and S. Mestelman (1994), "Emission trading with shares and coupons: a laboratory experiment", Energy Journal 15(2):185–211.

Muller, R.A., and S. Mestelman (1998), "What have we learned from emissions trading experiments", Managerial and Decision Economics 19(4–5):225–238.

Muller, R.A., S. Mestelman, J. Spraggon and R. Godby (2003), "Can double auctions control market power in emissions markets?", Journal of Environmental Economics and Management (forthcoming).

Plott, C.R. (1983), "Externalities and corrective policies in experimental markets", Economic Journal 93:106–127.

Plott, C.R. (1987), "Dimensions of parallelism: some policy applications of experimental methods", in: A.E. Roth, ed., Laboratory Experimentation in Economics (Cambridge University Press, Cambridge, UK).

Rhoads, T.A., and Shogren, J.F. (1999), "On Coasean bargaining with transaction costs", Applied Research, December.

Shogren, J.F. (1992), "An experiment on Coasean bargaining over ex ante lotteries and ex post rewards", Journal of Economic Behavior and Organization 17(1):153–169.

Shogren, J.F. (1993), "Experimental markets and environmental policy", Agricultural and Resource Economics Review 3(2):117–129.

Shogren, J.F. (1998), "Coasean bargaining with symmetric delay costs", Resource and Energy Economics 20(4):309–325.

Shogren, J.F., and C. Nowell (1991), "Economics and ecology: a comparison of experimental methodologies and philosophies", Ecological Economics.

Shogren, J., and S. Kask (1992), "Exploring the boundaries of the Coase theorem: efficiency and rationality given imperfect contract enforcement", Economics Letters 39:155–161.

Shogren, J.F., and T.M. Hurley (1999), "Experiments in environmental economics", in: J. van den Bergh, ed., Handbook of Environmental and Resource Economics (Edward Elgar, Cheltenham).

Smith, V.L. (1982), "Microeconomic systems as an experimental science", American Economic Review 72(5):923–955.

Smith, V.L., and A.W. Williams (1982), "The effects of rent asymmetries in experimental auction markets", Journal of Economic Behavior and Organization 3(1):99–116.

Smith, V.L., and A.W. Williams (1989), "The boundaries of competitive price theory: convergence, expectations, and transactions costs", in: L. Green and J. Kagel, eds., Advances in Behavioral Economics, Vol. 2 (Ablex Publishing, Norwood, NJ).

Soeberg, M. (2000), "Price expectations and international quota trading: an experimental evaluation", Environmental & Resource Economics 17:259–276.

Stavins, R.N. (1995), "Transaction costs and tradable permits", Journal of Environmental Economics and Management 29:133–148.

UN FCCC (1997), Kyoto Protocol to the United Nations Framework Convention on Climate Change (United Nations, New York).

Chapter 11

TECHNOLOGICAL CHANGE AND THE ENVIRONMENT

ADAM B. JAFFE

Department of Economics, Brandeis University and
National Bureau of Economic Research, Waltham, MA 02454-9110, USA

RICHARD G. NEWELL

Resources for the Future, Washington, DC 20036, USA

ROBERT N. STAVINS

John F. Kennedy School of Government, Harvard University, Cambridge, MA 02138, USA and
Resources for the Future, Washington, DC 20036, USA

Contents

Handbook of Environmental Economics, Volume 1, Edited by K.-G. Mäler and J.R. Vincent

Abstract

Environmental policy discussions increasingly focus on issues related to technological change. This is partly because the environmental consequences of social activity are frequently affected by the rate and direction of technological change, and partly because environmental policy interventions can themselves create constraints and incentives that have significant effects on the path of technological progress. This chapter summarizes current thinking on technological change in the broader economics literature, surveys the growing economic literature on the interaction between technology and the environment, and explores the normative implications of these analyses. We begin with a brief overview of the economics of technological change, and then examine theory and empirical evidence on invention, innovation, and diffusion and the related literature on the effects of environmental policy on the creation of new, environmentally friendly technology. We conclude with suggestions for further research on technological change and the environment.

Keywords

technological change, induced innovation, environmental policy, invention, diffusion

JEL classification: D24, D83, O14, O3, Q25, Q28, Q4

1. Introduction

In the last decade, discussions of environmental economics and policy have become increasingly permeated by issues related to technological change. An understanding of the process of technological change is important for two broad reasons. First, the environmental impact of social and economic activity is profoundly affected by the rate and direction of technological change. New technologies may create or facilitate increased pollution, or may mitigate or replace existing polluting activities. Further, because many environmental problems and policy responses thereto are evaluated over time horizons of decades or centuries, the cumulative impact of technological changes is likely to be large. Indeed, uncertainty about the future rate and direction of technological change is often an important sensitivity in "baseline" forecasts of the severity of environmental problems. In global climate change modeling, for example, different assumptions about autonomous improvements in energy efficiency are often the single largest source of difference among predictions of the cost of achieving given policy objectives [Weyant (1993), Energy Modeling Forum (1996)].

Second, environmental policy interventions themselves create new constraints and incentives that affect the process of technological change. These induced effects of environmental policy on technology may have substantial implications for the normative analysis of policy decisions. They may have quantitatively important consequences in the context of cost-benefit or cost-effectiveness analyses of such policies. They may also have broader implications for welfare analyses, because the process of technological change is characterized by externalities and market failures with important welfare consequences beyond those associated with environmental issues.

Our goals in this chapter are to summarize for environmental economists current thinking on technological change in the broader economics literature; to survey the growing literature on the interaction between technology and the environment; and to explore the normative implications of these analyses. This is a large task, inevitably requiring unfortunate but necessary omissions. In particular, we confine ourselves to the relationship between technology and problems of environmental pollution, leaving aside a large literature on technological change in agriculture and natural resources more broadly.[1] Because of the significant environmental implications of fossil fuel combustion, we include in our review some of the relevant literature on technological change and energy use.[2]

Section 2 provides a brief overview of the general literature on the economics of technological change. It is intended less as a true survey than as a checklist of issues

---

[1] See the recent surveys by Sunding and Zilberman (2000) and Ruttan (2000).

[2] Because our focus is *technological* change, we also exclude the growing literature on political and policy innovation and the evolution of social norms. See Chapters 8 ("The Political Economy of Environmental Policy") and 3 ("Property Rights, Public Goods, and the Environment").

that the interested reader can use to find entry points into the literature.[3] Section 3 discusses invention and innovation, including the idea of "induced innovation" whereby
environmental policy can stimulate the creation of new environmentally friendly technology. Section 4 focuses on issues related to technology diffusion. Section 5 provides
concluding observations and suggestions for future research.

2. Fundamental concepts in the economics of technological change

The literature pertaining to the economics of technological change is large and diverse.
Major sub-areas (with references to surveys related to those areas) include: the theory
of incentives for research and development [Tirole (1988), Reinganum (1989), Geroski
(1995)]; the measurement of innovative inputs and outputs [Griliches (1984, 1998)];
analysis and measurement of externalities resulting from the research process [Griliches
(1992), Jaffe (1998a)]; the measurement and analysis of productivity growth [Jorgen-
son (1990), Griliches (1998), Jorgenson and Stiroh (2000)]; diffusion of new technology [Karshenas and Stoneman (1995), Geroski (2000)]; the effect of market structure
on innovation [Scherer (1986), Sutton (1998)]; market failures related to innovation and
appropriate policy responses [Martin and Scott (2000)]; the economic effects of publicly
funded research [David, Hall and Toole (2000)]; the economic effects of the patent system [Jaffe (2000)]; and the role of technological change in endogenous macroeconomic
growth [Romer (1994), Grossman and Helpman (1994)]. In this section, we present a
selective overview designed to provide entry points into this large literature.

2.1. Schumpeter and the gale of creative destruction

The modern theory of the process of technological change can be traced to the ideas
of Josef Schumpeter (1942), who saw innovation as the hallmark of the modern capitalist system. Entrepreneurs, enticed by the vision of the temporary market power that
a successful new product or process could offer, continually introduce such products.
They may enjoy excess profits for some period of time, until they are displaced by subsequent successful innovators, in a continuing process that Schumpeter called "creative
destruction".

Schumpeter distinguished three steps or stages in the process by which a new, superior technology permeates the marketplace. *Invention* constitutes the first development
of a scientifically or technically new product or process.[4] Inventions may be patented,

---

[3] For surveys of other aspects of the economics of technological change, see Solow (1999) on neoclassical
growth theory, Grossman and Helpman (1995) on technology and trade, Evenson (1995) on technology and
development, and Reinganum (1989) on industrial organization theory of innovation and diffusion.

[4] The Schumpeterian "trichotomy" focuses on the commercial aspects of technological change. As discussed
in Section 3.1.2 below, the public sector also plays an important role. In addition, a non-trivial amount of
basic research – which one might think of as prior even to the invention stage – is carried out by private firms
[Rosenberg (1990)].

though many are not. Either way, most inventions never actually develop into an *innovation*, which is accomplished only when the new product or process is commercialized, that is, made available on the market.[5] A firm can innovate without ever inventing, if it identifies a previously existing technical idea that was never commercialized, and brings a product or process based on that idea to market. The invention and innovation stages are carried out primarily in private firms through a process that is broadly characterized as "research and development" (R&D). [6] Finally, a successful innovation gradually comes to be widely available for use in relevant applications through adoption by firms or individuals, a process labeled *diffusion*. The cumulative economic or environmental impact of new technology results from all three of these stages,[7] which we refer to collectively as the process of technological change.

2.2. *Production functions, productivity growth, and biased technological change*

The measurement of the rate and direction of technological change rests fundamentally on the concept of the transformation function,

$$T(Y, I, t) \leqslant 0, \tag{1}$$

where Y represents a vector of outputs, I represents a vector of inputs, and t is time. Equation (1) describes a production possibility frontier, that is, a set of combinations of inputs and outputs that are technically feasible at a point in time. Technological change is represented by movement of this frontier that makes it possible over time to use given input vectors to produce output vectors that were not previously feasible.

---

[5] More precisely, an invention may form the basis of a *technological* innovation. Economically important innovations need not be based on new technology, but can be new organizational or managerial forms, new marketing methods, and so forth. In this chapter, we use the word *innovation* as short-hand for the more precise *technological innovation*.

[6] Data regarding R&D expenditures of firms are available from the financial statements of publicly traded firms, if the expenditure is deemed "material" by the firm's auditors, or if the firm chooses for strategic reasons to report the expenditure [Bound et al. (1984)]. In the United States, the government carries out a "census" of R&D activity, and reports totals for broad industry groups [National Science Board (1998)]. Many industrialized countries now collect similar statistics, which are available through the Organization of Economic Cooperation and Development [OECD (2000)].

[7] Typically, for there to be environmental impacts of a new technology, a fourth step is required – utilization, but that is not part of the process of technological change *per se*. Thus, for example, a new type of hybrid motor vehicle engine might be *invented*, which emits fewer pollutants per mile; the same or another firm might commercialize this engine and place the *innovation* in new cars available for purchase on the market; individuals might purchase (or adopt) these cars, leading to *diffusion* of the new technology; and finally, by driving these cars instead of others (*utilization*), aggregate pollutant emissions might be reduced. Conversely, if higher efficiency and the resulting reduced marginal cost causes users to increase utilization, then the emissions reduction associated with higher efficiency may be partially or totally offset by higher utilization.

In most applications, separability and aggregation assumptions are made that make it possible to represent the economy's production technology with a production function,

$$Y = f(K, L, E; t),$$ (2)

where Y is now a scalar measure of aggregate output (for example, gross domestic product), and the list of inputs on the right-hand side of the production function can be made arbitrarily long. For illustrative purposes, we conceive of output as being made from a single composite of capital goods, K, a single composite of labor inputs, L, and a single composite of environmental inputs, E (for example, waste assimilation). Again, technological change means that the relationship between these inputs and possible output levels changes over time.

Logarithmic differentiation of Equation (2) with respect to time yields

$$y_t = A_t + \beta_{Lt} l_t + \beta_{Kt} k_t + \beta_{Et} e_t,$$ (3)

in which lower case letters represent the percentage growth rates of the corresponding upper case variable; the β's represent the corresponding logarithmic partial derivatives from Equation (2); and the t indicate that all quantities and parameters may change over time.[8] The term A_t corresponds to "neutral" technological change, in the sense that it represents the rate of growth of output if the growth rates of all inputs were zero. But the possibility that the β's can change over time allows for "biased" technological change, that is, changes over time in *relative* productivity of the various inputs.

Equations (2) and (3) are most easily interpreted in the case of process innovation, in which firms figure out more efficient ways to make existing products, allowing output to grow at a rate faster than inputs are growing. In principle, these equations also apply to product innovation. Y is a composite or aggregate output measure, in which the distinct outputs of the economy are each weighted by their relative value, as measured by their market price. Improved products will typically sell at a price premium, relative to lower quality products, meaning that their introduction will increase measured output even if the physical quantity of the new goods does not exceed the physical quantity of the old goods they replaced. In practice, however, product improvement will be included in measured productivity only to the extent that the price indices used to convert nominal GDP or other nominal output measures to real output measures are purged of the effects of product innovation. In general, official price indices and the corresponding real output measures achieve this objective only to a limited extent.

On its face, Equation (3) says nothing about the *source* of the productivity improvement associated with the neutral technological change term, A_t. If, however, all inputs and outputs are properly measured, and inputs (including R&D) yield only normal investment returns, then all endogenous contributions to output should be captured by

---

[8] This formulation can be considered a first-order approximation to an arbitrary functional form for Equation (2). Higher-order approximations can also be implemented.

returns to inputs, and there should be no "residual" difference between the weighted growth rates of inputs and the growth rate of output. The observation that the residual has been typically positive is therefore interpreted as evidence of some source of exogenous technological change.[9]

There is now a large literature on the measurement and explanation of the productivity residual. There are two basic approaches to the measurement of productivity. The "growth accounting" approach relies on neoclassical production theory under constant returns to scale for the proposition that the β's in Equation (3) are equal to the corresponding factor shares, and thereby calculates the A_t as an *arithmetic* residual after share-weighted input growth rates are subtracted from the growth rate of output [Denison (1979), U.S. Bureau of Labor Statistics (2000)]. The "econometric" approach estimates the parameters of Equation (3) from time series data and infers the magnitude of A_t as an *econometric* residual after the estimated effects of all measurable inputs on output have been allowed for [Jorgenson and Griliches (1967), Jorgenson and Stiroh (2000)]. In both of these approaches, much attention has focused on the difficulties of appropriately measuring both inputs and outputs [Jorgenson and Griliches (1967), Griliches (1994)]. This issue can be particularly problematic for the measurement of natural capital stocks, which can lead to bias in the productivity residual if they are ignored or mismeasured [see Dasgupta and Mäler (2000) and the chapter "National Income Accounts and the Environment" in this volume]. A particular focus has been understanding the slowdown in productivity growth in the 1970s and 1980s relative to the earlier postwar period, including the role played by rising energy prices in that slowdown [Berndt and Wood (1986), Jorgenson (1984)].

In many contexts, it is difficult to distinguish the effects of innovation and diffusion. We observe improvements in productivity (or other measures of performance) but do not have the underlying information necessary to separate such improvements into movements of the production frontier and movements of existing firms towards the frontier. A related issue, and one that is often significant for environment-related technological change, is that innovation can be undertaken either by the manufacturers or the users of industrial equipment. In the former case, the innovation must typically be *embodied* in new capital goods, and must then diffuse through the population of users via the purchase of these goods, in order to affect productivity or environmental performance. In the latter case, the innovation may take the form of changes in practices that are implemented with existing equipment. Alternatively, firms may develop new equipment for their own use, which they then may or may not undertake to sell to other firms. The fact that the locus of activity generating environment-related technological change can

---

[9] Fabricant (1954) was the first to observe that the growth of conventional inputs explained little of the observed growth in output in the twentieth century. This observation was elaborated by Abramowitz (1956), Kendrick (1956) and Solow (1957). The early writers were clear that the large "residual" of unexplained growth was "a measure of our ignorance" [Abramowitz (1956)] rather than a meaningful measure of the rate of technological progress. See Solow (1999) for a survey of neoclassical growth theory.

be supplying firms, using firms, or both, has important consequences for modeling the interaction of technological change and environmental policy.

The embodiment of new technology in new capital goods creates an ambiguity regarding the role played by technology *diffusion* with respect to Equations (2) and (3). One interpretation is that these equations represent "best practice," that is, what the economy would produce if all innovations made to date had fully diffused. In this interpretation, innovation would drive technological change captured in Equation (3); the issue of diffusion would then arise in the form of the presence of firms producing at points inside the production possibility frontier. Frontier estimation techniques [Aigner and Schmidt (1980)] or data envelopment methods [Fare, Grosskopf and Lovell (1994)] would be needed to measure the extent to which such sub-frontier behavior is occurring.[10] Alternatively, one can assume that the users of older equipment make optimal, informed decisions regarding when to scrap old machines and purchase newer ones that embody better technology. In this formulation, observed movements of the frontier – measured technological change – comprise the combined impacts of the invention, innovation and diffusion processes.

2.3. Technological change and endogenous economic growth

In the last two decades there has emerged a large macroeconomic literature that builds on the above concepts to produce models of overall economic growth based on technological change [Romer (1990, 1994), Grossman and Helpman (1994), Solow (2000)]. In these models, R&D is an *endogenous* equilibrium response to Schumpeterian profit incentives. Spillovers associated with this R&D generate a form of dynamic increasing returns, which allows an economy endogenously investing in R&D to grow indefinitely.[11] This stands in contrast to the older neoclassical growth model, in which *exogenous* technological change, in the presence of decreasing returns to investment in physical capital, typically yields an economy that tends towards a steady state in which income per capita does not grow.[12]

Endogenous growth theory has played an important role in re-introducing technological change – and the associated policy issues deriving from R&D market failures

---

[10] Boyd and McClelland (1999) and Boyd and Pang (2000) employ data envelopment analysis to evaluate the potential for improvements at paper and glass plants that increase productivity and reduce pollution.

[11] It is also possible to generate such endogenous growth through human capital investment [Lucas (1988)].

[12] Thus, in the literature, "endogenous technological change" and "induced technological change" refer to different concepts, even though the opposite of each is often described by the same phrase, that is, exogenous technological change. Endogenous technological change refers to the broad concept that technological change is the result of activities within the economic system, which are presumed to respond to the economic incentives of the system. Induced technological change refers to the more specific idea that changes in relative factor prices affect the rate and direction of innovation. In practice, papers that use the phrase "endogenous technological change" tend to focus on aggregate R&D expenditure and neutral technological change. Papers that used the phrase "induced technological change" or "induced innovation" tend to focus on the direction of R&D efforts and biases in technological change.

into discussions about economic growth [13] Modeling growth as a process driven by the endogenous creation and diffusion of new technology ought to have implications for important environmental issues such as sustainable development and global climate change. Its policy utility has been limited, however, by its relative lack of empirical foundation, and by the difficulty of linking the macroeconomic endogenous growth models to the microeconomic foundations of technological innovation and diffusion [Caballero and Jaffe (1993), Aghion and Howitt (1998)]. This remains an important area for future research.

3. Invention and innovation

As discussed in the introduction, if the imposition of environmental requirements can stimulate invention and innovation that reduces the (static) cost of complying with those requirements, this has profound implications for both the setting of environmental policy goals and the choice of policy instruments. Nonetheless, there has been some tendency to treat technology as a "black box" [Rosenberg (1982)]. For example, the production function/productivity growth paradigm described in Section 2 says little about what generates technological change. But following Schumpeter, there has been a line of theoretical and empirical analysis that has cast invention and innovation as a purposive economic activity, and has attempted to discern its determinants and effects. Milestones in this line of research are: Schmookler (1966), Mansfield (1968), Rosenberg (1982), Griliches (1984), Nelson and Winter (1982) and Scherer (1986).[14]

It is useful to identify two major strands of thought regarding the determinants of innovative activity. We call these two broad categories of modeling approaches the "induced innovation" approach and the "evolutionary" approach.[15] We now describe the induced innovation approach, while the evolutionary approach is discussed in Section 3.4.

3.1. The induced innovation approach

3.1.1. Neoclassical induced innovation

The recognition that R&D is a profit-motivated investment activity leads to the hypothesis that the rate and direction of innovation are likely to respond to changes in relative prices. Since environmental policy implicitly or explicitly makes environmental inputs

---

[13] See, for example, Jones and Williams (1998), and the symposium on "New Growth Theory" in the Winter 1994 issue of the *Journal of Economic Perspectives*.

[14] See also the survey by Thirtle and Ruttan (1987).

[15] In this section and Section 4, we focus separately on induced innovation and the economic forces driving diffusion. As noted above, however, the analytical distinction between innovation and diffusion is blurred in practice.

more expensive, the "induced innovation" hypothesis suggests an important pathway for the interaction of environmental policy and technology, and for the introduction of impacts on technological change as a criterion for evaluation of different policy instruments.

The induced-innovation hypothesis was first articulated by Sir John Hicks:

> "a change in the relative prices of the factors of production is itself a spur to invention, and to invention of a particular kind – directed to economizing the use of a factor which has become relatively expensive" [Hicks (1932, p. 124)].[16]

Analysis of this hypothesis has a long and somewhat tortured history in economics. Early empirical work was largely confined to aggregate data, and focused primarily on questions such as whether historical cross-country differences in wage levels could explain the location of development of labor-saving inventions [Thirtle and Ruttan (1987)].

Hicks did not link the induced-innovation hypothesis in a formal way to the research process, or to profit-maximizing R&D decisions by firms. This link was formalized in the 1960s by Ahmad (1966) and Kamien and Schwartz (1968), and developed further by Binswanger (1974). Binswanger and Ruttan (1978) summarize this literature. The general approach is to postulate a "meta" production function according to which investing in R&D changes the parameters of a production function such as Equation (2). Unfortunately, theoretical conclusions regarding the induced affect of changes in factor prices on the parameters of the production function are sensitive to the specification of the "meta" production function governing the research process.

Although formulated in terms of the R&D decisions of firms, this theory is nonetheless aggregate, because the result of the research process is change in the parameters of the aggregate production function. That is, "labor-saving" innovation in these models means a change in the parameters of Equation (2) that results in less labor being used. The model abstracts entirely from what kinds of new machines or processes might be yielding these changes. Further, because of the ambiguity described in Section 2.2 as to whether the production frontier does or does not encompass technology diffusion, there is really no distinction in these models between induced *innovation* and the effect of factor prices on the rate of technology *diffusion*.

A natural way to move the modeling of induced innovation to the microeconomic level is to recognize that factor-saving technological change comes about largely through the introduction of new capital goods that embody different input ratios. These input ratios can then be thought of as attributes or characteristics of the capital goods in the sense of Lancaster (1971). Thirtle and Ruttan (1987) provided a review of the non-environmental literature on induced innovation. Much of this work is in the agricultural

---

[16] Writing before Schumpeter, Hicks does not appear to use the word "invention" in the specific sense used by Schumpeter and adopted by later authors. Rather, Hicks uses it in a general sense encompassing both invention and innovation, as used today.

area in which excellent microdata has long provided fertile ground for empirical work on innovation and diffusion.[17] In general, available empirical analyses confirm that factor price changes are associated with factor-saving technological change.

3.1.2. Market failures and policy responses

Within the induced innovation approach, firms undertake an investment activity called "R&D" with the intention of producing profitable new products and processes. Decisions regarding the magnitude and nature of R&D activities are governed by firms' efforts to maximize their value, or, equivalently, to maximize the expected discounted present value of cash flows. In some applications, the output of R&D is explicitly modeled as "knowledge capital", an intangible asset that firms use together with other assets and other inputs to generate revenues.[18]

When viewed as an investment activity, R&D has important characteristics that distinguish it from investment in equipment or other tangible assets. First, although the outcome of any investment is uncertain to some extent, R&D investment appears to be qualitatively different. Not only is the variance of the distribution of expected returns much larger than for other investments, but much or even most of the value may be associated with very low-probability but very high value outcomes [Scherer, Harhoff and Kukies (2000)]. This skewness in the distribution of the outcomes of the research process has important implications for modeling firms' R&D decision making [Scherer and Harhoff (2000)]. In addition, the asset produced by the R&D investment process is specialized, sunk and intangible, so that it cannot be mortgaged or used as collateral. The combination of great uncertainty and intangible outcomes makes financing of research through capital market mechanisms much more difficult than for traditional investment. The difficulty of securing financing for research from outside sources may lead to under-investment in research, particularly for small firms that have less internally generated cash and/or less access to financial markets.

In addition to these financing difficulties, research investment differs from physical investment because the asset produced by the research process – new knowledge about how to make and do things – is difficult to exclude others from using. As first noted in the classic paper by Arrow (1962a), this means that the creator of this asset will typically fail to appropriate all or perhaps most of the social returns it generates. Much of this social return will accrue as "spillovers" to competing firms, to downstream firms that purchase the innovator's products, or to consumers [Griliches (1979, 1992), Jaffe (1986, 1998a)]. This "appropriability problem" is likely to lead to significant underinvestment by private firms in R&D, relative to the social optimum [Spence (1984)].[19]

---

[17] More recently, the availability of computerized firm-level data on R&D and patents has led to an increase in parallel analyses in the industrial sector.

[18] See Griliches (1979) for the seminal statement of this research approach. An example of a recent application measuring the knowledge capital of firms is Hall, Jaffe and Trajtenberg (2000).

[19] The recognition that the costs and benefits of R&D for the firm are affected by the appropriability problem and financing issues has led to a large literature on the effects of market structure on innovation. In the older

An important special case of the appropriability problem is created by "general purpose technologies" [Bresnahan and Trajtenberg (1995)]. GPTs are technologies that find use in many distinct application sectors within the economy, such as the electric motor, the steam engine, the internal combustion engine and now, the semiconductor and possibly the Internet. The development of such technologies increases the returns to R&D designed to incorporate them into the different applications sectors; development of such applications in turn increases the return to improving the GPT. Because of these dynamic feedback effects, GPTs may be an important factor in economic growth [Helpman (1998)]. The dynamic feedback between a GPT and its applications sectors also creates an important example of "path dependence", discussed in Section 4 below. With respect to the environment, whether the GPTs that drive a particular era are pollution-intensive or pollution-saving may have profound implications for the long-term environmental prognosis.

As a profit-motivated activity, R&D investment decisions are governed by the cost of R&D and its expected return. Theory and evidence suggest that the most important factors affecting the optimal level of R&D are the after-tax cost of R&D [Hall and Van Reenen (2000)], the size of the market [Schmookler (1966)], technological opportunity [Rosenberg (1982)], and appropriability conditions [Jaffe (1988)]. Each of these varies intrinsically across time, markets, and technologies, and also is affected by government policy. In particular, patents and other forms of intellectual property are used by firms to overcome the appropriability problem, although the effect of these institutions on investment in R&D or inventive activity has not been clearly demonstrated empirically [Jaffe (2000), Cohen, Nelson and Walsh (2000)].

As noted above, both the appropriability problem and the possibility of capital market failures in the financing of R&D lead to a presumption that laissez-faire levels of investment in innovation will be too low from a social perspective. There is, however, an offsetting *negative* externality that suggests that private R&D incentives may be *too*

---

literature, it was argued that both these problems would be overcome more easily by large firms and/or firms operating in concentrated industries characterized by market power. From these observations, it was hypothesized that innovation comes disproportionately from large firms and concentrated industries. This conjecture is known as the "Schumpeterian Hypothesis". After much debate about what the Schumpeterian Hypothesis really means, the volume of evidence seems to show that: (1) much innovation comes from large firms in moderately concentrated industries, if only because much economic activity comes from such firms; (2) truly competitive industries (for example, construction) perform little R&D; (3) beyond minimal size and concentration, there is little evidence of any monotonic relationship between innovation intensity and either size or concentration; and (4) innovation and market structure interact dynamically in a way that is not captured by an alleged causal influence of firm size and market concentration on innovation. For an extensive survey of this literature, see Cohen and Levin (1989). More recently, a large game-theoretic literature related to strategic R&D incentives has emerged [surveyed by Reinganum (1989)]. This literature has two strands. One views R&D or other innovative activities in a context of *continuous competition* in which, for example, marginal R&D investments result in marginal cost reductions or product improvements [for example, Dasgupta and Stiglitz (1980a), Levin and Reiss (1988), Spence (1984)]. The other R&D theory literature focuses on *patent races*, where firms compete to be the first to achieve a specific innovation goal [for example, Dasgupta and Stiglitz (1980b), Reinganum (1982), Fudenberg et al. (1983)].

great. R&D is a fixed cost that must, in equilibrium, be financed by the stream of quasi-rents it produces. The entry of another R&D competitor, or an increase in the R&D investment level of a competitor, reduces the expected quasi-rents earned by other R&D firms. This "rent-stealing" effect [Mankiw and Whinston (1986)] could, as a theoretical matter, lead to over-investment in R&D. This is analogous to the over-fishing of an open-access fishery by a competitive fishing industry.[20]

The empirical evidence suggests, however, that positive externalities associated with knowledge spillovers dominate the rent-stealing effect, leading to social rates of return to R&D substantially in excess of the private rates of return [Griliches (1992)]. In practice, virtually all industrialized countries engage in policies designed to encourage investment in innovation [Guellec and van Pottelsberghe (2000), Mowery and Rosenberg (1989)]. It is difficult to determine how well these policies do in moving R&D toward optimal levels. There is some evidence that social rates of return remain well above private levels [Griliches (1992), Jones and Williams (1998)], but there is also evidence that R&D subsidies drive up the wages of scientists enough to prevent significant increases in real R&D [Goolsbee (1998)]. This implies that the supply of scientists and engineers is relatively inelastic; whether such inelasticity could hold in the long run remains unresolved.

Policy can try to increase social investment in R&D by engaging in R&D in the public (and/or nonprofit) sector, or by trying to reduce the after-tax cost of R&D for private firms. R&D in the public sector and in universities is an important, though declining component of the overall research effort in the U.S. and other developed nations.[21] The evidence on the effectiveness of public research is mixed, partly because of the difficulty of measuring the output of the basic research process [Jaffe (1998b)], and partially because of the difficulty of determining the extent of complementarity or substitutability between public research investment and private investment [David, Hall and Toole (2000)]. Examples of successful government technology development (as opposed to research) have been particularly few [Cohen and Noll (1991)]. Nonetheless, public R&D may well play a particularly important role with respect to environment-related science and technology, since the external social benefits of environmentally benign technology are unlikely to be fully captured by private innovators.

---

[20] There is also a dynamic analogue to the tension between spillovers and rent-stealing. Over time, innovation may become cumulatively easier because subsequent inventors "stand on the shoulders" of those who came before; or it may become harder, because the pool of potential inventions is "fished out". In the 1980s, there was considerable interest in the idea that "fishing out" of invention potential may explain the productivity slowdown of the 1970s [Evenson (1991)]. But the surge in patenting and productivity growth rates in the 1990s has led to a fading of the fishing-out idea [Jaffe (2000)].

[21] Research performed in government labs, universities and other non-profit institutions is currently about one-fourth of all research performed in the U.S., versus three-quarters performed in the for-profit sector. In addition, some of the research *performed* by firms is *funded* by public money; altogether, over one-third of all R&D is funded by public sources [National Science Board (1998)]. This estimate excludes the implicit public subsidy for private research represented by the Research and Experimentation Tax Credit.

Government policy affects the after-tax cost of R&D via tax incentives [Hall and Van Reenen (2000)],[22] direct subsidies and grants for research [Klette, Møen and Griliches (2000), Trajtenberg (2000)], and also via educational policies that affect the supply of scientists and engineers [Romer (2000)]. Public policies can affect the market for new technologies via direct government purchase, subsidies for purchase or installation of products incorporating particular technologies, [Stoneman (1987)], and also disincentives against the adoption of competing technologies (pollution fees, for example). Finally, policies can affect the extent to which firms can successfully appropriate the returns to their research, by establishing the institutional environment of patent systems, employment relations, and antitrust or other competition policies.[23]

3.1.3. Empirical evidence on induced innovation in pollution abatement and energy conservation

The greatest challenge in testing the induced innovation hypothesis specifically with respect to environmental inducement is the difficulty of measuring the extent or intensity of inducement across firms or industries [Jaffe et al. (1995)]. Ideally, one would like to look at the relationship between innovation and the shadow price of pollution or environmental inputs. In practice, such shadow prices are not easily observed. Consequently, one must use proxies for this shadow price, such as characteristics of environmental regulations, expenditures on pollution abatement, or prices of polluting inputs (for example, energy). In the following paragraphs, we review, in turn, studies that have used each of these approaches.

There is a large literature on the impact of environmental regulation on productivity and investment.[24] To the extent that regulation inhibits investment and/or slows productivity growth, this can be viewed as indirect evidence suggesting that induced innovation effects are either small or are outweighed by other costs of regulation. Results of this type seem to be industry and methodology dependent. For measuring the characteristics of environmental regulations, studies have used expert judgements about relative regulatory stringency in different states [Gray and Shadbegian (1998)], number of enforcement actions [Gray and Shadbegian (1995)], attainment status with respect to environmental laws and regulations [Greenstone (2002)], and specific regulatory events

---

[22] The effect of taxation on R&D incentives is theoretically complex. On the one hand, any tax on profits derived from R&D drives a wedge between the before- and after-tax returns and hence discourages R&D investment. On the other hand, returns from R&D are taxed much more lightly than returns from investment in equipment and structures, both because of explicit R&D incentives, and also because R&D can be expensed rather than amortized. Thus *relative* to traditional investment, R&D is strongly tax-preferred.

[23] The primary explicit non-fiscal mechanism for encouraging innovation in industrialized countries is the patent system. Empirical evidence on the impact of patent protection on the rate of innovation is ambiguous. For a survey, see Jaffe (2000).

[24] See, for example, Gollop and Roberts (1983), Kolstad and Turnovsky (1998) and Yaisawarng and Klein (1994).

[Berman and Bui (1998)].[25] For example, Berman and Dui (1998) found significant productivity increases associated with air pollution regulation in the oil refining industry, but Gray and Shadbegian (1998) found that pollution abatement investment "crowds out" productive investment almost entirely in the pulp and paper industry. Greenstone (2002) found overall that air pollution regulation has a statistically significant but very small impact on overall costs, implying a small negative productivity impact.

Lanjouw and Mody (1996) showed a strong association between pollution abatement expenditures and the rate of patenting in related technology fields. Jaffe and Palmer (1997) examined the correlation between pollution expenditures by industry and indicators of innovation more broadly. They found that there is a significant correlation within industries over time between the rate of expenditure on pollution abatement and the level of R&D spending. They did not, however, find evidence of an effect of pollution control expenditure on overall patenting.

Evidence of inducement has also been sought by examining the response to changing energy prices. Newell (1997, Chapter 2) and Newell, Jaffe and Stavins (1999) examined the extent to which the energy efficiency of the menu of home appliances available for sale changed in response to energy prices between 1958 and 1993, using a model of induced innovation as changing characteristics of capital goods. Hicks formulated the induced innovation hypothesis in terms of factor prices. Newell, Jaffe and Stavins (1999) generalized this concept to include inducement by regulatory standards, such as labeling requirements that might increase the value of certain product characteristics by making consumers more aware of them. More generally, non-price regulatory constraints can fit within the inducement framework if they can be modeled as changing the shadow or implicit price that firms face in emitting pollutants. In their framework, the existing technology for making a given type of equipment at a point in time is identified in terms of vectors of characteristics (including cost of manufacture) that are feasible. The process of invention makes it possible to manufacture "models" (characteristics vectors) that were previously infeasible. Innovation means the offering for commercial sale of a model that was not previously offered for sale. Induced innovation is then represented as movements in the frontier of feasible models that reduce the cost of energy efficiency in terms of other attributes.

By constructing a series of dynamic simulations, they examined the effects of energy price changes and efficiency standards on average efficiency of the menu of products over time. They found that a substantial amount of the improvement was what may be described as autonomous (that is, associated with the passage of time), but significant amounts of innovation were also due to changes in energy prices and changes in energy-efficiency standards. They found that technological change in air conditioners was actually biased against energy efficiency in the 1960s (when real energy prices were falling), but that this bias was reversed after the two energy shocks of the 1970s.

---

[25] Of course, there is a parallel problem with respect to measurement of the rate of invention or innovation. See Griliches (1990) and Lanjouw and Schankerman (1999).

In terms of the efficiency of the average model offered, they found that energy efficiency in 1993 would have been about one-quarter to one-half lower in air conditioners and gas water heaters, if energy prices had stayed at their 1973 levels, rather than following their historical path. Most of the response to energy price changes came within less than five years of those changes.

Popp (2001, 2002) looked more broadly at energy prices and energy-related innovation. In the first paper, he found that patenting in energy-related fields increases in response to increased energy prices, with most of the effect occurring within a few years, and then fading over time. Popp attributed this fading to diminishing returns to R&D. In the second paper, he attempted to decompose the overall reduction in energy use that is associated with changing energy prices between the substitution effect – movements along a given production frontier – and the induced innovation effect – movement *of* the production frontier itself induced by the change in energy prices. Using energy-related patents as a proxy for energy innovation, he found that approximately one-third of the overall response of energy use to prices is associated with induced innovation, with the remaining two-thirds associated with factor substitution. Because energy patents are likely to measure energy innovation only with substantial error, one might interpret this result as placing a lower bound on the fraction of the overall response of energy use to changing prices that is associated with innovation.

3.2. *Effects of instrument choice on invention and innovation*

The effect of environmental policies on the development and spread of new technologies may, in the long run, be among the most important determinants of success or failure in environmental protection [Kneese and Schultze (1975)].[26] It has long been recognized that alternative types of environmental policy instruments can have significantly different effects on the rate and direction of technological change [Orr (1976)]. Environmental policies, particularly those with large economic impacts (for example, those intended to address global climate change) can be designed to foster rather than inhibit technological invention, innovation, and diffusion [Kemp and Soete (1990)].

3.2.1. *Categories of environmental policy instruments and criteria for comparison*

For purposes of examining the link between environmental policy instruments and technological change, policies can be characterized as either command-and-control or market-based approaches. Market-based instruments are mechanisms that encourage

---

[26] Whereas we focus in this section of the chapter on the effects of environmental policy instruments on technological change, it is also the case that exogenous technological change can differentially affect the performance of alternative environmental policy instruments. For example, technological change in monitoring and enforcement, such as improvements in remote-sensing of motor vehicle emissions, could render particular policy instruments that focus on emissions, rather than abatement equipment, more attractive.

behavior through market signals rather than through explicit directives regarding pollution control levels or methods. These policy instruments – such as pollution charges, subsidies, tradeable permits, and some types of information programs – have been described as "harnessing market forces". This is because if they are well designed and implemented, they encourage firms (and/or individuals) to undertake pollution control efforts that are in their own interests and that collectively meet policy goals.[27]

Conventional approaches to regulating the environment are often referred to as "command-and-control" regulations, since they allow relatively little flexibility in the means of achieving goals. Such regulations tend to force firms to take on similar shares of the pollution-control burden, regardless of the cost. Command-and-control regulations do this by setting uniform standards for firms, the most prevalent of which are performance- and technology-based standards. A performance standard sets a uniform control target for firms (emissions per unit of output, for example), while allowing some latitude in how this target is met. Technology-based standards specify the method, and sometimes the actual equipment, that firms must use to comply with a particular regulation. While even technology-based standards provide an incentive for innovation that reduces the cost of using specific technologies, performance standards allow a wider range of innovation, as long as standards are met at the plant level. In contrast, market-based instruments allow even greater flexibility in innovation possibilities, including flexibility in plant-level emissions.

Holding all firms to the same target can be expensive and, in some circumstances, counterproductive. While standards may effectively limit emissions of pollutants, they typically exact relatively high costs in the process, by forcing some firms to resort to unduly expensive means of controlling pollution. Because the costs of controlling emissions may vary greatly among firms, and even among sources within the same firm,[28] the appropriate technology in one situation may not be appropriate (cost-effective) in another.

All of these forms of intervention have the potential for inducing or forcing some amount of technological change, because by their very nature they induce or require firms to do things they would not otherwise do. Performance and technology standards can be explicitly designed to be "technology forcing", mandating performance levels that are not currently viewed as technologically feasible or mandating technologies that are not fully developed. One problem with these approaches, however, is that while regulators can typically assume that *some* amount of improvement over existing technology will always be feasible, it is impossible to know how much. Standards must either be made unambitious, or else run the risk of being ultimately unachievable, leading to great political and economic disruption [Freeman and Haveman (1972)].

Technology standards are particularly problematic, since they tend to freeze the development of technologies that might otherwise result in greater levels of control. Under

---

[27] See Chapter 9 ("Experience with Market-Based Environmental Policy Instruments").

[28] Control costs can vary enormously due to a firm's production design, physical configuration, inputs, age of assets, and other factors.

regulations that are targeted at technologies, as opposed to emissions levels, no financial incentive exists for businesses to exceed control targets, and the adoption of new technologies is discouraged. Under a "Best Available Control Technology" (BACT) standard, a business that adopts a new method of pollution abatement may be "rewarded" by being held to a higher standard of performance and thereby not benefit financially from its investment, except to the extent that its competitors have even more difficulty reaching the new standard [Hahn and Stavins (1991)]. On the other hand, if third parties can invent and patent better equipment, they can – in theory – have a ready market. Under such conditions, a BACT type of standard can provide a positive incentive for technology innovation. Unfortunately, as we note below, there has been very little theoretical or empirical analysis of such technology-forcing regulations.

In contrast with such command-and-control regulations, market-based instruments can provide powerful incentives for companies to adopt cheaper and better pollution-control technologies. This is because with market-based instruments, it pays firms to clean up a bit more if a sufficiently low-cost method (technology or process) of doing so can be identified and adopted.

In theory, the relative importance of the dynamic effects of alternative policy instruments on technological change (and hence long-term compliance costs) is greater in the case of those environmental problems which are of great magnitude (in terms of anticipated abatement costs) and/or very long time horizon.[29] Hence, the increased attention that is being given by scholars and by policy makers to the problem of global climate change[30] has greatly increased the prominence of the issues that are considered in this part of the chapter.

There are two principal ways in which environmental policy instruments can be compared with regard to their effects on technological change. First and foremost, scholars have asked – both with theoretical models and with empirical analyses – the most direct question: what effects do particular instruments have on the rate and direction of relevant technological change? In keeping with the Schumpeterian trichotomy identified above, such investigations can be carried out with reference to the pace of invention, innovation, or diffusion of new technologies.

It is also possible to ask whether environmental policies encourage *efficient* rates (and directions) of technological change, or more broadly, whether such policies result in overall economic efficiency (that is, whether the efficient degree of environmental protection is achieved). This second principal mode for comparison is linked more directly with criteria associated with welfare economics, but such comparisons have been

---

[29] Parry, Pizer and Fischer (forthcoming) showed that the importance of the welfare gains from cost-reducing technological change relative to the welfare gains from optimal pollution control using existing technology tends to be higher when marginal benefits are flatter, marginal costs are steeper (and optimal abatement is lower), the discount rate is lower, the rate of technological change is faster, and research costs are lower.

[30] For particular attention to the links between technological change and global climate policy, see: Jaffe, Newell and Stavins (1999).

made much less frequently than have direct assessments of technology effects. Within the limits of the existing literature, we consider both sets of criteria.[31]

Most of the work in the economics literature on the dynamic effects of environmental policy instruments on technological change has been theoretical, rather than empirical, and so we consider the theoretical literature first.

3.2.2. Theoretical analyses

Although, as we suggested above, decisions about technology commercialization are partly a demand-side function of anticipated sales (adoption), the relevant literature comparing the effects of alternative environmental policy instruments has given greater attention to the supply side, focusing on incentives for firm-level decisions to incur R&D costs in the face of uncertain outcomes.[32] Such R&D can be either inventive or innovative, but the theoretical literature in this area makes no particular distinction.

The earliest work that is directly relevant was by Magat (1978), who compared effluent taxes and CAC standards using an innovation possibilities frontier (IPF) model of induced innovation, where research can be used to augment capital or labor in a standard production function. He compared the output rate, effluent rate, output-effluent ratio, and bias (in terms of labor or capital augmenting technical change), but produced ambiguous results. Subsequently, Magat (1979) compared taxes, subsidies, permits, effluent standards, and technology standards, and showed that all but technology standards would induce innovation biased toward emissions reduction.[33] In Magat's model, if taxes and permits are set so that they lead to the same reduction in emissions as an effluent standard at all points in time, then the three instruments provide the same incentives to innovate.

A considerable amount of theoretical work followed in the 1980s. Although much of that work characterized its topic as the effects of alternative policy instruments on technology innovation, the focus was in fact on effects of policy on technology diffusion. Hence, we defer consideration of those studies to Section 4.3.1 of this chapter.

Taking a somewhat broader view than most economic studies, Carraro and Siniscalco (1994) suggested that environmental policy instruments should be viewed jointly with traditional industrial policy instruments in determining the optimal way to attain a given degree of pollution abatement. They showed that innovation subsidies can be used to attain the same environmental target, but without the output reductions that result from pollution taxes. Laffont and Tirole (1996a) examined how a tradeable permit system could – in theory – be modified to achieve desired incentive effects for technological change. They demonstrated that although spot markets for permits cannot induce the

---

[31] Enforceability of environmental regulations is another criteria for policy choice that it is rarely emphasized in the technology literature. See Macauley and Brennan (2001) for an evaluation of the potential role of remote sensing technology in the enforcement of environmental regulations.

[32] See Kemp (1997) for an overview of theoretical models of technology innovation.

[33] Technology standards provided no incentives for innovation whatsoever.

socially optimal degree of innovation, futures markets can improve the situation [Laffont and Tirole (1996a)].[34]

Cadot and Sinclair-Desgagne (1996) posed the following question: if a potentially regulated industry has private information on the costs of technological advances in pollution control (frequently a reasonable assumption), then since the industry has an incentive to claim that such technologies are prohibitively expensive (even if that is not the case), can the government somehow design an incentive scheme that will avoid the problems of this information asymmetry? The authors developed a solution to this game-theoretic problem. Not surprisingly, the scheme involves government issued threats of regulation (which diminish over time as the firm completes stages of technology development).

It was only recently that theoretical work followed up on Magat's attempt in the late 1970's to rank policy instruments according to their innovation-stimulating effects. Fischer, Parry and Pizer (forthcoming) found that an unambiguous ranking of market-based policy instruments was not possible. Rather, the ranking of policy instruments was shown by the authors to depend on the innovator's ability to appropriate spillover benefits of new technologies to other firms, the costs of innovation, environmental benefit functions, and the number of firms producing emissions.

The basic model consists of three stages. First, an innovating firm decides how much to invest in R&D by setting its marginal cost of innovation equal to the expected marginal benefits. Second, polluting firms decide whether or not to adopt the new technology, use an (inferior) imitation of it, or do nothing. Finally, firms minimize pollution control expenditures by setting their marginal costs equal to the price of pollution. Policy instruments affect the innovation incentives primarily through three effects: (1) an abatement cost affect, reflecting the extent to which innovation reduces the costs of pollution control; (2) an imitation effect, which weakens innovation incentives due to imperfect appropriability; and (3) an emissions payment effect, which can weaken incentives if innovation reduces firms' payments for residual emissions. There is some variation in this pattern depending on the instrument, as shown in Table 1, which summarizes the direction of the three effects under three alternative policy instruments. The ranking of instruments depends on the relative strength of these effects.

Table 1

Theoretical determinants of the incentives for innovation [Fischer, Parry and Pizer (forthcoming)]

Determinant	Emissions tax	Freely-allocated tradeable permits	Auctioned tradeable permits
Abatement cost effect	(+)	(+)	(+)
Imitation effect	(−)	(−)	(−)
Emissions payment effect	none	none	(+)

[34] In a subsequent analysis, Laffont and Tirole (1996b) examined the government's ability to influence the degree of innovative activity by setting the number of permits (and permit prices) in various ways in a dynamic setting.

In an analysis that is quite similar in its results to the study by Fischer, Parry and Pizer (forthcoming), Ulph (1998) compared the effects of pollution taxes and command-and-control standards, and found that increases in the stringency of the standard or tax had ambiguous effects on the level of R&D, because environmental regulations have two competing effects: a direct effect of increasing costs, which increases the incentives to invest in R&D in order to develop cost-saving pollution-abatement methods; and an indirect effect of reducing product output, which reduces the incentive to engage in R&D.[35] Carraro and Soubeyran (1996) compared an emission tax and an R&D subsidy, and found that an R&D subsidy is desirable if the output contractions induced by the tax are small or if the government finds output contractions undesirable for other reasons. Addressing the same trade-off, Katsoulacos and Xepapadeas (1996) found that a simultaneous tax on pollution emissions and subsidy to environmental R&D may be better suited to overcoming the joint market failure (negative externality from pollution and positive externality or spillover effects of R&D).[36]

Finally, Montero (2002) compared instruments under non-competitive circumstances, and found that the results are less clear than when perfect competition is assumed. He modeled a two-firm oligopoly facing environmental regulation in the form of emissions standards, freely-allocated permits, auctioned permits, and taxes. Firms can invest in R&D to lower their marginal abatement costs, and they can also benefit from spillover effects from the other firm's R&D efforts. In choosing whether and how much to invest in R&D in order to maximize profits, a firm must consider two effects of its investment choice: (1) the increase in profits due to a decrease in its abatement costs (less the R&D cost); and (2) the decrease in profits due to changes in the other firm's output, as a result of spillover from the first firm's R&D. The result is that standards and taxes yield higher incentives for R&D when the market is characterized by Cournot competition, but the opposite holds when the market is characterized by Bertrand competition.

3.2.3. Empirical analyses

There has been exceptionally little empirical analysis of the effects of alternative policy instruments on technology innovation in pollution abatement, principally because of the paucity of available data. One study by Bellas (1998) carried out a statistical analysis of the costs of flue gas desulfurization (scrubbing) installed at coal-fired power plants in the United States under the new-source performance standards of the 1970 and 1977 Clean Air Acts. Bellas failed to find any evidence of effects of scrubber vintage on cost, suggesting little technological change had taken place under this regulatory regime.

Although there has been very little analysis in the context of pollution-abatement technologies, there is a more extensive literature on the effects of alternative policy

---

[35] In addition, Ulph (1998) examined a situation where two firms produce identical products with two characteristics. If both firms innovate on the same characteristic, price competition will eliminate any gains from R&D; but consumer pressure can affect the direction of R&D by influencing the characteristic that firms focus on improving. See also: Ulph and Ulph (1996).

[36] See, also, Conrad (2000).

instruments on the innovation of energy-efficiency technologies, because data have been available. As described in Section 3.1.3, above, the innovation process can be thought of as affecting improvements in the characteristics of products on the market, and the process can be framed as the shifting inward over time of a frontier representing the tradeoffs between different product characteristics for the range of models available on the market. If one axis is the cost of the product and another axis is the energy flow associated with a product, that is, its energy intensity, then innovation is represented by inward shifts of the curve – greater energy efficiency at the same cost, or lower cost for given energy efficiency. With this approach, Newell, Jaffe and Stavins (1999) assessed the effects of changes in energy prices and in energy-efficiency standards in stimulating innovation. Energy price changes induced both commercialization of new models and elimination of old models. Regulations, however, worked largely through energy-inefficient models being dropped, since that is the intended effect of the energy-efficiency standards (models below a certain energy efficiency level may not be offered for sale).

A closely related approach to investigating the same phenomena is that of hedonic price functions. One hedonic study examined the effects of public policies in the context of home appliances. Greening, Sanstad and McMahon (1997) estimated the impacts of the 1990 and 1993 national efficiency standards on the quality-adjusted price of household refrigerator/freezer units. They found that quality-adjusted prices fell after the implementation of the energy efficiency standards, but such quality-adjusted price decreases are consistent with historical trends in refrigerator/freezer prices. Hence, one cannot rule out the possibility that the imposition of efficiency standards slowed the rate of quality-adjusted price decline.

Greene (1990) used data on fuel prices and fuel economy of automobiles from 1978 to 1989 to test the relative effectiveness of Corporate Average Fuel Economy (CAFE) Standards and gasoline prices in increasing fuel economy. He found that the big three U.S. firms faced a binding CAFE constraint, and for these firms compliance with CAFE standards had roughly twice the impact on fuel economy as did fuel prices. Japanese firms, however, did not face a binding CAFE constraint, and fuel prices had only a small effect. Luxury European manufactures seemed to base their fuel efficiency largely on market demand and often exceeded CAFE requirements. For these firms, neither the standards nor prices seemed to have much effects.

More recently, Pakes, Berry and Levinsohn (1993) investigated the effects of gasoline prices on the fuel economy of motor vehicles offered for sale, and found that the observed increase in miles per gallon (mpg) from 1977 onward was largely due to the consequent change in the mix of vehicles on the market. Fewer low-mpg cars were marketed, and more high-mpg cars were marketed. Subsequently, Berry, Kortum and Pakes (1996) combined plant-level cost data for the automobile industry and information on the characteristics of models that were produced at each plant to estimate a hedonic cost function – the supply-side component of the hedonic price function – finding that quality-adjusted costs generally *increased* over the period 1972–1982, thus coinciding with rising gasoline prices and emission standards.

Finally, Goldberg (1998) combined a demand-side model of discrete vehicle choice and utilization with a supply-side model of oligopoly and product differentiation to estimate the effects of CAFE standards on the fuel economy of the new car fleet. She found that automobile fuel operating costs have had a significant effect, although a gasoline tax of a magnitude that could match the effect of CAFE on fuel economy would have to be very large.

3.3. Induced innovation and optimal environmental policy

Though the magnitude of induced innovation effects remains uncertain, a few researchers have begun to explore the consequences of induced innovation for environmental policy. Section 3.2, above, addressed the important question of how considerations related to induced innovation affect the normative choice among different policy *instruments*. In this section, we consider the larger question of whether the possibility of induced innovation ought to change environmental policy *targets*, or the pace at which we seek to achieve them.

Intuitively, it seems logical that if environmental policy intervention induces innovation, this in some sense reduces the social cost of environmental intervention, suggesting that the optimal policy is more stringent than it would be if there were no induced innovation. This intuition contains an element of truth, but a number of complexities arise. First, one has to be careful what is meant by "reducing the cost of intervention". As shown by Goulder and Schneider (1999), if the policy intervention induces a reduction in the marginal cost of abatement, then any given policy target (for example, a particular aggregate emission rate or a particular ambient concentration) will be achieved at lower cost than it would without induced innovation. On the other hand, the lower marginal abatement cost schedule arising from induced innovation makes it socially optimal to achieve a greater level of pollution abatement. For a flat marginal social benefit function evaluated at the social optimum, or for any emission tax, this results in greater *total* expenditure on abatement even as the *marginal* abatement cost falls.

Another important issue is the general equilibrium effect of induced environmental innovation on innovation elsewhere in the economy [Schmalensee (1994)]. If inducement operates through increased R&D expenditure, then an issue arises as to the elasticity of supply of R&D inputs. To the extent that this supply is inelastic, then any induced innovation must come at the expense of other forms of innovation, creating an opportunity cost that may negate the "innovation offsets" observed in the regulated portion of the economy.[37] The general equilibrium consequences of these effects for welfare analysis depend on the extent of R&D spillovers or other market failures, and the magnitude of these distortions in the regulated firms or sectors relative to the rest of the economy [Goulder and Schneider (1999)].

---

[37] Goldberg (1998) provided evidence that the supply of R&D inputs (scientists and engineers) is relatively inelastic in the short run. It seems less likely that this supply is inelastic in the long run. See Romer (2000).

Goulder and Mathai (2000) looked at optimal carbon abatement in a dynamic setting, considering not only the optimal overall amount of abatement but also its timing.[38] In addition to R&D-induced innovation, they considered (in a separate model) reductions in abatement costs that come about via learning-by-doing. In the R&D model, there are two effects of induced innovation on optimal abatement: it reduces marginal abatement costs, which increases the optimal amount of abatement. But it also increases the cost of abatement today relative to the future, because of lower abatement costs in the future. The combination of these effects implies that with R&D-induced innovation, optimal abatement is lower in early years and higher in later years than it would otherwise be. In the learning-by-doing model, there is a third effect: abatement today lowers the cost of abatement in the future. This reinforces the tendency for cumulative optimal abatement to be higher in the presence of induced innovation, but makes the effect on optimal near-term abatement ambiguous.

Goulder and Mathai also considered the impact of innovation on the optimal tax rate. One might suppose that the potential for induced technological change justifies a higher environmental tax rate (or higher time-profile for an environmental tax), since in this setting environmental taxes have a dual role: discouraging emissions and triggering new technologies. Goulder and Mathai showed, however, that under typical conditions (a downward-sloping marginal damages curve) the presence of induced innovation implies a *lower* time-profile for the optimal environmental tax. The reason is that with induced innovation, a lower tax is all that is needed to achieve the desired abatement, even when the desired extent of abatement is higher.

Nordhaus (2000) introduced induced technological change into the "DICE" model of global climate change and associated economic activities. To calibrate the model, he needed parametric estimates of the private and social returns to fossil-fuel-related R&D. Using the existing R&D intensity of the fossil sector to derive these parameters, he found that the impact of induced innovation is modest. Essentially, the existing share of R&D investment in this sector is so small that even with large social returns the overall impact is modest. Indeed, comparing a model with induced innovation (but no factor substitution) with a model that has factor substitution but no induced innovation, he concluded that induced innovation has less effect than factor substitution on optimal emissions levels.

Overall, there is considerable ambiguity regarding the importance of induced innovation for the optimal stringency of environmental policy. Partly, this is because predictions depend on the magnitudes of parameters that are hard to measure. But, more fundamentally, if environmental policy affects the innovation process, and the innovation process is itself characterized by market failure, then this is a classic problem of the "second best". We know that robust results are generally hard to come by with respect to such problems. It will typically make a big difference whether we imagine optimizing R&D policy first, and then environmental policy, or vice versa, or if we imagine

---

[38] On the role of induced technological change in climate change modeling, see also Wigley, Richels and Edmonds (1996), Ha-Duong, Grubb and Hourcade (1997), and Grubb (1997).

simultaneous optimization in both realms, or if we assume that we are designing optimal environmental policy taking non-optimal R&D policy as given. Theory may be able to indicate the considerations that come into play, but is unlikely to provide robust prescriptions for policy.

3.4. The evolutionary approach to innovation

While viewing R&D as a profit-motivated investment activity comes naturally to most economists, the large uncertainties surrounding the outcomes of R&D investments make it very difficult for firms to make optimizing R&D decisions. Accordingly, Nelson and Winter (1982) used Herbert Simon's idea of boundedly rational firms that engage in "satisficing" rather than optimizing behavior [Simon (1947)] to build an alternative model of the R&D process. In this "evolutionary" model, firms use "rules of thumb" and "routines" to determine how much to invest in R&D, and how to search for new technologies. The empirical predictions of this model depend on the nature of the rules of thumb that firms actually use [Nelson and Winter (1982), Winter, Kaniovski and Dosi (2000)].

Because firms are not optimizing, a logical consequence of the evolutionary model is that it cannot be presumed that the imposition of a new external constraint (for example, a new environmental rule) necessarily reduces profits. There is at least the theoretical possibility that the imposition of such a constraint could be an event that forces a satisficing firm to rethink its strategy, with the possible outcome being the discovery of a new way of operating that is actually more profitable for the firm. This possibility of environmental regulation leading to a "win-win" outcome in which pollution is reduced and profits increased is discussed below.

3.4.1. Porter's "win-win" hypothesis

The evolutionary approach replaces optimizing firms with satisficing firms, and thereby admits greater scope for a variety of consequences when the firm's environment is modified. Satisficing firms may miss opportunities for increased profits simply because they do not look very hard for such opportunities as long as things are going reasonably well. An external shock such as a new environmental constraint can therefore constitute a stimulus to new search, possibly leading to discovery of previously undetected profit opportunities. This observation forms the basis for the normative observation that environmental regulation may not be as costly as we expect, because the imposition of the new constraint may lead to the discovery of new ways of doing things. In the limit, these new ways of doing things might actually be *more* profitable than the old ways, leading to an asserted "win-win" outcome.[39]

---

[39] Another related idea is that of "X-inefficiency" [Leibenstein (1966)].

In general, advocates of the "win-win" view of the consequences of environmental regulation seem unaware of the connection between their argument and the evolutionary school of technological change.[40] But the ideas are similar:

> It is sometimes argued that companies must, by the very notion of profit seeking, be pursuing all profitable innovation . . . In this view, if complying with environmental regulation can be profitable, in the sense that a company can more than offset the cost of compliance, then why is such regulation necessary?

> The possibility that regulation might act as a spur to [profitable] innovation arises because the world does not fit the Panglossian belief that firms always make optimal choices . . . [T]he actual process of dynamic competition is characterized by changing technological opportunities coupled with highly incomplete information, organizational inertia and control problems reflecting the difficulty of aligning individual, group and corporate incentives. Companies have numerous avenues for technological improvement, and limited attention [Porter and van der Linde (1995, pp. 98–99)].

Porter and other "win-win" theorists argued that in this non-optimizing world, regulation may lead to "innovation offsets" that "can not only lower the net cost of meeting environmental regulations, but can even lead to absolute advantages over firms in foreign countries not subject to similar regulations" [Porter and van der Linde (1995, p. 98)]. Of course, the fact that firms engage in non-optimizing behavior creates a *possibility* for profit improvements, without suggesting that such improvements would be the norm, would be systematic, or even likely. But win-win theorists propose several reasons why innovation offsets are likely to be common.

First, they argue that regulation provides a signal to companies about likely resource inefficiencies and potential technological improvements; that pollution is, by its very nature, indicative of resources being wasted, or at least not fully utilized. Regulation focuses attention on pollution, and such attention is likely to lead to the saving of resources, which will often lower costs. Second, regulation provides or requires the generation of information; since information is a public good it may be underprovided without such incentives. Third, regulation reduces uncertainty about the payoffs to investments in environmental innovation. There may be potential investments that are believed to be profitable in an expected value sense, and also deliver environmental benefits, but which are highly risky in the absence of regulation that ensures that the environmental benefits

---

[40] Neither Simon (1947) nor Nelson and Winter (1982) appear in the references of Porter and van der Linde (1995). Interestingly, Nelson and Winter themselves anticipated the connection. In their 1982 book, they say "In a regime in which technical advance is occurring and organizational structure is evolving in response to changing patterns of demand and supply, new nonmarket interactions that are not contained adequately by prevailing laws and policies are almost certain to appear, and old ones may disappear . . . The canonical 'externality' problem of evolutionary theory is the generation by new technologies of benefits and costs that old institutional structures ignore" (p. 368). See also Kemp and Soete (1990).

are also privately valuable. Regulation, in effect, provides "insurance" against the risk of investing in new technology, part of whose benefit cannot be internalized. Fourth, new technology that is initially more costly may produce long-run competitive advantage, because of learning-by-doing or other "first-mover" advantages, if other countries eventually impose similarly strict standards. Finally, regulation simply creates pressure. Such pressure plays an important role in the innovation process, "to overcome inertia, foster creative thinking and mitigate agency problems" [Porter and van der Linde (1995, p. 100)].

Porter and van der Linde (1995) provided numerous case studies of particular firms who developed or adopted new technology in response to regulation, and appear to have benefited as a result. It should be emphasized, however, that win-win theorists do not claim that all environmental regulations generate significant innovation offsets. Indeed, they emphasize that regulation must be properly designed in order to maximize the chances for encouraging innovation. Quantitative evidence is limited. Boyd and McClelland (1999) and Boyd and Pang (2000) employ data envelopment analysis to evaluate the potential at paper and glass plants for "win-win" improvements that increase productivity and reduce energy use or pollution. They find that the paper industry could reduce inputs and pollution by 2–8% without reducing productivity.

Generally, economists have been skeptical of the win-win theory [Palmer, Oates and Portney (1995)]. From a theoretical perspective, it is possible to model apparently inefficient firm behavior as the (second-best) efficient outcome of imperfect information and divergent incentives among managers or between owners and managers in a principal/agent framework.[41] From this perspective, the *apparent* inefficiency does not have normative implications. Since firms are doing the best they can given their information environment, it is unlikely that the additional constraints represented by environmental policy interventions would be beneficial.

On a more concrete level, it is not clear that pollution generally signals "waste"; most physical and biological processes have by-products of some sort, and whether the extent of such by-products is "wasteful" or not is inherently a question of prices and costs. More generally, firms' rationality is surely bounded, but that does not mean that unexploited profit opportunities are frequent. Palmer, Oates and Portney (1995) surveyed firms affected by regulation – including those cited by Porter and van der Linde as success stories – and found that most firms say that the net cost to them of regulation is, in fact, positive.

For regulation to have important informational effects, the government must have better information than firms have about the nature of environmental problems and their potential solutions. This seems questionable. Of course, the government may have better information about which environmental problems *it* considers most important, but it is not clear how conveying this type of information would produce win-win outcomes.

---

[41] For a survey, see Holmström and Tirole (1987).

As to overcoming inertia, most firms in today's world feel a lot of pressure, so it seems unlikely that the additional pressure of regulation is going to have beneficial stimulating effects on innovation. Finally, while it seems likely that environmental regulation will stimulate the innovation and diffusion of technologies that facilitate compliance, creation and adoption of new technology will typically require real resources, and have significant opportunity costs. The observation that the new technology is cost-saving on a forward-looking basis is not sufficient to conclude that the firm was made better off by being induced to develop and/or adopt the new technology.

Overall, the evidence on induced innovation and the win-win hypothesis seems to be a case of a "partially full glass" that analysts see as mostly full or mostly empty, depending on their perspective. This balance is summarized in Table 2.

Table 2
Overview of conclusions on induced innovation and the "win-win" hypothesis

Areas of agreement
Historical evidence indicates that a significant but not necessarily predominant fraction of innovation in the energy and environment area is induced.
Environmental regulation is likely to stimulate innovation and technology adoption that will facilitate environmental compliance.
Much existing environmental regulation uses inflexible mechanisms likely to stifle innovation; "incentive-based" mechanisms are likely to be more conducive to innovation.
Firms are boundedly rational so that external constraints can sometimes stimulate innovation that will leave the firm better off.
First-mover advantages may result from domestic regulation that correctly anticipates world-wide trends.

Areas of disagreement	
Win-win theory	*Neoclassical economics*
Widespread case-study evidence indicates significant "innovation offsets" are common.	Case studies are highly selective. Firms believe regulation is costly.
Innovation in response to regulation is evidence of offsets that significantly reduce or eliminate the cost of regulation.	When cost-reducing innovation occurs, the opportunity cost of R&D and management effort makes a true "win-win" outcome unlikely.
Pollution is evidence of waste, suggesting why cost-reducing innovation in response to regulation might be the norm.	Costs are costs; even if firms are not at the frontier, side-effects of pollution reduction could just as easily be bad as good.
Existing productivity or cost studies do not capture innovation offsets.	Existing productivity and cost studies suggest that innovation offsets have been very small.
There is much evidence of innovation offsets even though existing regulations are badly designed. This suggests that offsets from good regulation would be large.	Since there is agreement that bad regulations stifle innovation, the apparent beneficial effects of existing regulation only show that case studies can be very misleading.

4. Diffusion

4.1. Microeconomics of diffusion

From the mechanical reaper of the nineteenth century [David (1966)], through hybrid corn seed [Griliches (1957)], steel furnaces [Oster (1982)], optical scanners [Levin, Levin and Meisel (1987)] and industrial robots [Mansfield (1989)], research has consistently shown that the diffusion of new, economically superior technologies is a gradual process.[42] Typically, the fraction of potential users that has adopted a new technology follows a sigmoid or "S-shaped" path over time, rising only slowly at first, then entering a period of very rapid growth, followed by a slowdown in growth as the technology reaches maturity and most potential adopters have switched [Geroski (2000)].

The explanation for the apparent slowness of the technology diffusion process has been a subject of research in a variety of disciplines. Two main forces have been emphasized. First, potential technology adopters are heterogeneous, so that a technology that is generally superior will not be equally superior for all potential users, and may remain inferior to existing technology for some users for an extended period of time after its introduction. Second, adopting a new technology is a risky undertaking, requiring considerable information, both about the generic attributes of the new technology and about the details of its use in the particular application being considered. It takes time for information to diffuse sufficiently, and the diffusion of the technology is limited by this process of diffusion of information.

The two main models of the diffusion process each emphasize one of these two aspects of the process.[43] The *probit* or *rank* model, first articulated in an unpublished paper by Paul David (1969), posits that potential adopters are characterized by a distribution of value or returns associated with the new technology.[44] Because adoption is costly, at any moment in time there is a threshold point on this distribution, such that potential users with values at or above this threshold will want to adopt, and users for whom the value of the new technology is below this threshold will not want to adopt. Because the new technology will typically get cheaper and better as time passes after its initial introduction, this threshold will gradually move to the right, and eventually sweep out the entire distribution. If the distribution of underlying values is normal (or another single-peaked distribution with similar shape), this gradual movement of the threshold across the distribution will produce the typical S-shaped diffusion curve.

The other widely-used model is called the *epidemic* model [Griliches (1957), Stoneman (1983)]. The epidemic model presumes that the primary factor limiting diffusion is information, and that the most important *source* of information about a new technology is people or firms who have tried it. Thus technology spreads like a disease, with the

---

[42] See, also, Kemp (1997) for an overview of theoretical models of technology diffusion.

[43] For empirical examples that integrate the two models, see Trajtenberg (1990) and Kerr and Newell (forthcoming).

[44] This has sometimes been called the *rank* model since potential adopters can be ranked in terms of their potential benefits from adoption [Karshenas and Stoneman (1995)].

instigation of adoption being contact between the "infected" population (people who have already adopted) and the uninfected population. Denoting the fraction of the potential using population that has adopted as f, this leads to the differential equation $df/dt = \beta f(1 - f)$. Solution of this equation yields a logistic function, which has the characteristic S-shape. The parameter β captures the "contagiousness" of the disease, presumably related to the cost of the new technology and the degree of its superiority over the technology it replaces [Griliches (1957)].[45]

The probit model emphasizes adoption as the result of value-maximizing decisions by heterogeneous adopters. As such, at least in its basic form, it does not suggest that the slow diffusion of new technology is anything but optimal. In contrast, in the epidemic model each adopter generates a positive externality by transferring information to other potential adopters. This suggests that laissez-faire adoption rates may indeed be socially suboptimal. We return to this issue in Section 4.2.2 below.

Finally, we note an important issue of feedback from the diffusion process to the earlier stages of invention and innovation. The rate at which a technology diffuses determines in large part the rate at which its production volume grows. And as stated earlier, market size tends to be an important determinant of R&D effort and innovative activity, so that growing use increases the incentive for R&D to improve the product. Furthermore, if the production process is characterized by *learning by doing*, then quality may rise and production costs fall as production experience is accumulated. This possibility creates an additional source of positive externality associated with technology adoption, and may introduce dynamic increasing returns to scale for individual technologies. This issue is also discussed below in Section 4.1.1.

In the literature unrelated to environmental technology, both theory and empirical evidence are clear that technology diffusion rates depend on the strength of economic incentives for technology adoption. Both of the models discussed above predict that the present value of benefits from adoption and the initial adoption cost enter into decisions affecting the diffusion rate.[46] In the probit model, this net present value comparison determines the location of the adoption threshold that determines what fraction of potential adopters will adopt at a moment in time. In the epidemic model, this net present

---

[45] Both the probit and epidemic models typically focus on the fraction of the population that had adopted at a point in time. If one has individual-level data on adopters, one can take as the dependent variable the individual time until adoption. This leads to a duration or hazard model [Karshenas and Stoneman (1995), Rose and Joskow (1990)]. Kerr and Newell (forthcoming) employed a duration model to analyze technology adoption decisions by petroleum refineries during the phasedown of lead in gasoline, as discussed in Section 4.2.1 below.

[46] The fact that technology costs enter into the adoption decision demonstrates the close link between technology innovation and diffusion in both theory and reality. A key mechanism of diffusion is the gradual adoption of a new technology as its cost falls (and/or quality improves). Such cost and quality improvements represent innovation. Likewise, incentives for innovation will depend on the eventual demand for a new or improved product, that is, diffusion. This linkage also points to the difficulty of empirically distinguishing between technology innovation and diffusion since they depend on one another and also both depend on similar external incentives such as relative price changes.

value comparison determines the magnitude of the "contagiousness" parameter, which in turn determines the speed at which the technology spreads from adopters to previous non-adopters.

Empirical studies have addressed the influence on diffusion of factors such as firm size, R&D expenditure, market share, market structure, input prices, technology costs, firm ownership, and other institutional factors. The classic empirical study is by Griliches (1957), who showed that the rate of adoption of hybrid corn seed in different regions depended on the economic superiority of the new seed in that region. David (1966) showed that the first adopters of the mechanical reaper were larger farms, who benefited more from the decreased variable cost it permitted. Mansfield (1968) also found the rate of diffusion to depend on firm size (as do most studies), as well as the riskiness of the new technology and the magnitude of the investment required for adoption. [47]

4.1.1. Increasing returns and technology lock-in

Increasing returns to adoption – in the form of learning curves and positive adoption externalities – are a significant feature of market penetration processes for many technologies. Learning-by-doing describes how cumulative production experience with a product leads to reduced production costs, while learning-by-using captures how the value of a good increases for consumers as they gain experience using it. Positive adoption externalities arise when a non-user's probability of adoption is increased the greater the number of potential users who have already adopted [Berndt and Pindyck (2000)]. This could occur because of fad or herding effects, or because of "network externalities". Network externalities exist if a product is technologically more valuable to an individual user as other users adopt a compatible product (for example, telephone and computer networks). These phenomena can be critical to understanding the existing technological system, forecasting how that system might evolve, and predicting the potential effect of some policy or event.

Furthermore, increasing returns to adopting a particular technology or system have been linked with so-called technology "lock-in", in which a particular product, technical standard, production process, or service is produced by a market, and it is difficult to move to an alternative competing technology. Lock-in implies that, once led down a particular technological path, the barriers to switching may be prohibitive. This can be problematic if it would have been in the broader social interest to adopt a fundamentally different pattern of technological capacity. In turn, it raises the question of whether policy interventions – possibly involving central coordination and information assessment, direct technology subsidies, or publicly funded research, development, demonstration, and procurement programs – might avoid undesirable cases of technology lock-in by guiding technological paths in directions superior to those that would be taken by the free market.

---

[47] For further evidence and discussion, see the survey by Karshenas and Stoneman (1995).

A classic, although somewhat controversial, example given is the QWERTY keyboard layout [David (1985)]. As the story goes, the so-called Dvorak keyboard system is ergonomically more efficient than the standard layout. In other words, we could all type faster and better if we learned the Dvorak system. Unfortunately, the QWERTY system got there first, so to speak, and we may be stuck with it, due in part to network externalities and learning-by-doing.

Increasing returns are necessary but not sufficient conditions for persistent and undesirable lock-in. There must also be costs associated with maintaining parallel rival networks or "switching costs" associated with moving between systems (for example, cost of buying a new keyboard or learning to type on a new keyboard layout). The presence of these factors, however, in theory has the potential to lead to a market equilibrium in which a socially suboptimal standard or technology is employed. Nonetheless, an inefficient outcome need not necessarily result, and if it does it may not be lasting. Market forces will eventually tend to challenge the predominance of an inferior technology [see Ruttan (1997)].

A related characteristic of products or systems subject to increasing returns or "positive feedbacks" is that history can be critical. While other markets can often be explained by current demand and supply, markets subject to increasing returns may not be fully understandable without knowing the pattern of historical technology adoption. Work by Arthur (1989, 1990, 1994), David (1985, 1997) and others [Foray (1997)] on the importance of such "path dependence" have focused on the lasting role that chance historical events can play in leading market outcomes down one rather than another possible path. It is important to note that increasing returns and technology lock-in do not necessarily imply market failure. In cases where they may, the question becomes what policies, markets, or institutions, if any, can ameliorate undesirable technological paths or eventual lock-in.

We are far from having a well-established theoretical or empirical basis for when intervention is preferable to an unregulated market outcome or the form that intervention should take. (See Section 4.2.3 for applications to environmental and energy issues.) David (1997, p. 36) suggested that perhaps the most productive question to ask is "how can we identify situations in which it is likely that at some future time individuals really would be better off had another equilibria been selected" from the beginning. One thing that public policy can do, David suggested, is try to delay the market from irreversible commitments before enough information has been obtained about the likely implications of an early, precedent-setting decision.[48] One could construe current policy discussions surrounding certain biotechnology developments as potentially doing just that.

---

[48] See Majd and Pindyck (1989) for an analysis that explicitly treats learning-by-doing as an irreversible investment decision.

Network externalities. Besen and Farrell (1994), Katz and Shapiro (1994), and Liebowitz and Margolis (1994) together provide an overview of issues surrounding network effects. Several properties of "network markets" distinguish them from other markets, influence the strategies firms pursue, and may lead to market inefficiencies including oligopoly or monopoly. Network markets tend to tip; that is, the coexistence of incompatible products may be unstable, resulting in a single standard dominating the market. Two potential inefficiencies can also arise due to demand side coordination difficulties in the presence of network externalities: excess inertia (users wait too long to adopt a new technology) or excess momentum (users rush to an inferior technology to avoid being stranded) [see Farrell and Saloner (1985)]. The role of information is central; the possibility of locking into an inferior technology is greater when users have incomplete information, and it is expectations about the ultimate size of a network that is crucial to which technology dominates. The root problem is the difficulty of collective coordination in a decentralized process [David (1997)].

One way to address this coordination problem is through standards – that is, a particular technology chosen for universal adoption – which are often adopted by government or industry associations when network externalities are present (telephone signals, for example).[49] While standards may help avoid excess inertia and reduce users' search and coordination costs, they can also reduce diversity and may be subject to the strategy of a dominant firm. Katz and Shapiro (1994) point out that while network effects can lead to market inefficiencies, there are many possible market responses to these problems that do not necessarily involve government intervention. Furthermore, there is a question about whether the government has the proper incentives and information to improve the situation.

Learning-by-doing and learning-by-using. In early production stages, the manufacture of technologically complex products is fraught with difficulties. As a firm produces more and more of the product, however, it learns to produce it more efficiently and with higher levels of quality. Production experience leads to the rationalization of processes, reduced waste, and greater labor force expertise. When this is so, average production costs will tend to decrease over time and with increases in the firm's cumulative output, albeit at a decreasing rate. Alternative terms used to denote this characteristic learning pattern and related phenomena include "learning curve", "experience curve", "learning-by-doing", and "progress function". Learning-by-using, the demand-side counterpart of learning-by-doing, can complement and reinforce these learning effects as adoption increases with greater experience in use and increased productivity over time by the user [see Sunding and Zilberman (2000)].

A technology with an initial cost advantage can allow for pricing that increases market share. In turn, increased market share can lead to even greater learning, cost reductions, and competitive advantage – a virtuous circle for the firms producing the technology. Unfortunately, as with network effects, this persistent cost advantage can create

---

[49] See, for example, Katz and Shapiro (1986, 1994).

a kind of entry barrier if knowledge spillovers are incomplete. In the extreme, the cost advantage may completely deter or "lock out" the entry of new technologies or rival systems, at least for a time. Spence (1981) showed that the main factors affecting costs and competition in the presence of learning are the rate of learning, the extent of learning-induced cost decline relative to the market, the intertemporal pattern of demand (that is, demand elasticity and growth), and the degree of spillovers of learning to other firms.[50]

4.2. Diffusion of green technology

While the induced innovation literature focuses on the potential for environmental policy to bring forth new technology through innovation, there is also a widely-held view that significant reductions in environmental impacts could be achieved through more widespread diffusion of *existing* economically-attractive technologies, particularly ones that increase energy efficiency and thereby reduce emissions associated with fossil fuel combustion. For example, the report of the Interlaboratory Working Group (1997) compiled a comprehensive analysis of existing technologies that reportedly could reduce energy use and hence CO_2 emissions at low or even negative net cost to users. The observation that energy-efficient technologies that are cost-effective at current prices are diffusing only slowly dates back to the 1970s, having been identified as a "paradox" at least as far back as Shama (1983).

As discussed in Section 4.1, above, the observation of apparently slow diffusion of superior technology is not a surprise when viewed in historical context. Nonetheless, the apparent potential for emissions reductions associated with faster diffusion of existing technology raises two important questions. First, what is the theoretical and empirical potential for "induced diffusion" of lower-emissions technologies? Specifically, how do environmental policy instruments that implicitly or explicitly increase the economic incentive to reduce emissions affect the diffusion rate of these technologies?

A second and related question is the degree to which historical diffusion rates have been limited by market failures in the energy and equipment markets themselves [Jaffe and Stavins (1994)]. To the extent that diffusion has been and is limited by market failures, it is less clear that policies that operate by increasing the economic incentive to adopt such technology will be effective. On the other hand, if such market failures are important, then policies focused directly on correction of such market failures provide, at least in principle, opportunities for policy interventions that are social-welfare increasing, even without regard to any environmental benefit. Table 3 summarizes the potential influence on technology diffusion of many of the factors discussed in this section.

---

[50] See also Fudenberg and Tirole (1983) on strategic aspects of learning-by-doing and early work by Arrow (1962b). Yelle (1979), Dutton and Thomas (1984), Day and Montgomery (1983), and Argote and Epple (1990) together provide an excellent overview of the fairly large empirical literature on learning curves, which spans the fields of economics, marketing, and business administration.

Table 3
Factors influencing technology diffusion

Factor	Likely direction of effect on technology diffusion	Potential policy/ Institutional instrument
Increased relative price of resource conserved by the technology	(+)	tax on the resource
Decreased cost and/or increased quality of technology	(+)	technology subsidy
Inadequate information, uncertainty, and agency problems regarding benefits and costs of technology adoption	(−)	information dissemination, technology demonstration
Learning-by-doing and learning-by-using	(+)	technology demonstration and deployment, tax/subsidy
Network externalities	(?)	standards, planning, coordination
Characteristics of potential adopters	varied	flexible regulation

4.2.1. Effects of resource prices and technology costs

Kerr and Newell (forthcoming) used a duration model to analyze the influence of plant characteristics and the stringency and the form of regulation on technology adoption decisions by petroleum refineries during the leaded gasoline phasedown. They found that increased stringency (which raised the effective price of lead) encouraged greater adoption of lead-reducing technology. They also found that larger and more technically sophisticated refineries, which had lower costs of adoption, were more likely to adopt the new technology.

Rose and Joskow (1990) found a positive effect of fuel price increases on the adoption of a new fuel-saving technology in the U.S. electricity-generation sector, with the statistical significance of the effect depending on the year of the fuel price. In a tobit analysis of steel plant adoption of different furnace technologies, Boyd and Karlson (1993) found a significant positive effect of increases in a fuel's price on the adoption of technology that saves that fuel, although the magnitude of the effect was modest. For a sample of industrial plants in four heavily polluting sectors (petroleum refining, plastics, pulp and paper, and steel), Pizer et al. (2001) found that both energy prices and financial health were positively related to the adoption of energy-saving technologies.

Jaffe and Stavins (1995) carried out econometric analyses of the factors affecting the adoption of thermal insulation technologies in new residential construction in the United States between 1979 and 1988. They examined the dynamic effects of energy prices and technology adoption costs on average residential energy-efficiency technologies in new home construction.[51] They found that the response of mean energy efficiency to

---

[51] The effects of energy prices can be interpreted as suggesting what the likely effects of taxes on energy use would be, and the effects of changes in adoption costs can be interpreted as indicating what the effects of technology adoption subsidies would be. See Section 4.3.2.

energy price changes was positive and significant, both statistically and economically. Interestingly, they also found that equivalent percentage adoption cost changes were about three times as effective as energy price changes in encouraging adoption, although standard financial analysis would suggest they ought to be about equal in percentage terms. This finding offers confirmation for the conventional wisdom that technology adoption decisions are more sensitive to up-front cost considerations than to longer-term operating expenses.

Hassett and Metcalf (1995) found an even larger discrepancy between the effect of changes in installation cost (here coming through tax credits) and changes in energy prices. There are three interrelated possible explanations for this. One possibility is a behavioral bias that causes purchasers to focus more on up-front cost than they do on the lifetime operating costs of an investment. An alternative (but probably indistinguishable) view is that purchasers focus equally on both, but uncertainty about future energy prices makes them give less weight to the current energy price (which is only an indicator of future prices) than they do to the capital cost, which is known. A final interpretation might be that consumers actually have reasonably accurate expectations about future energy prices, and their decisions reflect those expectations, but our empirical proxies for their expectations are not correct.

For households and small firms, adoption of new technologies with significant capital costs may be constrained by inadequate access to financing. And in some countries, import barriers may inhibit the adoption of technology embodied in foreign-produced goods [Reppelin-Hill (1999)].

On the other hand, it must be acknowledged that it is impossible to generalize, particularly across countries. Nijkamp, Rodenburg and Verhoef (2001) presented the qualitative results of a survey of Dutch firms regarding their decisions on how much to invest in energy-efficient technologies. They found that general "barriers" to energy-efficient technology adoption – including the existence of alternative investments, low energy costs, and a desire to replace capital only when is fully depreciated – are more important than financial barriers and uncertainty about future technologies and prices.

4.2.2. Effects of inadequate information, agency problems, and uncertainty

As discussed in Section 4.1, above, information plays an important role in the technology diffusion process. There are two reasons why the importance of information may result in market failure. First, information is a public good that may be expected in general to be underprovided by markets. Second, to the extent that the adoption of the technology by some users is itself an important mode of information transfer to other parties, adoption creates a positive externality and is therefore likely to proceed at a socially suboptimal rate.[52] Howarth, Haddad and Paton (2000) explored the significance

---

[52] Transfer of useful information via technology adoption is a special case of the more general phenomenon of consumption externalities in technology adoption [Berndt and Pindyck (2000)]. If early adopters act randomly rather than on the basis of better information, then consumption externalities can result in socially excessive adoption or "herding" effects.

of inadequate information in inhibiting the diffusion of more efficient lighting equipment. Metcalf and Hassett (1999) compared available estimates of energy savings from new equipment to actual savings realized by users who have installed the equipment. They found that actual savings, while significant, were less than those promised by engineers and product manufacturers. Their estimate of the median realized rate of return is about 12%, which they found to be close to a discount rate for this investment implied by a CAPM analysis.

Also related to imperfect information are a variety of agency problems that can inhibit the adoption of superior technology. The agency problem can be either external or internal to organizations. An example of an external agency problem would be a landlord/tenant relationship, in which a tenant pays for utilities but the landlord makes decisions regarding which appliances to purchase, or vice versa. Internal agency problems can arise in organizations where the individual or department responsible for equipment purchase or maintenance differs from the individual or department whose budget covers utility costs.[53] DeCanio (1998) explored the significance of organizational factors in explaining firms' perceived returns to installation of energy-efficient lighting.[54]

Uncertainty is another factor that may limit the adoption of new technology [Geroski (2000)]. Such uncertainty is not a market failure, merely a fact of economic life. Uncertainty can be inherent in the technology itself, in the sense that its newness means that users are not sure how it will be perform [Mansfield (1968)]. For resource-saving technology, there is the additional uncertainty that the economic value of such savings depends on future resource prices, which are themselves uncertain. This uncertainty about future returns means that there is an "option value" associated with postponing the adoption of new technology [Pindyck (1991), Hassett and Metcalf (1995, 1996)].

Closely related to the issue of uncertainty is the issue of the discount rate or investment hurdle rate used by purchasers in evaluating the desirability of new technology, particularly resource-conserving technology. A large body of research demonstrates that purchasers *appear* to use relatively high discount rates in evaluating energy-efficiency investments [Hausman (1979), Ruderman, Levine and McMahon (1987), Ross (1990)]. The implicit or explicit use of relatively high discount rates for energy savings does not represent a market failure in itself; it is rather the manifestation of underlying aspects of the decision process including those just discussed. At least some portion of the discount rate premium is likely to be related to uncertainty, although the extent to which

---

[53] For a discussion of the implications of the separation of environmental decision-making in major firms from relevant economic signals, see: Hockenstein, Stavins and Whitehead (1997). A series of related case studies are provided by Reinhardt (2000).

[54] Agency problems are probably part of the basis for the hypothesis that energy-saving investments are ignored simply because energy is too small a fraction of overall costs to justify management attention and decisionmaking. This idea actually dates back to Alfred Marshall; one of his four laws of demand was "the importance of being unimportant". Marshall (1922) argued that inputs with small factor shares would receive little attention from firms and hence face inelastic factor demand curves.

the premium can be explained by uncertainty and option value is subject to debate [Hassett and Metcalf (1995, 1996), Sanstad, Blumstein and Stoft (1995)].[55] Capital market failures that make it difficult to secure external financing for these investments may also play a role.[56]

4.2.3. Effects of increasing returns

As described in Section 4.1, above, the presence of increasing returns in the form of learning effects, network externalities, or other positive adoption externalities presents the possibility that market outcomes for technologies exhibiting these features, including those with environmental consequences, may be inefficient. For example, the idea that we are "locked into" a fossil-fuel-based energy system is a recurring theme in policy discussions regarding climate change and other energy-related environmental problems. At a more aggregate level, there has been much discussion of the question of whether it is possible for developing countries to take less environmentally-damaging paths of development than have currently industrialized countries, for example by relying less on fossil fuels.[57]

While the empirical literature is quite thin, some studies have explored the issue of increasing returns and technology lock-in for competing technologies within the energy and environment arenas, including analysis of renewable energy and fossil fuels [Cowan and Kline (1996)], the internal combustion engine and alternatively-fueled vehicles [Cowan and Hulten (1996)], pesticides and integrated pest management [Cowan and Gunby (1996)], technologies for electricity generation [Islas (1997)], nuclear power reactor designs [Cowan (1990)], and the transition from hydrocarbon-based fuels [Kemp (1997)].

Energy and environment-related examples of empirical estimation of learning curves include work related to renewable energy and climate modeling [Nakicenovic (1996), Neij (1997), Grübler and Messner (1999), Grübler, Nakicenovic and Victor (1999)], nuclear reactors [Joskow and Rozanski (1979), Zimmerman (1982), Lester and McCabe (1993)], and electricity supply [Sharp and Price (1990)]. Although network externalities can be an important element of increasing returns, especially for information and communication technologies, their role in environmental technologies is less evident.

---

[55] Option values can arise for investments that can be postponed, and unless explicit account is taken of the option value, it will result in an increased effective hurdle rate for the investment.

[56] Shrestha and Karmacharya (1998) carried out an empirical analysis of the relative importance of various potential barriers to the adoption of fluorescent lighting in Nepal. They found that product information predicted adoption, but owner-occupancy and discount rates did not.

[57] See the survey by Evenson (1995) on technology and development. Also Chapter 5 ("Population, Poverty, and the Natural Environment") for a discussion of related issues, including the "environmental Kuznets curve".

4.3. Effects of instrument choice on diffusion

4.3.1. Theoretical analyses

The predominant theoretical framework for analyses of diffusion effects has been what could be called the "discrete technology choice" model: firms contemplate the use of a certain technology which reduces marginal costs of pollution abatement and which has a known fixed cost associated with it. While some authors have presented this approach as a model of "innovation",[58] it is more appropriately viewed as a model of adoption.

With such models, several theoretical studies have found that the incentive for the adoption of new technologies is greater under market-based instruments than under direct regulation [Zerbe (1970)],[59] Downing and White (1986), Milliman and Prince (1989), Jung, Krutilla and Boyd (1996). With the exception of Downing and White (1986), all of these studies examined the gross impacts of alternative policy instruments on the quantity of technology adoption.[60]

Theoretical comparisons *among* market-based instruments have produced only limited agreement. In a frequently-cited article, Milliman and Prince (1989) examined firm-level incentives for technology diffusion provided by five instruments: command-and-control; emission taxes; abatement subsidies; freely-allocated emission permits, and auctioned emission permits. Firm-level incentives for adoption in this representative-firm model were pictured as the consequent change in producer surplus. They found that auctioned permits would provide the largest adoption incentive of any instrument, with emissions taxes and subsidies second, and freely allocated permits and direct controls last. The Milliman and Prince (1989) study was criticized by Marin (1991) because of its assumption of identical firms, but it was subsequently shown that the results remain largely unchanged with heterogeneous abatement costs [Milliman and Prince (1992)].

In 1996, Jung, Krutilla and Boyd built on Milliman and Prince's basic framework for comparing the effects of alternative policy instruments, but rather than focusing on firm-level changes in producer surplus, they considered heterogeneous firms, and modeled the "market-level incentive" created by various instruments.[61] Their rankings echoed

---

[58] Downing and White (1986) and Malueg (1989) framed their work in terms of "innovation". Milliman and Prince (1989) used one model to discuss both diffusion and "innovation", the latter being defined essentially as the initial use of the technology by an "innovating" firm.

[59] Zerbe (1970) compared taxes, subsidies, and direct regulation (emissions standards). If a technology reduces emissions levels, rather than costs, taxes are still superior to direct regulation, but subsidies are not.

[60] Downing and White (1986) compared market-based instruments and command-and-control standards, and found for the case of small changes in emissions (so that the optimal pollution tax or permit quantity is unchanged) that all of the instruments except CAC standards would induce the socially optimal level of adoption. But for non-marginal emission changes, where the control authority does not modify the policy (tax or quantity of available permits), market-based instruments would induce too much diffusion, relative to the social optimum. Keohane (2001) demonstrated that the cost savings from adoption will always be greater under a market-based instrument than under an emissions rate standard that induces the same emissions.

[61] This measure is simply the aggregate cost savings to the industry as a whole from adopting the technology.

those of Milliman and Prince (1989): auctioned permits provided the greatest incentive, followed by taxes and subsidies, free permits, and performance standards.

Subsequent theoretical analyses [Parry (1998), Denicolò (1999), Keohane (1999)] clarified several aspects of these rankings. First, there is the question of relative firm-level incentives to adopt a new, cost-saving technology when the price of pollution (permit price or tax level) is endogenous. Milliman and Prince (1989), as well as Jung, Krutilla and Boyd (1996), argued that auctioned permits would provide greater incentives for diffusion than freely-allocated permits, because technology diffusion lowers the equilibrium permit price, bringing greater aggregate benefits of adoption in a regime where all sources are permit buyers (that is, auctions). But when technology diffusion lowers the market price for tradeable permits, all firms benefit from this lower price regardless of whether or not they adopt the given technology [Keohane (1999)]. Thus, if firms are price takers in the permit market, auctioned permits provide no more adoption incentive than freely-allocated permits.

The overall result is that both auctioned and freely-allocated permits are inferior in their diffusion incentives to emission tax systems (but superior to command-and-control instruments). Under tradeable permits, technology diffusion lowers the equilibrium permit price, thereby reducing the incentive for participating firms to adopt. Thus, a permit system provides a lower adoption incentive than a tax, assuming the two instruments are equivalent before diffusion occurs [Denicolò (1999), Keohane (1999)].[62]

More broadly, it appears that an unambiguous exhaustive ranking of instruments is not possible on the basis of theory alone. Parry (1998) found that the welfare gain induced by an emissions tax is *significantly* greater than that induced by tradable permits only in the case of very major innovations. Similarly, Requate (1998) included an explicit model of the final output market, and finds that whether (auctioned) permits or taxes provide stronger incentives to adopt an improved technology depends upon empirical values of relevant parameters.[63]

Furthermore, complete theoretical analysis of the effects of alternative policy instruments on the rate of technological change must include modeling of the government's response to technological change, because the degree to which regulators respond to technologically-induced changes in abatement costs affects the magnitude of the adoption incentive associated with alternative policy instruments.[64] Because technology diffusion presumably lowers the aggregate marginal abatement cost function, it results in a change in the efficient level of control. Hence, following diffusion, the optimal agency response is to set a more ambitious target. Milliman and Prince (1989) examined the incentives facing private industry, under alternative policy instruments, to oppose such

---

[62] The difference between diffusion incentives of permits and taxes/subsidies depends upon four conditions: (1) some diffusion occurs; (2) firms have rational expectations and recognize that diffusion lowers the permit price; (3) taxes and permits are equivalent *ex ante*; and (4) the level of regulation is fixed over the time horizon considered.

[63] See, also Parry (1995).

[64] See our discussion, above, of "Induced innovation and optimal environmental policy" in Section 3.3.

policy changes. Their conclusion was that firms would oppose optimal agency adjustment of the policy under all instruments *except taxes*. Under an emissions tax, the optimal agency response to cost-reducing technological change is to lower the tax rate (assuming convex damages); under a subsidy, the optimal response is to lower the subsidy; under tradeable permit systems, the optimal response is to decrease the number of available permits, and thereby drive up the permit price. Thus, firms have clear incentives to support the optimal agency response only under an emissions tax regime.

In a comparison of tradeable permits and pollution taxes, Biglaiser, Horowitz and Quiggin (1995) examined these instruments' ability to achieve the first-best outcome in a dynamic setting.[65] They found that effluent taxes can do so, but permits cannot. With an effluent tax, the optimal tax is presumably determined by marginal damages (which the authors assume to be constant), yielding a policy which is time consistent. Whether or not firms adopt a cost-saving technology, the government has no incentive to change the tax rate. From this perspective, however, tradeable permits are not time consistent, because the optimal number of permits in each period depends on both firms' costs, which are determined by all previous investments, and marginal damages. With constant marginal damages, and marginal abatement costs decreasing over time, the optimal number of permits should also be decreasing over time. Firms may internalize this, and thereby invest less than optimally in pollution control technology.

The result of Biglaiser, Horowitz and Quiggin (1995) depends, however, on the assumption of constant marginal damages. If marginal damages are not constant, then the optimal policy is determined by the interaction of marginal damages and marginal abatement costs for both taxes and permits. The result appears to be analogous to Weitzman's (1974) rule: if the marginal damage curve is relatively flat and there is uncertainty in marginal costs (from the regulator's perspective) due to potential innovation at the firm level, then a price instrument is more efficient.

4.3.2. Empirical analyses

Unlike the case of empirical analysis of the effects of alternative policy instruments on technology innovation (Section 3.2.3), where nearly all of the analysis focuses on energy-efficiency technologies, in the case of technology diffusion, there is a small, but significant literature of empirical analyses focused on pollution-abatement technologies *per se*.

One of the great successes during the modern era of environmental policy was the phasedown of lead in gasoline, which took place in the United States principally during the decade of the 1980s. The phasedown was accomplished through a tradeable permit system among refineries, whereby lead rights could be exchanged and/or banked for later use.[66] As noted in Section 4.2.1, Kerr and Newell (forthcoming) used a duration model to assess the effects of the phasedown program on technology diffusion.

---

[65] See, also, Biglaiser and Horowitz (1995).

[66] The tradeable permit system was also a great success. See Chapter 9 ("Experience with Market-Based Environmental Policy Instruments").

As theory suggests [Malueg (1989)], they found that the tradeable permit system provided incentives for more *efficient* technology adoption decisions, as evidenced by a significant divergence in the adoption behavior of refineries with low versus high compliance costs. Namely, the positive differential in the adoption propensity of expected permit sellers (i.e., low-cost refineries) relative to expected permit buyers (i.e., high-cost refineries) was significantly greater under market-based lead regulation compared to under individually binding performance standards.

Another prominent application of tradeable permit systems which has provided an opportunity for empirical analysis of the effects of policy instruments on technology diffusion is the sulfur dioxide allowance trading program, initiated under the U.S. Clean Air Act amendments of 1990. In an econometric analysis, Keohane (2001) found evidence of the way in which the increased flexibility of a market-based instrument can provide greater incentives for technology adoption. In particular, he found that the choice of whether or not to adopt a "scrubber" to remove sulfur dioxide – rather than purchasing (more costly) low-sulfur coal – was more sensitive to cost differences (between scrubbing and fuel-switching) under the tradeable permit system than under the earlier emissions rate standard.[67]

In an examination of the effects of alternative policy instruments for reducing oxygen-demanding water pollutants, Kemp (1998) found that effluent charges were a significant predictor of adoption of biological treatment by facilities. In earlier work, Purvis and Outlaw (1995) carried out a case study of EPA's permitting process for acceptable water-pollution control technologies in the U.S. livestock production sector. Those authors concluded that the relevant regulations encouraged the use of "time-tested" technologies that provided lower levels of environmental protection than other more innovative ones, simply because producers knew that EPA was more likely to approve a permit that employed the established approach.

Another body of research has examined the effects on technology diffusion of command-and-control environmental standards when they are combined with "differential environmental regulations". In many situations where command-and-control standards have been used, the required level of pollution abatement has been set at a far more stringent level for new sources than for existing ones.[68] There is empirical evidence that such differential environmental regulations have lengthened the time before plants were retired [Maloney and Brady (1988), Nelson, Tietenberg and Donihue (1993)]. Further, this dual system can actually worsen pollution by encouraging firms to keep older, dirtier plants in operation [Stewart (1981), Gollop and Roberts (1983), McCubbins, Noll and Weingast (1989)].

---

[67] Several additional research efforts on the sulfur dioxide allowance trading program are underway; a number of relevant hypotheses are described by Stavins (1998).

[68] It could be argued that new plants ought to have somewhat more stringent standards if their abatement costs are lower, although such standards should obviously be linked with actual abatement costs, not with the proxy of plant vintage.

In addition to economic incentives, direct regulation, and information provision some research has emphasized the role that "informal regulation" or community pressure can play in encouraging the adoption of environmentally clean technologies. For example, in an analysis of fuel adoption decisions for traditional brick kilns in Mexico, Blackman and Bannister (1998) suggested that community pressure applied by competing firms and local non-governmental organizations was associated with increased adoption of cleaner fuels, even when those fuels had relatively high variable costs.

Turning from pollution abatement to energy efficiency, the analysis by Jaffe and Stavins (1995), described above in Section 4.2.1, provided evidence of the likely effects of energy taxes and technology adoption subsidies on the adoption of thermal insulation technologies in new residential construction in the United States. Their findings suggest the response to energy taxes would be positive and significant, and that equivalent percentage technology cost subsidies would be about three times as effective as taxes in encouraging adoption, although standard financial analysis would suggest they ought to be about equal in percentage terms. These results were corroborated by the study of residential energy conservation investments by Hassett and Metcalf (1995), also described in Section 4.2.1, which suggested that tax credits for adoption would be up to eight times more effective than "equivalent" energy taxes.

Although empirical evidence from these two studies indicate that subsidies may be more effective than "equivalent" taxes in encouraging technology diffusion, it is important to recognize some disadvantages of such subsidy approaches. First, unlike energy prices, (energy-efficiency) adoption subsidies do not provide incentives to reduce utilization. Second, technology subsidies and tax credits can require large public expenditures per unit of effect, since consumers who would have purchased the product even in the absence of the subsidy still receive it. In the presence of fiscal constraints on public spending, this raises questions about the feasibility of subsidies that would be sizable enough to have desired effects.[69]

Given the attention paid to automobile fuel economy over the past two decades, it is not surprising that several hedonic studies of automobiles have addressed or focused on energy-efficiency, including Ohta and Griliches (1976) and Goodman (1983). Atkinson and Halvorsen (1984) found that the fuel efficiency of the new car fleet responds more than proportionally to changes in expected fuel prices. Using an analogue to the hedonic price technique, Wilcox (1984) constructed a quality-adjusted measure of automobile fuel economy over the period 1952–1980, finding that it was positively related to oil prices. Ohta and Griliches (1986) found that gasoline price changes over the period 1970–1981 could alone explain much of the observed change in related automobile characteristics.

---

[69] Mountain, Stipdonk and Warren (1989) attempted to assess the effects of relative prices on relevant technology diffusion in the Ontario manufacturing sector from 1962 to 1984. They found that fuel choices changed in response to changes in fuel prices, but given the nature of their analysis, they could not distinguish between product substitution and technology diffusion.

What about conventional command-and-control approaches? Jaffe and Stavins (1995) also examined the effects of more conventional regulations on technology diffusion, in the form of state building codes. They found no discernable effects. It is unclear to what extent this is due to inability to measure the true variation across states in the effectiveness of codes, or to codes that were in many cases not binding relative to typical practice. This is a reminder, however, that although price-based policies will always have some effect, typical command-and-control may have little effect if they are set below existing standards of practice.

In a separate analysis of thermal home insulation, this one in the Netherlands, Kemp (1997) found that a threshold model of diffusion (based on a rational choice approach) could not explain observed diffusion patterns. Instead, epidemic models provided a better fit to the data. Kemp also found that there was no significant effect of government subsidies on the adoption of thermal insulation by households.

Attention has also been given to the effects on energy-efficiency technology diffusion of voluntary environmental programs. Howarth, Haddad and Paton (2000) examined two voluntary programs of the U.S. Environmental Protection Agency, the Green Lights and Energy Star programs, both of which are intended to encourage greater private industry use of energy-saving technologies. A natural question from economics is why would firms carry out *additional* technology investments as part of a voluntary agreement? The authors respond that there are a set of agency problems that inhibit economically wise adoption of some technologies (see discussion of these issues in Section 4.2.2). For example, most energy-saving investments are small, and senior staff may rationally choose to restrict funds for small projects that cannot be perfectly monitored. The Green Lights program may be said to attempt to address this type of agency problem by providing information on savings opportunities at the level of the firm where decisions are made.[70]

Although the empirical literature on the effects of policy instruments on technology diffusion by no means settles all of the issues that emerge from the related theoretical studies, a consistent theme that runs through both the pollution-abatement and energy-efficiency empirical analyses is that market-based instruments are decidedly more effective than command-and-control instruments in encouraging the cost-effective adoption and diffusion of relevant new technologies.

5. Conclusion

In opening this chapter, we suggested that an understanding of the process of technological change is important for economic analysis of environmental issues for two broad

---

[70] Another potential explanation arises where the benefits and costs of a project are born by different units of a firm. Under such circumstances, projects that are good for the firm may not be undertaken. See discussion of this phenomenon in Chapter 9 ("Experience with Market-Based Environmental Policy Instruments").

reasons. First, the environmental impact of social and economic activity is greatly affected by the rate and direction of technological change. This linkage occurs because new technologies may either create or mitigate pollution, and because many environmental problems and policy responses are evaluated over timeframes in which the cumulative impact of technological changes is likely to be large.

The importance of the first link is manifest in determining the economic and environmental "baseline" against which to measure the impacts of proposed policies. That is, before we can discuss what we should or should not do about some environmental problem, we need to forecast how severe the problem will be in the absence of any action. Such forecasts are always based, in some way, on extrapolation of historical experience. Within that historical experience, the processes of technological change have been operating, often with significant consequences for the severity of environmental impacts. Forecasts for the future based on this historical experience depend profoundly on the relative magnitude of the effects of price-induced technological change, learning-by-doing and learning-by-using, public sector R&D, and exogenous technical progress. Sorting out these influences with respect to environmentally relevant technologies and sectors poses a major modeling and empirical challenge.

A particularly important aspect of this set of issues is the historical significance of "lock-in" phenomena for environmentally significant technologies. We understand the theory of increasing returns and other sources of path dependence, but we have little evidence regarding their quantitative importance. We know that it is theoretically possible, for example, that the dominant place of the internal combustion engine in our economy results from a combination of historical accidents and path dependence. But the actual magnitude of such effects, relative to the role played by the superior attractiveness of the technology to individual users, has enormous consequences for the question of whether developing nations will be able or likely to find a different path.

Another important area is in the conceptual and empirical modeling of how the various stages of technological change are interrelated, how they unfold over time, and the differential impact that various policies (for example, public-sector R&D, R&D subsidies to the private sector, environmental taxes, information programs) may have on each phase of technological change. We have reviewed the existing literature on various aspects of technology policy, but there has been relatively little empirical analysis of these policy options directed specifically at the development of environmentally beneficial technology.

There has been much debate surrounding the "win-win" hypothesis. Much of this debate has been explicitly or implicitly ideological or political. More useful would be detailed examinations regarding the kinds of policies and the kinds of private-sector institutions that are most likely to generate innovative, low-cost solutions to environmental problems.

This observation is a natural bridge to the second broad linkage between technology and environment, the effect of environmental policy interventions on the process of technological change. The empirical evidence to date is generally consistent with theoretical findings that market-based instruments for environmental protection provide better incentives than command-and-control approaches for the cost-effective diffusion of desirable, environmentally-friendly technologies. Further, empirical studies suggest that the response of technological change to relevant price changes can be surprisingly swift in terms of patenting activity and introduction of new model offerings – on the order of five years or less. Substantial diffusion can sometimes take considerably longer, depending on the rate of retirement of previously installed equipment. The longevity of much equipment reinforces the importance of taking a longer-term view toward improvements – on the order of decades. Existing empirical studies have also produced some results that may not be consistent with theoretical expectations, such as the finding from two independent analyses that the diffusion of energy-efficiency technologies is more sensitive to variation in adoption-cost than to commensurate energy price changes. Further theoretical and/or empirical work may resolve this apparent anomaly.

A variety of refutable hypotheses that emerge from theoretical models of alternative instruments have not been tested rigorously with empirical data. For example, the predictions from theory regarding the ranking of alternative environmental policy instruments is well-developed but much of the empirical analysis has focused on energy-efficient technologies, rather than pollution abatement technologies *per se*. The increased use of market-based instruments and performance-based standards has brought with it considerably more data with which hypotheses regarding the effects of policy instruments on technology innovation and diffusion can be tested.

The potential long-run consequences of today's policy choices create a high priority for broadening and deepening our understanding of the effects of environmental policy on innovation and diffusion of new technology. Unfortunately, these issues cannot be resolved at a purely theoretical level, or on the basis of aggregate empirical analyses. For both benefit-cost and cost-effectiveness analysis, we need to know the *magnitudes* of these effects, and these magnitudes are likely to differ across markets, technologies, and institutional settings. Thus, taking seriously the notion of induced technological change and its consequences for environmental policy requires going beyond demonstration studies that test whether or not such effects exist, to carry out detailed analyses in a variety of sectors in order to understand the circumstances under which they are large or small. This will require significant research attention from multiple methodological viewpoints over an extended period of time. But the alternative is continuing to formulate public policies with significant economic and environmental consequences without being able to take into account what is going on "inside the black box" of technological change.

Acknowledgements

We are grateful for valuable research assistance from Lori Snyder and helpful comments from Ernst Berndt, Karl-Göran Mäler, Lawrence Goulder, Nathaniel Keohane, Charles Kolstad, Ian Parry, Steven Polasky, David Popp, Vernon Ruttan, Manuel Trajtenberg, Jeffrey Vincent, and David Zilberman, but the authors alone are responsible for all remaining errors of omission and commission.

References

Abramowitz, M. (1956), "Resource and output trends in the U.S. since 1870", American Economic Review Papers and Proceedings 46:5–23.

Aghion, P., and P. Howitt (1998), Endogenous Growth Theory (MIT Press, Cambridge, MA).

Ahmad, S. (1966), "On the theory of induced innovation", Economic Journal 76:344–357.

Aigner D.J., and P. Schmidt (1980), "Specification and estimation of frontier production, profit and cost functions", Journal of Econometrics 13:1–138.

Argote, L., and D. Epple (1990), "Learning curves in manufacturing", Science 247:920–924.

Arrow, K.J. (1962a), "Economic welfare and the allocation of resources for invention", in: R. Nelson, ed., The Rate and Direction of Inventive Activity (Princeton University Press, Princeton, NJ).

Arrow, K.J. (1962b), "The economic implications of learning by doing", Review of Economic Studies 29:155–173.

Arthur, W.B. (1989), "Competing technologies, increasing returns, and lock-in by historical events", Economic Journal 99:116–131.

Arthur, W.B. (1990), "Positive feedbacks in the economy", Scientific American 262:92–95, 98–99.

Arthur, W.B. (1994), Increasing Returns and Path Dependence in the Economy, Economics, Cognition, and Society Series (University of Michigan Press, Ann Arbor, MI).

Atkinson, S.E., and R. Halvorsen (1984), "A new hedonic technique for estimating attribute demand: an application to the demand for automobile fuel efficiency", Review of Economics and Statistics 66:417–426.

Bellas, A.S. (1998), "Empirical evidence of advances in scrubber technology", Resource and Energy Economics 20:327–343.

Berman, E., and L. Bui (1998), "Environmental regulation and labor demand: evidence from the South Coast air basin", Working Paper No. 6776 (National Bureau of Economic Research, Cambridge, MA).

Berndt, E.R., and D.O. Wood (1986), "Energy rice shocks and productivity growth in US and UK manufacturing", Oxford Review of Economic Policy 2:1–31.

Berndt, E.R., and R.S. Pindyck (2000), "Consumption externalities and diffusion in pharmaceutical markets: antiulcer drugs," Working Paper No. 7772 (National Bureau of Economic Research, Cambridge, MA).

Berry, S., S. Kortum and A. Pakes (1996), "Environmental change and hedonic cost functions for automobiles", Proceedings of the National Academy of Sciences, U.S.A. 93:12731–12738.

Besen, S.M., and J. Farrell (1994), "Choosing how to compete: Strategies and tactics in standardization", Journal of Economic Perspectives 8:117–131.

Biglaiser, G., and J.K. Horowitz (1995), "Pollution regulation and incentives for pollution-control research", Journal of Economics and Management Strategy 3:663–684.

Biglaiser, G., J.K. Horowitz and J. Quiggin (1995), "Dynamic pollution regulation", Journal of Regulatory Economics 8:33–44.

Binswanger, H.P. (1974), "A microeconomic approach to innovation", Economic Journal 84:940–958.

Binswanger, H.P., and V.W. Ruttan (1978), Induced Innovation: Technology, Institutions, and Development (John Hopkins University Press, Baltimore, MD).

Blackman, A., and G.J. Bannister (1998), "Community pressure and clean technology in the informal sector: An econometric analysis of the adoption of propane by traditional Mexican brickmakers", Journal of Environmental Economics and Management 35(1):1–21.

Bound, J., C. Cummins, Z. Griliches, B. Hall and A. Jaffe (1984), "Who does R&D and who patents?", in: Z. Griliches, ed., R&D, Patents and Productivity (University of Chicago Press, Chicago, IL).

Boyd, G.A., and S.H. Karlson (1993), "The impact of energy prices on technology choice in the United States steel industry", The Energy Journal 14(2):47–56.

Boyd, G., and J. McClelland (1999), "The impact of environmental constraints on productivity improvement and energy efficiency an integrated paper and steel plants", The Journal of Economics and Environmental Management 38:121–146.

Boyd, G., and J. Pang (2000), "Estimating the linkage between energy efficiency and productivity", Energy Policy 28:289–296.

Bresnahan, T., and M. Trajtenberg (1995) "General purpose technologies: engines of growth?", Journal of Econometrics, Annals of Econometrics 65:83–108.

Caballero, R.J., and A.B. Jaffe (1993), "How high are the giants' shoulders: an empirical assessment of knowledge spillovers and creative destruction in a model of economic growth", in: O.J. Blanchard and S. Fischer, eds., NBER Macroeconomics Annual (MIT Press, Cambridge, MA) 15–74.

Cadot, O., and B. Sinclair-Desgagne (1996), "Innovation under the threat of stricter environmental standards", in: C. Carraro et al., eds., Environmental Policy and Market Structure (Kluwer Academic Publishers, Dordrecht) 131–141.

Carraro, C., and A. Soubeyran (1996), "Environmental policy and the choice of production technology", in: C. Carraro et al., eds., Environmental Policy and Market Structure (Kluwer Academic Publishers, Dordrecht) 151–180.

Carraro, C., and D. Siniscalco (1994), "Environmental policy reconsidered: The role of technology innovation", European Economic Review 38: 545–555.

Cohen, L., and R. Noll (1991), The Technology Pork Barrel (Brookings Institution, Washington, DC).

Cohen, W.M., and R.C. Levin (1989), "Empirical studies of innovation and market structure", in: R. Schmalensee and W. Willig, eds., Handbook of Industrial Organization (North-Holland, Amsterdam) Chapter 18.

Cohen, W.M., R.R. Nelson and J.P. Walsh (2000), "Protecting their intellectual assets: Appropriability conditions and why U.S. manufacturing firms patent (or not)", Working Paper No. 7552 (National Bureau of Economic Research, Cambridge, MA).

Conrad, K. (2000), "Energy tax and competition in energy efficiency: The case of consumer durables", Environmental and Resource Economics 15:159–177.

Cowan, R. (1990), "Nuclear power reactors: A study in technological lock-in", Journal of Economic History 50:541–567.

Cowan, R., and D. Kline (1996), The Implications of Potential 'Lock-In' in Markets for Renewable Energy (National Renewable Energy Laboratory, Golden, CO).

Cowan, R., and P. Gunby (1996), "Sprayed to death: Path dependence, lock-in and pest control strategies", Economic Journal 106:521–542.

Cowan, R., and S. Hulten (1996), "Escaping lock-in: The case of the electric vehicle", Technological Forecasting and Social Change 53:61–79.

Dasgupta, P., and K. Mäler (2000), "Net national product, wealth, and social well-being", Environment and Development Economics 5:69–93.

Dasgupta, P., and J. Stiglitz (1980a), "Industry structure and the nature of innovative activity", Economic Journal 90.266–293.

Dasgupta, P., and J. Stiglitz (1980b), "Uncertainty, industry structure and the speed of R&D", Bell Journal of Economics 11:1–28.

David, P. (1966), "The mechanization of reaping in the ante-bellum Midwest", in: H. Rosovsky, ed., Industrialization in Two Systems (Harvard University Press, Cambridge, MA) 3–39.

David, P. (1969), "A contribution to the theory of diffusion", Mimeo, in: David, P., Behind the Diffusion Curve (Westview Press, Boulder, CO) Chapters 1–3, forthcoming.

David, P., B. Hall and A. Toole (2000), "Is public R&D a complement or substitute for private R&D? A review of the econometric evidence", Research Policy 29:497–529.

David, P.A. (1985), "Clio and the economics of QWERTY", American Economic Review 75:332–337.

David, P.A. (1997), "Path dependence and the quest for historical economics: One more chorus in the ballad of QWERTY", Discussion Papers in Economic and Social History No. 20 (University of Oxford).

Day, G.S., and D.B. Montgomery (1983), "Diagnosing the experience curve", Journal of Marketing 47:22–58.

DeCanio, S.J. (1998), "The efficiency paradox: Bureaucratic and organizational barriers to profitable energy-saving investments", Energy Policy 26:441–454.

Denicolo, V. (1999), "Pollution-reducing innovations under taxes or permits", Oxford Economic Papers 51:184–199.

Denison, E. (1979), Accounting for Slower Economic Growth (Brookings Institution, Washington, DC).

Downing, P.B., and L.J. White (1986), "Innovation in pollution control", Journal of Environmental Economics and Management 13:18–29.

Dutton, J.M., and A. Thomas (1984), "Treating progress functions as a managerial opportunity", Academy of Management Review 9:235–247.

Energy Modeling Forum (1996), Markets for Energy Efficiency, EMF Report 13, Vol. 1 (Stanford University).

Evenson, R.E. (1991), "Patent data by industry: Evidence for invention potential exhaustion?", in: Technology and Productivity: The Challenge for Economic Policy (Organization for Economic Cooperation and Development, Paris) 233–248.

Evenson, R. (1995), "Technology change and technology strategy", in: J. Behrman and T.N. Srinivasan, eds., Handbook of Development Economics, Vol. 3A (North-Holland, Amsterdam).

Fabricant, S. (1954), Economic Progress and Economic Change (National Bureau of Economic Research, New York).

Fare, R., S. Grosskopf and C. Lovell (1994), Production Frontiers (Cambridge University Press, New York).

Farrell, J., and G. Saloner (1985), "Standardization, compatibility, and innovation", RAND Journal of Economics 16:70–83.

Fischer, C., I.W.H. Parry and W.A. Pizer (forthcoming), "Instrument choice for environmental protection when technological innovation is endogenous", Journal of Environmental Economics and Management.

Foray, D. (1997), "The dynamic implications of increasing returns: Technological change and path dependent inefficiency", International Journal of Industrial Organization 15:733–752.

Freeman, A.M., and R.H. Haveman (1972), "Clean rhetoric and dirty water", Public Interest 28:51–65.

Fudenberg, D., R. Gilbert, J. Stiglitz and J. Tirole (1983), "Preemption, leapfrogging and competition in patent races", European Economic Review 22:3–31.

Fudenberg, D., and J. Tirole (1983), "Learning by doing and market performance", Bell Journal of Economics 14:522–530.

Geroski, P.A. (1995), "Markets for technology: Knowledge, innovation and appropriability", in: P. Stoneman, ed., Handbook of the Economics of Innovation and Technological Change (Blackwell, Oxford) 90–131.

Geroski, P.A. (2000), "Models of technology diffusion", Research Policy 29:603–626.

Goldberg, P.K. (1998), "The effects of the corporate average fuel efficiency standards in the U.S.", Journal of Industrial Economics 46:1–3.

Gollop, F.M., and M.J. Roberts (1983), "Environmental regulations and productivity growth: The case of fossil-fueled electric power generation", Journal of Political Economy 91:654–674.

Goodman, A.C. (1983), "Willingness to pay for car efficiency: A hedonic price approach", Journal of Transport Economics 17:247–266.

Goolsbee, A. (1998), "Does government R&D policy mainly benefit scientists and engineers?", American Economic Review 88:298–302.

Goulder, L.H., and K. Mathai (2000), "Optimal CO_2 abatement in the presence of induced technological change", Journal of Environmental Economics and Management 39:1–38.

Goulder, L.H., and S.L. Schneider (1999), "Induced technological change and the attractiveness of CO_2 emissions abatement policies", Resource and Energy Economics 21:211–253.

Gray, W.B. and R.J. Shadbegian (1995), "Pollution abatement costs, regulation, and plant-level productivity", Working Paper No. 4994 (National Bureau of Economic Research).

Gray, W.B., and R.J. Shadbegian (1998), "Environmental regulation, investment timing, and technology choice", Journal of Industrial Economics 46:235–256.

Greene, D.L. (1990), "CAFE or price?: An analysis of the effects of federal fuel economy regulations and gasoline price on new car MPG, 1978–89", The Energy Journal 11(3):37–57.

Greening, L.A., A.H. Sanstad and J.E. McMahon (1997) "Effects of appliance standards on product price and attributes: An hedonic pricing model", Journal of Regulatory Economics 11:181–194.

Greenstone, M. (2002), "The impacts of environmental regulations on industrial activity: Evidence from the 1970 and 1977 Clean Air Act amendments and the census of manufactures", Journal of Political Economy 110(6):1175–1219.

Griliches, Z. (1957), "Hybrid corn: An exploration in the economics of technical change", Econometrica 48:501–522.

Griliches, Z. (1979), "Issues in assessing the contribution of research and development to productivity growth", Bell Journal of Economics 10:92–116.

Griliches, Z. (1984), R&D, Patents and Productivity (University of Chicago Press, Chicago, IL).

Griliches, Z. (1990), "Patent statistics as economic indicators: A survey", Journal of Economic Literature 28:1661–1707.

Griliches, Z. (1992), "The search for R&D spillovers", Scandinavian Journal of Economics 94:S29–S47.

Griliches, Z. (1994), "Productivity, R&D and the data constraint", American Economic Review 84:1–23.

Griliches, Z. (1998), R&D and Productivity, The Econometric Evidence (University of Chicago Press, Chicago, IL).

Grossman, G., and E. Helpman (1994), "Endogenous innovation in the theory of growth", Journal of Economic Perspectives 8:23–44.

Grossman, G., and E. Helpman (1995), Technology and trade, in: G.M. Grossman and K. Rogoff, eds., Handbook of International Economics, Vol. 3 (North-Holland, Amsterdam).

Grubb, M. (1997), "Technologies, energy systems, and the timing of CO_2 emissions abatement: An overview of economic issues", Energy Policy 25(2):159–172.

Grübler, A., and S. Messner (1999), "Technological change and the timing of mitigation measures", Energy Economics 20:495–512.

Grübler, A., N. Nakicenovic and D.G. Victor (1999), "Dynamics of energy technologies and global change", Energy Policy 27:247–280.

Guellec, D., and B. van Pottelsberghe (2000), "The impact of public R&D expenditure on business R&D", DSTI Working Paper (Organization for Economic Cooperation and Development, Paris).

Ha-Duong, M., M. Grubb and J.-C. Hourcade (1997), "Influence of socioeconomic inertia and uncertainty on optimal CO_2 abatement", Nature 390:270–273.

Hahn, R.W., and R.N. Stavins (1991), "Incentive-based environmental regulation: A new era from an old idea?" Ecology Law Quarterly 18:1–42.

Hall, B., and J. Van Reenen (2000), "How effective are fiscal incentives for R&D? A review of the evidence", Research Policy 29:449–469.

Hall, B., A. Jaffe and M. Trajtenberg (2000), "Market value and patent citations: A first look", Working Paper No. 7741 (National Bureau of Economic Research).

Hassett, K.A., and G.E. Metcalf (1995), "Energy tax credits and residential conservation investment: Evidence form panel data", Journal of Public Economics 57:201–217.

Hassett, K.A., and G.E. Metcalf (1996), "Can irreversibility explain the slow diffusion of energy saving technologies?", Energy Policy 24:7–8.

Hausman, J.A. (1979), "Individual discount rates and the purchase and utilization of energy-using durables", Bell Journal of Economics 10:33.

Helpman, E. (ed.) (1998), General Purpose Technologies and Economic Growth (MIT Press, Cambridge, MA).

Hicks, J. (1932), The Theory of Wages (Macmillan & Co, London).

Hockenstein, J.B., R.N. Stavins and B.W. Whitehead (1997), "Creating the next generation of market-based environmental tools", Environment 39(4):12–20, 30–33.

Holmström, B.R., and J. Tirole (1987), "The theory of the firm", in: R. Schmalensee and R. Willig, eds., Handbook of Industrial Organization, Vol. 1 (North-Holland, Amsterdam) 61–133.

Howarth, R.B., B.M. Haddad and B. Paton (2000), "The economics of energy efficiency: Insights from voluntary participation programs", Energy Policy 28:477–486.

Interlaboratory Working Group (1997), "Scenarios of carbon reductions: potential impacts of energy technologies by 2010 and beyond (Office of Energy Efficiency and Renewable Technologies, U.S. Department of Energy, Washington, DC).

Islas, J. (1997), "Getting round the lock-in in electricity generating systems: The example of the gas turbine", Research Policy 26:49–66.

Jaffe, A.B. (1980), "Technological opportunity and spillovers of R&D: Evidence from firms' patents, profits and market value", American Economic Review 76:984–1001.

Jaffe, A.B. (1988), "Demand and supply influences in R&D intensity and productivity growth", Review of Economics and Statistics 70:431–437.

Jaffe, A.B. (1998a), "The importance of 'spillovers' in the policy mission of the advanced technology program", Journal of Technology Transfer, Summer 1998:11–19.

Jaffe, A.B. (1998b), "Measurement issues", in: L.M. Branscomb and J.H. Keller, eds., Investing in Innovation (MIT Press, Cambridge, MA).

Jaffe, A.B. (2000), "The U.S. patent system in transition: policy innovation and the innovation process", Research Policy 29:531–558.

Jaffe, A.B., and K. Palmer (1997), "Environmental regulation and innovation: A panel data study", Review of Economics and Statistics 79:610–619.

Jaffe, A.B., and R.N. Stavins (1994), "The energy paradox and the diffusion of conservation technology", Resource and Energy Economics 16:91–122.

Jaffe, A.B. and R.N. Stavins (1995), "Dynamic incentives of environmental regulations: The effects of alternative policy instruments on technology diffusion", Journal of Environmental Economics & Management 29:S43–S63.

Jaffe, A.B., R.G. Newell and R.N. Stavins (1999), "Energy-efficient technologies and climate change policies: Issues and evidence", in: M. Toman, ed., Climate Change Economics and Policy (Resources for the Future, Washington, DC) 171–181.

Jaffe, A.B., S. Peterson, P. Portney and R.N. Stavins (1995), "Environmental regulation and the competitiveness of U.S. manufacturing: What does the evidence tell us?", Journal of Economic Literature 33:132–163.

Jones, C., and J. Williams (1998), "Measuring the social return to R&D", Quarterly Journal of Economics 113:1119–1135.

Jorgenson, D. (1984), "The role of energy in productivity growth", in: J.W. Kendrick, ed., International Comparisons of Productivity and the Causes of the Slowdown (Ballinger, Cambridge, MA) 279–323.

Jorgenson, D. (1990), "Productivity and economic growth", in: E. Berndt and J. Triplett, eds., Fifty Years of Economic Measurement (University of Chicago Press, Chicago, IL) 19–118.

Jorgenson, D., and Z. Griliches (1967), "The explanation of productivity change", Review of Economic Studies 34:249–283.

Jorgenson, D.W., and K.J. Stiroh (2000), "Raising the speed limit: U.S. economic growth in the information age", Brookings Papers on Economic Activity 1:125–211.

Joskow, P.L., and G.A. Rozanski (1979), "The effects of learning by doing on nuclear plant operating reliability", Review of Economics and Statistics 61:161–168.

Jung, C.H., K. Krutilla and R. Boyd (1996), "Incentives for advanced pollution abatement technology at the industry level: An evaluation of policy alternatives", Journal of Environmental Economics and Management 30:95–111.

Kamien, M.I., and N.L. Schwartz (1968), "Optimal induced technical change", Econometrica 36:1–17.

Karshenas, M., and P. Stoneman (1995), "Technological diffusion", in: P. Stoneman, ed., Handbook of the Economics of Innovation and Technological Change (Blackwell Publishers, Oxford).

Katsoulacos, Y., and A. Xepapadeas (1996), "Environmental innovation, spillovers and optimal policy rules", in: C. Carraro et al., eds., Environmental Policy and Market Structure (Kluwer Academic Publishers, Dordrecht) 143–150.

Katz, M.L., and C. Shapiro (1986), "Technology adoption in the presence of network externalities", Journal of Political Economy 94:822–841.

Katz, M.L., and C. Shapiro (1994), "Systems competition and network effects", Journal of Economic Perspectives 8:93–115.

Kemp, R. (1997), Environmental Policy and Technical Change (Edward Elgar, Cheltenham, UK).

Kemp, R. (1998), "The diffusion of biological waste-water treatment plants in the Dutch food and beverage industry", Environmental and Resource Economics 12:113–136.

Kemp, R., and L. Soete (1990), "Inside the 'Green Box': On the economics of technological change and the environment", in: Freeman, C. and L. Soete, eds., New Explorations in the Economics of Technological Change (Pinter, London).

Kendrick, J.W. (1956), Productivity Trends: Capital and Labor (National Bureau of Economic Research, New York).

Keohane, N.O. (1999), "Policy instruments and the diffusion of pollution abatement technology", Mimeo (Harvard University).

Keohane, N.O. (2001), "Essays in the economics of environmental policy", Ph.D. Dissertation (Harvard University), unpublished.

Kerr, S., and R.G. Newell (forthcoming), "Policy-induced technology adoption: Evidence from the U.S. lead phasedown", Journal of Industrial Economics.

Klette, T., J. Møen and Z. Griliches (2000), "Do subsidies to commercial R&D reduce market failures? Microeconometric evaluation studies", Research Policy 29:471–495.

Kneese, A., and C. Schultze (1975), Pollution, Prices, and Public Policy (Brookings Institution, Washington, DC).

Kolstad, C.D., and M.H.L. Turnovsky (1998), "Cost functions and nonlinear prices: Estimating a technology with quality-differentiated inputs", Review of Economics & Statistics 80:444–453.

Laffont, J., and J. Tirole (1996a), "Pollution permits and compliance strategies", Journal of Public Economics 62:85–125.

Laffont, J., and J. Tirole (1996b), "Pollution permits and environmental innovation", Journal of Public Economics 62:127–140.

Lancaster, K. (1971), Consumer Demand: A New Approach (Columbia University Press, New York).

Landjouw, J.O., and A. Mody (1996), "Innovation and the international diffusion of environmentally responsive technology", Research Policy 25:549–571.

Lanjouw, J., and M. Schankerman (1999), "The quality of ideas: Measuring innovation with multiple indicators," Working Paper No. 7345 (National Bureau of Economic Research).

Leibenstein, H. (1966), "Allocative efficiency as 'X-inefficiency' ", American Economic Review 56:392–415.

Lester, R.K., and M.J. McCabe (1993), "The effect of industrial structure on learning by doing in nuclear power plant operation", RAND Journal of Economics 24:418–438.

Levin, R.C., and P.C. Reiss (1988), "Cost-reducing and demand-creating R&D with spillovers", RAND Journal of Economics 19:538–556.

Levin, S., S. Levin and J. Meisel (1987), "A dynamic analysis of the adoption of a new technology: The case of optical scanners", Review of Economics and Statistics 69:12–17.

Liebowitz, S.J., and S.E. Margolis (1994), "Network externality: An uncommon tragedy", Journal of Economic Perspectives 8:133–150.

Lucas, R.E. (1988), "On the mechanics of economic development", Journal of Monetary Economics 22:3–42.

Macauley, M.K., and T.J. Brennan (2001), "Enforcing environmental regulation: Implications of remote sensing technology", in: P. Fishbeck and S. Farrow, eds., Improving Regulation: Cases in Environmental, Health, and Safety (Resources for the Future, Washington, DC).

Magat, W.A. (1978), "Pollution control and technological advance: A dynamic model of the firm", Journal of Environmental Economics and Management 5:1–25.

Magat, W.A. (1979), "The effects of environmental regulation on innovation", Law and Contemporary Problems 43:3–25.

Majd, S., and R.S. Pindyck (1989), "The learning curve and optimal production under uncertainty", Rand Journal of Economics 20:331–343.

Maloney, M.T., and G.L. Brady (1988), "Capital turnover and marketable pollution rights", Journal of Law and Economics 31:203–226.

Malueg, D.A. (1989), "Emission credit trading and the incentive to adopt new pollution abatement technology", Journal of Environmental Economics and Management 16:52–57.

Mankiw, N.G., and M.D. Whinston (1986), "Free entry and social inefficiency", RAND Journal of Economics 17:48–58.

Mansfield, E. (1968), Industrial Research and Technological Innovation (Norton, New York).

Mansfield, E. (1989), "Industrial robots in Japan and the USA", Research Policy 18:183–192.

Marin, A. (1991), "Firm incentives to promote technological change in pollution control: Comment", Journal of Environmental Economics and Management 21:297–300.

Marshall, A. (1922), Principles of Economics, Book 5 (Macmillan & Co, London) Chapter 6.

Martin, S., and J. Scott (2000), "The nature of innovation market failure and the design of public support for private innovation", Research Policy 29:437–448.

McCubbins, M.D., R.G. Noll and B.R. Weingast (1989), "Structure and process, politics and policy: Administrative arrangements and the political control of agencies", Virginia Law Review 75:431–482.

Metcalf, G.E., and K.A. Hassett (1999), "Measuring the energy savings from home improvement investments: Evidence from monthly billing data", The Review of Economics and Statistics 81:516–528.

Milliman, S.R., and R. Prince (1989), "Firm incentives to promote technological change in pollution control", Journal of Environmental Economics and Management 17:247–265.

Milliman, S.R., and R. Prince (1992), "Firm incentives to promote technological change in pollution control: Reply", Journal of Environmental Economics and Management 22:292–296.

Montero, J.P. (2002), "Market structure and environmental innovation", Journal of Applied Economics 5(2):293–325.

Mowery, D., and N. Rosenberg (1989), Technology and the Pursuit of Economic Growth (Oxford University Press, Oxford).

Mountain, D.C., B.P. Stipdonk and C.J. Warren (1989), "Technological innovation and a changing energy mix – a parametric and flexible approach to modeling Ontario manufacturing", The Energy Journal 10:139–158.

Nakicenovic, N. (1996), "Technological change and learning", in: N. Nakicenovic, W.D. Nordhaus, R. Richels and F.L. Toth, eds., Climate Change: Integrating Science, Economics, and Policy (International Institute for Applied Systems Analysis (IIASA), Laxenburg, Austria).

National Science Board (1998), Science and Engineering Indicators – 1998 (National Science Foundation, Washington), available online at http://www.nsf.gov/sbe/srs/seind98/frames.htm.

Neij, L. (1997), "Use of experience curves to analyze the prospects for diffusion and adoption of renewable energy technology", Energy Policy 23:1099–1107.

Nelson, R., and S. Winter (1982), An Evolutionary Theory of Economic Change (The Belknap Press of Harvard University Press, Cambridge, MA).

Nelson, R., T. Tietenberg and M. Donihue (1993), "Differential environmental regulation: Effects on electric utility capital turnover and emissions", Review of Economics and Statistics 75:368–373.

Newell, R.G., A.B. Jaffe and R.N. Stavins (1999), "The induced innovation hypothesis and energy-saving technological change", The Quarterly Journal of Economics 114:941–975.

Newell, R.G. (1997), Environmental Policy and Technological Change, Ph.D. Dissertation (Harvard University).

Nijkamp, P., C.A. Rodenburg and E.T. Verhoef (2001), "The adoption and diffusion of environmentally friendly technologies among firms", International Journal of Environmental Technology and Management 1(1/2): 87–103.

Nordhaus, W.D. (2000), "Modeling induced innovation in climate-change policy", Mimeo (Yale University).

Ohta, M., and Z. Griliches (1976), "Automobile prices revisited: Extensions of the hedonic hypothesis", in: N.E. Terleckyj, ed., Household Production and Consumption (National Bureau of Economic Research, New York).

Ohta, M., and Z. Griliches (1986), "Automobile prices and quality: Did the gasoline price increases change consumer tastes in the U.S.?", Journal of Business & Economic Statistics 4:187–198.

Organization for Economic Cooperation and Development (2000), Basic Science and Technology Statistics: 1999 Edition (Organization for Economic Cooperation and Development, Paris).

Orr, L. (1976), "Incentive for innovation as the basis for effluent charge strategy", American Economic Review 66:441–447.

Oster, S. (1982), "The diffusion of innovation among steel firms: The basic oxygen furnace", The Bell Journal of Economics 13:45–56.

Pakes, A., S. Berry and J.A. Levinsohn (1993), "Applications and limitations of some recent advances in empirical industrial organization: Prices indexes and the analysis of environmental change", American Economic Review 83:240–246.

Palmer, K., W.E. Oates and P.R. Portney (1995), "Tightening environmental standards: The benefit-cost or the no-cost paradigm?" Journal of Economic Perspectives 9:119–132.

Parry, I.W., W.A. Pizer and C. Fischer (forthcoming), "How large are the welfare gains from technological innovation induced by environmental policies?", Journal of Regulatory Economics.

Parry, I.W.H. (1995), "Optimal pollution taxes and endogenous technological progress", Resource and Energy Economics 17:69–85.

Parry, I.W.H. (1998), "Pollution regulation and the efficiency gains from technological innovation", Journal of Regulatory Economics 14:229–254.

Pindyck, R. (1991), "Irreversibility, uncertainty, and investment", Journal of Economic Literature 29:1110–1152.

Pizer, W.A., W. Harrington, R.J. Kopp, R.D. Morgenstern and J. Shih (2001), "Technology adoption and aggregate energy efficiency", Resources for the Future Discussion Paper 01-21 (Resources for the Future, Washington, DC).

Popp, D. (2001), "The effect of new technology on energy consumption", Resource and Energy Economics 23(4):215–239.

Popp, D. (2002), "Induced innovation and energy prices", American Economic Review, 92(1):339–349.

Porter, M.E., and C. van der Linde (1995), "Toward a new conception of the environment-competitiveness relationship", Journal of Economic Perspectives 9:97–118.

Purvis, A., and J. Outlaw (1995), "What we know about technological innovation to achieve environmental compliance: Policy issues for an industrializing animal agriculture sector", American Journal of Agricultural Economics 77:1237–1243.

Reinganum, J. (1982), "A dynamic game of R&D: patent protection and competitive behavior", Econometrica 50:671–688.

Reinganum, J. (1989), "The timing of innovation: Research, development and diffusion", in: R. Schmalensee and R. Willig, eds., Handbook of Industrial Organization, Vol. 1 (North-Holland, Amsterdam) 850–908.

Reinhardt, F.L. (2000), Down to Earth: Applying Business Principles to Environmental Management (Harvard Business School Press, Boston, MA).

Reppelin-Hill, V. (1999), "Trade and environment: An empirical analysis of the technology effect in the steel industry", Journal of Environmental Economics and Management 38:283–301.

Requate, T. (1998), "Incentives to innovate under emission taxes and tradeable permits", European Journal of Political Economy 14:139–165.

Romer, P. (2000), "Should the government subsidize supply or demand for scientists and engineers", in: A. Jaffe et al., eds., Innovation Policy and the Economy (MIT Press, Cambridge, MA).

Romer, P.M. (1990), "Endogenous technical change", Journal of Political Economy 98:S71–S102.

Romer, P.M. (1994), "The origins of endogenous growth", Journal of Economic Perspectives 8:3–22.

Rose, N., and P. Joskow (1990), "The diffusion of new technologies: Evidence from the electric utility industry", RAND Journal of Economics 21:354–373.

Rosenberg, N. (1982), Inside the Black Box (Cambridge University Press, Cambridge, UK).

Rosenberg, N. (1990), "Why do firms do basic research (with their own money)?", Research Policy 19:165–174.

Ross, M. (1990), "Capital budgeting practices of twelve large manufacturers", in: P. Cooley, ed., Advances in Business Financial Management (Dryden Press, Chicago, IL) 157–170.

Ruderman, H., M.D. Levine and J.E. McMahon (1987), "The behavior of the market for energy efficiency in residential appliances including heating and cooling equipment", The Energy Journal 8:101–123.

Ruttan, V.W. (1997), "Induced innovation, evolutionary theory and path dependence: Sources of technical change", Economic Journal 107:1520–1529.

Ruttan, V.W. (2000), "Technology, resources and environment", in: V.W. Ruttan, ed., Technology, Growth and Environment: An Induced Innovation Perspective (University of Minnesota, St. Paul, MN).

Sanstad, A., C. Blumstein and S. Stoft (1995), "How high are option values in energy-efficiency investments", Energy Policy 23:739–743.

Scherer, F. (1986), Innovation and Growth, Schumpeterian Perspectives (The MIT Press, Cambridge, MA).

Scherer, F.M., and D. Harhoff (2000), "Technology policy for a world of skew-distributed outcomes", Research Policy 29:559–566.

Scherer, F., D. Harhoff and J. Kukies (2000), "Uncertainty and the size distribution of rewards from technological innovation", Journal of Evolutionary Economics 10:175–200.

Schmalensee, R. (1994), "The costs of environmental protection", in: M.B. Kotowski, ed., Balancing Economic Growth and Environmental Goals (American Council for Capital Formation, Center for Policy Research, Washington, DC) 55–75.

Schmookler, J. (1966), Invention and Economic Growth (Harvard University Press, Cambridge, MA).

Schumpeter, J. (1942), Capitalism, Socialism and Democracy (Harper, New York).

Shama, A. (1983), "Energy conservation in U.S. buildings, solving the high potential/low adoption paradox from a behavioral perspective", Energy Policy 11:148–168.

Sharp, J.A., and D.H.R. Price (1990), "Experience curve models in the electricity supply industry", International Journal of Forecasting 6:531–540.

Shrestha, R.M., and B.K. Karmacharya (1998), "Testing of barriers to the adoption of energy-efficient lamps in Nepal", The Journal of Energy and Development 23(1):71–82.

Simon, H.A. (1947), Administrative Behavior: A Study of Decision-Making Processes in Administrative Organization (Macmillan Co., New York).

Solow, R.M. (1957), "Technical change and the aggregate production function", Review of Economics and Statistics 39:312–320.

Solow, R.M. (1999), "Neoclassical growth theory", in: J.B. Taylor and M. Woodford, eds., Handbook of Macroeconomics, Vol. 1A (North-Holland, Amsterdam).

Solow, R.M. (2000), Growth Theory: An Exposition (Oxford University Press, New York).

Spence, A.M. (1981), "The learning curve and competition", Bell Journal of Economics 12:49–70.

Spence, A.M. (1984), "Cost reduction, competition and industry performance", Econometrica 52(1):101–121.

Stavins, R.N. (1998), "What have we learned from the grand policy experiment: Lessons from SO$_2$ allowance trading", Journal of Economic Perspectives 12(3):69–88.

Stewart, R.B. (1981), "Regulation, innovation, and administrative law: A conceptual framework", California Law Review 69:1256–1270.

Stoneman, P. (1983), The Economic Analysis of Technological Change (Oxford University Press, Oxford, UK).

Stoneman, P. (1987), The Economic Analysis of Technology Policy (Oxford University Press, Oxford, UK).

Sunding, D., and D. Zilberman (2000), "The agricultural innovation process: Research and technology adoption in a changing agricultural industry", in: B. Gardner and G.C. Rausser, eds., Handbook of Agricultural and Resource Economics (North-Holland, Amsterdam), forthcoming.

Sutton, J. (1998), Technology and Market Structure (The MIT Press, Cambridge, MA).

Thirtle, C.G., and V.W. Ruttan (1987), "The role of demand and supply in the generation and diffusion of technical change", Fundamentals of Pure and Applied Economics, Vol. 21 (Harwood Academic Publishers, New York).

Tirole, J. (1988), The Theory of Industrial Organization (The MIT Press, Cambridge, MA).

Trajtenberg, M. (1990), Economic Analysis of Product Innovation: The Case of CT Scanners (Harvard University Press, Cambridge, MA).

Trajtenberg, M. (2000), "R&D Policy in Israel: An Overview and Reassessment", Working Paper No. 7930 (National Bureau of Economic Research).

Ulph, A., and D. Ulph (1996), "Trade, strategic innovation and strategic environmental policy – a general analysis", in: C. Carraro et al., eds., Environmental Policy and Market Structure (Kluwer Academic Publishers, Dordrecht) 181–208.

Ulph, D. (1998), "Environmental policy and technological innovation", in: C. Carraro and D. Siniscalaco, eds., Frontiers of Environmental Economics (Edward Elgar, Cheltensham, UK), forthcoming.

U.S. Bureau of Labor Statistics (2000), Multifactor Productivity Trends, 1998 (U.S. Bureau of Labor Statistics, Washington, DC), available at http://www.bls.gov/news.release/prod3.toc.htm.

Weitzman, M.L. (1974), "Prices vs. quantities", Review of Economic Studies 41:477–491.

Weyant, J.P. (1993), "Costs of reducing global carbon emissions", Journal of Economic Perspectives 7:27–46.

Wigley, T.M.L., R. Richels and J.A. Edmonds (1996), "Economic and environmental choices in the stabilization of atmospheric CO_2 concentrations", Nature 379(18):240–243.

Wilcox, J. (1984), "Automobile fuel efficiency: measurement and explanation", Economic Inquiry 22:375–385.

Winter, S.G., Y.M. Kaniovski and G. Dosi (2000), "Modeling industrial dynamics with innovative entrants", Structural Change and Economic Dynamics 11:255–293.

Yaisawarng, S., and J.D. Klein (1994), "The effects of sulfur dioxide controls on productivity change in the U.S. electric power industry", Review of Economics & Statistics 76:447–460.

Yelle, L.E. (1979), "The learning curve: Historical review and comprehensive survey", Decision Sciences 10:302–328.

Zerbe, R.O. (1970), "Theoretical efficiency in pollution control", Western Economic Journal 8:364–376.

Zimmerman, M.B. (1982), "Learning effects and the commercialization of new energy technologies: The case of nuclear power", Bell Journal of Economics 13:297–310.

AUTHOR INDEX

n indicates citation in footnote.

SUBJECT INDEX

HANDBOOKS IN ECONOMICS

1. HANDBOOK OF MATHEMATICAL ECONOMICS (in 4 volumes)
   Volumes 1, 2 and 3 edited by Kenneth J. Arrow and Michael D. Intriligator
   Volume 4 edited by Werner Hildenbrand and Hugo Sonnenschein

2. HANDBOOK OF ECONOMETRICS (in 6 volumes)
   Volumes 1, 2 and 3 edited by Zvi Griliches and Michael D. Intriligator
   Volume 4 edited by Robert F. Engle and Daniel L. McFadden
   Volume 5 edited by James J. Heckman and Edward Leamer
   Volume 6 is in preparation (editors James J. Heckman and Edward Leamer)

3. HANDBOOK OF INTERNATIONAL ECONOMICS (in 3 volumes)
   Volumes 1 and 2 edited by Ronald W. Jones and Peter B. Kenen
   Volume 3 edited by Gene M. Grossman and Kenneth Rogoff

4. HANDBOOK OF PUBLIC ECONOMICS (in 4 volumes)
   Edited by Alan J. Auerbach and Martin Feldstein

5. HANDBOOK OF LABOR ECONOMICS (in 5 volumes)
   Volumes 1 and 2 edited by Orley C. Ashenfelter and Richard Layard
   Volumes 3A, 3B and 3C edited by Orley C. Ashenfelter and David Card

6. HANDBOOK OF NATURAL RESOURCE AND ENERGY ECONOMICS
   (in 3 volumes). Edited by Allen V. Kneese and James L. Sweeney

7. HANDBOOK OF REGIONAL AND URBAN ECONOMICS (in 4 volumes)
   Volume 1 edited by Peter Nijkamp
   Volume 2 edited by Edwin S. Mills
   Volume 3 edited by Paul C. Cheshire and Edwin S. Mills
   Volume 4 is in preparation (editors J. Vernon Henderson and Jacques-François Thisse)

8. HANDBOOK OF MONETARY ECONOMICS (in 2 volumes)
   Edited by Benjamin Friedman and Frank Hahn

9. HANDBOOK OF DEVELOPMENT ECONOMICS (in 4 volumes)
   Volumes 1 and 2 edited by Hollis B. Chenery and T.N. Srinivasan
   Volumes 3A and 3B edited by Jere Behrman and T.N. Srinivasan

10. HANDBOOK OF INDUSTRIAL ORGANIZATION (in 3 volumes)
    Volumes 1 and 2 edited by Richard Schmalensee and Robert R. Willig
    Volume 3 is in preparation (editors Mark Armstrong and Robert H. Porter)

11. HANDBOOK OF GAME THEORY with Economic Applications (in 3 volumes)
    Edited by Robert J. Aumann and Sergiu Hart

FORTHCOMING TITLES

HANDBOOK OF EXPERIMENTAL RESULTS ECONOMICS
Editors Charles Plott and Vernon L. Smith

HANDBOOK ON THE ECONOMICS OF GIVING, RECIPROCITY AND ALTRUISM
Editors Serge-Christophe Kolm and Jean Mercier Ythier

HANDBOOK ON THE ECONOMICS OF ART AND CULTURE
Editors Victor Ginsburgh and David Throsby

HANDBOOK OF ECONOMIC GROWTH
Editors Philippe Aghion and Steven N. Durlauf

HANDBOOK OF LAW AND ECONOMICS
Editors A. Mitchell Polinsky and Steven Shavell

HANDBOOK OF ECONOMIC FORECASTING
Editors Graham Elliott, Clive W.J. Granger and Allan Timmermann

HANDBOOK OF THE ECONOMICS OF EDUCATION
Editors Eric Hanushek and Finis Welch

All published volumes available